T0203120

Lecture Notes in Computer Science 14426

The series Lecture Notes in Computer Science (LNCS), including its subseries Lecture Notes in Artificial Intelligence (LNAI) and Lecture Notes in Bioinformatics (LNBI), has established itself as a medium for the publication of new developments in computer science and information technology research, teaching, and education.

LNCS enjoys close cooperation with the computer science R & D community, the series counts many renowned academics among its volume editors and paper authors, and collaborates with prestigious societies. Its mission is to serve this international community by providing an invaluable service, mainly focused on the publication of conference and workshop proceedings and postproceedings. LNCS commenced publication in 1973.

Qingshan Liu · Hanzi Wang · Zhanyu Ma ·
Weishi Zheng · Hongbin Zha · Xilin Chen ·
Liang Wang · Rongrong Ji
Editors

Pattern Recognition and Computer Vision

6th Chinese Conference, PRCV 2023
Xiamen, China, October 13–15, 2023
Proceedings, Part II

Editors
Qingshan Liu
Nanjing University of Information Science and Technology
Nanjing, China

Zhanyu Ma
Beijing University of Posts and Telecommunications
Beijing, China

Hongbin Zha
Peking University
Beijing, China

Liang Wang
Chinese Academy of Sciences
Beijing, China

Hanzi Wang
Xiamen University
Xiamen, China

Weishi Zheng
Sun Yat-sen University
Guangzhou, China

Xilin Chen
Chinese Academy of Sciences
Beijing, China

Rongrong Ji
Xiamen University
Xiamen, China

ISSN 0302-9743 ISSN 1611-3349 (electronic)
Lecture Notes in Computer Science
ISBN 978-981-99-8431-2 ISBN 978-981-99-8432-9 (eBook)
https://doi.org/10.1007/978-981-99-8432-9

This Springer imprint is published by the registered company Springer Nature Singapore Pte Ltd.
The registered company address is: 152 Beach Road, #21-01/04 Gateway East, Singapore 189721, Singapore

Preface

Welcome to the proceedings of the Sixth Chinese Conference on Pattern Recognition and Computer Vision (PRCV 2023), held in Xiamen, China.

PRCV is formed from the combination of two distinguished conferences: CCPR (Chinese Conference on Pattern Recognition) and CCCV (Chinese Conference on Computer Vision). Both have consistently been the top-tier conference in the fields of pattern recognition and computer vision within China's academic field. Recognizing the intertwined nature of these disciplines and their overlapping communities, the union into PRCV aims to reinforce the prominence of the Chinese academic sector in these foundational areas of artificial intelligence and enhance academic exchanges. Accordingly, PRCV is jointly sponsored by China's leading academic institutions: the Chinese Association for Artificial Intelligence (CAAI), the China Computer Federation (CCF), the Chinese Association of Automation (CAA), and the China Society of Image and Graphics (CSIG).

PRCV's mission is to serve as a comprehensive platform for dialogues among researchers from both academia and industry. While its primary focus is to encourage academic exchange, it also places emphasis on fostering ties between academia and industry. With the objective of keeping abreast of leading academic innovations and showcasing the most recent research breakthroughs, pioneering thoughts, and advanced techniques in pattern recognition and computer vision, esteemed international and domestic experts have been invited to present keynote speeches, introducing the most recent developments in these fields.

PRCV 2023 was hosted by Xiamen University. From our call for papers, we received 1420 full submissions. Each paper underwent rigorous reviews by at least three experts, either from our dedicated Program Committee or from other qualified researchers in the field. After thorough evaluations, 522 papers were selected for the conference, comprising 32 oral presentations and 490 posters, giving an acceptance rate of 37.46%. The proceedings of PRCV 2023 are proudly published by Springer.

Our heartfelt gratitude goes out to our keynote speakers: Zongben Xu from Xi'an Jiaotong University, Yanning Zhang of Northwestern Polytechnical University, Shutao Li of Hunan University, Shi-Min Hu of Tsinghua University, and Tiejun Huang from Peking University.

We give sincere appreciation to all the authors of submitted papers, the members of the Program Committee, the reviewers, and the Organizing Committee. Their combined efforts have been instrumental in the success of this conference. A special acknowledgment goes to our sponsors and the organizers of various special forums; their support made the conference a success. We also express our thanks to Springer for taking on the publication and to the staff of Springer Asia for their meticulous coordination efforts.

We hope these proceedings will be both enlightening and enjoyable for all readers.

October 2023

Qingshan Liu
Hanzi Wang
Zhanyu Ma
Weishi Zheng
Hongbin Zha
Xilin Chen
Liang Wang
Rongrong Ji

Organization

General Chairs

Hongbin Zha	Peking University, China
Xilin Chen	Institute of Computing Technology, Chinese Academy of Sciences, China
Liang Wang	Institute of Automation, Chinese Academy of Sciences, China
Rongrong Ji	Xiamen University, China

Program Chairs

Qingshan Liu	Nanjing University of Information Science and Technology, China
Hanzi Wang	Xiamen University, China
Zhanyu Ma	Beijing University of Posts and Telecommunications, China
Weishi Zheng	Sun Yat-sen University, China

Organizing Committee Chairs

Mingming Cheng	Nankai University, China
Cheng Wang	Xiamen University, China
Yue Gao	Tsinghua University, China
Mingliang Xu	Zhengzhou University, China
Liujuan Cao	Xiamen University, China

Publicity Chairs

Yanyun Qu	Xiamen University, China
Wei Jia	Hefei University of Technology, China

Local Arrangement Chairs

Xiaoshuai Sun	Xiamen University, China
Yan Yan	Xiamen University, China
Longbiao Chen	Xiamen University, China

International Liaison Chairs

Jingyi Yu	ShanghaiTech University, China
Jiwen Lu	Tsinghua University, China

Tutorial Chairs

Xi Li	Zhejiang University, China
Wangmeng Zuo	Harbin Institute of Technology, China
Jie Chen	Peking University, China

Thematic Forum Chairs

Xiaopeng Hong	Harbin Institute of Technology, China
Zhaoxiang Zhang	Institute of Automation, Chinese Academy of Sciences, China
Xinghao Ding	Xiamen University, China

Doctoral Forum Chairs

Shengping Zhang	Harbin Institute of Technology, China
Zhou Zhao	Zhejiang University, China

Publication Chair

Chenglu Wen	Xiamen University, China

Sponsorship Chair

Yiyi Zhou	Xiamen University, China

Exhibition Chairs

Bineng Zhong	Guangxi Normal University, China
Rushi Lan	Guilin University of Electronic Technology, China
Zhiming Luo	Xiamen University, China

Program Committee

Baiying Lei	Shenzhen University, China
Changxin Gao	Huazhong University of Science and Technology, China
Chen Gong	Nanjing University of Science and Technology, China
Chuanxian Ren	Sun Yat-Sen University, China
Dong Liu	University of Science and Technology of China, China
Dong Wang	Dalian University of Technology, China
Haimiao Hu	Beihang University, China
Hang Su	Tsinghua University, China
Hui Yuan	School of Control Science and Engineering, Shandong University, China
Jie Qin	Nanjing University of Aeronautics and Astronautics, China
Jufeng Yang	Nankai University, China
Lifang Wu	Beijing University of Technology, China
Linlin Shen	Shenzhen University, China
Nannan Wang	Xidian University, China
Qianqian Xu	Key Laboratory of Intelligent Information Processing, Institute of Computing Technology, Chinese Academy of Sciences, China
Quan Zhou	Nanjing University of Posts and Telecommunications, China
Si Liu	Beihang University, China
Xi Li	Zhejiang University, China
Xiaojun Wu	Jiangnan University, China
Zhenyu He	Harbin Institute of Technology (Shenzhen), China
Zhonghong Ou	Beijing University of Posts and Telecommunications, China

Contents – Part II

Character Recognition

Fundamental Theory of Computer Vision

3D Vision and Reconstruction

Deep Stereo Matching with Superpixel Based Feature and Cost

Kai Zeng[1], Hui Zhang[2(✉)], Wei Wang[3(✉)], Yaonan Wang[1], and Jianxu Mao[1]

[1] School of Electrical and Information Engineering, Hunan University, Changsha, China

[2] School of Robotics, Hunan University, Changsha, China
zhanghuihby@126.com

[3] School of Architecture and Planning, Hunan University, Changsha, China
wangwei10731@hnu.edu.cn

Abstract. Previous stereo methods achieved state-of-the-art performances but are still difficult to handle the well-known edge-fattening issue at depth discontinuity regions. In this paper, we proposed differentiable superpixel-based feature and cost aggregation (DSFCA) networks for stereo matching. More specifically, we generated the superpixel maps using the simple linear iterative clustering (SLIC) method and used the efficient feature extraction network to extract the dynamic scale unary features of stereo pair images. Next, we exploited the edge or contour feature information of superpixel maps to aggregate the unary features using the differentiable superpixel-based feature aggregation (DSFA) module. The aggregated features better represent the similarity of stereo pair image feature maps and produce high-quality initial cost volumes. Furthermore, the matching cost is also aggregated by the proposed differentiable superpixel-based cost aggregation (DSCA) module. The experimental results demonstrate that the proposed method outperforms previous methods on the KITTI 2012, KITTI 2015, and SceneFlow datasets.

Keywords: Stereo Matching · Superpixels · Feature Aggregation · Cost Aggregation

This work was supported in part by the National Key Research and Development Program of China under Grant 2021ZD0114503; in part by the Major Research plan of the National Natural Science Foundation of China under Grant 92148204; in part by the National Natural Science Foundation of China under Grants 62027810, 62303171, 62293512, 62133005; in part by the Hunan Provincial Natural Science Foundation under Grants 2023JJ40159; in part by Hunan Leading Talent of Technological Innovation under Grant 2022RC3063; in part by Hunan Key Research and Development Program under Grants 2022GK2011, 2023GK2051; in part by the China Post-Doctoral Science Foundation under Grant 2022M721098; in part by Special funding support for the construction of innovative provinces in Hunan Province under Grant 2021GK1010; in part by Major project of Xiangjiang Laboratory under Grant 22XJ01006.

Q. Liu et al. (Eds.): PRCV 2023, LNCS 14426, pp. 3–15, 2024.
https://doi.org/10.1007/978-981-99-8432-9_1

1 Introduction

Stereo matching is a crucial pre-processing step in many computer vision applications, including autonomous driving, robot navigation, and 3D object or scene reconstruction. The rectified stereo-pair images are given and then estimated the disparity of the corresponding pixels between the left and right images. The traditional stereo-matching method usually computes a correlation score at all disparity levels to determine where optimal matching occurs. Afterward, the post-processing step is followed to refine the predicted disparities.

Recently, deep learning-based methods have demonstrated a strong potential for use in stereo matching. It wants to replace the traditional stereo matching pipeline steps by using the convolutional layer. For example, Yang et al. [29] proposed a WaveletStereo network to learn the wavelet coefficients for the disparity map of stereo matching. The low-frequency wavelet predictors focus on learning the global contextual information of the disparity map. Moreover, the high-frequency wavelet predictors focus on generating the details of the disparity map. In comparison to traditional methods, deep learning-based approaches have several advantages; however, they do have some potential shortcomings that limit performance improvements for stereo matching depth estimation.

Under harsh imaging conditions-such as repeated or lost texture patterns, object occlusion, and color/lighting noise-it is difficult for existing deep stereo matching networks to extract more efficient unary features from stereo pair images to construct the matching cost volume. Moreover, the scale variation with depth changes cannot successfully be mitigated. Therefore, it is still a challenge to efficiently extract the unary feature maps with good similarity.

Furthermore, the well-known edge-fattening issue is still the challenge that prevents existing methods from attaining optimum performance. Conventional techniques used include the segment-based method [27] or superpixel method [15] for matching cost aggregation to handle this issue. For stereo-pair images, accurate segmentation results at the boundary or contour positions reduced the disparity error and rectified the mismatch phenomenon at these positions. However, existing deep stereo methods often solve this issue using data-driven techniques that predict the optimal disparity at the boundary or contour positions by the matching cost aggregation network performance.

In this paper, we propose a deep model-driven stereo matching framework with differentiable superpixel-based features and cost aggregation. Our network introduces an ElasticNet [23] to extract the dynamic scale features for cost volume construction. Moreover, we use the DSFA method to facilitate unary feature map aggregation. The DSFA module exploits the edge or contour feature information to guide the unary feature map aggregation, which better represents the similarity of stereo pair features. Furthermore, we use the DSFA method to guide the cost volume aggregation through the same strategy. We also conducted various experiments to verify our method and compared it with popular benchmarks for disparity map estimation. The experimental results demonstrate that the proposed method outperforms previous approaches on the KITTI 2012, KITTI 2015, and SceneFlow datasets.

2 Related Works

2.1 Traditional Methods

For the traditional stereo methods, many researchers have taken the superpixel information into account for stereo matching by exploiting the edge or contour information of stereo images to improve the performance of depth estimation. For example, Jiao et al. [9] proposed a hybrid-superpixel-based disparity refinement method for stereo matching. They extracted the boundary or edge information from the color images and then used it to further refine the disparity map by removing the boundary-inconsistent regions. Gouveia et al. [6] proposed the superpixel-based confidence disparity estimation approach for stereo matching. They extracted a representative set of features from the disparity map and the superpixel fitting maps, training the random forest classifier for disparity estimation. Kim et al. [11] used a similar concept and extracted robust confidence features from superpixel maps, which determined the confidence measure of stereo matching. Yan et al. [22] proposed a segment-based method for stereo matching to refine the disparity of occlusion handling, where the stereo image is over-segmented into the superpixels maps. The disparity plane is fitted using an improved random sample consensus within each superpixel.

2.2 Learning Based Stereo Method

With the flourishing of deep learning, learning-based methods have increased the success of stereo-matching tasks. For example,

Shen et al. [18] proposed the cascade and fused stereo matching network (CFNet), which handled the large domain differences issue with fused cost volume representation and alleviate the unbalanced disparity distribution phenomenon by the cascade cost volume representation.

More recently, the performance of deep stereo networks has improved through the application of human-design differentiable operation to guide the matching cost aggregated at a finer level. For example,

Yang et al. [29] proposed a WaveletStereo network to learn the wavelet coefficients for the disparity map of stereo matching. The low-frequency wavelet predictors focus on learning the global contextual information of the disparity map. Moreover, the high-frequency wavelet predictors focus on generating the details of the disparity map.

Xu et al. [24] proposed the real-time bilateral grid network for deep stereo matching, which presented the edge-preserving cost volume up-sampling module with a learned bilateral grid to guide low-resolution cost volume up-sampling to high-quality cost volume.

These methods perform well in terms of stereo disparity estimation.

However, the disparities between smooth regions and detailed regions are difficult to accurately estimate simultaneously. Furthermore, a more efficient stereo-matching network architecture is also required to further improve the performance by alleviating the edge-fattening issue at discontinuities.

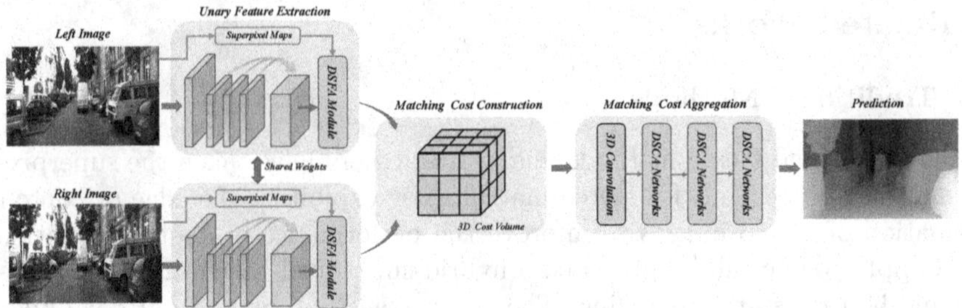

Fig. 1. The flowchart of our proposed superpixel-based feature and cost aggregation network for deep stereo matching.

3 Methods

In this paper, we proposed a differentiable superpixel-based feature and cost aggregation network for deep stereo matching. The flow diagram of the proposed method is shown in Fig. 1.

3.1 Unary Feature Extraction Networks

Both traditional and learning-based approaches suffer from drawbacks related to scale variation in computer vision applications, particularly in the field of stereo matching. In this paper, we introduce a more efficient feature extraction network, called, ElasticNet [23], with the improvement of feature representation ability by learning a dynamic scale policy in training processing. These extracted features could better reflect the similarity of stereo pair unary feature maps.

Our feature extraction network used Resnet-50 [8] as the benchmark architecture, modifying it to extract the unary feature maps utilizing an elastic convolution block. That network extracted multi-level unary features to concatenate the high-dimensional representations for matching cost volume construction. Those high-dimensional features include significant amounts of contextual information and reflect the similarity of left and right feature maps better.

3.2 Superpixel-Based Feature Aggregation

The unary feature extraction network's performance determined how well the similarity of stereo pair feature maps was represented. However, it faces some remains some challenges for stereo matching, such as poor imaging conditions, scale variation, and edge/contour information preservation issues, which must be solved to improve stereo matching performance.

In this paper, we propose a model-driven technique for feature aggregation and enhancement. The proposed DSFA method focuses more on the boundary or contour position feature aggregation and enhancement. Multi-level features are extracted and strengthen, reflecting the similarities between stereo pair images better. Let $F^A(\cdot)$ denote the aggregated feature maps and $F(\cdot)$ denote the raw feature maps. The differentiable superpixel-based feature aggregation model is represented as follows:

$$F^A(p) = \frac{1}{N_f}\sum_q W(p,q)\,S(p,q)\,F(q) \tag{1}$$

where q is the pixel within the user-aggregated support region R_f centered on pixel p, N_f is the regularization term that denotes the number of pixels in R_f, and the number of the support region superpixel label is the same as that of the region center superpixel label. $W(p,q)$ is the space similarity weight function and $S(p,q)$ is the truncated threshold function, which are defined as follows:

$$W(p,q) = e^{-\frac{(p-q)^2}{2\cdot\sigma^2}} \tag{2}$$

$$S(p,q) = \begin{cases} 1, & L(p) = L(q) \\ 0, & L(p) \neq L(q) \end{cases} \tag{3}$$

where $L(p)$ is the superpixel label of pixel p.

Specifically, the gradients of aggregated features $F^A(p)$ with respect to the raw features $F(q)$ will be computed as follows:

$$\frac{\partial F^A(p)}{\partial F(q)} = \frac{1}{N_f}W(p,q)\,S(p,q) \tag{4}$$

Therefore, the gradient of loss function F with respect to the raw features $F(q)$ will be computed as follows:

$$\frac{\partial F}{\partial F(q)} = \frac{\partial F}{\partial F^A(p)}\frac{\partial F^A(p)}{\partial F(q)} \tag{5}$$

3.3 Matching Cost Volume Computation

Once the unary feature maps are extracted, they are used to build the cost volume. Some researchers have used the full correlation method to construct the cost volume with the inner product of two feature maps at each disparity level [13,19,28] . Other scholars have used approaches that construct the cost volume by concatenating the left features with corresponding right feature maps at each disparity level [1,10,31]. Furthermore, GwcNet [7] evenly divides all feature map channels into several groups and constructs the cost volume in each corresponding feature group using the correlation method across all disparity levels. Based on these studies, we use a hybrid method that aims to combine the advantages of concatenation cost volume and correlation cost volume. Given the unary feature maps of stereo pair images, we constructed the combined matching cost volume for cost aggregation networks.

3.4 Superpixel-Based Cost Aggregation

The basis of this paper is the same as that for superpixel-based feature aggregation; therefore, we used a model-driven technique with differentiable superpixel-based cost aggregation (DSCA) to better preserve the edge or contour information and reduce disparity prediction error at edge positions. Let $C_d^A(\cdot)$ denote the aggregated cost volume and $C_d(\cdot)$ denote the raw cost volume. Then, the DSCA model can be defined as follows:

$$C_d^A(p) = \frac{1}{N_c} \sum_q W(p,q)\, S(p,q)\, C_d(q) \tag{6}$$

where q is a pixel within user-aggregated support region R_c centered on pixel p, N_c is the regularization term that denotes the number of pixels in R_c, and the number of support region superpixel labels is the same as that of the region center superpixel label. $W(p,q)$ is the space similarity weight function and $S(p,q)$ is the truncated threshold function. The definitions of these functions are the same as those for the aforementioned DSFA module.

Specifically, the gradients of aggregated cost volume $C_d^A(p)$ with respect to the raw cost volume $C_d(q)$ will be computed as follows:

$$\frac{\partial C_d^A(p)}{\partial C_d(q)} = \frac{1}{N_c} W(p,q)\, S(p,q) \tag{7}$$

Therefore, the gradients of loss function F with respect to the raw cost volume $C_d(q)$ will be computed as follows:

$$\frac{\partial F}{\partial C_d(q)} = \frac{\partial F}{\partial C_d^A(p)} \frac{\partial C_d^A(p)}{\partial C_d(q)} \tag{8}$$

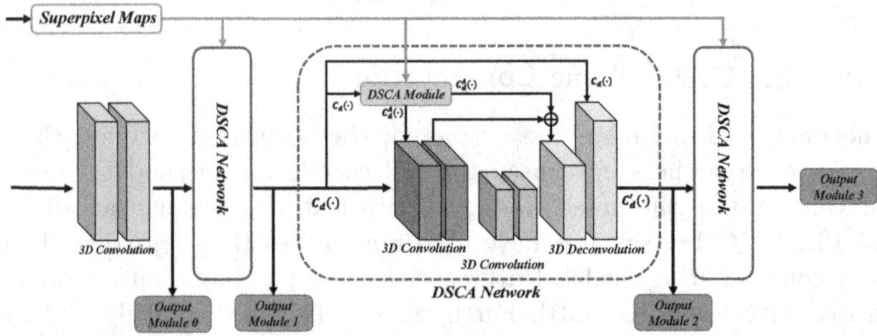

Fig. 2. The flowchart of superpixel based cost aggregation framework. The cost volume is guided aggregated by multiple stacked DSCA network.

3.5 Superpixel-Based Cost Aggregation Networks

Intuitively, the edge or contour feature information plays an important role in the accurate disparity prediction. In this paper, we proposed the differentiable superpixel-based cost aggregation (DSCA) network, which exploits the edge or contour feature information to guide the matching cost aggregation. The DSCA network comprises an encoder-decoder architecture with several 3D convolution layers. The cost volume is aggregated with the correspondence superpixel map by using the DSCA module. The aggregated cost volume will residual conjunction to the different stages (encoding and decoding stage) through the $\times 2$ down-sampling operation. The framework of DSCA networks is showing in Fig. 2.

Furthermore, we used the stacked hourglass DSCA networks to aggregate more contextual information for the matching cost volume. It comprises multiple encoder-decoder architectures, consisting of repeated top-down/bottom-up processing for cost aggregation. Each cost aggregation block is connected to the output module for disparity map prediction. Multiple outputs are trained with different weights and learned using a coarse-to-fine strategy.

4 Experiments

To explore our proposed DSFCA-Net, we have trained and tested it on different datasets, including Sceneflow [16], KITTI 2015 [17], and KITTI 2012 [5] datasets. We constructed the ablation study on Sceneflow dataset and trained with 20 epochs. The initial learning rate of that benchmark training is also set at 0.001, and it changed at $10, 14$, and 18 epochs by down-scaled 2 times.

The pre-trained model of DSFCA-Net on Sceneflow datasets is fine-tuned and trained with 300 epochs for KITTI benchmarks. The initial learning rate is also set at 0.001, and it changed at 200 epochs by down-scaled 10 times. The DSFCA-Net is implemented using the popular deep learning platform `PyTorch`. It trained with the `Adam` [12] optimizer, by setting $\beta_1 = 0.9$, $\beta_2 = 0.999$. There are two `Nvidia Tesla V100` GPUs for training our proposed network architecture, and fixed the network batch size to 8.

The metric of Sceneflow dataset is used as the mean average disparity error of pixels, called the end-point error (EPE) metric. The evaluation server of KITTI 2015 computes the percentage number of disparity errors pixels for background (*D1-bg*), foreground (*D1-fg*), or all pixels (*D1- all*) in non-occluded or all regions. For the KITTI 2012 benchmark, the evaluation server computes averages of bad pixels for all non-occluded (Out-Noc) and all occluded (Out-All) pixels.

The loss function is defined as follows:

$$L = \sum_{i=1}^{N_{out}} \eta_i \cdot \texttt{Smooth}_{L1}\left(G^* - \hat{G}_i\right). \tag{9}$$

where N_{out} denoted the output disparity numbers, G^* denoted groundtruth. η_i denoted the training weights with $\eta_0 = 0.5$, $\eta_1 = 0.5$, $\eta_2 = 0.7$, $\eta_3 = 1.0$., for the i-th prediction disparity map $\hat{G}_i$. $\texttt{Smooth}_{L1}(x)$ operator is defined as:

$$\texttt{Smooth}_{L1}(x) = \begin{cases} \frac{1}{2} \cdot x^2, & |x| < 1 \\ |x| - \frac{1}{2}, & |x| \geq 1 \end{cases}. \tag{10}$$

4.1 Ablation Study on SceneFlow Dataset

In the ablation study, we first analyzed the performance of our feature extraction networks. Without using the DSFA and DSCA modules, we achieves an EPE error of 0.7351. In comparison to other state-of-the-art methods' performance that shown in Table 3, it is clear that our feature network architecture is more efficient and with the improvement performance.

We also investigate the performance of the DSFA and DSCA modules. With only the DSFA module, we obtained the EPE performance gain of 0.7153, with an aggregation kernel size of 5, and a sigma value of 0.5. Meanwhile, we also obtained the better EPE performance of 0.7113 by only applying the DSCA module under the parameters aggregation kernel size 3 and sigma value 0.5. As the aggregate kernel size of DSFA and DSCA modules increases, the overall performance of our network model deteriorates slightly. Within the large aggregate kernel region, there is still some noise information to influence our network benchmark. Although the noise information is filtered out by the same superpixel label. Instinctively, we conducted the performance comparsion of the combination of DSFA and DSCA modules. Our method obtained the EPE performance 0.7002 with the aggregation kernel size 5 and sigma value 0.5 of DSFA model, and the aggregation kernel size 7 and sigma value 0.5 of DSCA model (Table 1).

Table 1. The ablation study on the SceneFlow.

Methods	Baseline	DSFA	DSCA	> 1px (%)	> 2px (%)	> 3px (%)	EPE (px)
DSFCA-Net	✓			7.7236	4.2694	3.1436	0.7351
	✓	✓		7.6109	4.1272	3.0637	0.7153
	✓		✓	7.5813	4.1412	3.0473	0.7113
	✓	✓	✓	7.5873	4.1566	3.0239	0.6966

Furthermore, we conducted the ablation study of the performance of superpixel maps changes. The ablation study about it is shown in the bottom part of Table 2. With the superpixel's segments and compactness increasing, the performance of our method is slightly worse. Over-aggressive segmentation of superpixel maps will destroy the texture and structure of the scene or target. There are some disparity prediction results of our DSFCA network on the SceneFlow dataset as shown in Fig. 3. Those results were obtained from our method DSFCA with different superpixel maps.

Finally, to objectively show the performance gain of our method, we used publically released stereo matching codes GwcNet [7] to analyze the performance of the proposed module. By applying the DSFA module, GwcNet has obtained much more performance gain and got better performance with aggregation kernel size 5 and sigma value 0.5 that reduced the end-point error from 0.7663 to 0.7341. We are with ablation study the DSCA module for the GwcNet method. It has reduced the end-point error from 0.7663 to 0.7301 with cost aggregation kernel

Table 2. The ablation study of superpixel changes and generalization on the SceneFlow dataset. The major parameters set include the number of segments N_S and compactness N_C of superpixel map generation.

Methods	Superpixel		DSFA	DSCA	> 1px (%)	> 2px (%)	> 3px (%)	EPE (px)
	N_S	N_C						
DSFCA-Net	–	–	–	–	7.7236	4.2694	3.1436	0.7351
	600	10	✓	✓	7.3625	4.1136	2.9782	0.6925
	1000	15	✓	✓	7.5873	4.1566	3.0239	0.6966
	2000	20	✓	✓	7.5895	4.1579	3.0372	0.7062
GwcNet-gc [7]	–	–	–	–	8.0315	4.4812	3.3107	0.7663
	1000	15	✓	–	7.6681	4.2607	3.1443	0.7341
	1000	15	–	✓	7.7430	4.2647	3.1449	0.7301

size 5 and sigma value 0.5. Hence, these ablation study results for the GwcNet show the performance gain of our proposed module and verify its universality and performance improvements over other network architectures. Those detailed experiment results are reported in Table 2.

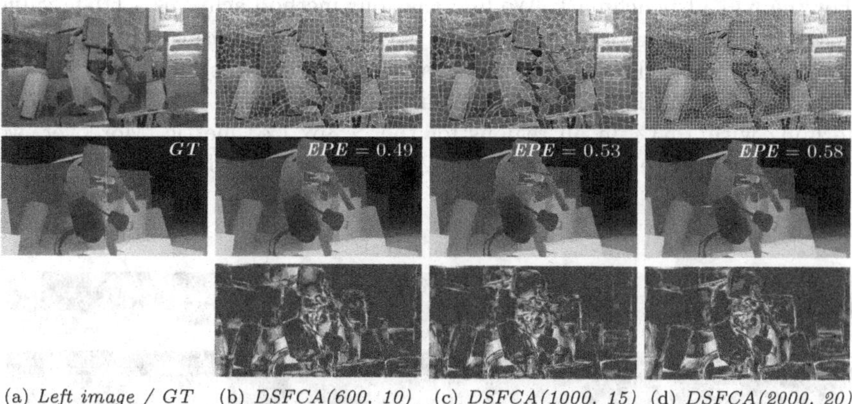

(a) *Left image / GT* (b) *DSFCA(600, 10)* (c) *DSFCA(1000, 15)* (d) *DSFCA(2000, 20)*

Fig. 3. Some disparity prediction results on the SceneFlow dataset. The illustrations from top to bottom are the left image or superpixel maps, the groundtruth or prediction disparity result, and the error maps. The *DSFCA* (1000, 15) donated using the superpixel maps obtained with the number of segment 1000 and compactness 15.

4.2 The Benchmark Evaluation Results

As shown in Table 3, it is the quantitative comparison results of the EPE metric for start-of-the-art stereo methods on the Sceneflow dataset. The DSFCA-Net obtained the 0.69px end-point error and outperforms most of the previous stereo-matching methods on the Sceneflow dataset.

Table 3. Comparative results with top-performing methods on different benchmarks.

Methods	Sceneflow	KITTI 2012		KITTI 2015			Times(s)
	EPE (px)	> 3px (%)		All Pixels (%)			
		Noc	All	D1-bg	D1-fg	D1-all	
ACFNet(2020) [30]	0.87	**1.17**	1.54	1.51	3.80	1.89	–
CamliFlow(2022) [14]	–	–	–	**1.48**	3.46	1.81	0.118
PVStereo(2022) [2]	–	1.20	1.61	1.50	3.43	1.82	0.05
CFNet(2021) [18]	–	1.23	1.58	1.54	3.56	1.88	0.18
DPCTF-S(2021) [4]	–	1.31	1.72	1.71	3.34	1.98	0.12
SANet(2022) [3]	0.86	1.32	1.70	1.64	3.61	1.93	0.32
GwcNet(2019) [7]	0.77	1.32	1.70	1.74	3.93	2.11	0.38
HITNet [20]	**0.43**	1.41	1.89	1.74	3.20	1.98	0.02
AANet(2020) [25]	0.83	1.91	2.42	1.65	3.96	2.03	0.14
LMNet(2022) [26]	1.02	–	–	2.81	4.39	3.24	0.0156
DSFCA-Net	0.69	1.19	**1.53**	1.52	**2.90**	**1.75**	0.28

For the KITTI benchmark, We find that our method achieves a D1-fg value of 2.90% in the foreground regions for KITTI 2015 dataset, which outperforms those published best-performing approaches such as ACFNet [30] and CFNet [18]. For the KITTI 2012 benchmark, our method achieves the best-performing with 1.53% in all regions for the > 3px (%) metric. Moreover, we showed the qualitative results on KITTI 2015 and 2012 benchmark in Fig. 4 and Fig. 5.

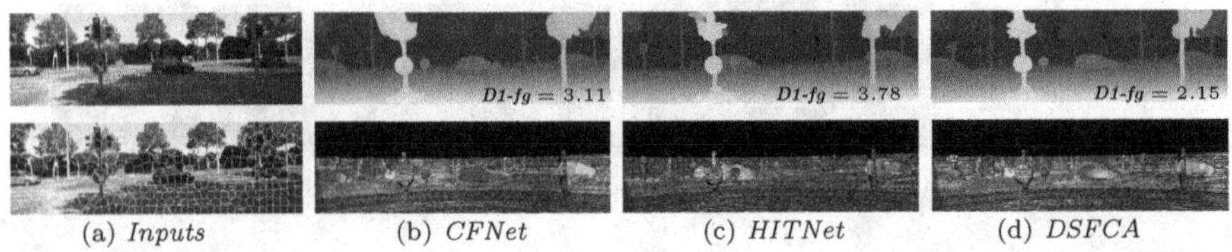

Fig. 4. Results of disparity estimation for KITTI 2015 test images. The left input image and its superpixel map will be showing in the left panel. For each input image, the disparity result maps predicted by (b) CFNet [18], (c) HITNet [21], and (d) our method are illustrated together above their error maps.

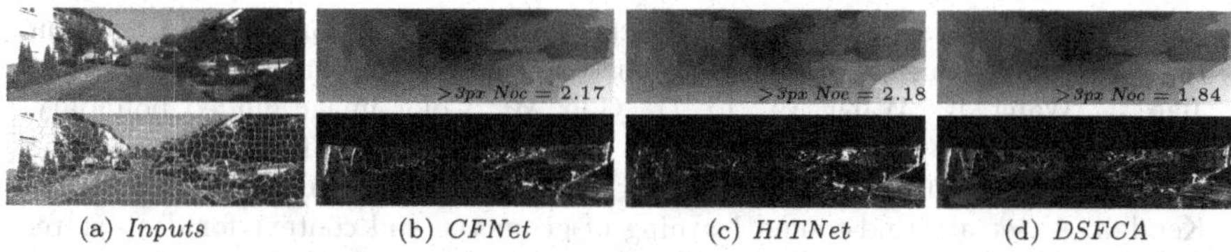

(a) *Inputs* (b) *CFNet* (c) *HITNet* (d) *DSFCA*

Fig. 5. Results of disparity estimation for KITTI 2012 test images. The left input image and its superpixel map will be showing in the left panel. For each input image, the disparity result maps predicted by (b) CFNet [18], (c) HITNet [21], and (d) our method are illustrated together above their error maps.

5 Conclusion

In this manuscript, we proposed superpixel-based feature and cost aggregation networks for deep stereo matching, which used the differentiable superpixel-based feature and cost aggregation module to guide the unary feature maps and matching cost volume aggregation. We used the SLIC method to generate the superpixel maps of stereo-pair images. The unary feature maps of stereo pair images are extracted and aggregated using the corresponding superpixel maps. The aggregate features represented the similarity of the left and right feature maps well and produced a high-quality initial cost volume. Then, the cost volume was aggregated using the differentiable superpixel-based cost aggregation model with the 3D hourglass aggregation network. Finally, the proposed hierarchical architecture was trained in a coarse-to-fine manner.

References

1. Chang, J.R., Chen, Y.S.: Pyramid stereo matching network. In: Proceedings of the IEEE Conference on Computer Vision and Pattern Recognition, pp. 5410–5418 (2018)
2. Chen, W., Jia, X., Wu, M., Liang, Z.: Multi-dimensional cooperative network for stereo matching. IEEE Robot. Automat. Lett. **7**(1), 581–587 (2022). https://doi.org/10.1109/LRA.2021.3130984
3. Chong, A.X., Yin, H., Wan, J., Liu, Y.T., Du, Q.Q.: Sa-net: scene-aware network for cross-domain stereo matching. Appl. Intell. 1–14 (2022)
4. Deng, Y., Xiao, J., Zhou, S.Z., Feng, J.: Detail preserving coarse-to-fine matching for stereo matching and optical flow. IEEE Trans. Image Process. **30**, 5835–5847 (2021). https://doi.org/10.1109/TIP.2021.3088635
5. Geiger, A., Lenz, P., Urtasun, R.: Are we ready for autonomous driving? the Kitti vision benchmark suite. In: 2012 IEEE Conference on Computer Vision and Pattern Recognition, pp. 3354–3361. IEEE (2012)
6. Gouveia, R., Spyropoulos, A., Mordohai, P.: Confidence estimation for superpixel-based stereo matching. In: 2015 International Conference on 3D Vision (3DV) (2015)
7. Guo, X., Yang, K., Yang, W., Wang, X., Li, H.: Group-wise correlation stereo network. In: Proceedings of the IEEE Conference on Computer Vision and Pattern Recognition, pp. 3273–3282 (2019)

8. He, K., Zhang, X., Ren, S., Sun, J.: Deep residual learning for image recognition. In: IEEE Conference on Computer Vision and Pattern Recognition (2016)
9. Jiao, J., Wang, R., Wang, W., Li, D., Gao, W.: Color image guided boundary-inconsistent region refinement for stereo matching. In: IEEE Transactions on Circuits and Systems for Video Technology, pp. 1155–1159 (2015)
10. Kendall, A., et al.: End-to-end learning of geometry and context for deep stereo regression. In: Proceedings of the IEEE International Conference on Computer Vision, pp. 66–75 (2017)
11. Kim, S., Min, D., Kim, S., Sohn, K.: Feature augmentation for learning confidence measure in stereo matching. In: IEEE Transactions on Image Processing A Publication of the IEEE Signal Processing Society, no. 12, p. 1 (2017)
12. Kingma, D., Ba, J.: Adam: a method for stochastic optimization. Comput. Sci. (2014)
13. Liang, Z., et al.: Learning for disparity estimation through feature constancy. In: Proceedings of the IEEE Conference on Computer Vision and Pattern Recognition, pp. 2811–2820 (2018)
14. Liu, H., Lu, T., Xu, Y., Liu, J., Li, W., Chen, L.: Camliflow: bidirectional camera-lidar fusion for joint optical flow and scene flow estimation. In: 2022 IEEE/CVF Conference on Computer Vision and Pattern Recognition (CVPR), pp. 5781–5791 (2022). https://doi.org/10.1109/CVPR52688.2022.00570
15. Lu, J., Li, Y., Yang, H., Min, D., Eng, W., Do, M.N.: Patchmatch filter: edge-aware filtering meets randomized search for visual correspondence. In: IEEE Transactions on Pattern Analysis and Machine Intelligence, no. 9, p. 1 (2017)
16. Mayer, N., et al.: A large dataset to train convolutional networks for disparity, optical flow, and scene flow estimation, pp. 4040–4048 (2016)
17. Menze, M., Geiger, A.: Object scene flow for autonomous vehicles. In: Proceedings of the IEEE Conference on Computer Vision and Pattern Recognition, pp. 3061–3070 (2015)
18. Shen, Z., Dai, Y., Rao, Z.: Cfnet: cascade and fused cost volume for robust stereo matching. In: IEEE/CVF Conference on Computer Vision and Pattern Recognition (CVPR) (2021)
19. Song, X., Zhao, X., Hu, H., Fang, L.: EdgeStereo: a context integrated residual pyramid network for stereo matching. In: Jawahar, C.V., Li, H., Mori, G., Schindler, K. (eds.) ACCV 2018. LNCS, vol. 11365, pp. 20–35. Springer, Cham (2019). https://doi.org/10.1007/978-3-030-20873-8_2
20. Tankovich, V., Häne, C., Zhang, Y., Kowdle, A., Fanello, S., Bouaziz, S.: Hitnet: hierarchical iterative tile refinement network for real-time stereo matching. In: CVPR (2021)
21. Tankovich, V., Häne, C., Zhang, Y., Kowdle, A., Fanello, S., Bouaziz, S.: Hitnet: Hierarchical iterative tile refinement network for real-time stereo matching. In: 2021 IEEE/CVF Conference on Computer Vision and Pattern Recognition (CVPR), pp. 14357–14367 (2021). https://doi.org/10.1109/CVPR46437.2021.01413
22. Tingman, Y., Yangzhou, G., Zeyang, X., Qunfei, Z.: Segment-based disparity refinement with occlusion handling for stereo matching. IEEE Trans. Image Process. (2019)
23. Wang, H., Kembhavi, A., Farhadi, A., Yuille, A.L., Rastegari, M.: Elastic: improving cnns with dynamic scaling policies. In: Proceedings of the IEEE/CVF Conference on Computer Vision and Pattern Recognition, pp. 2258–2267 (2019)
24. Xu, B., Xu, Y., Yang, X., Jia, W., Guo, Y.: Bilateral grid learning for stereo matching network. In: Proceedings of the IEEE/CVF Conference on Computer Vision and Pattern Recognition (CVPR) (2021)

25. Xu, H., Zhang, J.: Aanet: adaptive aggregation network for efficient stereo matching. In: IEEE/CVF Conference on Computer Vision and Pattern Recognition (CVPR), pp. 1956–1965 (2020). https://doi.org/10.1109/CVPR42600.2020.00203
26. Xue, Y., Zhang, D., Li, L., Li, S., Wang, Y.: Lightweight multi-scale convolutional neural network for real time stereo matching. Image Vis. Comput. **124**, 104510 (2022)
27. Yan, T., Gan, Y., Xia, Z., Zhao, Q.: Segment-based disparity refinement with occlusion handling for stereo matching. IEEE Trans. Image Process. **28**(8), 3885–3897 (2019). https://doi.org/10.1109/TIP.2019.2903318
28. Yang, G., Zhao, H., Shi, J., Deng, Z., Jia, J.: Segstereo: exploiting semantic information for disparity estimation. In: Proceedings of the European Conference on Computer Vision, pp. 636–651 (2018)
29. Yang, M., Wu, F., Li, W.: Waveletstereo: learning wavelet coefficients of disparity map in stereo matching. In: IEEE/CVF Conference on Computer Vision and Pattern Recognition (CVPR), pp. 12882–12891 (2020). https://doi.org/10.1109/CVPR42600.2020.01290
30. Zhang, Y., Chen, Y., Bai, X., Yu, S., Yang, K.: Adaptive unimodal cost volume filtering for deep stereo matching. Proc. AAAI Conf. Artif. Intell. **34**(7), 12926–12934 (2020)
31. Zhong, Y., Li, H., Dai, Y.: Open-world stereo video matching with deep RNN. In: Proceedings of the European Conference on Computer Vision (ECCV), pp. 101–116 (2018)

GMA3D: Local-Global Attention Learning to Estimate Occluded Motions of Scene Flow

Zhiyang Lu[1] and Ming Cheng[2(✉)]

[1] Institute of Artificial Intelligence, Xiamen University, Xiamen 361005, China
zhiyang@stu.xmu.edu.cn
[2] School of Informatics, Xiamen University, Xiamen 361005, China
chm99@xmu.edu.cn

Abstract. Scene flow plays a pivotal role in the 3D perception task by capturing motion information between consecutive frames. However, occlusion phenomena are commonly observed in dynamic scenes, whether from the sparsity data sampling or real-world occlusion. In this paper, we focus on addressing occlusion issues in scene flow by the semantic self-similarity and motion consistency of the moving objects. We propose a GMA3D module based on the transformer framework, which utilizes local and global semantic similarity to infer the motion information of occluded points from the motion information of local and global non-occluded points respectively, and then uses the Offset Aggregator to aggregate them. Our module is the first to apply the transformer-based architecture to gauge the scene flow occlusion problem on point clouds. Experiments show that our GMA3D can solve the occlusion problem in the scene flow, especially in the real scene. We evaluated the proposed method on the occluded version of point cloud datasets and get state-of-the-art results on the real-world KITTI dataset. To testify that GMA3D is still beneficial to non-occluded scene flow, we also conducted experiments on non-occluded version datasets and achieved promising performance on FlyThings3D and KITTI. The code is available at https://github.com/O-VIGIA/GMA3D.

Keywords: Scene flow estimation · Deep learning · Point clouds · Local-global attention

1 Introduction

Scene flow [22], a fundamental low-level motion information for the purpose of 3D dynamic scene understanding, computes the motion field between consecutive frames. This underlying motion information finds wide applications in various scene understanding downstream tasks related to autonomous driving, robotic navigation, and so on. The previous methods use RGB images to estimate the scene flow. However, with the advances in 3D sensors, it is easy to obtain point cloud data. PointNet [16] and PointNet++ [17] pioneered the direct extraction of features for raw point clouds, and then the deep learning networks [3,17, 27] in the field of point clouds continued to emerge. These works provide the

Q. Liu et al. (Eds.): PRCV 2023, LNCS 14426, pp. 16–27, 2024.
https://doi.org/10.1007/978-981-99-8432-9_2

necessary conditions for the scene flow task. Combined with these frameworks, many neural network architectures suitable for scene flow [1,2,7–9,23,26,28] are proposed, which have better performance than the traditional optimization-based methods. However, despite their successes on non-occluded datasets, these methods encounter challenges when inferring motion information for occluded objects.

In the scene flow task, the occluded points exist in the first frame (source) point cloud. We define it as a set of points without corresponding points and/or corresponding patches in the second frame (target). Furthermore, we divide occluded points into two categories: the first category has non-occluded points in local areas of the first frame point cloud, and these points are called local occluded points. The second kind of point is global occluded points, where there are no non-occluded points in their local areas. The previous method calculates the corresponding scene flow through the feature matching between two frames, which can well infer the scene flow of non-occluded points in the first frame, because such points have corresponding matching patches in the second frame, and the motion information can be deduced through the cross-correlation between the point clouds of the two frames. However, the occluded points have no corresponding matching patches in the second point cloud, so it is incapable to infer the motion information by cross-correlation. In contrast, humans often employ self-correlation when inferring the motion of occluded objects in dynamic scenes. For example, without considering collision, we can infer the motion information of the occluded head of the same vehicle from the tail. Therefore, the self-correlation of motion is very significant to solve the occlusion problem in scene flow.

Ouyang et al. integrated scene flow estimation with occlusion detection to infer motion information of occluded points [13]. While effective for local small-scale occlusions, this approach falls short in handling local large-scale and global occlusions. Moreover, it relies on point occlusion annotations, which are challenging to obtain from the real world. Jiang et al. [6] introduced a global motion aggregation (GMA) module based on transformers to estimate the motion information of occluded pixels in optical flow. Inspired by this, we propose GMA3D, which incorporates the transformer [21] into scene flow. By leveraging the self-similarity of point cloud features, we aggregate non-occluded motion features and extract the motion information of occluded points. However, previous approaches only focus on the global perspective of motion features and fail to consider the local consistency of motion, potentially resulting in erroneous motion estimation for local occlusion points.

To address these issues, we present the Local-Global Similarity Map (LGSM) module to calculate the local-global semantic similarity map and then employ the Offset Aggregator (OA) to aggregate motion information based on self-similarity. For local occlusion, we deduce the motion information of the occluded points from their local non-occluded neighbors based on local motion consistency. As far as global occlusion points, we apply the global semantic features to aggregate occluded motion features from non-occluded points. We utilize these local and

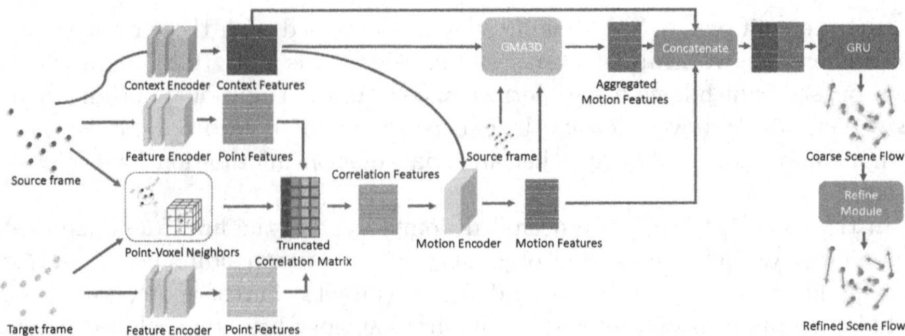

Fig. 1. The overall pipeline of our proposed framework. Our network is based on the successful PV-RAFT [26] architecture. The input of the GMA3D module is the context features and motion features of the point cloud in the first frame, and the output is the motion features aggregated locally and globally. These aggregated motion features are concatenated with context features and the original motion features, and then the concatenated features are fed into GRU for residual flow estimation, which is finally refined by the refine module.

global aggregated motion features to augment the successful PV-RAFT [26] framework and achieve state-of-the-art results in occluded scene flow estimation.

The key contributions of our paper are as follows. Firstly, we introduce GMA3D, a transformer-based framework specifically designed to tackle motion occlusion in the context of scene flow. This framework can be seamlessly integrated into any existing scene flow network, allowing for the inference of occluded motion. Within the GMA3D framework, we propose the LGSM module, which effectively leverages the self-consistency of motion information from both local and global perspectives. This module enhances the overall accuracy and robustness of motion estimation by considering multiple viewpoints. Additionally, we incorporate the Offset Aggregator, a mechanism that aggregates the motion features of non-occluded points with self-similarity to the occluded points. This aggregation process further improves the quality and completeness of motion inference for occluded regions. Experiments have shown that our GMA3D module has attained exceptional results in scene flow tasks and exhibits strong generalization on real-world point cloud data, whether in the case of occluded or non-occluded datasets.

2 Related Work

2.1 Motion Occlusion of Scene Flow

There are few techniques to address the occlusion problem of scene flow. Self-Mono-SF [4] utilizes self-supervised learning with 3D loss function and occlusion reasoning to infer the motion information of occlusion points in monocular scene flow. [5] combines occlusion detection, depth, and motion boundary estimation to infer occlusion points and scene flow. PWOC-3D [18] constructs a compact CNN architecture to predict scene flow in stereo image sequences and proposes a self-supervised strategy to produce the occlusion map for improving the accuracy of flow estimation. OGSF [13] presents the same backbone network to optimize scene flow and occlusion detection, then integrate the occlusion detection results with the cost volume between two frame point clouds, changing the cost volume of occlusion points to 0. Estimation&Propagation [25] utilizes a two-stage method to estimate scene flow, first for unobstructed points and then employing flow propagation for occluded points. However, they all rely on occlusion mask ground truth in the scene flow dataset. This paper will propose a module designed to estimate motion information of occluded points, which can be seamlessly integrated into any scene flow network architecture.

3 Methodology

3.1 Problem Statement

We consider scene flow as a 3D motion estimation task. It inputs two consecutive frames of point cloud data $PC_t = \{pc_t^i \in \mathbb{R}^3\}_{i=1}^N$ and $PC_{t+1} = \{pc_{t+1}^j \in \mathbb{R}^3\}_{j=1}^M$ and outputs the 3D vector $Flow = \{f_i \in \mathbb{R}^3\}_{i=1}^N$ of each point in the first frame of PC_t to indicate how to move to the corresponding position of the second frame.

3.2 Background

The backbone architecture of our GMA3D module is PV-RAFT [26]. The overall network diagram is shown in Fig. 1. For completeness, we will briefly introduce the PV-RAFT model. PV-RAFT adopts the point-voxel strategy to calculate the cost volume of the source point cloud. At the point level, the KNN method is used to find the points in the neighborhood of the target point cloud for short-distance displacement. At the voxel level, the points in the target point cloud are voxelized based on the source point cloud to capture the long-distance displacement. Then, it sends the point cloud context features together with the cost volumes into the GRU-based iteration module to estimate the residual flow. Finally, the flow features are smoothed in refine module. However, PV-RAFT removes the occlusion points when processing datasets, so it is unable to address the occlusion problem in scene flow.

3.3 Overview

We utilize the LGSM (Local-Global Similarity Map) module to quantify the semantic similarity between the local and global features of the point cloud in the first frame. Within the LGSM module, we employ a linear model with shared weights to establish mappings between the context features, query feature map, and key feature map. Next, we apply softmax and Multi-Layer Perception

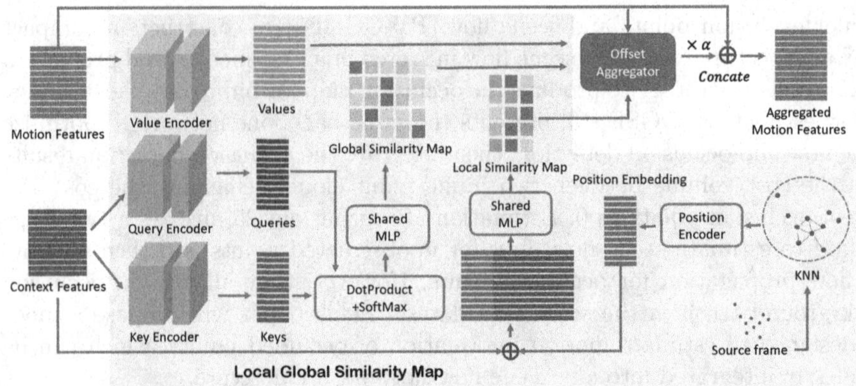

Fig. 2. Illustration of GMA3D module details. We use the LGSM module to map the local and global semantic similarity of the point cloud in the first frame. In the LGSM module, we map the context features to the query feature map and key feature map by a linear model with shared weights. Next, the attention map produced by the dot product is applied with softmax and Muti-Layer Perception(MLP) to generate the global self-similarity map. Then, the local similarity map is calculated by utilizing a Local GNN to model the positional relations in Euclidean space among the first point clouds. Finally, the local and global semantic similarity map are weighted sum with the motion features projected from the value encoder through the Offset Aggregator to output local and global aggregated motion features.

(MLP) to the attention map produced by the dot product, generating the global self-similarity map. Then, we calculate the local similarity map using the Local Graph Neural Network (GNN) [19] to model positional relations in Euclidean space among the first frame. Finally, we combine the maps with motion features projected from the value encoder through the Offset Aggregator to produce local and global aggregated motion features. These aggregated local-global motion features are concatenated with the original motion features and context features and then fed into the GRU module for iterative estimation of the scene flow. The detailed diagram of our GMA3D module is demonstrated in Fig. 2.

3.4 Mathematical Formulation

Let q, k, v be the *query*, *key* and *value* projection operators respectively, the formula is as follows:

$$q(\mathbf{x}_i) = \mathbf{Q}_m \mathbf{x}_i \tag{1}$$

$$k(\mathbf{x}_j) = \mathbf{K}_m \mathbf{x}_j \tag{2}$$

$$v(\mathbf{y}_j) = \mathbf{V}_m \mathbf{y}_j \tag{3}$$

In these formulas, $\mathbf{x} = \{\mathbf{x}_i\}_{i=1}^N \in \mathbb{R}^{N \times D_c}$ denote the context features and $\mathbf{y} = \{\mathbf{y}_j\}_{j=1}^N \in \mathbb{R}^{N \times D_m}$ indicate the motion features, where N is the number of the source point cloud, D_c and D_m refer to the dimension of context features

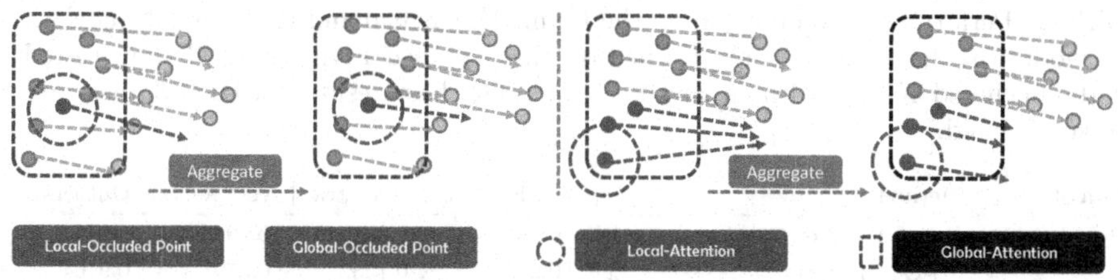

Fig. 3. Local-Global attention mechanism for GMA3D in local occluded points (left) and global occluded points (right)

and motion features respectively. Moreover, $\mathbf{Q}_m, \mathbf{K}_m \in \mathbb{R}^{D_c \times D_{q,k}}$ is a shared learnable linear projection and $\mathbf{V}_m \in \mathbb{R}^{D_m \times D_m}$.

First, we project the contextual information to the query map and key map employing $q(x)$ and $k(x)$, and compute the local similarity map by the Local-GNN [19] and global similarity map by the function $f(x, y)$ separately. Then, we map the motion features into value features by $v(y)$ and generate local and global aggregated motion features by semantic similarity maps, respectively.

$$\mathbf{g}_{local} = \sum_{j \in N(x_i)} \psi([\psi(pc_{t+1}^j - pc_t^i), \mathbf{x}_j, \mathbf{x}_i]))v(\mathbf{y}_j), \tag{4}$$

$$\mathbf{g}_{global} = \sum_{j \in P_{source}} f(q(\mathbf{x}_i), k(\mathbf{x}_j))v(\mathbf{y}_j). \tag{5}$$

Here $N(x_i)$ is the set of local neighborhood points of x_i captured by KNN, ψ is approximated with an MLP followed by an activation layer, '[,]' denotes the concatenation operator and f represents the operation given by

$$\bar{x}_{i,j} = \frac{\exp\left(\mathbf{x}_i^\top \mathbf{x}_j\right)}{\sum_{j=1}^{N} \exp\left(\mathbf{x}_i^\top \mathbf{x}_j\right)}, \tag{6}$$

$$f\left(\mathbf{x}_i, \mathbf{x}_j\right) = MLP\left(\frac{\bar{x}_{i,j}}{\sum_k \bar{x}_{i,k}}\right). \tag{7}$$

Finally, we apply the Offset Aggregator to get the local and global aggregated motion information and add it to the original motion information according to the learnable coefficient to obtain the final output.

$$\mathbf{g}_{offset} = h_{l,b,r}(\mathbf{y} - (\mathbf{g}_{local} + \mathbf{g}_{global})), \tag{8}$$

$$\tilde{y_i} = y_i + \alpha(\mathbf{g}_{offset}), \tag{9}$$

where $h_{l,b,r}$ refers to linear model, batch-norm and relu.

Table 1. Performance comparison on the FlyingThings3Do and KITTIo datasets. All the models in the table are only trained on the occluded Flyingthings3D and tested on the occluded KITTI without any fine-tune. The best results for each dataset are marked in bold.

Dataset	Method	Sup.	EPE3D(m)↓	Acc Strict↑	Acc Relax↑	Outliers↓
FlyThings3Do	PointPWC-Net [28]	Self	0.3821	0.0489	0.1936	0.9741
	3D-OGFLow [14]	Self	0.2796	0.1232	0.3593	0.9104
	PointPWC-Net [28]	Full	0.1552	0.4160	0.6990	0.6389
	SAFIT [20]	Full	0.1390	0.4000	0.6940	0.6470
	OGSF [13]	Full	0.1217	0.5518	0.7767	0.5180
	FESTA [24]	Full	0.1113	0.4312	0.7442	–
	3D-OGFLow [14]	Full	0.1031	0.6376	0.8240	0.4251
	Estimation&Propagation [25]	Full	0.0781	0.7648	0.8927	0.2915
	Bi-PointFlowNet [1]	Full	0.0730	0.7910	0.8960	0.2740
	WM [23]	Full	**0.0630**	0.7911	0.9090	0.2790
	GMA3D(Ours)	Full	0.0683	**0.7917**	**0.9171**	**0.2564**
KITTIo	PointPWC-Net [28]	Self	0.3821	0.0489	0.1936	0.9741
	3D-OGFLow [14]	Self	0.2796	0.1232	0.3593	0.9104
	SCOOP [8]	Self	0.0470	**0.9130**	0.9500	0.1860
	PointPWC-Net [28]	Full	0.1180	0.4031	0.7573	0.4966
	SAFIT [20]	Full	0.0860	0.5440	0.8200	0.3930
	OGSF [13]	Full	0.0751	0.7060	0.8693	0.3277
	FESTA [24]	Full	0.0936	0.4485	0.8335	–
	3D-OGFLow [14]	Full	0.0595	0.7755	0.9069	0.2732
	Estimation&Propagation [25]	Full	0.0458	0.8726	0.9455	0.1936
	Bi-PointFlowNet [1]	Full	0.0650	0.7690	0.9060	0.2640
	WM [23]	Full	0.0730	0.8190	0.8900	0.2610
	GMA3D(Ours)	Full	**0.0385**	0.9111	**0.9652**	**0.1654**

4 Experiments

4.1 Datasets

Following previous methods [7,13,15,26], we trained our model on the FlyThings3D [10] dataset and tested it on both FlyThings3D and KITTI [11,12] datasets respectively. Currently, there are two different approaches to processing these datasets, so we compare them separately on datasets generated by these different processing methods. The first method is derived from [2], in which the occluded point and some difficult points are removed. Following [15], we call this version of these datasets FlyThings3Ds and KITTIs. Another way to obtain the scene flow datasets comes from [9], where the information on occluded points is preserved. We refer to the second version of these datasets as FlyThings3Do and KITTIo. But unlike the previous method [7,9,15,26], we trained on all points including the occlusion points in the datasets of the occluded version to demonstrate that our GMA3D module can be used to solve the occlusion problem in the scene flow.

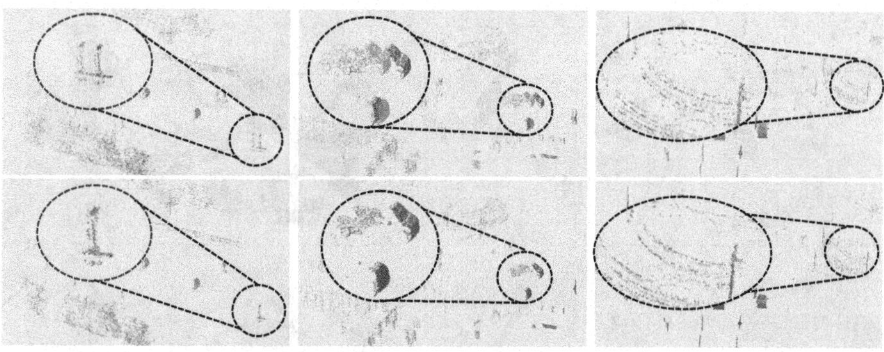

Fig. 4. Qualitative results on the KITTIo dataset of occluded version. Top: source point cloud (green) and target point cloud (red). Bottom: warped point cloud (green) utilizing the estimated flow of source point cloud and target point cloud (red). (Color figure online)

Table 2. Performance comparison on the FlyingThings3Ds and KITTIs datasets. All the models in the table are only trained on the non-occluded Flyingthings3D in a supervised manner and tested on the non-occluded KITTI without fine-tune. Our results for each dataset are marked in bold.

Dataset	Method	Sup.	EPE3D(m)↓	Acc Strict↑	Acc Relax↑	Outliers↓
FlyThings3Ds	PointPWC-Net [28]	Full	0.0588	0.7379	0.9276	0.3424
	FLOT [15]	Full	0.0520	0.7320	0.9270	0.3570
	PV-RAFT(baseline) [26]	Full	0.0461	0.8169	0.9574	0.2924
	GMA3D(Ours)	Full	**0.0397**	**0.8799**	**0.9727**	**0.2293**
KITTIs	PointPWC-Net [28]	Full	0.0694	0.7281	0.8884	0.2648
	FLOT [15]	Full	0.0560	0.7550	0.9080	0.2420
	PV-RAFT(baseline) [26]	Full	0.0560	0.8226	0.9372	0.2163
	GMA3D(Ours)	Full	**0.0434**	**0.8653**	**0.9692**	**0.1769**

4.2 Evaluation Metrics

Adhere to the previous methods [1,2,9,15,23,26], we still employed traditional evaluation operators to compare the performance of our GMA3D module, including EPE3D(m), Acc strict, Acc relax, and Outliers:

- **EPE3D(m)**: $\|f_{pred} - f_{gt}\|_2$. Average of the end-point-error at each point.
- **Acc Strict**: the percentage of points whose **EPE3D(m)** $< 0.05m$ or relative error $< 5\%$.
- **Acc Relax**: the percentage of points whose **EPE3D(m)** $< 0.1m$ or relative error $< 10\%$.
- **Outliers**: the percentage of points whose **EPE3D(m)** $> 0.3m$ or relative error $> 30\%$.

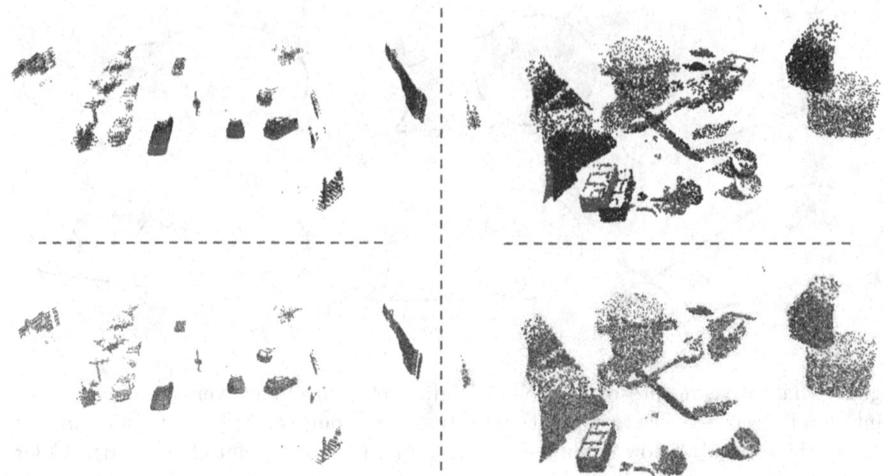

Fig. 5. Qualitative results on the KITTIs dataset(left) and FlyThings3Ds dataset (right) of non-occluded version. Top: source point cloud (blue) and target point cloud (red). Bottom: warped point cloud (green) utilizing the estimated flow of source point cloud and target point cloud (red). (Color figure online)

4.3 Performance on FT3Do and KITTIo with Occlusion

We conducted a comparative analysis between our GMA3D module and existing methods [1,8,9,13,15,23,25,28] on the datasets FT3Do and KITTIo. The KITTIo dataset contains 150 real-world LiDAR scenes. We trained the GMA3D module on n = 8192 points with an initial learning rate of 0.001 for 45 epochs, 12 iterations, and the refine module for 10 epochs, 32 iterations. The detailed results of the comparison are shown in Table 1. In the synthetic FT3Do dataset, the performance of our GMA3D model is basically the same as that of the state-of-the-art [25] method. However, GMA3D has a stronger generalization ability, which performs well on the KITTIo without any fine-tuning. And [13,14,25] relies on the ground truth of the occluded mask, which is challenging to obtain in the real world. In contrast, our GMA3D only depends on the 3D coordinates of the point clouds and exhibits greater competitiveness in the real world. Figure 4 visualizes the effect of GMA3D on scene flow estimation of occlusion points in the KITTIo dataset.

4.4 Performance on FT3Ds and KITTIs Without Occlusion

We also compared the results obtained for GMA3D on datasets FT3Ds and KITTIs with previous methods [15,26,28], and detailed comparison results are shown in Table 2. We increase the number of GRU iterations to 12 and epochs to 45 in training unlike the baseline [26], so the model can better integrate the original motion information and the motion information obtained by self-similarity aggregation. Experiments show that our GMA3D module achieves

Table 3. Ablation Studies of GMA3D on the KITTIo dataset

Method	EPE3D(m) ↓
Backbone w/o GMA3D	0.1084
GMA3D w/o Offset Aggregator (original aggregator = MLP)	0.0412
GMA3D w/o Offset Aggregator and Local Similarity Map	0.0882
GMA3D w/o Offset Aggregator and Global Similarity Map	0.0803
GMA3D (full, with Offset Aggregator and Local-Global Similarity Map)	**0.0385**

promising results and outperforms the baseline in terms of EPE by 13.9% and 22.5% on FT3Ds and KITTIs datasets respectively, which demonstrates that GMA3D can still produce more beneficial solutions to the non-occlusion scene while solving the occlusion problem, as revealed in Fig. 5.

Through our experiments, we concluded that there are two reasons why GMA3D improves the performances on the non-occlusion version of the datasets: First, from the analysis of occlusion above, we infer that the farthest point sampling algorithm may cause the deletion of local matching areas between two consecutive point clouds, leading to the occurrence of hidden occlusion. Secondly, our GMA3D module aggregates local and global motion information through self-similarity, which not only can smooth local motion but also decrease the motion inconsistency of local areas in the dynamic scene.

4.5 Ablation Studies

We conducted experiments on KITTIo datasets to testify to the effectiveness of various modules in the GMA3D, including the Offset Aggregator and LGSM module. We gradually add these modules to GMA3D, and the final results are shown in Table 3. From Table 3, we can deduce that each module plays an essential role in GMA3D. First, The model does not perform well when the Offset Aggregator is not introduced. This is because the original transformer is designed for the domain of natural language processing. However, there are many differences between natural language and point clouds, so it is unable to be directly applied to point clouds. Secondly, we find that only focusing on the global motion information will produce poor results. With the Local-Global Similarity Map Module, GMA3D can improve accuracy by aggregating motion features from local and global aspects, respectively.

5 Conclusion

In this study, we introduce GMA3D, a novel approach to address motion occlusion in scene flow. GMA3D leverages a local-global motion aggregation strategy to infer motion information for both local and global occluded points within the source point cloud. Furthermore, GMA3D can enforce local motion consistency for moving objects, which proves advantageous in estimating scene flow for non-occluded points as well. Experimental results obtained from datasets involving both occluded and non-occluded scenarios demonstrate the superior performance and generalization capabilities of our GMA3D module.

References

1. Cheng, W., Ko, J.H.: Bi-PointFlowNet: bidirectional learning for point cloud based scene flow estimation. In: Avidan, S., Brostow, G., Cissé, M., Farinella, G.M., Hassner, T. (eds.) ECCV 2022. LNCS, vol. 13688, pp. 108–124. Springer, Cham (2022). https://doi.org/10.1007/978-3-031-19815-1_7
2. Gu, X., Wang, Y., Wu, C., Lee, Y.J., Wang, P.: Hplflownet: hierarchical permutohedral lattice flownet for scene flow estimation on large-scale point clouds. In: Proceedings of the IEEE/CVF Conference on Computer Vision and Pattern Recognition, pp. 3254–3263 (2019)
3. Guo, M.H., Cai, J.X., Liu, Z.N., Mu, T.J., Martin, R.R., Hu, S.M.: PCT: point cloud transformer. Comput. Vis. Media **7**(2), 187–199 (2021)
4. Hur, J., Roth, S.: Self-supervised monocular scene flow estimation. In: Proceedings of the IEEE/CVF Conference on Computer Vision and Pattern Recognition (CVPR) (June 2020)
5. Ilg, E., Saikia, T., Keuper, M., Brox, T.: Occlusions, motion and depth boundaries with a generic network for disparity, optical flow or scene flow estimation. In: Proceedings of the European Conference on Computer Vision (ECCV), pp. 614–630 (2018)
6. Jiang, S., Campbell, D., Lu, Y., Li, H., Hartley, R.: Learning to estimate hidden motions with global motion aggregation. In: Proceedings of the IEEE/CVF International Conference on Computer Vision, pp. 9772–9781 (2021)
7. Kittenplon, Y., Eldar, Y.C., Raviv, D.: Flowstep3d: model unrolling for self-supervised scene flow estimation. In: Proceedings of the IEEE/CVF Conference on Computer Vision and Pattern Recognition, pp. 4114–4123 (2021)
8. Lang, I., Aiger, D., Cole, F., Avidan, S., Rubinstein, M.: Scoop: self-supervised correspondence and optimization-based scene flow. In: Proceedings of the IEEE/CVF Conference on Computer Vision and Pattern Recognition, pp. 5281–5290 (2023)
9. Liu, X., Qi, C.R., Guibas, L.J.: Flownet3d: learning scene flow in 3d point clouds. In: Proceedings of the IEEE/CVF Conference on Computer Vision and Pattern Recognition, pp. 529–537 (2019)
10. Mayer, N., et al.: A large dataset to train convolutional networks for disparity, optical flow, and scene flow estimation. In: Proceedings of the IEEE Conference on Computer Vision and Pattern Recognition, pp. 4040–4048 (2016)
11. Menze, M., Geiger, A.: Object scene flow for autonomous vehicles. In: Proceedings of the IEEE Conference on Computer Vision and Pattern Recognition, pp. 3061–3070 (2015)
12. Menze, M., Heipke, C., Geiger, A.: Joint 3d estimation of vehicles and scene flow. ISPRS Ann. Photogram. Remote Sens. Spatial Inf. Sci. **2**, 427 (2015)
13. Ouyang, B., Raviv, D.: Occlusion guided scene flow estimation on 3d point clouds. In: Proceedings of the IEEE/CVF Conference on Computer Vision and Pattern Recognition, pp. 2805–2814 (2021)
14. Ouyang, B., Raviv, D.: Occlusion guided self-supervised scene flow estimation on 3d point clouds. In: 2021 International Conference on 3D Vision (3DV), pp. 782–791. IEEE (2021)

15. Puy, G., Boulch, A., Marlet, R.: Flot: scene flow on point clouds guided by optimal transport. In: European Conference on Computer Vision, pp. 527–544. Springer (2020)
16. Qi, C.R., Su, H., Mo, K., Guibas, L.J.: Pointnet: deep learning on point sets for 3d classification and segmentation. In: Proceedings of the IEEE Conference on Computer Vision and Pattern Recognition, pp. 652–660 (2017)
17. Qi, C.R., Yi, L., Su, H., Guibas, L.J.: Pointnet++: deep hierarchical feature learning on point sets in a metric space. Adv. Neural Inf. Process. Syst. **30** (2017)
18. Saxena, R., Schuster, R., Wasenmuller, O., Stricker, D.: Pwoc-3d: deep occlusion-aware end-to-end scene flow estimation. In: 2019 IEEE Intelligent Vehicles Symposium (IV), pp. 324–331. IEEE (2019)
19. Shi, W., Rajkumar, R.: Point-GNN: graph neural network for 3d object detection in a point cloud. In: Proceedings of the IEEE/CVF Conference on Computer Vision and Pattern Recognition (CVPR) (June 2020)
20. Shi, Y., Ma, K.: Safit: segmentation-aware scene flow with improved transformer. In: 2022 International Conference on Robotics and Automation (ICRA), pp. 10648–10655. IEEE (2022)
21. Vaswani, A., et al.: Attention is all you need. Adv. Neural Inf. Process. Syst. **30** (2017)
22. Vedula, S., Baker, S., Rander, P., Collins, R., Kanade, T.: Three-dimensional scene flow. In: Proceedings of the Seventh IEEE International Conference on Computer Vision, vol. 2, pp. 722–729. IEEE (1999)
23. Wang, G., et al.: What matters for 3D scene flow network. In: Avidan, S., Brostow, G., Cissé, M., Farinella, G.M., Hassner, T. (eds.) ECCV 2022. LNCS, vol. 13693, pp. 38–55. Springer, Cham (2022). https://doi.org/10.1007/978-3-031-19827-4_3
24. Wang, H., Pang, J., Lodhi, M.A., Tian, Y., Tian, D.: Festa: flow estimation via spatial-temporal attention for scene point clouds. In: Proceedings of the IEEE/CVF Conference on Computer Vision and Pattern Recognition, pp. 14173–14182 (2021)
25. Wang, K., Shen, S.: Estimation and propagation: scene flow prediction on occluded point clouds. IEEE Robot. Automat. Lett. **7**, 12201–12208 (2022)
26. Wei, Y., Wang, Z., Rao, Y., Lu, J., Zhou, J.: Pv-raft: point-voxel correlation fields for scene flow estimation of point clouds. In: Proceedings of the IEEE/CVF Conference on Computer Vision and Pattern Recognition, pp. 6954–6963 (2021)
27. Wu, W., Qi, Z., Fuxin, L.: Pointconv: deep convolutional networks on 3d point clouds. In: Proceedings of the IEEE/CVF Conference on Computer Vision and Pattern Recognition, pp. 9621–9630 (2019)
28. Wu, W., Wang, Z.Y., Li, Z., Liu, W., Fuxin, L.: PointPWC-Net: cost volume on point clouds for (self-)spervised scene flow etimation. In: Vedaldi, A., Bischof, H., Brox, T., Frahm, J.-M. (eds.) ECCV 2020. LNCS, vol. 12350, pp. 88–107. Springer, Cham (2020). https://doi.org/10.1007/978-3-030-58558-7_6

Diffusion-Based 3D Object Detection with Random Boxes

Xin Zhou[1], Jinghua Hou[1], Tingting Yao[1], Dingkang Liang[1], Zhe Liu[1], Zhikang Zou[2], Xiaoqing Ye[2], Jianwei Cheng[3], and Xiang Bai[1(✉)]

[1] Huazhong University of Science and Technology, Wuhan, China
{xzhou03,xbai}@hust.edu.cn
[2] Baidu Inc., Beijing, China
[3] JIMU Intelligent Technology Co., Ltd., Wuhan, China

Abstract. 3D object detection is an essential task for achieving autonomous driving. Existing anchor-based detection methods rely on empirical heuristics setting of anchors, which makes the algorithms lack elegance. In recent years, we have witnessed the rise of several generative models, among which diffusion models show great potential for learning the transformation of two distributions. Our proposed Diff3Det migrates the diffusion model to proposal generation for 3D object detection by considering the detection boxes as generative targets. During training, the object boxes diffuse from the ground truth boxes to the Gaussian distribution, and the decoder learns to reverse this noise process. In the inference stage, the model progressively refines a set of random boxes to the prediction results. We provide detailed experiments on the KITTI benchmark and achieve promising performance compared to classical anchor-based 3D detection methods.

Keywords: 3D object detection · Diffusion models · Proposal generation

1 Introduction

3D object detection, a fundamental task in computer vision, aims to regress the 3D bounding boxes and recognize the corresponding category from the point clouds. It is widely used in autonomous driving as a core component for 3D scene understanding. However, due to the intrinsic sparsity and irregularity of the derived point clouds, building high-accuracy LiDAR-based 3D object detection methods is challenging.

Recently, the mainstream approaches can be divided into two categories according to the representation formats of the point clouds: point-based [36,47] and voxel-based methods [9,19,28,37,42,49]. Although point-based methods have achieved reliable performance on object localization, they are challenging to handle large-scale point clouds scenarios due to high computation costs of point clouds sampling and grounding operations [32]. In contrast, the voxel-based manners can convert irregular raw point clouds into regular voxel grid format and implement efficient feature extraction on voxels by highly optimized 3D sparse convolution operations. Thus, to trade off the performance and efficiency, the network in this paper is mainly based on the voxel representation. However, existing voxel-based methods [19,28,42,44,49] still rely on empirical heuristics to set anchor sizes or center radii, which might not be an elegant strategy, as shown

Q. Liu et al. (Eds.): PRCV 2023, LNCS 14426, pp. 28–40, 2024.
https://doi.org/10.1007/978-981-99-8432-9_3

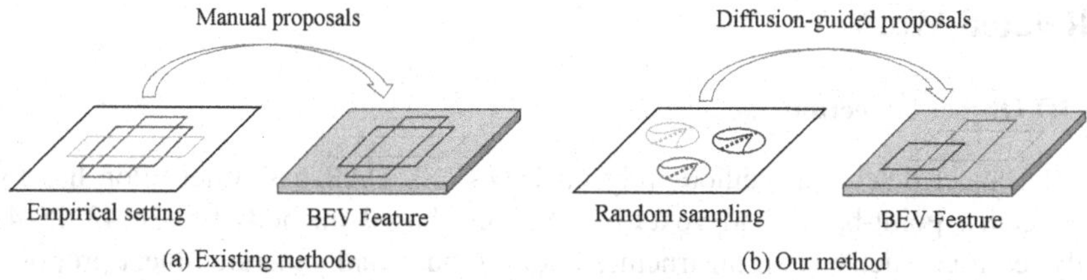

Fig. 1. Compared with existing anchor-based 3D object detection paradigms. (a) Manual set proposals; (b) Diffusion-guided proposals (ours). The existing methods rely on manual anchors for prediction, while ours requires only Gaussian noises.

in Fig. 1(a). This leads us to ask *whether it is feasible to recover predictive boxes from more succinct random boxes directly*.

Fortunately, we have witnessed the rise of diffusion models [15,39], a probabilistic modeling paradigm that demonstrates its potential in many 2D vision tasks (e.g., image generation [13,25,33], segmentation [1,3,6] and 2D object detection [5]). Among these, DDPM [15] stands out as a seminal advancement. It treats the process of image generation as a Markov process, wherein Gaussian noise is intentionally introduced to the target distribution. Through the training of a dedicated network, this noise is subsequently removed, facilitating the restoration of the pristine underlying structure. By learning the intricate mapping from a Gaussian distribution to the target distribution, diffusion models inherently possess the intrinsic capability to denoise ideal data. Despite their resounding success in 2D vision tasks, the untapped potential of diffusion models in proposal generation for 3D object detection remains unexplored.

In this paper, we present a framework named Diff3Det to explore the feasibility of generative models for 3D object detection. Specifically, the model adds Gaussian noise with a controlled variance schedule to the ground truth boxes in the training stage to obtain noisy boxes, as shown in Fig. 1(b). These noisy boxes are then used to extract Region of Interest (RoI) features from the BEV feature map, which does not need to set the manual anchor. The detection decoder then incorporates these features and time planes to predict offsets between the noisy and ground truth boxes. As a result, the model can recover the ground truth boxes from the noisy ones. We reverse the learned diffusion process during inference to generate bounding boxes that fit a noisy prior distribution to the learned distribution over the bounding boxes.

Our main contributions of this paper can be summarized in two folds as follows:

- We present a framework named Diff3Det that explores the feasibility of generative models for 3D object detection, achieving promising performance compared with the popular 3D object detectors, and demonstrating the potential of diffusion models in 3D vision tasks.
- We design several simple strategies to select proposal boxes to address the sparsity of point cloud data and 3D features to improve the usability of the methods. In addition, we propose an optimized noise variance scheduling for diffusion models that can be better adapted to the 3D object detection task.

2 Related Work

2.1 3D Object Detection

Most 3D object detection methods rely on LiDAR. LiDAR-based detection has two main streams: point-based and voxel-based. Point-based methods [20,36,43,45–47] directly learn geometry from unstructured point clouds and generate object proposals. However, these methods often have insufficient learning capacity and limited efficiency. In contrast, VoxelNet [49] converts the irregular point clouds into regular voxel grids. To improve computational efficiency, some methods [9,10,14,28,42] leverage highly optimized 3D submanifold sparse convolutional networks. Due to the assumption of no repeating objects in height in autonomous driving, some methods [19,34] only voxelize in the plane and apply 2D convolution to further improve computational efficiency. Although voxel-based methods are computationally efficient and better suited for feature extraction, they inevitably introduce information loss. Researchers [30,35] utilize both point-based and voxel-based representations for further learning.

Different from LiDAR-based 3D object detection, image-based methods can significantly reduce sensor costs. Image-based methods [22,26,41] for 3D object detection estimate depth and detect objects from 2D images, but the performance of these methods is still limited. To overcome this challenge, multimodal-based 3D object detection methods [2,16,21,27] combine precise geometric information from LiDAR with rich semantic information from images, resulting in state-of-the-art performance.

Here, we notice that many previous approaches [19,28,42,49] still require manual selection of anchor boxes in advance for subsequent proposal generation, which largely depends on the human experience. To the best of our knowledge, more elegant ways to generate proposals are still under-explored.

2.2 Diffusion Models in Vision

Diffusion is a physical model aimed at minimizing the spatial concentration difference. In the computer vision field, Diffusion [15,39] is a probabilistic model that uses a forward process to transform the initial distribution into a normal distribution and train a network to reverse the noise. The diffusion model has shown promising results in many tasks, including image generation [13,25,33], segmentation [1,3,6] and depth estimation [11,17]. Recently, DiffusionDet [5] extends the diffusion process into generating detection box proposals, showing that the prospects for applying diffusion models to detection tasks are bright. Inspired by DiffusionDet, we explore the application of Diffusion models on proposals generation in 3D object detection.

3 Proposed Method

In this section, we first revisit the diffusion model in Sect. 3.1. Then, we introduce the overall of our method (Sect. 3.2) and details of the proposed proposal generator (Sect. 3.3) as well as training and inference processes separately.

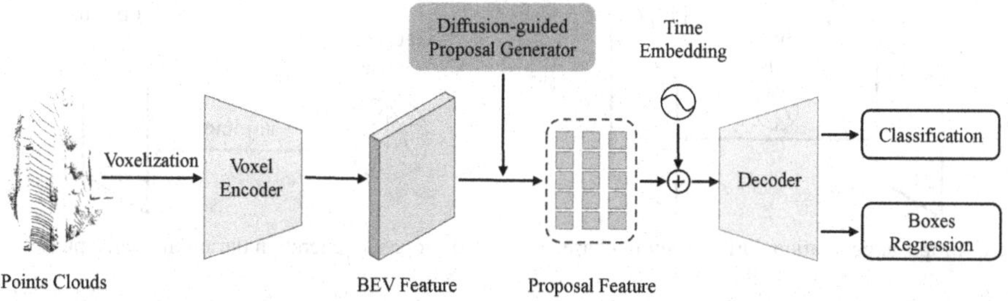

Fig. 2. The pipeline of our method. The point clouds are fed to a 3D encoder for generating BEV features. Then, the diffusion-guided proposal generator generates some random proposals on the BEV. Finally, the detection decoder consumes BEV proposal features and time embeddings to predict the detection results.

3.1 A Revisit of Diffusion Models

The diffusion model [15,39] is a powerful generative model that generates high-quality samples from Gaussian noise, whose pipeline consists of forward and backward processes. Specifically, it builds the forward process by gradually adding noise to a sample and transforming the sample into a latent space with an increasing noise, which follows the Markov chain. The forward process is formulated as follows:

$$q\left(x_t \mid x_0\right)=\mathcal{N}\left(x_t ; \sqrt{\bar{\alpha}_t} x_0,\left(1-\bar{\alpha}_t\right) I\right), \tag{1}$$

$$\bar{\alpha}_t:=\prod_{i=0}^t \alpha_i=\prod_{i=0}^t\left(1-\beta_i\right), \tag{2}$$

where x_0, x_t, and β_i represent the sample, latent noisy sample, and noise variance schedule, respectively. During training, a neural network $f_\theta(x_t, t)$ is trained to predict the original sample x_0 from the noisy sample x_t at each time step t by minimizing the ℓ_2 loss between the predicted and original sample.

$$\mathcal{L}_{train}=\frac{1}{2}\left\|f_\theta(x_t, t)-x_0\right\|^2 . \tag{3}$$

The model reconstructs the original data sample from the noisy sample at the inference stage by iteratively applying the updating rule in reverse order.

In our work, we attempt to apply the diffusion model to the 3D object detection task. We consider the ground-truth bounding boxes as x_0, where $x_0 \in \mathbb{R}^{N\times 5}$. A network $f_\theta(x_t, t, x)$ is trained to predict x_0 from noisy boxes x_t by the corresponding point clouds features x.

3.2 Overall

Our Diff3Det consists of a diffusion-guided proposal generator, an encoder, and a decoder, as shown in Fig. 2. The diffusion-guided proposal generator generates corrupted boxes x_t by adding Gaussian noise on the ground truth boxes. The encoder, a 3D voxel backbone [42], is utilized to extract the features of point clouds. The decoder aims

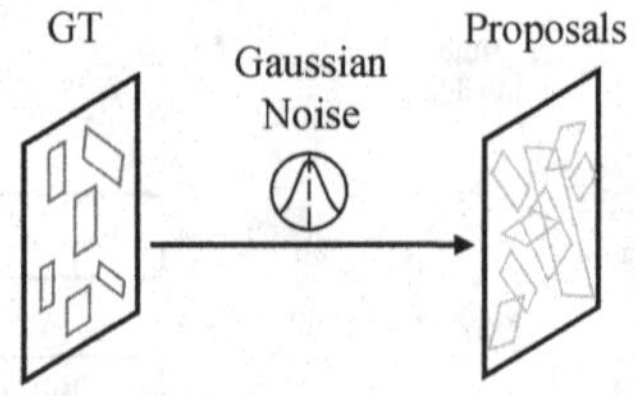

(a) Proposal generation during training phase

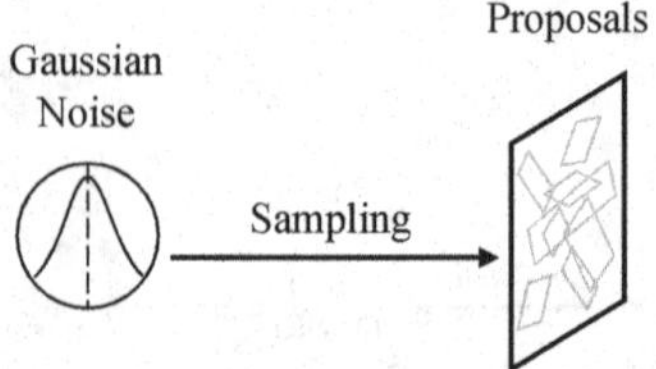

(b) Proposal generation during inference phase

Fig. 3. Diffusion-guided Proposal Generator. Our diffusion-guided proposal generator generates proposals in the training (a) and inference (b) phases by adding Gaussian noise on the ground truth and sampling from Gaussian noise.

to predict the original ground truth boxes by the corrupted boxes x_t and corresponding region of interest (RoI) features. Specifically, we utilize the dynamic head [40], extended to 3D object detection. Our approach does not rely on the learnable query and learnable embedding for dynamic convolution prediction but instead adopts randomly selected proposal boxes, temporal noise levels, and RoI features.

3.3 Diffusion-Guided Proposal Generator

Bird's eye view (BEV) is an effective representation for 3D object detection. Therefore, our method uses the BEV boxes (cx, cy, dx, dy, θ) for the diffusion process. For constructing the initial boxes x_0, we repeat the ground truth boxes to N and normalize them between 0 and 1. A signal scaling factor controls the diffusion process's signal-to-noise ratio (SNR) [7]. Then, we generate the corrupted boxes x_t by adding Gaussian noise to x_0, which is formulated as:

$$x_t = \sqrt{\bar{\alpha}_t} x_0 + \sqrt{1 - \bar{\alpha}_t} \varepsilon, \tag{4}$$

where $\varepsilon \sim \mathcal{N}(0, I_5)$, $t = \text{randint}(1, T_{\text{max}})$, $\bar{\alpha}_t$ is the same as [39]. The maximum time T_{max} is set to an integer (e.g., 1000).

As shown in Fig. 3(a), the diffusion-guided proposal generator generates proposal boxes from the ground truth during training. Firstly, a proposal box with no point makes it tough to recover the target. We adopt a resampling operation by calculating the number of points m in each proposal box. If $m < \eta$, remove the boxes and resample random boxes. Repeat this loop until every proposal box with at least η points in it. Moreover, we find that the quality of the proposal boxes is the key to the success of our method, so we adopt simple ways to refine the proposal boxes, which will be explained in detail in the following sections.

Correlation Coefficient on Size. It is clear that in the real world, there is a definite relationship between the width and length of a 3D detection box. However, the two independent random distributions of width and length will produce unrealistic proposals, as shown in Fig. 4(a). For this reason, it is inappropriate to consider the size (w, l)

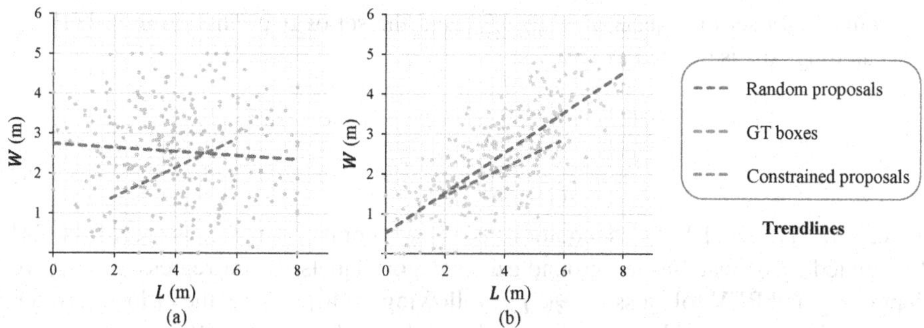

Fig. 4. Distribution of Random proposals vs. Constrained proposals. The distribution of our constrained proposals is more correlated with GT than random proposals.

as two independent and identically distributed random variables. Therefore, we introduce a correlation coefficient to restrict the box size for those resampled boxes and the boxes used during inference.

$$W = \rho L + \sqrt{1 - \rho^2} X, \tag{5}$$

where $L, X \overset{\text{i.i.d.}}{\sim} \mathcal{N}(0, 1)$, and we set the value of correlation coefficient $\rho = 0.8$. After generating the random vector (W, L), we scale them to the ranges $(0, w)$ and $(0, l)$ as the width and length of the proposals. We set $w = 8, l = 5$ to satisfy the target. As shown in Fig. 4(b), after the correlation coefficient constraint, the distribution of generated proposals more correlated with ground truth.

Dynamic Time Step. In the early training phase, recovering samples from seriously corrupted samples with high noise levels is difficult, which harms the final performance. Therefore, we propose a sine schedule to control the time step range, where the noise gradually increases in training. Specifically, n is the total training epoch number, x is the current epoch index, and T is the maximum time to reach. The maximum time on one training epoch T_{max} can be calculated as:

$$T_{max} = \begin{cases} T \left\lfloor \sin\left(\frac{\cos^{-1}\left(\frac{\omega}{T}\right)}{\sigma n} x + \sin^{-1}\left(\frac{\omega}{T}\right) \right) \right\rfloor & , x < \sigma n \\ T & , x \geq \sigma n \end{cases} \tag{6}$$

where ω and σ are hyperparameters that control the initial time steps at the first epoch and control the training time-point to reach the maximum time T, respectively. We empirically set $\omega = 5$ and $\sigma = 0.5$.

3.4 Loss Function

Some methods [4,40] minimize the time-consuming post-processing operation by the bipartite matching. Therefore, we extend the bipartite matching from 2D to 3D. Given the ground truth set of objects $y = \{y_i\}_{i=1}^{M}$ and the set of N prediction $\hat{y} = \{\hat{y}_i\}_{i=1}^{N}$. The matching cost is defined as follows:

$$\mathcal{C}_{\text{match}} = \lambda_{cls} \cdot \mathcal{L}_{cls} + \lambda_{reg} \cdot \mathcal{L}_{reg} + \lambda_{IoU} \cdot \mathcal{L}_{BEV\_IoU} \tag{7}$$

$$\mathcal{C} = \underset{i \in \mathrm{M}, j \in \mathrm{N}}{\arg\min}\, \mathcal{C}_{\text{match}}(\hat{y}_i, y_j), \tag{8}$$

where, λ_{cls}, λ_{reg}, and λ_{IoU} are coefficients of each component. $\mathcal{L}_{cls}$ is focal loss [24] of predicted classifications and ground truth category labels. As for regression loss, we adopt the ℓ_1 and BEV IoU loss $\mathcal{L}_{BEV\_IoU}$ following [4,40]. $\mathcal{L}_{reg}$ is the ℓ_1 loss between the normalized predicted boxes and ground truth boxes following [49].

The training loss consists of classification, regression, and IoU, applied only to the matched pairs. The IoU loss adopts the rotated 3D DIoU loss [48] denoted as $\mathcal{L}_{DIoU}$.

$$\mathcal{L} = \lambda_{cls} \cdot \mathcal{L}_{cls} + \lambda_{reg} \cdot \mathcal{L}_{reg} + \lambda_{IoU} \cdot \mathcal{L}_{DIoU}, \tag{9}$$

where the λ_{cls}, λ_{reg}, and λ_{IoU} represent the weight of corresponding loss, which is the same as the parameters in Eq. 7. We set $\lambda_{cls} = 2$, $\lambda_{reg} = 5$, and $\lambda_{IoU} = 2$.

3.5 Inference Phase

The inference procedure of Diff3Det is a denoising process from noise to object boxes. As shown in Fig. 3(b), Diff3Det progressively refines its predictions from boxes sampled in Gaussian distribution. In each sampling step, the random boxes or the estimated boxes from the last step are fed to the decoder to predict the results in the current stage. The proposal boxes for the next step can be computed by the formula [39]:

$$\boldsymbol{x}_{t-s} = \sqrt{\alpha_{t-s}} \left(\frac{\boldsymbol{x}_t - \sqrt{1-\alpha_t}\epsilon_\theta^{(t)}(\boldsymbol{x}_t)}{\sqrt{\alpha_t}} \right) + \sqrt{1 - \alpha_{t-s} - \sigma_t^2} \cdot \varepsilon_\theta^{(t)}(\boldsymbol{x}_t) + \sigma_t \varepsilon_t, \tag{10}$$

$$\sigma_t = \sqrt{\frac{1 - \alpha_t/\alpha_{t-s}}{(1-\alpha_{t-s})/(1-\alpha_t)}}, \tag{11}$$

where $\boldsymbol{x}_t, \boldsymbol{x}_{t-s}$ represent the proposal boxes in two adjacent steps, $\varepsilon_\theta^{(t)}(\boldsymbol{x}_t)$ is the predicted offsets by the decoder, and ε_t is the Gaussian noises. The number of sampling steps is allowed to be equal to or higher than 1, and the s is the starting time level (i.e., 1000) divided by sampling steps. Besides, the multiple iterations will lead to redundant boxes requiring an added NMS to filter them.

4 Results and Analysis

4.1 Dataset

Our experiments are conducted on the KITTI dataset [12], which is split into 3717 training and 3769 validation samples. We use the average precision (AP) metric, where the IoU threshold is set to 0.7 for the car category. All experiments are conducted on the car category with easy, moderate, and hard three levels.

Table 1. 3D object detection results are evaluated on the KITTI validation set with AP calculated by 11 recall positions. We report the average precision of 3D boxes (AP_{3D}) and bird's eye view (AP_{BEV}) for the car category.

Method	Modality	AP_{3D} ($IoU = 0.7$)			AP_{BEV} ($IoU = 0.7$)		
		Easy	Mod.	Hard	Easy	Mod.	Hard
MV3D [8]	RGB + LiDAR	71.29	62.68	56.56	86.55	78.10	76.67
ContFuse [23]	RGB + LiDAR	82.54	66.22	64.04	88.81	85.83	77.33
AVOD-FPN [18]	RGB + LiDAR	84.40	74.44	68.65	–	–	–
F-PointNet [31]	RGB + LiDAR	83.76	70.92	63.65	88.16	84.02	76.44
PointPillars [19]	LiDAR only	79.76	77.01	74.77	–	–	–
VoxelNet [49]	LiDAR only	81.97	65.46	62.85	89.60	84.81	78.57
SECOND [42]	LiDAR only	87.43	76.48	69.10	**89.95**	87.07	79.66
TANet [28]	LiDAR only	**88.21**	77.85	75.62	–	–	–
Ours (step = 1)	LiDAR only	87.84	**77.90**	76.07	89.81	88.24	86.68
Ours (step = 4)	LiDAR only	87.38	77.71	**76.44**	89.87	**88.31**	**87.05**

4.2 Implementation Details

The voxelization range is set to $[0m, 70.4m]$ for X axis, $[-40m, 40m]$ for Y axis, and $[-3m, 1m]$ for Z axis. The voxel size is set to $(0.05m, 0.05m, 0.1m)$. We adopt standard data augmentation techniques [19,36,37,42], including GT sampling, flipping, rotation, scaling, and more. The Diff3Det is trained on 2 NVIDIA RTX 3090 GPUs with batch size 32. We adopt the AdamW [29] optimizer with a one-cycle learning rate policy.

4.3 Main Results

The main results are shown in Table 1, where we compare the proposed Diff3Det with classic methods. Our approach achieves better performance compared with the representative anchor-based methods [19,42,49]. Specifically, one-step Diff3Det outperforms SECOND [42] by 1.42% and 6.97% on the moderate and hard levels. Besides, our method exceeds PointPillars [19] 0.89% on the moderate level, 1.3% on the hard level, and 8.08% on the easy level. Qualitative results of one-step Diff3Det are shown in Fig. 5.

When using the multi-step sampling approach commonly used in Diffusion models (i.e., step = 4), the performance improvement is mainly in the hard level of AP_{3D} (0.37%). We argue the main reason is that with the increase in sampling steps, the decoder generates more detection boxes, which is beneficial for detecting difficult samples. However, the large number of boxes may confuse post-processing because of similar predicted classification scores, which causes slight performance damage. The influence of sampling steps will be discussed in the next section.

Table 2. Ablation study of each component in Diff3Det. We gradually add the designed proposal refinements methods by setting random boxes during training as our baseline.

Component	Easy	Mod.	Hard	mAP
Baseline	82.62	74.21	73.56	76.80
+ Corrupted proposals from GT	84.32 (+1.70)	76.17 (+1.96)	74.41 (+0.85)	78.30 (+1.50)
+ Resample	85.31 (+0.99)	76.31 (+0.14)	74.48 (+0.08)	78.70 (+0.40)
+ Size correlation	86.14 (+0.83)	76.80 (+0.49)	75.29 (+0.81)	79.41 (+0.71)
+ Dynamic time step	**87.84** (+1.70)	**77.90** (+1.10)	**76.07** (+0.78)	**80.61** (+1.20)

4.4 Ablation Studies

The diffusion-guided proposal generator is the key to Diff3Det. This section explores how it affects performance with extensive ablation studies.

Proposed Components. To illustrate the effectiveness of the proposed components in the diffusion-guided proposal generator, we conduct ablation studies on our Diff3Det as shown in Table 2. Here, our baseline is Diff3Det directly using proposals sampled from a Gaussian distribution for training and inference. When adding boxes corrupted from the ground truth during training, there is a performance gain with mAP of 1.5% over the baseline. Then, we observe that some of the boxes may not have point clouds when selecting initial proposals. Thus, we propose a resampling procedure to ensure each proposal box contains at least several points (*e.g.,* 5 points). This resampling operation further brings an improvement of 0.99% on the easy level. Besides, we adopt a size correlation strategy to control the aspect of the size of 3D boxes, which is beneficial to capturing more effective 3D objects. This strategy also brings a performance improvement of 0.71% mAP, which demonstrates the importance of proposal quality. Finally, different from using the fixed time step in most diffusion models, we propose a new dynamic time step to make the whole learning process easier, which produces superior performance with mAP of 80.61% for all three levels.

Table 3. Effect of sampling steps in test.

Steps	AP_{40} (IoU= 0.7)			
	Easy	Mod.	Hard	mAP
1	89.29	79.91	75.48	81.56
2	**89.56**	80.26	76.35	82.06
4	89.45	**80.86**	**77.41**	**82.57**
6	89.43	80.28	76.76	82.16
8	88.76	80.24	77.24	82.08

Sampling Steps. Diffusion models [15,39] often employ iterative inference to obtain the target distribution. Therefore, we also utilize multiple sampling steps in the test. As shown in Table 3, the average precision (AP) calculated by 40 recall positions exhibits varying degrees of improvement. Notably, the performance at the moderate level is enhanced by 0.95%, while the hard level shows an improvement of 1.93% when comparing sampling step=4 to step=0. However, we find that some metrics decreased with calculated by 11 recall positions in

Table 4. Ablation study of hyperparameters.

(a) Ablation on scale					(b) Ablation on proposal number					(c) Ablation on η in Resample				
Scale	Easy	Mod.	Hard	mAP	N	Easy	Mod.	Hard	mAP	η	Easy	Mod.	Hard	mAP
0.1	15.36	14.66	14.35	14.79	100	86.59	77.09	75.75	79.81	1	86.44	77.22	75.78	79.81
1.0	83.87	74.50	73.44	77.27	300	**87.84**	**77.90**	**76.07**	**80.61**	5	**87.84**	**77.90**	**76.07**	**80.61**
2.0	**87.84**	**77.90**	**76.07**	**80.61**	500	86.25	76.91	75.53	79.63	10	86.69	77.38	76.05	80.04

Fig. 5. Qualitative results of Diff3Det on the KITTI validation set. We show the prediction boxes (red) and ground truth boxes (green). (Color figure online)

Table 1. We believe it is due to the fact that recall 40 metrics are more accurate [38] and provide a better reflection of the effectiveness of the iterative approach. We want to assure readers that the slight performance drop in the recall 11 metrics should not lead to any misunderstanding.

Hyperparameters. We perform ablation studies for several important sets of hyperparameters, as shown in Table 4. For the signal scale, which is used to control the signal-to-noise ratio (SNR) of the diffusion process, we empirically find that the value set to 2.0 yields the highest average precision (AP) performance (Table 4(a)). For the proposal number N, the performance achieves best when its value is set to 300 (Table 4(b)). The parameter η controls the minimum number of point clouds for each proposal, which can effectively improve the quality of proposals. We can observe that the best result is achieved when its value is set to 5 (Table 4(c)).

4.5 Limitation

The primary limitation is that the proposed method poses difficulty for the decoder to regress prediction from random boxes, leading to a relatively slow convergence speed. Besides, there remains scope for improving the performance of our approach. In the future, we would like to explore fast converging diffusion-based 3D object detection.

5 Conclusion

In this paper, we propose a generative-based 3D object detection framework, Diff3Det, by viewing 3D object detection as a denoising diffusion process from noisy boxes to object boxes. Our key idea is to utilize the diffusion method to avoid the empirical heuristics setting of anchors. We hope that our method will provide new insights into the application of generative methods on 3D vision tasks.

Acknowledgement. This work was supported by the National Science Fund for Distinguished Young Scholars of China (Grant No. 62225603) and the National Undergraduate Training Projects for Innovation and Entrepreneurship (202310487020).

References

1. Amit, T., Nachmani, E., Shaharbany, T., Wolf, L.: Segdiff: image segmentation with diffusion probabilistic models. arXiv preprint arXiv:2112.00390 (2021)
2. Bai, X., et al.: Transfusion: robust lidar-camera fusion for 3d object detection with transformers. In: CVPR (2022)
3. Brempong, E.A., Kornblith, S., Chen, T., Parmar, N., Minderer, M., Norouzi, M.: Denoising pretraining for semantic segmentation. In: CVPR (2022)
4. Carion, N., Massa, F., Synnaeve, G., Usunier, N., Kirillov, A., Zagoruyko, S.: End-to-end object detection with transformers. In: Vedaldi, A., Bischof, H., Brox, T., Frahm, J.-M. (eds.) ECCV 2020. LNCS, vol. 12346, pp. 213–229. Springer, Cham (2020). https://doi.org/10.1007/978-3-030-58452-8_13
5. Chen, S., Sun, P., Song, Y., Luo, P.: Diffusiondet: diffusion model for object detection. arXiv preprint arXiv:2211.09788 (2022)
6. Chen, T., Li, L., Saxena, S., Hinton, G., Fleet, D.J.: A generalist framework for panoptic segmentation of images and videos. arXiv preprint arXiv:2210.06366 (2022)
7. Chen, T., Zhang, R., Hinton, G.: Analog bits: generating discrete data using diffusion models with self-conditioning. arXiv preprint arXiv:2208.04202 (2022)
8. Chen, X., Ma, H., Wan, J., Li, B., Xia, T.: Multi-view 3d object detection network for autonomous driving. In: CVPR (2017)
9. Chen, Y., Liu, J., Zhang, X., Qi, X., Jia, J.: Voxelnext: fully sparse voxelnet for 3d object detection and tracking. In: CVPR (2023)
10. Deng, J., Shi, S., Li, P., Zhou, W., Zhang, Y., Li, H.: Voxel r-cnn: towards high performance voxel-based 3d object detection. In: AAAI (2021)
11. Duan, Y., Guo, X., Zhu, Z.: Diffusiondepth: diffusion denoising approach for monocular depth estimation. arXiv preprint arXiv:2303.05021 (2023)
12. Geiger, A., Lenz, P., Urtasun, R.: Are we ready for autonomous driving? the kitti vision benchmark suite. In: CVPR (2012)
13. Gu, S., et al.: Vector quantized diffusion model for text-to-image synthesis. In: CVPR (2022)
14. Han, J., Wan, Z., Liu, Z., Feng, J., Zhou, B.: Sparsedet: towards end-to-end 3d object detection. In: VISAPP (2022)
15. Ho, J., Jain, A., Abbeel, P.: Denoising diffusion probabilistic models. In: NeurIPS (2020)
16. Huang, T., Liu, Z., Chen, X., Bai, X.: EPNet: enhancing point features with image semantics for 3D object detection. In: Vedaldi, A., Bischof, H., Brox, T., Frahm, J.-M. (eds.) ECCV 2020. LNCS, vol. 12360, pp. 35–52. Springer, Cham (2020). https://doi.org/10.1007/978-3-030-58555-6_3
17. Ji, Y., et al.: DDP: diffusion model for dense visual prediction. arXiv preprint arXiv:2303.17559 (2023)
18. Ku, J., Mozifian, M., Lee, J., Harakeh, A., Waslander, S.L.: Joint 3d proposal generation and object detection from view aggregation. In: IROS (2018)

19. Lang, A.H., Vora, S., Caesar, H., Zhou, L., Yang, J., Beijbom, O.: Pointpillars: fast encoders for object detection from point clouds. In: CVPR (2019)
20. Li, J., Liu, Z., Hou, J., Liang, D.: Dds3d: dense pseudo-labels with dynamic threshold for semi-supervised 3d object detection. In: ICRA (2023)
21. Li, X., et al.: Logonet: towards accurate 3d object detection with local-to-global cross-modal fusion. In: CVPR (2023)
22. Li, Y., et al.: Bevdepth: acquisition of reliable depth for multi-view 3d object detection. In: AAAI (2023)
23. Liang, M., Yang, B., Wang, S., Urtasun, R.: Deep continuous fusion for multi-sensor 3D object detection. In: Ferrari, V., Hebert, M., Sminchisescu, C., Weiss, Y. (eds.) ECCV 2018. LNCS, vol. 11220, pp. 663–678. Springer, Cham (2018). https://doi.org/10.1007/978-3-030-01270-0_39
24. Lin, T.Y., Goyal, P., Girshick, R., He, K., Dollár, P.: Focal loss for dense object detection. In: ICCV (2017)
25. Liu, N., Li, S., Du, Y., Torralba, A., Tenenbaum, J.B.: Compositional visual generation with composable diffusion models. In: Avidan, S., Brostow, G., Cissé, M., Farinella, G.M., Hassner, T. (eds.) ECCV 2022. LNCS, vol. 13677, pp. 423–439. Springer, Cham (2022). https://doi.org/10.1007/978-3-031-19790-1_26
26. Liu, Y., Wang, T., Zhang, X., Sun, J.: PETR: position embedding transformation for multi-view 3D object detection. In: Avidan, S., Brostow, G., Cissé, M., Farinella, G.M., Hassner, T. (eds.) ECCV 2022. LNCS, vol. 13687, pp. 531–548. Springer, Cham (2022). https://doi.org/10.1007/978-3-031-19812-0_31
27. Liu, Z., Huang, T., Li, B., Chen, X., Wang, X., Bai, X.: Epnet++: cascade bi-directional fusion for multi-modal 3d object detection. IEEE Trans. Pattern Anal. Mach. Intell. (2022)
28. Liu, Z., Zhao, X., Huang, T., Hu, R., Zhou, Y., Bai, X.: Tanet: robust 3d object detection from point clouds with triple attention. In: AAAI (2020)
29. Loshchilov, I., Hutter, F.: Decoupled weight decay regularization. arXiv preprint arXiv:1711.05101 (2017)
30. Noh, J., Lee, S., Ham, B.: Hvpr: hybrid voxel-point representation for single-stage 3d object detection. In: CVPR (2021)
31. Qi, C.R., Liu, W., Wu, C., Su, H., Guibas, L.J.: Frustum pointnets for 3d object detection from rgb-d data. In: CVPR (2018)
32. Qi, C.R., Yi, L., Su, H., Guibas, L.J.: Pointnet++: Deep hierarchical feature learning on point sets in a metric space. In: NeurIPS (2017)
33. Rombach, R., Blattmann, A., Lorenz, D., Esser, P., Ommer, B.: High-resolution image synthesis with latent diffusion models. In: CVPR (2022)
34. Shi, G., Li, R., Ma, C.: PillarNet: real-time and high-performance pillar-based 3D object detection. In: Avidan, S., et al. (eds.) ECCV 2022. LNCS, vol. 13670, pp. 35–52. Springer, Cham (2022). https://doi.org/10.1007/978-3-031-20080-9_3
35. Shi, S., et al.: Pv-rcnn: point-voxel feature set abstraction for 3d object detection. In: CVPR (2020)
36. Shi, S., Wang, X., Li, H.: Pointrcnn: 3d object proposal generation and detection from point cloud. In: CVPR (2019)
37. Shi, S., Wang, Z., Shi, J., Wang, X., Li, H.: From points to parts: 3D object detection from point cloud with part-aware and part-aggregation network. IEEE Trans. Pattern Anal. Mach. Intell. (2020)
38. Simonelli, A., Bulo, S.R., Porzi, L., López-Antequera, M., Kontschieder, P.: Disentangling monocular 3d object detection. In: ICCV (2019)
39. Song, J., Meng, C., Ermon, S.: Denoising diffusion implicit models. In: ICLR (2021)
40. Sun, P., et al.: Sparse r-cnn: end-to-end object detection with learnable proposals. In: CVPR (2021)

41. Xiong, K., et al.: Cape: camera view position embedding for multi-view 3d object detection. In: CVPR (2023)
42. Yan, Y., Mao, Y., Li, B.: Second: sparsely embedded convolutional detection. Sensors (2018)
43. Yang, Z., Sun, Y., Liu, S., Jia, J.: 3dssd: point-based 3d single stage object detector. In: CVPR (2020)
44. Yin, T., Zhou, X., Krahenbuhl, P.: Center-based 3d object detection and tracking. In: CVPR (2021)
45. Zhang, D., et al.: Sam3d: zero-shot 3d object detection via segment anything model. arXiv preprint arXiv:2306.02245 (2023)
46. Zhang, D., et al.: A simple vision transformer for weakly semi-supervised 3d object detection. In: ICCV (2023)
47. Zhang, Y., Hu, Q., Xu, G., Ma, Y., Wan, J., Guo, Y.: Not all points are equal: learning highly efficient point-based detectors for 3d lidar point clouds. In: CVPR (2022)
48. Zhou, D., et al.: Iou loss for 2d/3d object detection. In: 3DV (2019)
49. Zhou, Y., Tuzel, O.: Voxelnet: end-to-end learning for point cloud based 3d object detection. In: CVPR (2018)

Blendshape-Based Migratable Speech-Driven 3D Facial Animation with Overlapping Chunking-Transformer

Jixi Chen[1], Xiaoliang Ma[1], Lei Wang[2(✉)], and Jun Cheng[2]

[1] College of Computer Science and Software Engineering, Shenzhen University, Shenzhen, China
chenjixi2022@email.szu.edu.cn, maxiaoliang@szu.edu.cn

[2] Shenzhen Institute of Advanced Technology, Chinese Academy of Sciences, Shenzhen, China
{lei.wang1,jun.cheng}@siat.ac.cn

Abstract. Speech-driven 3D facial animation has attracted an amount of research and has been widely used in games and virtual reality. Most of the latest state-of-the-art methods employ Transformer-based architecture with good sequence modeling capability. However, most of the animations produced by these methods are limited to specific facial meshes and cannot handle lengthy audio inputs. To tackle these limitations, we leverage the advantage of blendshapes to migrate the generated animations to multiple facial meshes and propose an overlapping chunking strategy that enables the model to support long audio inputs. Also, we design a data calibration approach that can significantly enhance the quality of blendshapes data and make lip movements more natural. Experiments show that our method performs better than the methods predicting vertices, and the animation can be migrated to various meshes.

Keywords: Speech-driven 3D Facial Animation · Blendshapes · Digital Human

1 Introduction

Over the past few years, there have been many studies in the exploration of 3D digital humans. This innovative technique aims to produce a digital model of the human body that emulates reality with a high degree of accuracy. A crucial component of this innovative technology is speech-driven 3D facial animation, which allows for realistic facial expressions to be generated based on audio input. This technique has emerged as a critical element in the development of 3D digital human technology, with potential benefits spanning a diverse range of applications, including virtual reality, film production, games, and education.

Supplementary Information The online version contains supplementary material available at https://doi.org/10.1007/978-981-99-8432-9_4.

Q. Liu et al. (Eds.): PRCV 2023, LNCS 14426, pp. 41–53, 2024.
https://doi.org/10.1007/978-981-99-8432-9_4

We focus on generating 3D face animations directly from speech rather than 2D talking videos like Wav2Lip [22]. In speech-driven 3D facial animation, recent state-of-the-art works [4,9,11,23,31] are mesh-based and learning-based, *i.e.*, using a deep learning model to regress the coordinate offsets of mesh vertices of each frame relative to the template mesh. For example, the representative work FaceFormer [9] is a Transformer-based architecture that uses Wav2Vec 2.0 [2] to encode audio, fuses audio features with the mesh via cross-attention, and ultimately predicts the mesh vertices on each frame. Despite the vivid effects, the animations are limited to a specific face model and cannot receive lengthy audio input due to the complexity of self-attention.

To address the first issue mentioned above, we utilize ARKit Blendshapes, an authoritative standard for blendshapes introduced by Apple Inc., to represent facial expressions rather than the vertices of the specific facial mesh. Blendshapes, or Morph Target Animation, is a technique of mesh morphing through vertices interpolation that can control a lot of mesh vertices using a small number of weights. Similar to Facial Action Coding System (FACS) [7], ARKit Blendshapes decomposes facial expressions into a combination of different movements of face parts such as brow, eye, cheek, mouth, and jaw, for a total of 52 expression weights. With this set of weights and baseline expressions corresponding to the weights one by one, it is possible to combine them linearly to get new facial expressions. This idea of controlling the mesh with a few parameters is widely used in 3D Morphable Model (3DMM) [6,16,19] represented by FLAME [16]. However, most of the parameters of 3DMM are exclusive to that model and have not become the accepted standard for facial expressions like ARKit Blendshapes. Many publicly available face meshes currently support ARKit Blendshapes, and you can add support for your face meshes by simply deforming the mesh to generate a set of baseline expressions according to ARKit documentation.

In addition, we adopt an overlapping chunking strategy to support long audio training and inference. A current state-of-the-art method for 3D facial animation, FaceFormer [9], has a complexity of $O(n^2)$, while the training datasets VOCAset [4] and BIWI [10] both have audios within 5 s each, greatly circumventing the complexity problem of self-attention. However, working with long audios is quite common in practical applications. We employ a chunking strategy where adjacent chunks overlap to ensure that boundaries obtain sufficient context, which seems very simple but helpful for 3D facial animation. It enables our Transformer-based method to be trained successfully on the BEAT [17] dataset with an average audio length of 60 s.

Meanwhile, we propose a data calibration method for blendshapes to calibrate the lip closure and amplify the lip movements. To alleviate the congestion around the median in the lower face of blendshapes, we stretch the data within a specified range centered around the median, ensuring a more even distribution of blendshapes. Our experiments have yielded promising results, demonstrating that our calibrations applied to the BEAT dataset result in animations that exhibit enhanced vividness and significantly improve the data quality.

To summarize, we make the following contributions. (1) We first use ARKit Blendshapes to represent facial expressions in speech-driven 3D facial animation. Compared to vertex coordinates, animations represented by blendshapes are more versatile and portable. (2) We employ an overlapping chunking mechanism for our Transformer-based method. It can also be adopted into other Transformer-based approaches in 3D facial animation without modifying the architecture. (3) We propose a data calibration method to enhance low-quality blendshapes data, such as amplifying the amplitude of lip movements.

2 Related Work

Speech-Driven 3D Facial Animation. Early works based on linguistic features of speech [5,8,25] focus on establishing a set of mapping rules between phonemes and visemes [14] to produce procedural animation. Although these methods can control facial movements based on articulation, they require complex procedures and lack a principled way to animate the entire face. Recent works based on deep learning [4,9,11,23,26,31] focus on directly predicting the coordinates of face mesh vertices from high-quality 3D captured data. Karras *et al.* [11] first propose a convolutional network to directly predict the offsets of the vertex coordinates corresponding to the facial motions from the audio. VOCA [4] encodes the audio with DeepSpeech [1] and adds subject encoding to the CNN-based decoder, enabling it to autonomously select the speaking styles. MeshTalk [23] decomposes audio-relevant and audio-irrelevant face movements into a classified latent space and auto-regressively predicts the representation of audio in the latent space during the inference. FaceFormer [9] encodes long-term audio contexts with Transformer [29] and devises two biased attention mechanisms for generalization. CodeTalker [31] casts speech-driven facial animation as a code query task in a finite proxy space with Transformer-based VQ-VAE. Although the animations generated by these works are vivid, they are constrained by certain facial mesh and cannot generate migratable facial animations. The blendshape weights can be easily applied to any facial mesh, only requiring a pre-defined expression deformation for each parameter.

Talking Head Generation. Most works [22,24] use 2D landmarks to synthesize 2D talking videos. Suwajanakorn *et al.* [24] use 19h videos of Obama's speeches for training, and generate the results through an LSTM-based network. Wav2Lip [22] trains a network that determines whether the sound is in sync with the lips as a discriminator, forcing the generator to produce accurate lip movements. Recent state-of-the-art methods [13,27,30] use the facial 3DMM [16] to replace 2D landmarks as intermediate representations to guide video synthesis. Audio2Head [30] designs a head pose predictor by modeling rigid 6D head movements with a motion-aware recurrent neural network. LipSync3D [13] proposes efficient learning of data for personalized talking heads, focusing on the pose and light normalization. While these methods also generate speech-driven facial animation, our work focuses on synthesizing 3D facial animation, which is much more informative and can be combined with 3D digital humans.

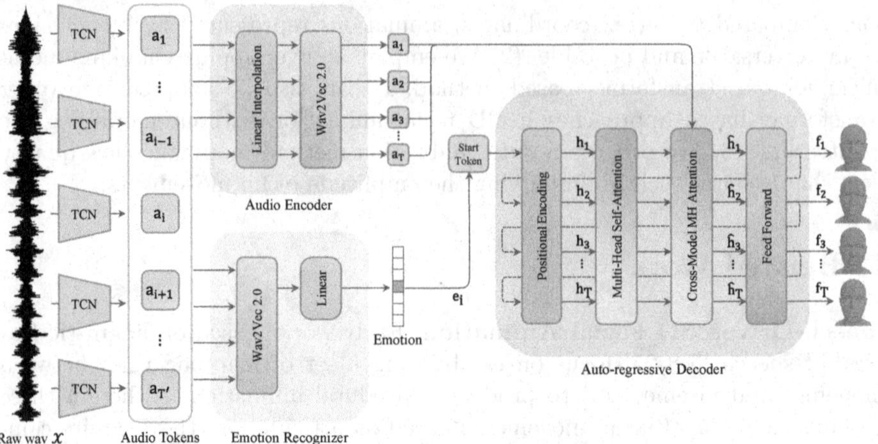

Fig. 1. The architecture of our method. The encoder and recognizer take audio as input to obtain audio representation and emotion vector, respectively. The decoder takes the emotion vector as the start token and auto-regressively predicts the blendshape weights of each frame. The blendshapes predicted by the model can be visualized with several facial meshes.

3 Method

We aim to predict ARKit blendshape weights directly from speech, which differs from existing speech-driven 3D facial animation methods that predict vertices. Specifically, we construct a Transformer-based model that learns to extract speech sequence representations and auto-regressively synthesize blendshape weights frame from audio features for each animation. Additionally, we introduce a special chunking strategy for our Transformer model to support long audio inputs. In the following sections, we will describe our model architecture, chunking strategy, and training objectives.

3.1 Model Architecture

Our Transformer-based architecture consists of three components (see Fig. 1): an audio encoder, an emotion recognizer, and an auto-regressive decoder.

Audio Encoder. Our speech encoder utilizes a self-supervised pre-trained speech model, Wav2Vec 2.0 [2], consisting of a CNN-based audio feature extractor and a Transformer-based encoder. The audio feature extractor converts the raw speech waveform into feature vectors through a temporal convolutions network (TCN), while the transformer encoder contextualizes the audio features to obtain latent speech representations. However, when considering facial animation as a sequence-to-sequence task, the audio representation should match the sampling frequency of the motion data. In this way, the audio and motion data can be aligned by cross-attention of the decoder as they need to be aligned.

Given target frames $\mathbf{T}$, the audio encoder transfers raw waveform audio $\mathcal{X}$ into a latent representation $\mathbf{A_T} = (\mathbf{a}_1, ..., \mathbf{a_T})$.

Emotion Recognizer. The emotion recognizer aims to recognize emotions from the input speech $\mathcal{X}$, which enables the model to learn the facial expression features associated with each emotion, thus enhancing the vividness of the generated animations. We design the recognizer as a lightweight Wav2Vec 2.0 model with a classification head. Like the audio encoder, its backbond encodes the raw audio into a latent discretized representation. But only the output corresponding to the CLS token is fed into a two-layer linear classification head to classify the audio into one of the $\mathbf{n}$ emotions. The recognized result is a $\mathbf{n}$-dimensional one-hot vector, which will be mapped to an emotion embedding $\mathbf{e_i}(1 \leq \mathbf{i} \leq \mathbf{n})$, the start token for the decoder. Compared to the audio encoder, the emotion recognizer can use a more lightweight model to reduce parameters while maintaining satisfactory performance.

Auto-regressive Decoder. Following FaceFormer [9], our decoder adopts a Transformer-based architecture, which takes latent audio representations $\mathbf{A_T}$ as input and produces blendshape weights $\mathbf{F_T} = (\mathbf{f}_1, ..., \mathbf{f_T})$ in an auto-regressive manner. Specifically, the decoder auto-regressively predicts every blendshapes frame $\mathbf{f_t}$ considering past blendshapes frames $\mathbf{F_{t-1}}$ and speech features $\mathbf{A_t}$. We first inject emotion embedding $\mathbf{e_i}$ and temporal positional information into input $\mathbf{F_{t-1}}$, obtaining intermediate representations $\mathbf{H_t} = (\mathbf{h}_1, ..., \mathbf{h_t})$:

$$\mathbf{h_t} = \begin{cases} \mathbf{e_i} + \mathrm{PPE}(\mathbf{t}), & \mathbf{t} = 1 \\ \mathrm{L}(\mathbf{f_{t-1}}) + \mathbf{e_i} + \mathrm{PPE}(\mathbf{t}), & 1 < \mathbf{t} \leq \mathbf{T} \end{cases}, \tag{1}$$

where $\mathrm{L}(\mathbf{f_{t-1}})$ is a linear projection mapping blendshapes into the hidden dimension of the decoder and $\mathrm{PPE}(\mathbf{t})$ is periodic positional encoding [9]. The emotion embedding $\mathbf{e_i}$ intends to learn special facial features related to emotions to improve the animation. Afterward, the intermediate representation $\mathbf{H_t}$ is fed into the multi-head self-attention block and cross-modal multi-head attention block, allowing each frame to consider its previous context and audio information, respectively. The intermediate representation $\hat{\mathbf{H}}_\mathbf{t}$ after two attention blocks is then passed through a feed forward network and a linear projection to map the hidden dimension to the blendshapes, resulting in the final blendshapes frame $\mathbf{f_t}$.

3.2 Chunking-Transformer

Since the emergence of Transformer [29], various methods have been proposed to reduce complexity, such as locality-sensitive hashing attention [12] and shifted windows [18]. However, these complex mechanisms are designed to ensure the integrity of global information, but speech-driven 3D facial animation is more concerned with the short-term than the global context. Specifically, the current frame is most influenced by the preceding few seconds, and information from too far back in time does not significantly contribute to predicting the current animation frame. Therefore, feeding the model lengthy audio at once does not

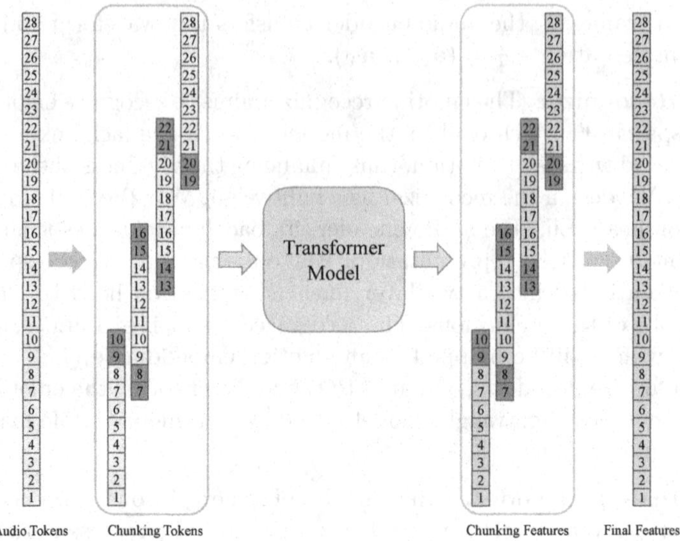

Fig. 2. The overlapping chunking strategy for long audio input. It shows the process when chunking audio into four chunks, with longer audio requiring more intermediate chunks.

make the animation more realistic because the model focuses almost exclusively on information within a short window. Instead, it will incur an unacceptable performance overhead, especially for the Transformer-based model.

Based on the prior knowledge mentioned above, we propose an effective chunking strategy for Transformer in speech-driven 3D facial animation. Specifically, we introduce overlapping regions for adjacent chunks and remove the output of these overlapping regions when concatenating (see Fig. 2). Our approach offers several advantages, including its simplicity and non-intrusive modification. It only requires adding chunking and concatenating at the beginning and end of the model, respectively, without modifying the network structure.

3.3 Training Objective

To train our model effectively, we define a comprehensive set of training losses:

$$\mathcal{L}_{\text{total}} = \lambda_{\text{reg}} \cdot \mathcal{L}_{\text{reg}} + \lambda_{\text{vel}} \cdot \mathcal{L}_{\text{vel}} + \lambda_{\text{emo}} \cdot \mathcal{L}_{\text{emo}}, \tag{2}$$

where $\mathcal{L}_{\text{reg}}$ defines a blendshapes regression loss, $\mathcal{L}_{\text{vel}}$ defines a velocity loss for smoothing, and $\mathcal{L}_{\text{emo}}$ defines an emotion loss.

Regression Loss. Inspired by Audio2Face [28], we adopt a Huber loss to supervise the regression of blendshape weights:

$$\mathcal{L}_{\text{reg}} = \begin{cases} \frac{1}{2}(y-\hat{y})^2, & \text{if } |(y-\hat{y})| < \delta \\ \delta((y-\hat{y}) - \frac{1}{2}\delta), & \text{otherwise} \end{cases}, \tag{3}$$

where δ is the threshold of Huber loss and also a tunable hyper-parameter. Considering the distribution of blendshapes, we set it to 0.1.

Velocity Loss. Relying solely on the regression loss will result in significant temporal instabilities due to the similarity between consecutive frames. To improve temporal consistency, we introduce a velocity loss similar to [4]:

$$\mathcal{L}_{\text{vel}} = \sum_{t=2}^{\text{T}} \sum_{s=1}^{\text{S}} \left\| (y_{t,s} - y_{t-1,s}) - (\hat{y}_{t,s} - \hat{y}_{t-1,s}) \right\|^2, \tag{4}$$

where $y_{t,s}$ represents the predicted blendshape weights at time t in s-th dimension (with a total of S dimensions) and $\hat{y}_{t,s}$ denotes the ground truth.

Emotion Loss. To enhance the quality of the generated animations, we augment our model with an emotion recognizer that performs sentiment analysis on the speech. To train the emotion recognizer, we employ a cross-entropy (CE) loss:

$$\mathcal{L}_{\text{emo}} = -\sum_{c=1}^{\text{M}} y_c \log(p_c), \tag{5}$$

where y_c denotes the ground truth label for emotion class c, and p_c represents the predicted probability assigned to class c.

4 Experiments

4.1 Experimental Setup

BEAT Dataset. We select the BEAT [17] dataset in our experiment, which stands out as it provides blendshape weights instead of raw 3D captured mesh for representing facial expressions. It comprises rich information from 30 speakers engaging in 118 self-talk sessions, with eight emotional annotations and an average length of one minute, for 76 h of data. In our experiments, we downsample the original blendshape weights sampled at 60fps to 30fps. Moreover, we focus on seven specific emotions: *anger, disgust, fear, happiness, neutral, sad*, and *surprise*, and exclude the *contempt* due to its challenging differentiation. After filtering, we obtain a final dataset comprising 1646 instances. We randomly divide the dataset into an 8:1:1 ratio, yielding the BEAT-Train, BEAT-Valid, and BEAT-Test subsets.

Implementation Details. In our experiments, we employ the wav2vec2-xlsr-53 [3] on the audio encoder and the wav2vec2-base-960h [2] on the emotion recognizer while randomly initializing the auto-regressive decoder. In our chunking strategy, we divide the 16 kHz audio into 10-second chunks with 2-second overlapping region between adjacent chunks. This setup enables the model to perform inference on audio sequences up to 9 min on a 12 GB GPU. In the model training, we utilize the AdamW optimizer [21] for all weights except biases, and cosine schedule [20] for learning rate decay. We set the batch size to 2 and the learning rate to 1e-5. We train the model for a maximum of 50 epochs, implementing early stopping to prevent overfitting. The training process takes approximately 50 h using a single Titan Xp.

4.2 Data Calibration

We found many poor-quality blendshapes data in BEAT during our experiments, which hindered our model training. However, since blendshapes contain semantic information, we can make targeted adjustments to improve the data quality.

Lip Closure Calibration. We observe some offsets in the weight *jawOpen* controlling lip closure in almost all data of BEAT. To address this, we perform max-min normalization on the *jawOpen* and then remap it back to the original range. As shown in Fig. 3a, the *jawOpen* in the initial state is close to 0.1, which causes the subsequent lips to be unclosed. As a comparison, the smaller values of *jawOpen* after calibration are close to 0 (see Fig. 3b) to achieve proper lip closure. Note that *jawOpen* is also amplified, so we only need to concentrate on the changes in smaller values.

Lip Amplitude Amplification. In addition to lip closure, almost all the weights controlling the lower face exhibit excessive oscillations within a narrow range, forcing the model to fit the data by repeatedly jittering. To this end, we devise a lip movements amplifier that preserves the original trends while significantly enhancing the lip amplitude. Supposing w_i represents one of the lower face blendshape weights (27 in total, ranging from *jawForward* to *mouthUpperUpRight*), we can formularize our amplifier as follows:

$$w_i = \begin{cases} \mu \cdot w_i, & \text{if} \, |w_i - \sigma| \leq \delta \ \& \ w_i \geq \sigma \\ \frac{1}{\mu} \cdot w_i, & \text{if} \, |w_i - \sigma| \leq \delta \ \& \ w_i < \sigma \\ w_i, & \text{otherwise} \end{cases}, \tag{6}$$

where σ denotes the median of w_i, δ represents the amplification threshold and μ is the amplification factor. In our experiments, we set $\delta = 0.1$ and $\mu = 1.2$. As shown in Fig. 3, the original data (see Fig. 3c) exhibit overcrowding and small oscillation around the median, while the calibrated data (see Fig. 3d) are more evenly distributed and have a greater lip amplitude while retaining the original trend.

4.3 Evaluation Results

To our knowledge, there exist no works that directly predict ARKit blendshape weights. Therefore, for quantitative evaluation, we have to convert the blendshape weights to several facial meshes and compare them with previous works of predicting vertices.

Facial Meshes. We utilize ICT-Face [15] and BEAT-Face [17] as the carriers of blendshape weights. ICT-Face is a highly realistic 3DMM toolkit with 26,719 vertices, while BEAT-Face is the default facial mesh in ARKit with 1,220 vertices. In our experiments, we consider ICT-Face as the benchmark and BEAT-Face as a reference because the blendshapes animation of ICT-Face is more vivid. To match the facial size with VOCA [4], we scaled down ICT-Face by a factor of

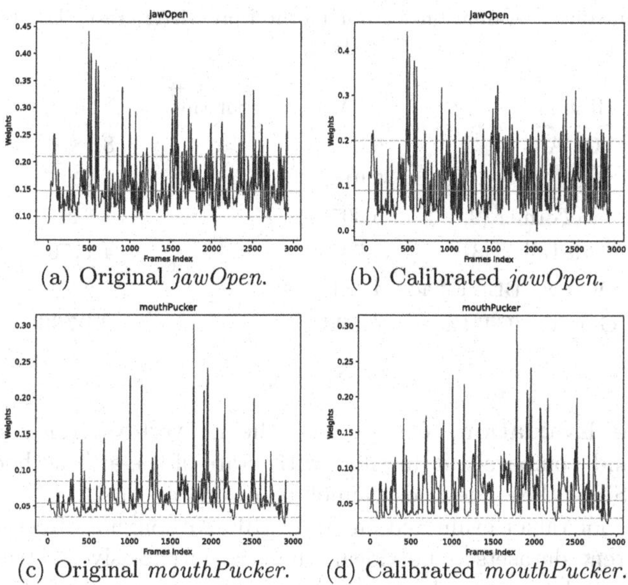

(a) Original *jawOpen*. (b) Calibrated *jawOpen*.

(c) Original *mouthPucker*. (d) Calibrated *mouthPucker*.

Fig. 3. Comparison between the original and calibrated data for *jawOpen* and *mouthPucker*. The red line is the respective median. Figures (a) and (c) depict the original data. Figures (b) and (d) demonstrate the calibrated data. (Color figure online)

100 and BEAT-Face by a factor of 500. Only when the mesh scales are aligned can meaningful quantitative results be obtained, as it directly determines the magnitude of the evaluation results. Note that even for the same blendshape weights, there may be variations in the evaluation metrics calculated on different meshes. Therefore, we can only make a quantitative comparison as relatively fair as possible, to reflect the performance objectively.

Evaluation Metrics. Following CodeTalker [31], we utilize the L2 vertex error and face dynamics deviation (FDD) as evaluation metrics. In contrast to CodeTalker, where the vertex error only acts on the lips and the FDD only acts on the upper face, we employ both metrics to the entire face to avoid ambiguous vertices partitioning. The L2 vertex error calculates the maximum L2 distance between the predicted and ground truth vertices on each frame. By applying this metric to the entire face rather than just the lips, we consider more vertices resulting in higher errors. The FDD measures the difference between the average standard deviation of the L2 norms on each vertex and its corresponding ground truth over time. Although applying FDD to the whole face results in a smaller mean, it also includes the lips, the area with the maximum standard deviation, because the most frequent facial movements are concentrated in this region. Overall, it is relatively fair that both metrics act on the whole face, which is not favorable to us.

Table 1. Quantitative evaluation for our method on BEAT-Test. The results of the compared methods are taken from CodeTalker.

Method	L2 Vertex Error (mm) ↓	FDD ↓
VOCA [4]	6.5563	8.1816
MeshTalk [23]	5.9181	5.1025
FaceFormer [9]	5.3077	4.6408
CodeTalker [31]	4.7914	4.1170
Ours w/ BEAT-Face	4.8310	5.5832
Ours w/ ICT-Face	**4.5933**	**2.9786**

Quantitative Evaluation. We compute the L2 vertex error and FDD on BEAT-Test and compare our method with state-of-the-art methods predicting vertices. Since these methods cannot be evaluated on BEAT, we take the results from CodeTalker evaluated on BIWI [10]. Although a direct comparison between different datasets is not ideal, there is no excessive distribution bias between datasets because each dataset is captured from real-life faces and the same language has a high similarity in lip motions. As long as the mesh scales are consistent, such a comparison is still credible. According to Table 1, our method achieves the best results on ICT-Face, indicating accurate facial movements and good temporal stability in the predicted blendshapes animations. However, BEAT-Face exhibits worse temporal stability due to its exaggerated amplitude of blendshapes animations. Note that due to the absence of similar methods predicting blendshapes, we have to adopt this compromise quantitative evaluation scheme. While the numerical comparison across datasets lacks direct comparability, it demonstrates that our method is at least not inferior to others.

Qualitive Evaluation. We also conduct a qualitative evaluation to demonstrate the effectiveness of our data calibration. After data calibration, we can observe a significant increase in lip amplitude, and the lips are closed more effectively in several keyframes (see Fig. 4). In contrast, ground truth data without data calibration cannot achieve lip closure in most cases, and the model trained with such data is hardly realistic. As a reference, FaceFormer [9] represents the expected lip states without data issues. Comparing our results after data calibration with it reveals similar lip states, indicating the effectiveness of our data calibration approach in improving low-quality blendshapes data.

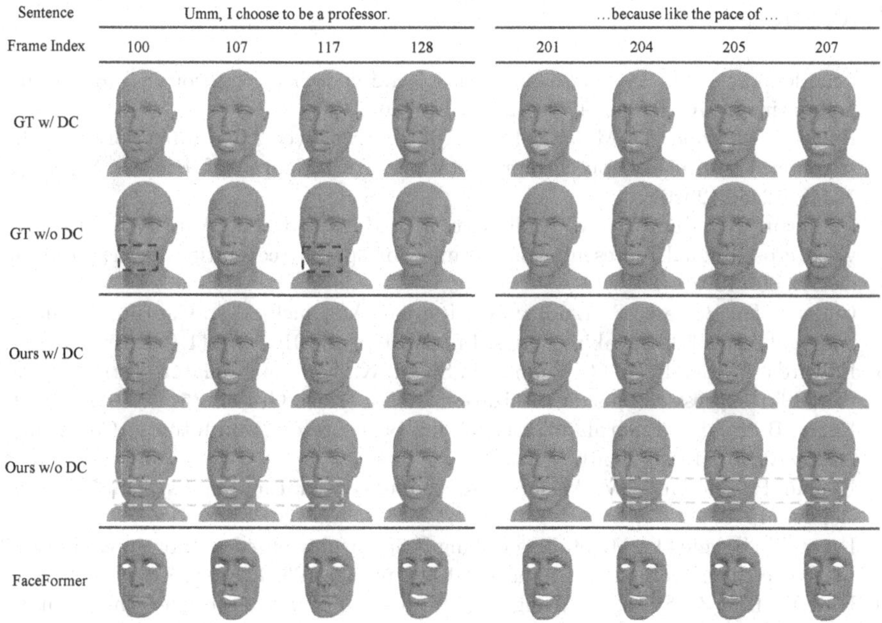

Fig. 4. Qualitive evaluation of data calibration. DC stands for data calibration, and Frame Index refers to the index of Ours w/ DC in audio. We highlight the too-small lip amplitude and lip closure problem of the original data.

5 Conclusion

We present a migratable Transformer-based approach for speech-driven 3D facial animation. Our method leverages the versatility of blendshapes, allowing for the convenient transfer of the generated animations to multiple facial meshes. Additionally, we introduce an overlapping chunking strategy for long audio within the Transformer framework without modifying the network architecture. Furthermore, we propose a data calibration approach that effectively enhances the low-quality blendshapes data. We believe that migratable blendshapes animation will continue to be a crucial technology in digital humans, as it can be easily integrated with other 3D human technologies.

Acknowledgement. This work was supported in part by the Shenzhen Technology Project (JCYJ20220531095810023), National Natural Science Foundation of China (61976143, U21A20487), Guangdong-Hong Kong-Macao JointLaboratory of Human-Machine Intelligence-Synergy Systems (2019B121205007).

References

1. Amodei, D., et al.: Deep speech 2: end-to-end speech recognition in English and mandarin. In: ICML, pp. 173–182. PMLR (2016)
2. Baevski, A., Zhou, Y., Mohamed, A., Auli, M.: wav2vec 2.0: a framework for self-supervised learning of speech representations. Adv. Neural Inf. Process. Syst. **33**, 12449–12460 (2020)
3. Conneau, A., Baevski, A., Collobert, R., Mohamed, A., Auli, M.: Unsupervised cross-lingual representation learning for speech recognition. arXiv preprint arXiv:2006.13979 (2020)
4. Cudeiro, D., Bolkart, T., Laidlaw, C., Ranjan, A., Black, M.J.: Capture, learning, and synthesis of 3d speaking styles. In: CVPR, pp. 10101–10111 (2019)
5. Edwards, P., Landreth, C., Fiume, E., Singh, K.: Jali: an animator-centric viseme model for expressive lip synchronization. ACM Trans. Graph. **35**(4), 1–11 (2016)
6. Egger, B., et al.: 3d morphable face models-past, present, and future. ACM Trans. Graph. **39**(5), 1–38 (2020)
7. Ekman, P., Friesen, W.V.: Facial action coding system. Environ. Psychol. Nonverb. Behav. (1978)
8. Ezzat, T., Poggio, T.: Miketalk: a talking facial display based on morphing visemes. In: Proceedings Computer Animation 1998, pp. 96–102. IEEE (1998)
9. Fan, Y., Lin, Z., Saito, J., Wang, W., Komura, T.: Faceformer: speech-driven 3d facial animation with transformers. In: CVPR, pp. 18770–18780 (2022)
10. Fanelli, G., Gall, J., Romsdorfer, H., Weise, T., Van Gool, L.: A 3-d audio-visual corpus of affective communication. IEEE Trans. Multim. **12**(6), 591–598 (2010)
11. Karras, T., Aila, T., Laine, S., Herva, A., Lehtinen, J.: Audio-driven facial animation by joint end-to-end learning of pose and emotion. ACM Trans. Graph. **36**(4), 1–12 (2017)
12. Kitaev, N., Kaiser, Ł., Levskaya, A.: Reformer: the efficient transformer. arXiv preprint arXiv:2001.04451 (2020)
13. Lahiri, A., Kwatra, V., Frueh, C., Lewis, J., Bregler, C.: Lipsync3d: data-efficient learning of personalized 3d talking faces from video using pose and lighting normalization. In: CVPR, pp. 2755–2764 (2021)
14. Lewis, J.: Automated lip-sync: background and techniques. J. Vis. Comput. Animat. **2**(4), 118–122 (1991)
15. Li, R., et al.: Learning formation of physically-based face attributes. In: CVPR, pp. 3410–3419 (2020)
16. Li, T., Bolkart, T., Black, M.J., Li, H., Romero, J.: Learning a model of facial shape and expression from 4d scans. ACM Trans. Graph. **36**(6), 194–1 (2017)
17. Liu, H., et al.: BEAT: a large-scale semantic and emotional multi-modal dataset for conversational gestures synthesis. In: Avidan, S., Brostow, G., Cissé, M., Farinella, G.M., Hassner, T. (eds.) ECCV 2022. LNCS, vol. 13667, pp. 612–630. Springer, Cham (2022). https://doi.org/10.1007/978-3-031-20071-7_36
18. Liu, Z., et al.: Swin transformer: hierarchical vision transformer using shifted windows. In: ICCV, pp. 10012–10022 (2021)
19. Loper, M., Mahmood, N., Romero, J., Pons-Moll, G., Black, M.J.: Smpl: a skinned multi-person linear model. ACM Trans. Graph. **34**(6), 1–16 (2015)
20. Loshchilov, I., Hutter, F.: SGDR: stochastic gradient descent with warm restarts. arXiv preprint arXiv:1608.03983 (2016)
21. Loshchilov, I., Hutter, F.: Decoupled weight decay regularization. arXiv preprint arXiv:1711.05101 (2017)

22. Prajwal, K., Mukhopadhyay, R., Namboodiri, V.P., Jawahar, C.: A lip sync expert is all you need for speech to lip generation in the wild. In: Proceedings of the 28th ACM International Conference on Multimedia, pp. 484–492 (2020)
23. Richard, A., Zollhöfer, M., Wen, Y., De la Torre, F., Sheikh, Y.: Meshtalk: 3d face animation from speech using cross-modality disentanglement. In: ICCV, pp. 1173–1182 (2021)
24. Suwajanakorn, S., Seitz, S.M., Kemelmacher-Shlizerman, I.: Synthesizing obama: learning lip sync from audio. ACM Trans. Graph. **36**(4), 1–13 (2017)
25. Taylor, S.L., Mahler, M., Theobald, B.J., Matthews, I.: Dynamic units of visual speech. In: Proceedings of the 11th ACM SIGGRAPH/Eurographics Conference on Computer Animation, pp. 275–284 (2012)
26. Thambiraja, B., Habibie, I., Aliakbarian, S., Cosker, D., Theobalt, C., Thies, J.: Imitator: personalized speech-driven 3d facial animation. arXiv preprint arXiv:2301.00023 (2022)
27. Thies, J., Elgharib, M., Tewari, A., Theobalt, C., Nießner, M.: Neural voice puppetry: audio-driven facial reenactment. In: Vedaldi, A., Bischof, H., Brox, T., Frahm, J.-M. (eds.) ECCV 2020. LNCS, vol. 12361, pp. 716–731. Springer, Cham (2020). https://doi.org/10.1007/978-3-030-58517-4_42
28. Tian, G., Yuan, Y., Liu, Y.: Audio2face: generating speech/face animation from single audio with attention-based bidirectional LSTM networks. In: ICME, pp. 366–371. IEEE (2019)
29. Vaswani, A., et al.: Attention is all you need. Adv. Neural Inf. Process. Syst. **30** (2017)
30. Wang, S., Li, L., Ding, Y., Fan, C., Yu, X.: Audio2head: audio-driven one-shot talking-head generation with natural head motion. arXiv preprint arXiv:2107.09293 (2021)
31. Xing, J., Xia, M., Zhang, Y., Cun, X., Wang, J., Wong, T.T.: Codetalker: speech-driven 3d facial animation with discrete motion prior. In: CVPR, pp. 12780–12790 (2023)

FIRE: Fine Implicit Reconstruction Enhancement with Detailed Body Part Labels and Geometric Features

Junzheng Zhang, Xipeng Chen, Keze Wang, Pengxu Wei, and Liang Lin(✉)

School of Computer Science, Sun Yat-sen University, Guangzhou, China
{zhangjzh8,chenxp37}@mail2.sysu.edu.cn, weipx3@mail.sysu.edu.cn,
linliang@ieee.org

Abstract. While considerable progress has been made in 3D human reconstruction, existing methodologies exhibit performance limitations, mainly when dealing with complex clothing variations and challenging poses. In this paper, we propose an innovative approach that integrates fine-grained body part labels and geometric features into the reconstruction process, addressing the aforementioned challenges and boosting the overall performance. Our method, Fine Implicit Reconstruction Enhancement (FIRE), leverages blend weight-based soft human parsing labels and geometric features obtained through Ray-based Sampling to enrich the representation and predictive power of the reconstruction pipeline. We argue that incorporating these features provides valuable body part-specific information, which aids in improving the accuracy of the reconstructed model. Our work presents a subtle yet significant enhancement to current techniques, pushing the boundaries of performance and paving the way for future research in this intriguing field. Extensive experiments on various benchmarks demonstrate the effectiveness of our FIRE approach in terms of reconstruction accuracy, particularly in handling challenging poses and clothing variations. Our method's versatility and improved performance render it a promising solution for diverse applications within the domain of 3D human body reconstruction.

Keywords: Human reconstruction · Implicit function · Parametric model

1 Introduction

Clothed Human Reconstruction represents a pivotal research direction due to its widespread applications encompassing fields such as virtual reality, augmented reality, virtual try-on, telepresence, *etc.* Conventional methods have primarily relied on exact multi-view scanning systems, which significantly constricts their potential applicability across diverse scenarios. Therefore, 3D human reconstruction from a single image has gained significant attention. Although substantial

Q. Liu et al. (Eds.): PRCV 2023, LNCS 14426, pp. 54–65, 2024.
https://doi.org/10.1007/978-981-99-8432-9_5

progress has been made in this field [4,12,22,29,30], accurately reconstructing complex and diverse human body shapes and poses from monocular images remains a challenging task. This is primarily due to the inherent ambiguity of recovering 3D information from 2D data, the wide range of human body shapes and poses, as well as the presence of occlusions and varying lighting conditions.

In recent years, implicit function-based methods [8,9,11,25,26,31] have attained substantial success, demonstrating a marked capability to reconstruct intricate details such as hair and clothing. However, these methods often encounter challenges in producing accurate reconstruction results when confronted with a diversity of human poses. To address this, the incorporation of parametric human body models has been proposed to enhance the reconstruction results by leveraging them as prior knowledge. Many parametric models based methods [2,3,10,14,28,32,34] have achieved great results. The most prevalently employed parametric models are SMPL [15] and SMPL-X [21].

In this paper, our proposed FIRE addresses these challenges by enhancing the existing implicit function learning framework with advanced geometric and semantic features extracted from the SMPL model. Following the ICON [32] approach, our FIRE incorporates an iterative optimization strategy for refining the estimation of the SMPL model and the normal maps. Inspired by S-PIFu [2], we use Ray-based Sampling (RBS) to extract rich geometric features from the parametric human body model. Moreover, we employ blend weight-based labels as a proxy for soft human parsing labels, which provide fine-grained part-level information and significantly improve the model's comprehension of the input data. As such, FIRE effectively exploits the information from both paradigms, resulting in a more accurate and robust clothed human reconstruction model. By integrating these features into the implicit function learning framework, FIRE is capable of producing highly detailed and accurate 3D human reconstructions.

Our primary contributions are two-fold:

1) Our proposed approach, FIRE, to our knowledge, is the first to integrate Ray-based Sampling (RBS) based features within an advanced implicit function learning framework. This amalgamation bridges the strengths of these approaches, leading to significant enhancements in the accuracy and detail of 3D human reconstructions.
2) Our model, FIRE, introduces an efficient method for exploiting the high-level geometric and semantic cues embedded in parametric human body models. The ability to leverage these cues greatly boosts the model's understanding of the human form, thereby providing a robust solution to the long-standing challenges of pose and clothing diversity.

We train FIRE on THuman2.0 [39] dataset and evaluate FIRE on CAPE [17] dataset. We conduct extensive experiments and both quantitative results and qualitative results demonstrate the effectiveness of our approach in handling diverse body poses. Furthermore, FIRE exhibits robust generalizability, demonstrating a marked capability for effective deployment within real-world scenarios.

2 Related Work

2.1 Implicit Function Based Human Reconstruction

Deep learning methods for 3D human reconstruction based on implicit functions [6,18,20] primarily learn a relationship between a 3D coordinate and a corresponding value. Using the marching cubes algorithm [16], we can then derive the meshes by setting a threshold to this value. The strength of implicit function-based methods lies in their capacity to portray intricate geometrical details without being bound by static topology or resolution. This is an advantage over other representations such as voxels or meshes. In addition, these methods offer a balanced computational load.

A notable contribution in this field is PIFu [25], which aligns the 3D occupancy prediction with 2D pixels. It shows effectiveness in reconstructing both the shape and appearance of individuals wearing various types of clothes. To overcome the resolution limitation of the image feature extractor, PIFuHD [26] employs a multi-level architecture and obtains improved performance. HumanNeRF [30] introduces NeRF [19] and the motion field mapping from the observed to canonical space for 3D human reconstruction. SelfRecon [12] combines implicit and explicit representation to take advantage of both representations. PHORHUM [1] successfully reconstructs the mesh along with albedo and shading information. Other significant works include SuRS [23], which integrates a super-resolution technique to recover detailed meshes from low-resolution images, and FOF [7], which proposes the Fourier Occupancy Field for real-time monocular human reconstruction, offering high-quality results at 30 frames per second.

2.2 Parametric Human Models Based Reconstruction

Though deep implicit functions excel in depicting detailed 3D shapes, they may create disembodied limbs or non-human shapes when confronted with diverse human poses or severe occlusion. To mitigate this issue, there is growing interest in integrating parametric human models with implicit functions.

The ARCH model [11] designs a semantic space and a semantic deformation utilizing a parametric 3D body estimator. A subsequent model, ARCH++ [9], enhances ARCH by jointly estimating point occupancy in both posed and canonical spaces. The SNARF model [4] learns a forward deformation field without direct supervision, capitalizing on linear blend skinning. The ClothWild [37] model employs two separate networks to recover body and clothing. SHARP [38] introduces a combination of a parametric body prior with a non-parametric peeled depth map representation for clothed human reconstruction. FITE [14] combines the strengths of both implicit and explicit representations and introduces diffused skinning to aid in template training, particularly for loose clothing. PaMIR [34] introduces a depth-ambiguity-aware training loss and a body reference optimization technique to enhance the consistency between the parametric model and the implicit function. ICON [32] leverages local features to boost generalizability and implements an iterative optimization strategy. Finally, models such as IntegratedPIFu [3] and S-PIFu [2] propose to utilize human parsing and depth information to assist with reconstruction.

3 Methodology

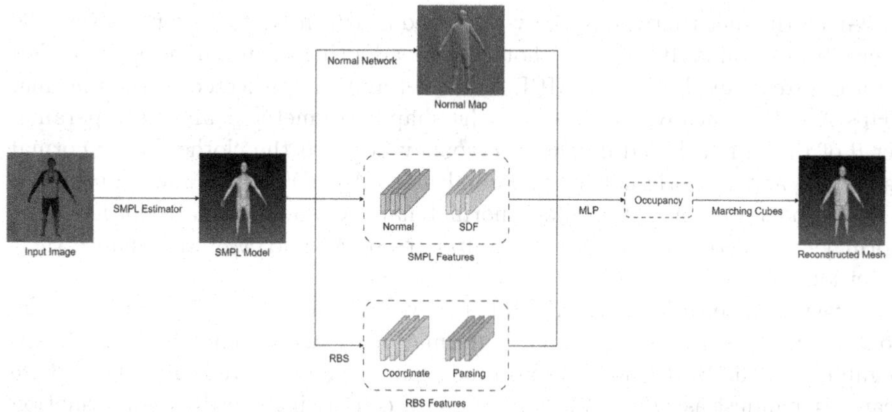

Fig. 1. This is the overview of our Fine Implicit Reconstruction Enhancement process. Starting with an input image, we first generate its corresponding SMPL model. We then use this SMPL model to help estimate the normal map and extract various features. Additionally, we employ Ray-based Sampling to draw out RBS features. All these features are then fed into a Multilayer Perceptron (MLP) to predict occupancy. Finally, we apply the marching cubes algorithm to obtain the reconstructed mesh.

Given a single image, our objective is to recreate the underlying 3D structure of a clothed human, while keeping the intricate details visible in the image. Our model is built upon the implicit function-based method, integrating a parametric human model. Furthermore, we employ Ray-based Sampling to gather coordinate information and blend weight-based labels. With these modules in place, we manage to attain excellent reconstruction results.

3.1 Overview of Our Framework

Figure 1 provides an illustration of our Fine Implicit Reconstruction Enhancement (FIRE) process. Starting with an input image denoted as $\mathcal{I}$, we employ the SMPL estimator [40] represented by $\mathbf{E}$. This estimator is used to compute the SMPL shape parameter, $\beta \in \mathbb{R}^{10}$, and the pose parameter, $\theta \in \mathbb{R}^{3\times K}$. These parameters collectively define the SMPL model expressed as $\mathcal{M}(\beta, \theta)$. Subsequently, we utilize the PyTorch3D [41] differentiable renderer $\mathbf{R}$, which is responsible for rendering normal maps of the SMPL body, $\mathcal{N}_B$, both on the front and back sides. Then, we apply the normal network $\mathbf{G}$ to predict the clothed-body normal maps $\mathcal{N}_C$, which are inferred directly from $\mathcal{N}_B$. This procedure is shown in Eq. 1.

$$\begin{aligned} \mathcal{N}_C &= \mathbf{G}(\mathcal{N}_B), \\ \mathcal{N}_B &= \mathbf{R}(\mathcal{M}), \\ \mathcal{M}(\beta, \theta) &= \mathbf{E}(\mathcal{I}) \end{aligned} \tag{1}$$

Given the fact that the initially predicted SMPL model might not always be accurate, we follow ICON's method [32] by adopting an iterative optimization approach to refine both the SMPL model $\mathcal{M}$ and the projected clothed normal maps $\mathcal{N}_C$. Specifically, we fine-tune the shape parameter β and pose parameter θ of the SMPL based on the discrepancy between the clothed-body normal maps $\mathcal{N}_C$ and the original body normal maps $\mathcal{N}_B$. With the newly optimized SMPL model, we render fresh body normal maps $\mathcal{N}B$ and estimate clothed-body normal maps $\mathcal{N}_C$. Through iterative updates of $\mathcal{M}$ and $\mathcal{N}_C$, we enhance their precision.

Drawing inspiration from S-PIFu [2], we recognize that the SMPL model contains more than just coordinate information. Thus, we introduce Ray-based Sampling to the SMPL model to obtain what we term as RBS features. These features manifest as a set of 2D feature maps, comprising coordinate information $\mathcal{C}$ and blend weight-based labels $\mathcal{B}$. Given the improved accuracy of our refined SMPL model, the extracted features hold considerable value. In addition, we also incorporate two SMPL features: the SMPL SDF $\mathcal{S}$ and the SMPL normal $\mathcal{N}$. The SDF $\mathcal{S}$ represents the signed distance of the query point to the SMPL body, whereas the SMPL normal is the barycentric surface normal of the body point closest to the query point. With these features at our disposal, we can predict the occupancy field using a Multilayer Perceptron (MLP). Subsequently, we employ the marching cubes algorithm [16] to extract meshes from the 3D occupancy space. The entire pipeline of our framework is succinctly encapsulated in Eq. 2, where $\mathcal{V}$ refers to the occupancy value.

$$\mathcal{V} = f\left(\mathcal{N}_C, \mathcal{B}, \mathcal{C}, \mathcal{N}, \mathcal{S}\right) \tag{2}$$

3.2 Ray-Based Sampling Features

Coordinates Information. In the Ray-based Sampling procedure, we initially deploy parallel rays toward the SMPL, with each ray corresponding to a specific pixel location. A weak-perspective camera is employed, rendering rays that are perpendicular to the SMPL mesh. Upon striking a face, each ray continues its trajectory, penetrating through. Importantly, each ray documents the faces it penetrates. For instance, if a ray penetrates only through a particular body part of the mesh, it will record the two faces it has traversed - the front and back sides of the body part. This process allows us to obtain the coordinates of the vertices forming the penetrated faces. Subsequently, we record the average coordinate of these vertices in 2D feature maps.

We assume that a single ray will intersect no more than six faces, thus necessitating the use of 18 feature maps to record the x, y, and z coordinates of each intersection point. In the aforementioned example, the values of six feature maps at the pixel corresponding to the ray will be updated, while all other feature maps maintain zero values. In situations where a ray intersects more than two faces, say four, we interpret this as occlusion. In such circumstances, the coordinate information feature maps will register the coordinates of the vertices forming the intersected faces. This implies that 12 feature maps at this pixel will carry non-zero values. This collected information is subsequently conveyed to the Multilayer Perceptron. Given this occlusion data, the MLP is thus better equipped to predict the occupancy field accurately.

Blendweight-Based Labels. In addition, to coordinate information, we integrate blend weight-based labels to serve as soft human parsing information. In the SMPL model, blend weights are utilized to determine how each vertex moves in relation to the skeletal joints. Each vertex in the model is attached to one or more joints with a corresponding blend weight. When a specific joint moves or rotates, the vertices associated with that joint move correspondingly. The extent of this movement is governed by the blend weights. While the pose and shape of the SMPL model are subject to change, the blend weights for each vertex largely remain constant.

Directly assigning discrete human parsing labels to vertices of the SMPL model has its limitations, as a single vertex may be associated with multiple human body parts. Therefore, in accordance with S-PIFu [2], we employ blend weight-based labels as a form of soft human parsing information. The blend weight of a vertex can be used to determine its degree of association with each joint, thereby providing human parsing information. Through the use of Ray-based Sampling, we generate a set of 2D feature maps comprising blend weight-based labels for each vertex. These feature maps offer insights into the relationships between vertices and aid the Multilayer Perceptron in more accurately predicting the occupancy field.

4 Experiments

4.1 Datasets

Training Dataset. We train our model on THuman2.0 [39] dataset. This dataset consists of high-quality scans, primarily featuring young Chinese adults. For training, we employ 500 human meshes from THuman2.0, with each mesh having corresponding fits for both the SMPL and SMPL-X models. Furthermore, for each mesh, we generate 36 RGB images from varying yaw angles using a weak-perspective camera.

Testing Dataset. In line with previous research [32], we evaluate our model using the CAPE [17] dataset. This dataset includes both posed and unposed clothed bodies for each frame, alongside corresponding SMPL and SMPL-X fits. For testing purposes, we generate 3 RGB images from distinct yaw angles for each mesh, utilizing a weak-perspective camera. This dataset is not used during training, allowing us to evaluate the model's capacity for generalization.

4.2 Evaluation Metrics

Chamfer Distance. The Chamfer distance is a popular metric for measuring the dissimilarity between two point sets. We sample points uniformly on the ground-truth mesh and our reconstructed mesh to produce a set of points and the Chamfer distance is calculated based on these point sets. This provides a measure of how similar the shapes of the two meshes are, irrespective of their positioning, orientation, or scaling in 3D space.

"P2S" Distance. Point-to-surface (P2S) distance is a common method to measure the similarity between a 3D point and a 3D surface. Given a 3D point and a 3D surface, the point-to-surface distance is the minimum Euclidean distance from the point to any point on the surface. This metric gives a more accurate representation of the difference between a point set and a surface, as compared to point-to-point metrics like the Chamfer distance, as it considers the entire surface rather than just a set of discrete points.

Normals Difference. The normals of a mesh are vectors that are perpendicular to the surface of the mesh at each vertex. They play an important role in rendering the mesh and can significantly affect the appearance of the resulting 3D model, especially in terms of shading and lighting. The normals difference metric provides a measure of the geometric accuracy of the reconstructed mesh, in terms of both shape and orientation of the surface.

4.3 Evaluation Results

Quantitative Results. We compare our proposed method with the existing state-of-the-art clothed human reconstruction methods on CAPE [17] dataset. As shown in Table 1, we find that our model achieves the lowest error when utilizing global features generated by 2D convolutional filters. With a Chamfer distance of 0.891, a Point-to-surface distance of 0.854, and a Normals difference of 0.054, our method significantly outperforms other methods by large margins. When we employ local features, our method achieves the best performance for Point-to-surface distance, measuring at 0.901. As for the other evaluation metrics, our method performs on par with the state-of-the-art approaches.

Qualitative Results. In this paper, we present a qualitative comparison of our approach, Fine Implicit Reconstruction Enhancement (FIRE), with other leading 3D human reconstruction methods such as PIFu [25], PaMIR [34], and

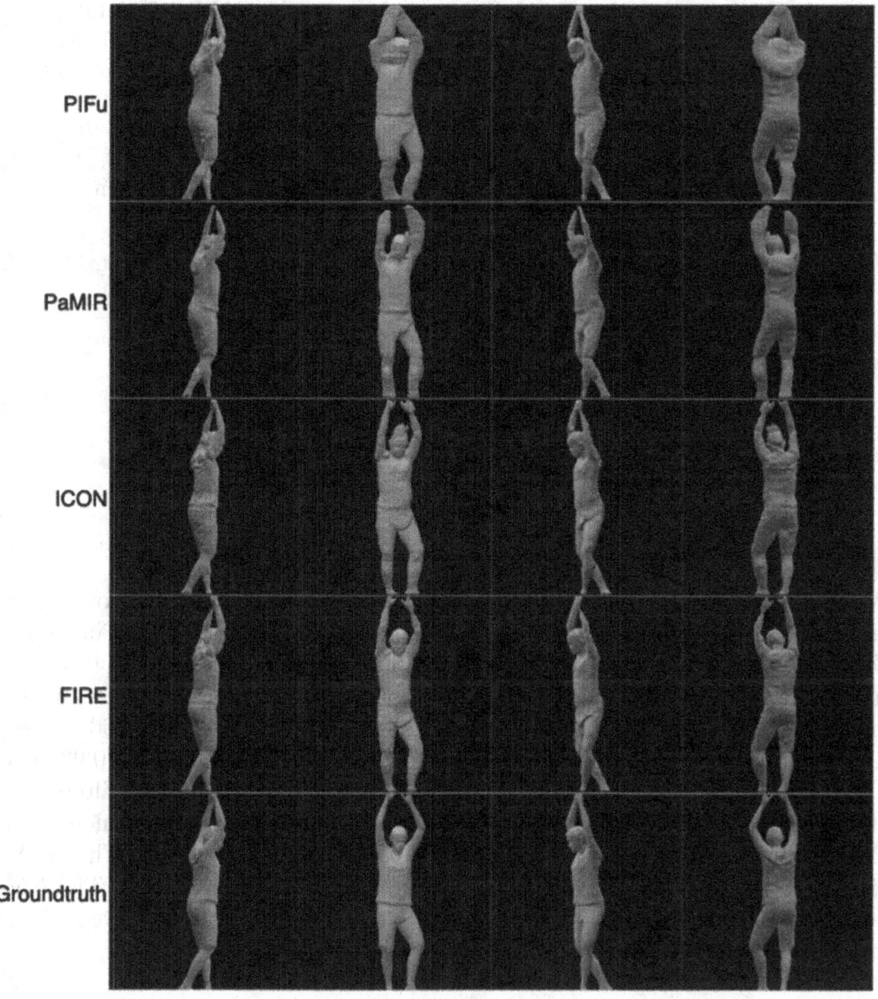

Fig. 2. Qualitative evaluation with state-of-the-art methods. Our model achieves clean and detailed 3D reconstruction of clothed human.

ICON [32]. This comparison, conducted on the CAPE dataset, is illustrated in Fig. 2. Our method, FIRE, demonstrates superior performance, producing a more detailed and realistic 3D human mesh with subtle features like fingers and cloth wrinkles. In addition, our model successfully prevents the generation of floating artifacts, a common issue seen in other methods. We believe this improvement is due to the incorporation of Ray-based Sampling features, as they provide information on the presence of a subject within a specific space. Overall, FIRE successfully captures the accurate structure of the human body, generating a clean and realistic 3D human mesh.

Table 1. Comparison with state-of-the-art methods on the CAPE [17] dataset. The results from PIFu [25], PIFuHD [26], and PaMIR [34] are reported by ICON [32], and these models were trained on AGORA [35] and THuman [36] datasets. We reimplemented ICON [32] and trained it on THuman2.0 [39] dataset. Additionally, both 'ICON + filter' and 'Ours + filter' utilized the hourglass model [27] to extract features.

Methods	Chamfer distance	"P2S" distance	Normals difference
PIFu	3.627	3.729	0.116
PIFuHD	3.237	3.123	0.112
PaMIR	2.122	1.495	0.088
ICON	1.034	0.960	0.064
FIRE	1.037	0.901	0.075
ICON + filter	1.016	0.933	0.068
FIRE + filter	**0.891**	**0.854**	**0.054**

4.4 Ablation Study

We conduct ablation experiments to verify the effectiveness of each component of FIRE on the CAPE dataset with Chamfer distance, "P2S" distance, and Normals difference, shown in Table 2. As illustrated in Table 2, when local features are used, disabling coordinates information and blend weight-based labels increases the Chamfer distance error, going from 1.037 to 1.048 and 1.100, respectively. Similarly, the "P2S" distance error rises, moving from 0.901 to 0.934 and 1.028. On the other hand, when we use global features obtained from filters, disabling coordinates information and blend weight-based labels see an increase in Chamfer distance error, from 0.891 to 0.928 and 0.950, respectively. The "P2S" distance error also increases, moving from 0.854 to 0.880 and 0.903. The Normals difference rises from 0.054 to 0.056 and 0.058. These findings assert the importance of every component in our model.

Table 2. Ablation study on the CAPE [17] dataset. We analyze the effects of two key components: coordinates information (C) and blend weight-based labels (B). Both C and B are extracted using Ray-based Sampling.

Methods	Chamfer distance	"P2S" distance	Normals difference
FIRE w/o C	1.048	0.934	0.075
FIRE w/o B	1.100	1.028	0.069
FIRE	1.037	0.901	0.075
FIRE + filter w/o C	0.928	0.880	0.056
FIRE + filter w/o B	0.950	0.903	0.058
FIRE + filter	**0.891**	**0.854**	**0.054**

5 Conclusion

In this paper, we introduced the Fine Implicit Reconstruction Enhancement (FIRE) model, designed to reconstruct 3D meshes of clothed humans from a single image. Our approach incorporates Ray-based Sampling features and blend weight-based labels within an iterative optimization framework, enhancing the utilization of information extracted from the parametric human model. The experimental results obtained from public benchmarks validate the efficacy and robustness of our model. Looking forward, we aim to integrate multi-view information to further enhance the performance of our model, paving the way for its application in real-world scenarios.

References

1. Alldieck, T., Zanfir, M., Sminchisescu, C.: Photorealistic monocular 3d reconstruction of humans wearing clothing. In: CVPR, pp. 1506–1515 (2022)
2. Chan, K., Lin, G., Zhao, H., Lin, W.: S-pifu: integrating parametric human models with pifu for single-view clothed human reconstruction. Adv. Neural. Inf. Process. Syst. **35**, 17373–17385 (2022)
3. Chan, K.Y., Lin, G., Zhao, H., Lin, W.: Integratedpifu: integrated pixel aligned implicit function for single-view human reconstruction. In: Computer Vision-ECCV 2022: 17th European Conference, Tel Aviv, Israel, 23–27 October 2022, Proceedings, Part II, pp. 328–344. Springer (2022). https://doi.org/10.1007/978-3-031-20086-1_19
4. Chen, X., Zheng, Y., Black, M.J., Hilliges, O., Geiger, A.: Snarf: Differentiable forward skinning for animating non-rigid neural implicit shapes. In: Proceedings of the IEEE/CVF International Conference on Computer Vision, pp. 11594–11604 (2021)
5. Kavan, L., Collins, S., Žára, J., O'Sullivan, C.: Skinning with dual quaternions. In: Proceedings of the 2007 Symposium on Interactive 3D Graphics and Games, pp. 39–46 (2007)
6. Chen, Z., Zhang, H.: Learning implicit fields for generative shape modeling. In: CVPR, pp. 5939–5948 (2019)
7. Feng, Q., Liu, Y., Lai, Y.K., Yang, J., Li, K.: Fof: learning fourier occupancy field for monocular real-time human reconstruction. arXiv:2206.02194 (2022)
8. He, T., Collomosse, J., Jin, H., Soatto, S.: Geo-pifu: geometry and pixel aligned implicit functions for single-view human reconstruction. Adv. Neural. Inf. Process. Syst. **33**, 9276–9287 (2020)
9. He, T., Xu, Y., Saito, S., Soatto, S., Tung, T.: Arch++: animation-ready clothed human reconstruction revisited. In: Proceedings of the IEEE/CVF International Conference on Computer Vision, pp. 11046–11056 (2021)
10. Huang, Y., et al.: One-shot implicit animatable avatars with model-based priors. arXiv:2212.02469 (2022)
11. Huang, Z., Xu, Y., Lassner, C., Li, H., Tung, T.: Arch: animatable reconstruction of clothed humans. In: CVPR, pp. 3093–3102 (2020)
12. Jiang, B., Hong, Y., Bao, H., Zhang, J.: Selfrecon: self reconstruction your digital avatar from monocular video. In: CVPR, pp. 5605–5615 (2022)
13. Joo, H., Simon, T., Sheikh, Y.: Total capture: a 3d deformation model for tracking faces, hands, and bodies. In: CVPR, pp. 8320–8329 (2018)

14. Lin, S., Zhang, H., Zheng, Z., Shao, R., Liu, Y.: Learning implicit templates for point-based clothed human modeling. In: Computer Vision-ECCV 2022: 17th European Conference, Tel Aviv, Israel, October 23–27, 2022, Proceedings, Part III, pp. 210–228. Springer (2022). https://doi.org/10.1007/978-3-031-20062-5_13
15. Loper, M., Mahmood, N., Romero, J., Pons-Moll, G., Black, M.J.: Smpl: a skinned multi-person linear model. ACM Trans. Graph. (TOG) **34**(6), 1–16 (2015)
16. Lorensen, W.E., Cline, H.E.: Marching cubes: a high resolution 3d surface construction algorithm. ACM siggraph computer graphics **21**(4), 163–169 (1987)
17. Ma, Q., et al.: Learning to dress 3d people in generative clothing. In: CVPR, pp. 6469–6478 (2020)
18. Mescheder, L., Oechsle, M., Niemeyer, M., Nowozin, S., Geiger, A.: Occupancy networks: learning 3d reconstruction in function space. In: CVPR, pp. 4460–4470 (2019)
19. Mildenhall, B., Srinivasan, P.P., Tancik, M., Barron, J.T., Ramamoorthi, R., Ng, R.: Nerf: representing scenes as neural radiance fields for view synthesis. Commun. ACM **65**(1), 99–106 (2021)
20. Park, J.J., Florence, P., Straub, J., Newcombe, R., Lovegrove, S.: Deepsdf: learning continuous signed distance functions for shape representation. In: CVPR, pp. 165–174 (2019)
21. Pavlakos, G., et al.: Expressive body capture: 3d hands, face, and body from a single image. In: CVPR, pp. 10975–10985 (2019)
22. Peng, S., et al.: Neural body: implicit neural representations with structured latent codes for novel view synthesis of dynamic humans. In: CVPR, pp. 9054–9063 (2021)
23. Pesavento, M., Volino, M., Hilton, A.: Super-resolution 3d human shape from a single low-resolution image. In: Computer Vision-ECCV 2022: 17th European Conference, Tel Aviv, Israel, 23–27 October 2022, Proceedings, Part II, pp. 447–464. Springer (2022). https://doi.org/10.1007/978-3-031-20086-1_26
24. Romero, J., Tzionas, D., Black, M.J.: Embodied hands: modeling and capturing hands and bodies together. arXiv:2201.02610 (2022)
25. Saito, S., Huang, Z., Natsume, R., Morishima, S., Kanazawa, A., Li, H.: Pifu: pixel-aligned implicit function for high-resolution clothed human digitization. In: Proceedings of the IEEE/CVF International Conference on Computer Vision, pp. 2304–2314 (2019)
26. Saito, S., Simon, T., Saragih, J., Joo, H.: Pifuhd: multi-level pixel-aligned implicit function for high-resolution 3d human digitization. In: CVPR, pp. 84–93 (2020)
27. Jackson, A.S., Manafas, C., Tzimiropoulos, G.: 3D human body reconstruction from a single image via volumetric regression. In: Leal-Taixé, L., Roth, S. (eds.) ECCV 2018. LNCS, vol. 11132, pp. 64–77. Springer, Cham (2019). https://doi.org/10.1007/978-3-030-11018-5_6
28. Wang, K., Zheng, H., Zhang, G., Yang, J.: Parametric model estimation for 3d clothed humans from point clouds. In: ISMAR, pp. 156–165. IEEE (2021)
29. Wang, S., Mihajlovic, M., Ma, Q., Geiger, A., Tang, S.: Metaavatar: learning animatedly clothed human models from few depth images. Adv. Neural. Inf. Process. Syst. **34**, 2810–2822 (2021)
30. Weng, C.Y., Curless, B., Srinivasan, P.P., Barron, J.T., Kemelmacher-Shlizerman, I.: Humannerf: free-viewpoint rendering of moving people from monocular video. In: CVPR, pp. 16210–16220 (2022)
31. Xiong, Z., et al.: Pifu for the real world: a self-supervised framework to reconstruct dressed humans from single-view images. arXiv:2208.10769 (2022)
32. Xiu, Y., Yang, J., Tzionas, D., Black, M.J.: Icon: implicit clothed humans obtained from normals. In: CVPR, pp. 13286–13296. IEEE (2022)

33. Xu, H., Bazavan, E.G., Zanfir, A., Freeman, W.T., Sukthankar, R., Sminchisescu, C.: Ghum & ghuml: generative 3d human shape and articulated pose models. In: CVPR, pp. 6184–6193 (2020)
34. Zheng, Z., Yu, T., Liu, Y., Dai, Q.: Pamir: Parametric model-conditioned implicit representation for image-based human reconstruction. IEEE Trans. Pattern Anal. Mach. Intell. **44**(6), 3170–3184 (2021)
35. Patel, P., Huang, C.H.P., Tesch, J., Hoffmann, D.T., Tripathi, S., Black, M.J.: Agora: avatars in geography optimized for regression analysis. In: CVPR, pp. 13468–13478 (2021)
36. Zheng, Z., Yu, T., Wei, Y., Dai, Q., Liu, Y.: Deephuman: 3d human reconstruction from a single image. In: Proceedings of the IEEE/CVF International Conference on Computer Vision. pp. 7739–7749 (2019)
37. Moon, G., Nam, H., Shiratori, T., Lee, K.M.: 3d clothed human reconstruction in the wild. In: Computer Vision-ECCV 2022: 17th European Conference, Tel Aviv, Israel, 23–27 October 2022, Proceedings, Part II, pp. 184–200. Springer (2022). https://doi.org/10.1007/978-3-031-20086-1_11
38. Jinka, S.S., Srivastava, A., Pokhariya, C., Sharma, A., Narayanan, P.: Sharp: shape-aware reconstruction of people in loose clothing. Int. J. Comput. Vision **131**(4), 918–937 (2023)
39. Yu, T., Zheng, Z., Guo, K., Liu, P., Dai, Q., Liu, Y.: Function4d: real-time human volumetric capture from very sparse consumer rgbd sensors. In: CVPR, pp. 5746–5756 (2021)
40. Kocabas, M., Huang, C.H.P., Hilliges, O., Black, M.J.: Pare: part attention regressor for 3d human body estimation. In: Proceedings of the IEEE/CVF International Conference on Computer Vision, pp. 11127–11137 (2021)
41. Ravi, N., Reizenstein, J., Novotny, D., Gordon, T., Lo, W.Y., Johnson, J., Gkioxari, G.: Accelerating 3d deep learning with pytorch3d. arXiv:2007.08501 (2020)

Sem-Avatar: Semantic Controlled Neural Field for High-Fidelity Audio Driven Avatar

Xiang Zhou[1], Weichen Zhang[1], Yikang Ding[1], Fan Zhou[1], and Kai Zhang[1,2](✉)

[1] Shenzhen International Graduate School, Tsinghua University, Beijing, China
{zhoux21,zwc21,dyk20,zhouf21}@mails.tsinghua.edu.cn, zhangkai@sz.tsinghua.edu.cn

[2] Research Institute of Tsinghua, Pearl River Delta, Beijing, China

Abstract. In this paper, we tackle the audio-driven avatar challenge by fitting a semantic controlled neural field to a talking-head video. While existing methods struggle with realism and head-torso inconsistency, our novel end-to-end framework, semantic controlled neural field (Sem-Avatar) sucessfully overcomes the above problems, delievering high-fidelity avatar. Specifically, we devise a one-stage audio-driven forward deformation approach to ensure head-torso alignment. We further propose to use semantic mask as a control signal for eye opening, lifting the naturalness of the avatar to another level. We train our framework via comparing the rendered avatar to the original video. We further append a semantic loss which leverages human face prior to stabilize training. Extensive experiments on public datasets demonstrate Sem-Avatar's superior rendering quality and lip synchronization, establishing a new state-of-the-art for audio-driven avatars.

Keywords: Avatar Synthesis · Talking Head · Neural Rendering

1 Introduction

Synthesizing audio-driven avatar is a trending and challenging topic that faces increasing demands in real-world applications. For instance, AR/VR technologies, human digitization and metaverse. Early approaches tried to solve this problem by complicated camera capturing [3], professional artists were also required to polish the generated model [5]. These methods demand highly expensive equipments and are not applicable for users in daily life. With the witnessed success of 3D parametric face model, e.g., 3D Morphable Models [2], researchers tried to solve this problem by using parametric models as intermediates. However, due to the fixed topology of the parametric face models, they are incapable of reconstructing hair or accessories like glasses for avatars.

The recent success of implicit-based rendering receives considerable attention from computer vision researchers. Implicit approaches use a continuous function

X. Zhou and W. Zhang—Equal Contribution

Q. Liu et al. (Eds.): PRCV 2023, LNCS 14426, pp. 66–78, 2024.
https://doi.org/10.1007/978-981-99-8432-9_6

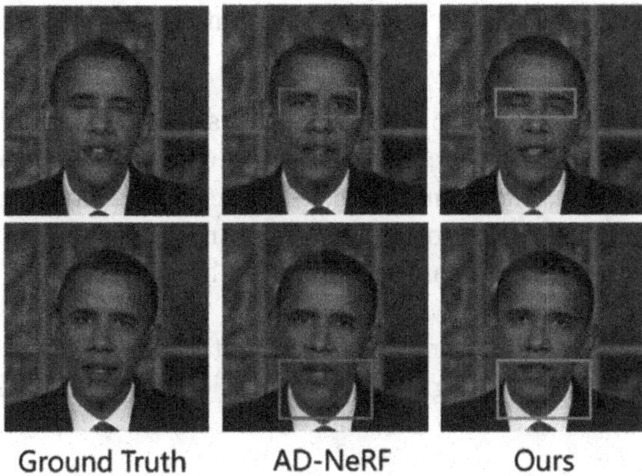

Fig. 1. Audio driven results. We show that our approach successfully addressed the inconsistency between head and torso which AD-NeRF [8] struggles to deal with (second row). The semantic eye control module further improves the fidelity of our audio-driven results (first row).

to fit the scene by optimizing to a set of calibrated input images. Differentiable rendering is then applied to accomplish the generation of the scene image. Among implicit approaches, NeRF [13] is the most widely-adopted approach for its stunning rendering quality and the ability to synthesize novel view images. Owing to this, many follow-up methods [6,8,11,17] proposed to use NeRF to synthesize realistic audio-driven avatar. NeRF-based avatar cleverly treats head pose changes via inversely changing the camera pose (*e.g.* the head looking up can be interpreted as the camera moving down), but the torso part has to be rendered separately because it has a different pose. Despite their promising rendering results, they struggle to maintain consistency between head and torso. The severe drawback largely degrades the audience's perception, indicating that there is still a long way to synthesize realistic audio-driven avatar.

In this paper, to tackle this challenge and open gate for high-fidelity audio-driven human avatar synthesis, our novel framework Sem-Avatar is hereby proposed. Sem-Avatar incorporates audio features into a flame-guided neural field and deform the avatar instead of changing the camera, this shift allows us to render the avatar as a whole and completely solve the inconsistency issue. Specifically, given a video sequence and the related audio track of a talking human, Sem-Avatar will first encode the audio into features. Together with blendshape coefficients and lbs weights, the audio features will deform the avatar in canonical space represented as an occupancy field. By having the deformed avatar, we can render the audio-driven human avatar via a texture field. We further employ semantic mask to control the eye region for higher fidelity. We briefly show the performance of our framework by comparing with AD-NeRF [8] in Fig. 1.

We evaluate our Sem-Avatar both qualitatively and quantitatively on public benchmark datasets. These comparisons demonstrate the performance of Sem-

Avatar w.r.t. the state-of-the-art methods. In summary, the contributions of this paper are:

- We propose a novel implicit neural field framework for avatar that can, by directly leveraging the encoded audio features, produce satisfying audio-driven results and maintain motion consistency between the head and torso.
- We employ semantic mask to explicitly control the opening and closing of eyes. This technique significantly improves the realness and naturalness of our synthesized avatar.
- We conduct extensive experiments to evaluate the effectiveness of our proposed approach, establishing better results over previous state-of-the-art approaches.

2 Related Works

2.1 Model-Based Audio-Driven Avatar

Audio-driven animation aims to generate photo-realistic video that is in sync with the driving audio. As mapping audio directly to high-dimensional facial movement is a challenging task. Researchers typically chose to map audio to expression parameters for 3D morphable face model that is in relatively low-dimensional space. NVP [19] first pretrained the audio to expression module on a large dataset then finetuned the module on target video to fit personal talking style. AudioDVP [9] utilized pretrained audio feature and achieved good driving results under much less data. Apart from using parametric face models as intermediate representations, another choice for researcher is facial landmarks. MakeitTalk [23] utilized audio to regress content embedding and speaker identity embedding separately, aiming for person-specific talking style. LSP [12] used manifold projection to extract audio features, then used the audio features to regress head poses which together with audio features predicts movements of landmarks and is used to produce real-time photo-realistic talking avatar. Wav2lip [16], on the other hand, used GAN-based method to directly synthesize the talking portrait image.

2.2 Implicit-Based Avatar

The recently proposed NeRF-based avatar, different from model-based methods, use a multi-layer-perceptron(MLP) that regresses density and radiance to represent the avatar and a condition vector to animate the avatar. Nerfies [14] and HyperNeRF [15] are the first to use NeRF [13] to build 4D avatars, for each frame, they use a latent code to deform the frame space back to canonical space that is shared by all frames. The density and radiance of points in frame space can then be retrieved by querying its canonical correspondence. But Nerfies and HyperNeRF avatar are not animatable because the per-frame latent code does not provide meaningful control signals. To resolve this, Nerface [6] uses expression parameters from 3DMM as condition for NeRF to drive the avatar. Different from Nerface, AD-NeRF [8] uses audio feature as a condition to NeRF and animates the avatar according to the audio feature. Neural head avatar [7] used a FLAME model to represent the avatar and use a SIREN-MLP [18] to render the avatar for the better rendering quality of implicit rendering. IMAvatar [22] employed a forward-deformation approach to generalize the reconstructed avatar on unseen expressions and poses.

3 Sem-Avatar

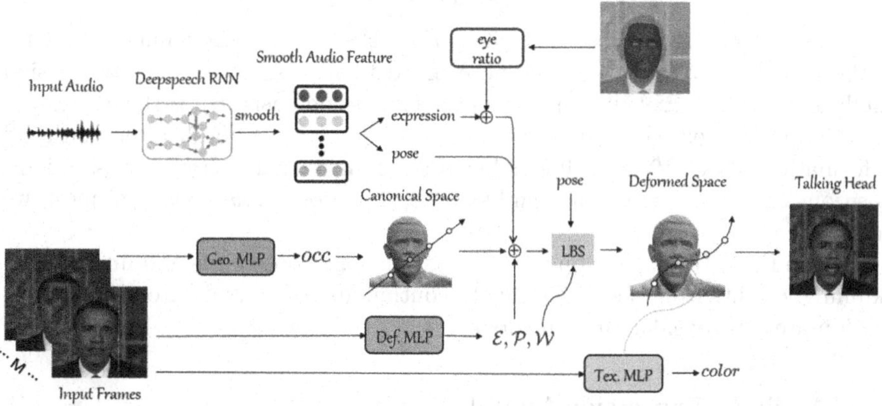

Fig. 2. The architecture of our proposed Sem-Avatar. Given a talking portrait video, we train MLPs for geometry and texture to represent the avatar in canonical space, and a deformation MLP to deform the avatar from canonical space. Given a segment of audio, it is mapped to expression and pose parameters to synthesize the talking portrait via deformation field.

We present a semantic-controlled semi-implicit neural field for audio-driven avatar (Sem-Avatar) that synthesize high-fidelity audio-driven avatar. See Fig. 2 for an overview of our framework. In the following sections we will introduce the framework in detail. First, we will brief on the basic knowledge of the proposed Sem-Avatar. Then we will explain how we encode the audio feature. We will then describe our neural implicit field including the occupancy field and texture field for generating the canonical face model, followed by the deformation field that deforms canonical space to observation space. Finally, we introduce our semantic eye control module.

3.1 Basic Knowledge

FLAME. In order to deform the avatar, we base our deformation process on FLAME [10], a 3D morphable face model, whose deformation process goes like this: given an average face mesh $\overline{T} \in \mathbb{R}^{3N}$ where N is the number of vertex, FLAME will first perform non-rigid deformation by multiplying expression coefficient with expression basis and pose coefficient with pose basis, followed by rigid deformation calculated by linear blend skinning(LBS). LBS takes the rotation vector of the four joints(neck, jaw and both eyes) and the global rotation vector of the full body, weight these rotation vectors with linear blend skinning weight and perform rigid transformation for the whole face.

We can formulate the deformation procedure as:

$$T_P(\boldsymbol{\theta}, \boldsymbol{\psi}) = \overline{\boldsymbol{T}} + B_P(\boldsymbol{\theta}; \mathcal{P}) + B_E(\boldsymbol{\psi}; \mathcal{E}), \tag{1}$$

$$M(\boldsymbol{\theta}, \boldsymbol{\psi}) = W\left(T_P(\boldsymbol{\theta}, \boldsymbol{\psi}), J(\boldsymbol{\psi}), \boldsymbol{\theta}, \mathcal{W}\right), \tag{2}$$

where $\theta \in \mathbb{R}^{|\theta|}$ is the pose coefficient, $\mathcal{P} \in \mathbb{R}^{3N \times 9K}$ (K is the number of joints) is the pose basis and B_p denotes pose blendshape. $\psi \in \mathbb{R}^{|\psi|}$ is the expression coefficient and $\mathcal{E} \in \mathbb{R}^{3N \times |\psi|}$ represents the expression basis, B_E is the expression blendshape. Added together and they form T_P, the face mesh after non-rigid deformation. $\mathcal{W} \in \mathbb{R}^{K \times N}$ is linear blend skinning weights, $J(\psi)$ regresses joint positions and W is the linear blend skinning function. $M(\theta, \psi)$ is final mesh we get.

Unlike FLAME, we redefine the expression basis, pose basis and linear blend skinning weights from discrete space to continuous space, and extend the deformation area to consider the full torso.

3.2 Audio to Expression Module

Given the audio, we accomplish the mapping from audio to expression space with our audio to expression module. Specifically, we first extract audio features using Deepspeech model [1]. Here we use Deepspeech model pretrained on a large-scale multi-lingual audio dataset. This procedure will provide us with a 29-dimensional feature for every 20ms of audio. To acquire the audio feature for each frame in the input video, we aggregate the features from its neighbouring 16 segments to obtain the final audio feature $\mathbf{A} \in \mathbb{R}^{16 \times 29}$.

Given the audio feature $\mathbf{A}$, a network consisting of 4-layer 1-D temporal convolution layers, is employed to map it to expression ψ and jaw pose θ_{jaw}, which is also related to audio but for simplicity we will omit it from here. In order to smooth the expression parameters from neighbouring frames to avoid jittering phenomenon, we adopt a self-attention module that takes as input a window size of 8 with the current frame in the center and outputs a weighted sum of the neighbouring expression. Note that we did not separately train this audio to expression module using FLAME expression as ground truth, but rather train it in an end-to-end fashion. As only part of FLAME expression is correlated with audio, directly using audio to regress all the expression can cause confusion and lead to poor audio driving performance.

3.3 Avatar Represented in Neural Field

We represent the avatar using 2 MLPs, one for regressing occupancy value representing the avatar geometry, and one for regressing RGB value to render the avatar. This disentanglement of the two attributes brings better geometry and rendering performance.

$$\mathcal{F}_{\theta_o} : (\mathbf{A}, x_c) \longrightarrow occ. \tag{3}$$

where $\mathbf{A}$ stands for the extracted audio feature, $x_c \in \mathbb{R}^3$ is the querying coordinate in canonical space, $occ \in [0, 1]$ represents the probability whether x_c is occupied by the reconstructed avatar and θ_o are the learnable parameters. We appended audio as one of the inputs to achieve audio-driven setup and more importantly, to account for topological changes during driving that cannot be explained by a fixed canonical space.

We use a neural texture MLP to assign the avatar with color. The MLP will predict the RGB color $c \in \mathbb{R}^3$ of a 3D point $\mathbf{x_c}$ in the canonical space. We can formulate the texture function as follows:

$$\mathcal{F}_{\theta_t} : (x_c, n_d, \theta, \psi) \longrightarrow c \tag{4}$$

where n_d is the normal value, obtained by normalizing the gradient from occupancy field. θ represents pose coefficients and ψ is the expression coefficients.

3.4 Animation of Neural Implicit Avatar

Based on the neural occupancy field and neural texture field we have our canonical space avatar. Our deformation network can animate the neural implicit face avatar by deforming the points $\mathbf{x_c}$ in canonical space to observed space $\mathbf{x_d}$ in a way that resembles FLAME but in continuous space.

For each 3D point $\mathbf{x_c}$ in the canonical space, we will first predict the expression basis $\mathcal{E}$, pose basis $\mathcal{P}$ and LBS weights $\mathcal{W}$:

$$\mathcal{F}_{\theta_d} : (x_c) \longrightarrow \mathcal{E}, \mathcal{P}, \mathcal{W}, \tag{5}$$

where the basis vectors $\mathcal{E}$ and $\mathcal{P}$ will help conduct the non-rigid body transformation. LBS weights will help perform the rigid-body transformation.

Specifically, the non-rigid transformation is performed by adding the product of expression coefficient, pose coefficient and their respective basis vectors.

$$x_{nr} = x_c + B_P(\boldsymbol{\theta}; \mathcal{P}) + B_E(\boldsymbol{\psi}; \mathcal{E}) \tag{6}$$

LBS weights are then applied to x_{nr}. So that the coordinates of the deformed points x_d can be formulated as:

$$x_d = W\left(x_{nr}, J(\boldsymbol{\psi}), \boldsymbol{\theta}, \mathcal{W}\right) \tag{7}$$

It will form the final deformed avatar.

Fig. 3. A semantic segmentation of the avatar.

3.5 Semantic Eye Control

Since FLAME does not model the opening and closing of eyes, our neural implicit avatar cannot perform any eye movements. This limitation inherited from FLAME seriously degrades the realness of the avatar, as a result we get an avatar whose eyes keeps open for the whole time, rendering it highly unreal. To address this problem, here we propose a novel method by using semantic map as an explicit control signal.

In the preprocessing stage, we use an off-the-shelf semantic parsing method [21] to get the semantic segmentation of the avatar as shown in Fig. 3. We count the total number of pixels from the eye region and formulate a value called eye ratio, we normalize this eye ratio and append it to our expression coefficient.

During training, the deformation module gradually learns to interpret this dimension as an eye opening metric and we realize the control of eye.

3.6 Training Objectives

We adopt photometric loss, which calculates the L_2 distance from generated pixels to ground-truth pixels, to train our neural implicit avatar.

$$L_{rgb} = \frac{1}{|P|} \sum_{p \in P_{in}} ||I_{pred}[p] - I_{gt}[p]||_2^2, \tag{8}$$

where p refers to the pixels whose ray hits the surface. To improve geometry and the rendering results, we add an additional mask loss for the rays that does not hit surface:

$$\mathcal{L}_M = \frac{1}{|P|} \sum_{p \in P \setminus P^{in}} CE\left(O_p, F_{\theta_o}\left(x_c^\star\right)\right), \tag{9}$$

where CE stands for cross entropy loss imposed upon ground truth occupancy O_p and the predicted occupancy $F_{\theta_o}(x_c^\star))$. Since these rays do not hit the surface, we choose the point on the ray that is nearest to surface, x^*, to predict the occupancy value.

We further append a semantic loss, which leverages semantic prior from the aforementioned semantic segmentation to stabilize the training process. It goes like:

$$\mathcal{L}_{sem} = \frac{1}{|P|} \sum_{p \in P_{face}} ||x_c^p - \overline{T}_p||_2^2 \tag{10}$$

For pixels that belong to face semantic region, we find its 3D canonical correspondence x_c^p and calculate its distance to the nearest FLAME vertex $\overline{T}_p$. The intuition behind this loss is the face of the implicit avatar should resemble FLAME mesh. Note that we cannot impose this loss on non-face region because FLAME mesh doesn't possess these regions. In practice, we empirically find this loss conducive to training.

4 Experiments

Datasets. Different from previous study that requires a large training set to train an audio-to-expression module, a 3–5 minute video is sufficient for training our model. Specifically, we choose one publicly-released video from AD-NeRF [8] and one video from LSP [12]. For each of the videos, we select 80% of the frames as our training set and 20% as the testset.

Evaluation Metrics. To evaluate the results quantitatively, we choose Peak Signal-to-Noise Ratio(PSNR), Structural Similarity(SSIM) as metrics to evaluate the image quality. Additionally, we employ Landmark Distance(LMD) and Synchronization Confidence score(Sync) as metrics to evaluate the synchronization between audio and lip.

Implementation Details. We implement our framework with PyTorch. The networks are trained via Adam optimizer with a learning rate of 0.0002. We train our model for 40 epochs with a batch size of 8 on 8 NVIDIA Tesla V100s. The model is trained at 256 × 256 resolution on a 5-minute dataset and it takes around 40 h to converge.

Experimental Settings. We compare our methods with 1) AudioDVP [20], state-of-the-art approach for model-based audio driven avatar 2) Wav2lip [16], the approach that produces SOTA lip synchronization results 3) AD-NeRF [8], the existing SOTA approach for implicit-based audio-driven avatar.

4.1 Comparison with Other Methods

Quantitative Comparisons. We demonstrate our quantitative results in Table 1 and Table 2. Since model-based approaches use audio to drive only the mouth region and copy the rest from the original video. We crop only the lower-face region and resize it into the same size for fair comparison. We compare with AD-NeRF [8] under full resolution setting because it can also drive the full avatar.

Table 1. Quantitative results under cropped setting on public datasets. We compare our approach against recent SOTA methods.

	Dataset A [8]				Dataset B [12]			
Methods	PSNR↑	SSIM↑	LMD↓	Sync↑	SNR↑	SSIM↑	LMD↓	Sync↑
Ground Truth	N/A	1.000	0	8.2	N/A	1.000	0	8.5
AudioDVP [20]	18.8	0.614	16.6	4.2	16.1	0.561	21.8	4.8
Wav2Lip [16]	24.9	0.807	2.16	**7.8**	24.1	0.861	**1.11**	**10.7**
AD-NeRF [8]	25.1	0.801	1.81	5.2	24.5	0.862	1.75	6.1
Ours	**26.3**	**0.819**	**1.69**	5.7	**25.3**	**0.875**	1.61	6.3

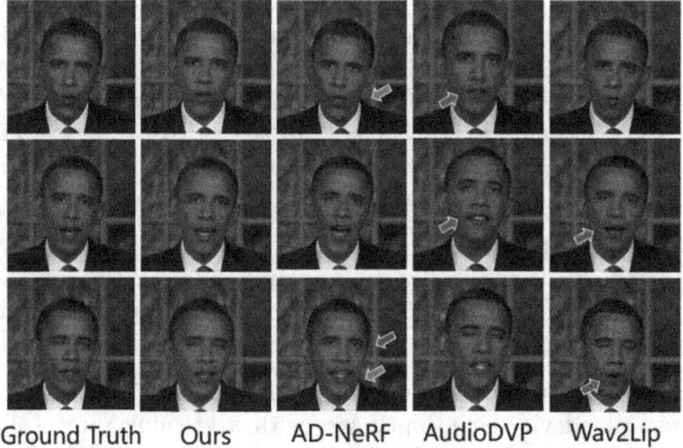

Fig. 4. Comparison with other approaches. The driven results show that our approach can generate the most realistic talking head. The orange arrow indicates a separation of head and torso, blue arrow indicates an eye mismatch, as for the red arrow, it represents a lip error.

Under cropped setting, our method achieves SOTA results on PSNR, SSIM and LMD metrics in dataset A. It proves that our method can generate photometric talking avatar with fine-grained details. Note that Wav2Lip [16] achieves the highest Sync score, owing to the fact they used a pretrained Syncnet [4] in their training process. Despite high lip synchronization score, Wav2lip generates unnatural talking avatar, the mouth region is blurry and we can see clear boundary surrounding the mouth region. What's more, generating 3D-aware avatar is out of bound for wav2lip. Under full resolution setting, our method beats AD-NeRF on all metrics. Because AD-NeRF did not utilize eye semantic information, we conduct another experiment that disabled the semantic control module for fair comparison. As is shown in Table 2, under both setting our approach beats AD-NeRF, proving that we successfully tackle the inconsistency between head and torso.

Fig. 5. A comparison of our method against AD-NeRF [8]. We enalrge part of the image to stress the inconsistency caused by AD-NeRF.

Table 2. Quantitative results under full resolution setting. We compare our approach against AD-NeRF [8]

Methods	PSNR↑	SSIM↑	LMD↓	Sync↑
Ground Truth	N/A	1.000	0	8.2
AD-NeRF [8]	25.4	0.877	2.44	5.2
Ours w/o sem control	27.1	0.902	1.67	**5.7**
Ours	**27.8**	**0.911**	**1.41**	**5.7**

Qualitative Comparisons We conduct qualitative experiments by comparing key-frames from the rendering results of each method. Results are shown in Fig. 4, where AD-NeRF frequently encounters the aforementioned inconsistency problem, AudioDVP [20]'s lip movements do not synchronize well with driving audio. Wav2lip [16], despite achieving the highest score on Syncnet metric, its lip region is highly unnatural. Our method achieves the subtle balance between lip synchronization and image quality, producing the most realistic talking-head. Enlarged comparisons with AD-NeRF [8] are shown in Fig. 5.

5 Ablation Study

Audio to Expression Module. In our framework, we choose to train the audio to expression module in an end-to-end manner with no supervision imposed on expression or basis. Therefore, we conduct ablation experiments under these 2 settings. (1) With expression supervision: We train the audio to expression module with tracked FLAME expression as ground truth, in this manner we regress audio feature directly to FLAME expression coefficients. According to Table 3, we can inspect a decline in lip synchronization metric, meaning this setup is not suitable for generating lip-sync expression. (2) With basis supervision: We set the ground truth for the predicted expression basis to be the expression basis from the nearest vertex in FLAME. This also results in a drop for lip synchronization metric as shown in Table 3. The two experiments suggest that while we are able use audio to drive the avatar in resemblance to FLAME, it does not achieve the optimal result. The reason behind is both the tracked FLAME expression and the pretrained FLAME expression basis can be inaccurate, leading to increased fallacies. Using audio feature directly brings us the best result.

Semantic Loss. We also test our module without semantic loss to demonstrate the effect of the proposed loss. The result is shown on Table 3, we find the semantic loss beneficial to the final rendering results and audio synchronization. What's more, semantic loss stabilizes the training process, effectively preventing the frequent collapse of training before imposing this loss.

Table 3. Ablation Study Results

Methods	PSNR↑	SSIM↑	LMD↓	Sync↑
Exp. Supervision	25.1	0.866	2.77	4.3
BS. Supervision	26.6	0.881	1.96	5.3
Ours w/o sem loss	26.3	0.876	2.51	5.1
Ours	**27.8**	**0.911**	**1.41**	**5.7**

Semantic Eye Control Module. To prove the effectiveness of the semantic eye control module. We conduct an ablation experiment without the semantic control module. We crop the eye region for comparison on image fidelity and the result is shown in Table 4. We can observe that the semantic eye control module, by explicitly controlling eye movements, brings better fidelity and higher realness to the driven results.

Table 4. Ablation Study Results on Semantic Control Module

Methods	PSNR↑	SSIM↑
w/o semantic control	27.00	0.833
Ours	**29.56**	**0.913**

6 Conclusion

In this paper, we present a novel framework Semantic Controlled Neural Field for Audio Driven Avatar (**Sem-Avatar**) that utilizes explicit model FLAME [10] to guide the deformation of implicit audio avatar in one stage and completely tackles the inconsistency problem, bringing a high-quality audio driven avatar. We then propose to use semantic information to guide the eye region of the avatar, which is previously uncontrollable for FLAME model. Effectiveness of our method is proven by extensive experiments, suggesting our approach can synthesize a highly authentic audio-driven avatar.

Acknowledgments. This work was supported by the Key-Area Research and Development Program of Guangdong Province, under Grant 2020B0909050003.

References

1. Amodei, D., et al.: Deep speech 2: end-to-end speech recognition in English and mandarin. In: International Conference on Machine Learning (2015)
2. Blanz, V., Vetter, T.: A morphable model for the synthesis of 3d faces. In: Proceedings of the 26th Annual Conference on Computer Graphics and Interactive Techniques, pp. 187–194 (1999)
3. Cao, Y., Tien, W.C., Faloutsos, P., Pighin, F.: Expressive speech-driven facial animation. ACM Trans. Graph. (TOG) **24**(4), 1283–1302 (2005)
4. Chung, J.S., Zisserman, A.: Out of time: automated lip sync in the wild. In: Asian Conference on Computer Vision (2016)
5. Edwards, P., Landreth, C., Fiume, E., Singh, K.: Jali: an animator-centric viseme model for expressive lip synchronization. ACM Trans. Graph. (TOG) **35**(4), 1–11 (2016)
6. Gafni, G., Thies, J., Zollhöfer, M., Nießner, M.: Dynamic neural radiance fields for monocular 4d facial avatar reconstruction. In: Computer Vision and Pattern Recognition (2021)
7. Grassal, P.W., Prinzler, M., Leistner, T., Rother, C., Nießner, M., Thies, J.: Neural head avatars from monocular rgb videos. In: Proceedings of the IEEE/CVF Conference on Computer Vision and Pattern Recognition, pp. 18653–18664 (2022)
8. Guo, Y., Chen, K., Liang, S., Liu, Y.J., Bao, H., Zhang, J.: Ad-nerf: audio driven neural radiance fields for talking head synthesis. In: Proceedings of the IEEE/CVF International Conference on Computer Vision, pp. 5784–5794 (2021)
9. Ji, X., et al.: Audio-driven emotional video portraits. In: Proceedings of the IEEE/CVF Conference on Computer Vision and Pattern Recognition, pp. 14080–14089 (2021)
10. Li, T., Bolkart, T., Black, M.J., Li, H., Romero, J.: Learning a model of facial shape and expression from 4d scans. ACM Trans. Graph. (2017)
11. Liu, X., Xu, Y., Wu, Q., Zhou, H., Wu, W., Zhou, B.: Semantic-aware implicit neural audio-driven video portrait generation. In: Computer Vision-ECCV 2022: 17th European Conference, Tel Aviv, Israel, 23–27 October 2022, Proceedings, Part XXXVII, pp. 106–125. Springer (2022). https://doi.org/10.1007/978-3-031-19836-6_7
12. Lu, Y., Chai, J., Cao, X.: Live speech portraits: real-time photorealistic talking-head animation. ACM Trans. Graph. (TOG) **40**(6), 1–17 (2021)
13. Mildenhall, B., Srinivasan, P.P., Tancik, M., Barron, J.T., Ramamoorthi, R., Ng, R.: NeRF: representing scenes as neural radiance fields for view synthesis. In: Vedaldi, A., Bischof, H., Brox, T., Frahm, J.-M. (eds.) ECCV 2020. LNCS, vol. 12346, pp. 405–421. Springer, Cham (2020). https://doi.org/10.1007/978-3-030-58452-8_24
14. Park, K., et al.: Nerfies: deformable neural radiance fields. In: International Conference on Computer Vision (2020)
15. Park, K., et al.: Hypernerf: a higher-dimensional representation for topologically varying neural radiance fields. arXiv preprint arXiv:2106.13228 (2021)
16. Prajwal, K.R., Mukhopadhyay, R., Namboodiri, V.P., Jawahar, C.V.: A lip sync expert is all you need for speech to lip generation in the wild. ACM Multimedia (2020)

17. Shen, S., Li, W., Zhu, Z., Duan, Y., Zhou, J., Lu, J.: Learning dynamic facial radiance fields for few-shot talking head synthesis. In: Computer Vision-ECCV 2022: 17th European Conference, Tel Aviv, Israel, 23–27 October 2022, Proceedings, Part XII, pp. 666–682. Springer (2022). https://doi.org/10.1007/978-3-031-19775-8_39
18. Sitzmann, V., Martel, J., Bergman, A., Lindell, D., Wetzstein, G.: Implicit neural representations with periodic activation functions. Adv. Neural. Inf. Process. Syst. **33**, 7462–7473 (2020)
19. Thies, J., Elgharib, M., Tewari, A., Theobalt, C., Nießner, M.: Neural voice puppetry: audio-driven facial reenactment. In: Vedaldi, A., Bischof, H., Brox, T., Frahm, J.-M. (eds.) ECCV 2020. LNCS, vol. 12361, pp. 716–731. Springer, Cham (2020). https://doi.org/10.1007/978-3-030-58517-4_42
20. Wen, X., Wang, M., Richardt, C., Chen, Z.Y., Hu, S.M.: Photorealistic audio-driven video portraits. In: International Symposium on Mixed and Augmented Reality (2020)
21. Yu, C., Wang, J., Peng, C., Gao, C., Yu, G., Sang, N.: BiSeNet: bilateral segmentation network for real-time semantic segmentation. In: Ferrari, V., Hebert, M., Sminchisescu, C., Weiss, Y. (eds.) ECCV 2018. LNCS, vol. 11217, pp. 334–349. Springer, Cham (2018). https://doi.org/10.1007/978-3-030-01261-8_20
22. Zheng, Y., Abrevaya, V.F., Bühler, M.C., Chen, X., Black, M.J., Hilliges, O.: Im avatar: implicit morphable head avatars from videos. In: Proceedings of the IEEE/CVF Conference on Computer Vision and Pattern Recognition, pp. 13545–13555 (2022)
23. Zhou, Y., Han, X., Shechtman, E., Echevarria, J., Kalogerakis, E., Li, D.: Makelttalk: speaker-aware talking-head animation. ACM Trans. Graph. (TOG) **39**(6), 1–15 (2020)

Depth Optimization for Accurate 3D Reconstruction from Light Field Images

Xuechun Wang, Wentao Chao, and Fuqing Duan(✉)

School of Artificial Intelligence, Beijing Normal University, Beijing 100875, China
fqduan@bnu.edu.cn

Abstract. Because the light field camera can capture both the position and direction of light simultaneously, it enables us to estimate the depth map from a single light field image and subsequently obtain the 3D point cloud structure. However, the reconstruction results based on light field depth estimation often contain holes and noisy points, which hampers the clarity of the reconstructed 3D object structure. In this paper, we propose a depth optimization algorithm to achieve a more accurate depth map. We introduce a depth confidence metric based on the photo consistency of the refocused angular sampling image. By utilizing this confidence metric, we detect the outlier points in the depth map and generate an outlier mask map. Finally, we optimize the depth map using the proposed energy function. Experimental results demonstrate the superiority of our method compared to other algorithms, particularly in addressing issues related to holes, boundaries, and noise.

Keywords: Light field · Depth map · Optimization · 3D reconstruction

1 Introduction

Three-dimensional reconstruction is a process that converts 2D images into a three-dimensional structure. Currently, 3D reconstruction is extensively applied in various fields, such as architectural design [3], cultural heritage preservation [17], and medical image reconstruction [6]. Depth maps, point clouds, voxels, and grids are commonly used representations in 3D reconstruction. Since a single view often lacks complete information, most reconstruction algorithms rely on multiple views to achieve accurate results.

The light field camera [11] records both the position and orientation information of the light rays, which provides rich information for 3D reconstruction [7]. Many depth estimation methods for 3D reconstruction from light field images have been proposed. For example, Zhang et al. [18] proposed the spinning parallelogram operator (SPO) to estimate the optimal orientation of the corresponding line in the epipolar image (EPI), where the occlusion is seen as a discrete

This work is supported by National Key Research and Development Project Grant, Grant/Award Number: 2018AAA0100802.

Q. Liu et al. (Eds.): PRCV 2023, LNCS 14426, pp. 79–90, 2024.
https://doi.org/10.1007/978-981-99-8432-9_7

labeling problem. Han et al. [4] proposed a vote cost to estimate the depth map by counting the number of pixels whose color deviation was less than an adaptive threshold.

These depth estimation methods can estimate an accurate depth map if the light field image has simple scenes. However, issues such as occlusion, noise, and specular reflection often lead to depth maps with holes, noise, and blurred boundaries, as depicted in Fig. 1. In this paper, we collectively identify these holes, noise, and blurry boundary points as outlier points. Therefore, obtaining a more precise depth map becomes a crucial step in the process of 3D reconstruction.

(a) Central view image (b) Ground truth (c) The estimated depth map

Fig. 1. The holes, noise and blurry boundary in the estimated depth map. (a) The central view image of light field image **StillLife**. (b) The groundtruth of image **StillLife**. (c) The estimated depth map using *OACC* [15].

To solve these problems, we propose a new optimization method to obtain the accurate depth map. The main contributions are as follow:

- We propose a universal light field depth optimization method that only requires a light field image and a depth map as input.
- By combining the photo consistency of refocused angular sampling image and an adaptive threshold, we define a depth confidence to detection the outlier points.
- We introduce an energy function based on the depth confidence and outlier mask to optimize the depth map, especially at holes, noise, and fuzzy boundaries.

2 Related Work

An accurate depth map plays a vital role in faithful 3D reconstruction. Consequently, numerous researchers have proposed depth estimation methods and post-processing techniques for optimizing the depth map.

Depth Map Estimation. Zhang et al. [18] proposed the spinning parallelogram operator (SPO) to estimate the optimal orientation of the corresponding line in EPI, where the occlusion is seen as a discrete labeling problem. However, the SPO cannot handle complex occlusions. Han et al. [4] analysed the color deviations from the non-central view pixels to the central view pixel, when refocused to different depth labels. They found the occluded view pixels have the large deviations when refocused to correct depth. Hence, they set an adaptive threshold and counted the number of pixels whose color deviation was less than this threshold and proposed a vote cost. The depth map is estimated via vote cost using the obtained number. However, it will fail in scenes with severe noise, because the adaptive threshold is affected by noise. There are also many deep learning methods proposed to estimate the depth map. Tsai et al. [13] proposed an attention-based views selection module to fuse multi-view information and extracted more context information of the image by a spatial pyramid pooling for depth estimation. Wang et al. [15] constructed an occlusion-aware cost by modulating pixels from different view images and integrated pixels via using the convolutions with specifically designed dilation rates. They aggregated these costs to estimate depth map.

However, the aforementioned methods often struggle to perform well in scenarios with complex occlusion, specular reflection regions, and textureless regions. Consequently, the resulting estimated depth map may contain holes and exhibit significant errors along the boundaries.

Depth Map Optimization. Perra et al. [9] directly processed the depth map obtained from the Lytro software by histogram stretching, and then performed edge contour extraction on the depth map. Finally, the edge contour map and depth map are fused together and converted into a 3D point cloud. Zhao et al. [19] used superpixels to segment the central view image to detect the corresponding positions of outliers in the depth map, and used bilateral filtering to correct and optimize the depth map. After the 3D point cloud is generated from the optimized depth map, the statistical filtering method is used to further eliminate outliers. Peng et al. [8] estimated the depth maps of all sub-views of the light field image. Based on the point cloud fusion of Kinect, they used the volume integration algorithm to fuse the depth maps of multiple views to obtain a 3D point cloud. However, due to the short baseline of the light field camera, the obtained 3D point cloud will have large noise and error. Farhood et al. [1] performed histogram stretching and histogram equalization on depth maps to enhance the contrast of depth maps. Then, the multi-modal edge information is estimated based on the correspondence matching relationship between the depth map and the sub-aperture image to further optimize the depth map. Finally, the optimized depth map is converted into a 3D point cloud. However, this method cannot deal with the scene with more occlusion. Galea et al. [2] used a statistical outlier removal filter to detect the outlier points and then used the consistency between the light field sub-views to correct the location of the outlier points. However, this method needs to fit a plane with a large number of points, so it is not always efficient.

The aforementioned methods solely rely on edge detection results from the central view image as a reference, subsequently employing optimization algorithms to address outlier points. However, they tend to overlook the influence of noise and holes during the optimization process. In contrast, our proposed outlier point detection algorithm introduces weight adjustment within the optimization algorithm itself, thereby effectively reducing the impact of error propagation during optimization.

3 The Proposed Method

In this section, we will describe the proposed depth map optimization algorithm in detail. The complete algorithm framework is shown in Fig. 2.

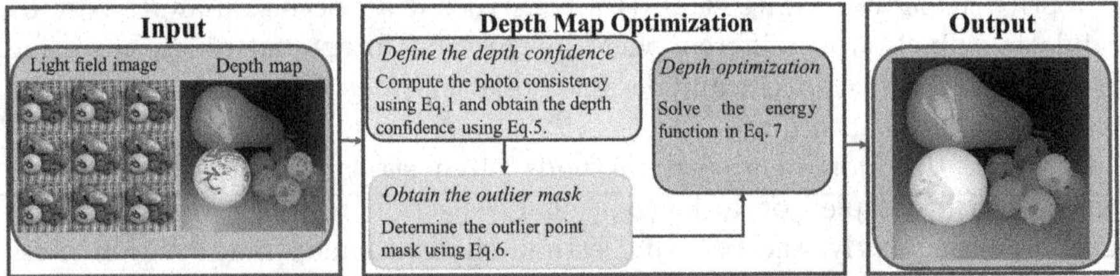

Fig. 2. Overview of the proposed depth map optimization method. The proposed method for depth map optimization is a universal approach that requires only a light field image and a depth map as input. It can be divided into two main parts: the definition of depth confidence for detecting outlier points, and the minimization function that assigns different weights to points based on their corresponding confidence.

3.1 Outliers Detection

The conventional joint bilateral filtering optimization algorithm optimizes based on the pixels within the neighborhood without considering the accuracy of the depth values of these pixels. Consequently, it fails to achieve improved optimization results for outlier points. Therefore, to obtain a more accurate depth map, our approach focuses on first detecting outlier points in the depth map.

A 4D light field image can be represented as $L(x, y, u, v)$ in Fig. 3 (a), where (x, y) means the spatial coordinate in the sub-aperture image, and (u, v) means the angular coordinate. Hence, an angular sampling image $A_\alpha^{\boldsymbol{p}}$ of the point $\boldsymbol{p} = (x, y)$ can be generated by extracting the pixels from the sheared light field image $L_\alpha(x, y, u, v)$, as shown in Fig. 3 (b), where α means the estimated depth value of the point $\boldsymbol{p} = (x, y)$. If the point $\boldsymbol{p}$ is focused on a proper depth, its angular sampling image will exhibit the highest photo consistency. Figure 3 (c) shows the point $\boldsymbol{p}$ is focused on a incorrect depth, the angular sampling image has chaotic

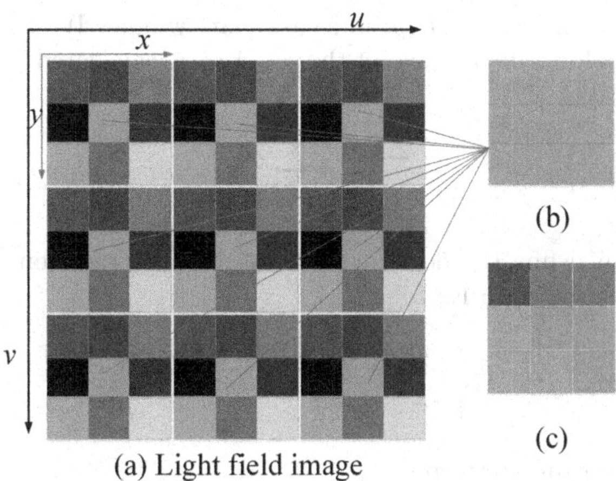

Fig. 3. (a) The light field image and $p = (x, y)$ is the gray point. (b) The angular sampling image of point $p = (x, y)$ focused in the proper depth. (c) The angular sampling image of point $p = (x, y)$ focused in the incorrect depth. (Color figure online)

color distributions. Then we can measure the photo consistency $\rho_\alpha(x, y)$ in the image A^p_α according to the following equation

$$\rho_\alpha(x, y) = \frac{1}{N} \sum_{u,v} D_\alpha(x, y, u, v) \tag{1}$$

where N represents the number of pixels in the image A^p_α, and $D_\alpha(x, y, u, v)$ is the color difference between the non-central pixel and the central pixel in the image A^p_α.

$$D_\alpha(x, y, u, v) = \|A^p_\alpha(u, v) - A^p_\alpha(u_c, v_c)\| \tag{2}$$

where $\|.\|$ means L2 norm.

Since different light field images exhibit varying color distributions, the photo consistency $\rho_\alpha(x, y)$ of the angular sampling image A^p_α can be influenced by these color distributions. To enhance the universality of the algorithm, we introduce an adaptive threshold to evaluate the photo consistency $\rho_\alpha(x, y)$ of the angular sampling image A^p_α. Previous studies have demonstrated that the occlusion edges in the spatial domain align with those in the angular sampling image when correctly refocused [14]. Based on this observation, we consider the color distribution in the spatial domain as the guiding information for the angular sampling image. Therefore, we construct a local square window W_{UV} centered at $p = (x, y)$ in the central view image, where $U * V$ is same as the angular resolution of the light field image. We compute the color difference $E(p)$ between point p and each point $q = (s, t)$ in local square window W_{UV},

$$E(p) = \|L(s, t, 0, 0) - L(x, y, 0, 0)\|, q \in W_{UV} \tag{3}$$

All the color difference values $E(p)$ in local square window W_{UV} are denoted as a set A_{UV}. Therefore, the average of the set A_{UV} is calculated as the adaptive threshold. That is

$$\Delta\varepsilon = \frac{1}{k}\sum_{i=1}^{k} A_i \tag{4}$$

where $k = U * V$.

Then, we can define the depth confidence measure according to Eq. 1 and Eq. 4, which is as shown in Eq. 5.

$$M_{conf}(p) = \begin{cases} \mu^2(1-\exp(-\frac{(\rho_\alpha-\Delta\varepsilon)^2}{2\mu^2})) & \rho_\alpha \geq \Delta\varepsilon \\ 1 & \rho_\alpha < \Delta\varepsilon \end{cases} \tag{5}$$

where μ is the control parameter.

It is obvious that we can determine the outlier points using the depth confidence in Eq. 5. That is we have the outlier point mask of the estimated depth map.

$$M_{out}(p) = \begin{cases} 0 & M_{conf}(p) = 1 \\ 1 & M_{conf}(p) < 1 \end{cases} \tag{6}$$

If $M_{out}(p) = 1$, it means the point p is a outlier point.

3.2 Depth Map Optimization

Most algorithms employ joint bilateral filtering [12] to optimize the depth map, which incorporates both spatial and image information of pixels. The algorithm assigns relatively higher weights to the category to which the point belongs and then performs a weighted summation within the neighborhood to generate the final result. By simultaneously considering the spatial and image characteristics, joint bilateral filtering aims to achieve more accurate and visually pleasing depth maps. That is

$$\boldsymbol{D}_p = \frac{\sum_{q\in\Omega} w_c(p,q)w_r(p,q)\tilde{\boldsymbol{D}}_p}{\sum_{q\in\Omega} w_c(p,q)w_r(p,q)} \tag{7}$$

where $\boldsymbol{D}_p$ is the depth after optimization, $\tilde{\boldsymbol{D}}_p$ is the original depth, $w_r(p,q) = \exp(-\frac{\|q-p\|^2}{2\sigma_r^2})$ and $w_c(p,q) = \exp(-\frac{\|\boldsymbol{I}_q-\boldsymbol{I}_p\|^2}{2\sigma_c^2})$

However, when a significant number of outlier points are present in the depth map, joint bilateral filtering struggles to effectively handle cases with numerous outlier points within the neighborhood. To address this issue, we introduce different weight assignments for points during the optimization process in order to minimize the influence of outlier points on error propagation to other points. Building upon the depth confidence defined in Sect. 3.1, we propose an optimization energy function. By assigning appropriate weights to each point, we can enhance the accuracy and reliability of the optimization process, particularly in the presence of outlier points. That is

$$E = \alpha E_D + \beta E_S \tag{8}$$

where α and β are the weight of each term. The first term E_D controls the impact of the initial depth value on the depth optimization based on the depth confidence. That is

$$E_D = \sum_{p \in \boldsymbol{I}} M_{conf}(p) \left\| D(p) - \tilde{D}(p) \right\|^2 \tag{9}$$

The smooth regularization term E_s uses the adjacent depth values to optimize the depth of point p, which is defined as Eq. 10

$$E_s = \sum_{p \in \boldsymbol{I}} \sum_{q \in \boldsymbol{\Omega}} M_{conf}(q) w_r(p, q) w_c(p, q) \| D(p) - D(q) \|^2 \tag{10}$$

where the size of $\boldsymbol{\Omega}$ is $U * V$.

To prevent the presence of multiple outliers within the same neighborhood, we employ an eight-connected search method. If the count of outlier points, denoted as t, within the neighborhood of a given point exceeds a predetermined threshold τ, we expand the neighborhood. This approach helps ensure that outlier points are adequately accounted for and reduces the potential impact of clustering outliers in close proximity. Then, the second term E_s can represented as follow

$$E_s = \begin{cases} \sum\limits_{p \in \boldsymbol{I}} \sum\limits_{q \in \boldsymbol{\Omega}} M_{conf}(q) w_r(p, q) w_c(p, q) \| D(p) - D(q) \|^2 & t < \tau \\ \sum\limits_{p \in \boldsymbol{I}} \sum\limits_{q \in \boldsymbol{\Omega}'} M_{conf}(q) w_r(p, q) w_c(p, q) \| D(p) - D(q) \|^2 & t \geq \tau \end{cases} \tag{11}$$

where $\boldsymbol{\Omega}' = (U+1) * (V+1)$. Then the minimization function can be iteratively optimized to obtain the final optimized result.

4 Experimental Results

In this section, we conduct a comprehensive set of experiments on both synthetic and real light field images to evaluate the performance of the proposed method. The synthetic datasets, which include ground truth depth maps, are obtained from Wanner et al. [16] and the 4D Light Field Benchmark [5]. The real images are captured using Lytro Illum cameras [10]. To evaluate the estimated depth results on synthetic datasets, we use the bad pixel metric with a threshold of 0.07 [5]. The initial depth maps are estimated using two methods including a traditional method Han et al. (*OAVC*) [4] and a deep learning method Wang et al. (*OACC*) [15]. To distinguish from the optimization algorithms, italics indicate the depth maps from *OACC* and *OAVC* in the following comparison experiments.

To evaluate the quality of the optimized depth map, the proposed method is compared with the three methods proposed by Zhao et al. (SPBO) [19], Galea et al. (DNPC) [2] and Farhood et al. (EHPC) [1].

Fig. 4. The outlier points detection comparison using SPBO [19] and Ours based on the depth maps obtained by *OACC* [15] and *OAVC* [4].

Table 1. The accuracy of outlier detection using SPBO [19] and Ours on the datasets [16].

		Papillon	Medieval	StillLife	Mona	Buddha	Buddha2	Hourse	Average
OACC	SPBO	79.2	90.2	89.3	87.8	81.5	83.3	80.3	84.5
	Ours	89.3	92.6	97.3	92.1	82.8	85.6	87.1	**89.5**
OAVC	SPBO	82.4	91.3	93.7	90.2	83.6	88.8	89.9	88.6
	Ours	93.6	95.5	97.9	93.5	85.8	89.9	91.1	**92.5**

4.1 Outliers Detection Evaluation

In this part, we will compare the detection results of outlier points with SPBO [19]. Since the dataset contains ground truth, we consider the points in the depth maps obtained from *OAVC* [4] and *OACC* [15] that deviate from the ground truth by more than 0.07 as outlier points. Figure 4 illustrates the outlier points in the depth map, represented in red. The comparison results are also shown in Fig. 4. From Fig. 4, we can find that our method can detect more outlier points more accurately.

To quantitatively evaluate the performance of the proposed method in outlier point detection, we calculate the accuracy of the outlier point detection, and the results are shown in Table 1. From Table 1, we can find that the average detection accuracy of our method on the depth map results of both *OACC* [15] and *OAVC* [4] is higher than that of SPBO [19]. The above experimental results demonstrate the effectiveness and robustness of the proposed detection method.

4.2 Depth Optimization

We compare the optimization rsesults of the depth maps used by our method with those by Zhao et al. (SPBO) [19], Galea et al. (DNPC) [2] and Farhood et al. (EHPC) [1]. The badpixel errors of the images in two datasets [5,16] are

Table 2. The badpixel errors of all the methods on the two datasets [5,16].

Image	*OACC*	SPBO	DNPC	EHPC	Ours	*OAVC*	SPBO	DNPC	EHPC	Ours
Papillon	*23.6*	12.3	18.5	21.7	7.17	*14.4*	7.71	10.5	13.6	5.58
Medieval	*15.4*	6.68	8.91	13.3	2.27	*10.9*	5.52	7.38	9.79	0.78
StillLife	*12.7*	5.17	7.96	10.8	2.11	*5.97*	4.33	5.12	5.43	1.72
Mona	*9.56*	5.54	6.83	7.81	4.75	*6.01*	4.26	5.25	5.93	2.21
Buddha	*3.33*	2.06	2.98	3.03	1.68	*1.78*	1.43	1.61	1.58	1.22
Buddha2	*28.6*	11.4	18.6	23.3	8.86	*11.7*	8.56	9.89	10.8	7.08
Hourse	*13.7*	7.21	9.42	11.5	4.54	*5.45*	4.92	5.07	5.34	3.18
Average	*15.3*	7.19	10.5	13.1	**4.48**	*8.03*	5.25	6.41	7.49	**3.11**
Cotton	*0.49*	0.47	0.48	0.48	0.38	*2.55*	1.96	2.16	2.34	1.01
Boxes	*13.3*	11.2	12.4	12.7	6.09	*16.1*	13.4	15.5	15.5	8.21
Prami.	*0.11*	0.11	0.11	0.11	0.08	*0.83*	0.65	0.72	0.76	0.25
Bakgm.	*4.81*	4.33	4.61	4.57	2.29	*3.12*	2.83	2.96	3.06	2.67
Dots	*1.65*	1.58	1.65	1.65	1.43	*69.1*	11.6	19.7	23.4	8.84
Dino	*0.97*	0.93	0.97	1.04	0.85	*3.94*	3.52	3.91	3.89	1.94
Sideb.	*3.35*	3.24	3.35	3.33	2.78	*12.4*	7.30	9.35	10.8	3.37
Strip.	*2.92*	2.25	2.85	2.92	1.28	*29.31*	11.3	18.8	23.6	4.97
Average	*4.07*	3.01	3.31	3.35	**1.91**	*12.5*	6.57	9.14	10.4	**3.91**

calculated as shown in Table 2. As can be seen from Table 2, the average badpixel errors of our method is decreased compared with the best result of the other three methods. In addition, we can see that the average errors of our method are the smallest both on the depth maps obtained by *OAVC* [4] and *OACC* [15]. It is certain that our method performs better than these methods on almost all datasets.

We display estimated depth maps for the scene *Medieval* from Wanner's dataset [16] as in Fig. 5. From the ground truth, we can see that the buildings in the image *Medieval* have clear straight edges and some occlusion regions. As shown the depth map of *OACC* [15] in Fig. 5, it has many holes in the wall and too much noise. The results optimized by the three algorithms including SPBO [19], (DNPC) [2] and (EHPC) [1] still have noise and holes. The result obtained using our method is smoother. For the depth map of *OAVC* [4], it generates the blurry boundaries in the window frame of image *Medieval.* Compared with the three methods, the boundaries of the window frame in our method is more straight and shape. In general, it can be found from Fig. 5 that compared with the three comparison algorithms, our method can better deal with noise, holes, and fuzzy boundaries.

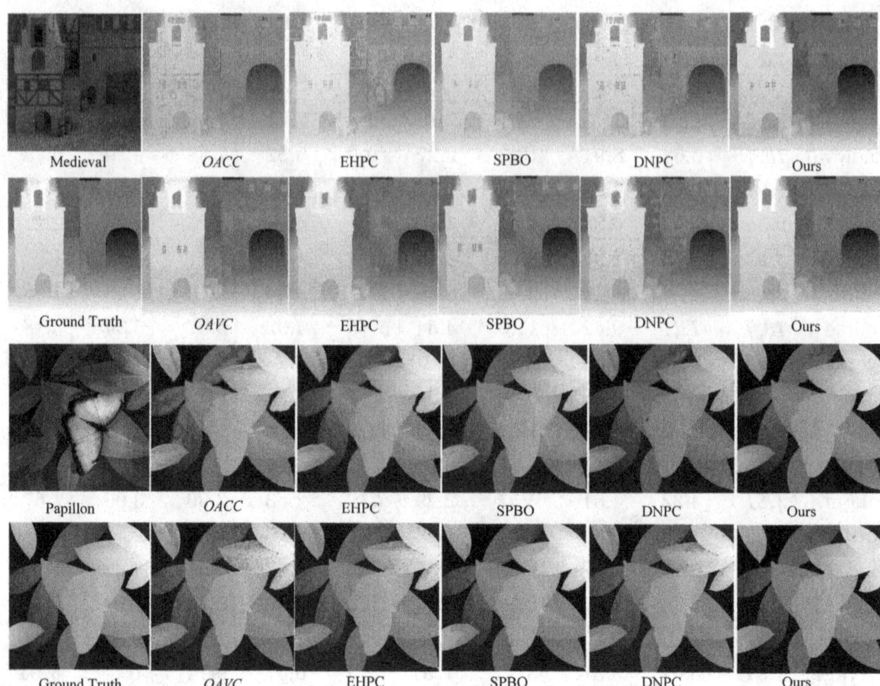

Fig. 5. The depth optimization comparison using SPBO [19], Galea et al. (DNPC) [2], Farhood et al. (EHPC) [1] and Ours based on the depth maps obtained by *OACC* [15] and *OAVC* [4]. The light field images are from the synthetic datasets [16] with ground truth.

We also have carried out a lot of experiments to evaluate the performance of our method in the real scenes. Since the real light field images do not have ground truth, we just conduct qualitative analysis by visual depth results. We estimate the depth maps of real scenes taken by Lytro Illum light field cameras [10]. The depth optimization results are shown in Fig. 6.

The first scene contains lots of fuzzy boundaries in the depth maps of *OACC* [15] and *OAVC* [4]. The comparison results can not recovery the fuzzy boundaries, and our optimization result has a clearer and more complete structure. The last two scenes contain specular reflections, so there are a lot of holes and noise in the depth maps of *OACC* [15] and *OAVC* [4]. From Fig. 6, it can be clearly found that the optimization results of the three comparison algorithms are poor, and they cannot deal with a large number of holes and noise at the same time. The experimental results once again demonstrate the robustness and effectiveness of our method in dealing with holes, noise, and fuzzy boundaries.

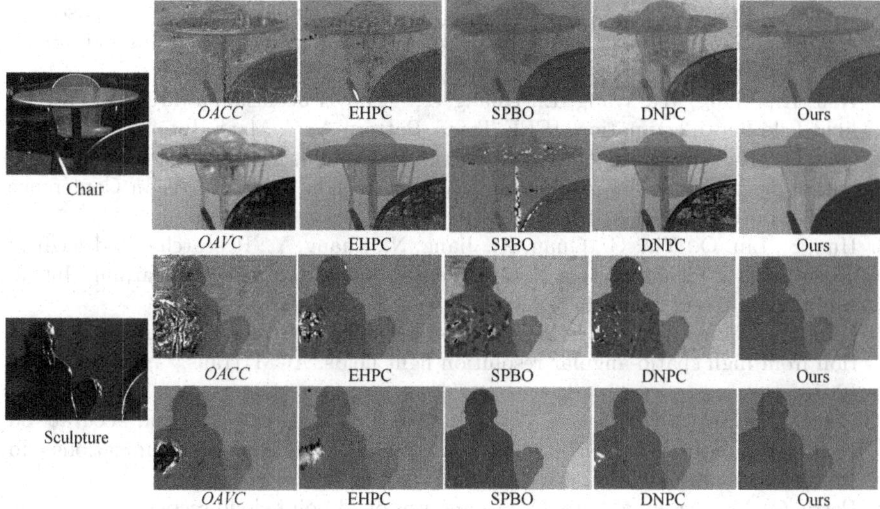

Fig. 6. The depth optimization comparison using SPBO [19], Galea et al. (DNPC) [2], Farhood et al. (EHPC) [1] and Ours based on the initial depth maps obtained by *OACC* [15] and *OAVC* [4]. The light field images are from the real datasets [10].

5 Conclusion

In this paper, we present a novel depth map optimization method to obtain the accurate 3D reconstruction for light field images. Firstly, we define a depth confidence according to the photo consistency of the refocused angular sampling image. Using this confidence measure, we detect the outlier points in the depth map and generate an outlier mask map. Finally, the depth map is optimized using the proposed energy function. The proposed method can detect the outlier points accurately and well handle the depth optimization at the boundary and holes. Compared with the optimization works, experimental results demonstrate that the proposed method has better performance on both synthetic and real light field datasets, especially for holes, noise. The proposed method can be applicated in all the Lambertian surface and a scene with a very small proportion of specular reflections. Because the proposed method is based on the photo consistency of Lambertian, the proposed method cannot accurately deal with large specular reflection regions. In the future, we will consider depth map optimization for scenes with specular reflections.

References

1. Farhood, H., Perry, S., Cheng, E., Kim, J.: Enhanced 3d point cloud from a light field image. Remote Sens. **12**(7), 1125 (2020)
2. Galea, C., Guillemot, C.: Denoising of 3d point clouds constructed from light fields. In: ICASSP 2019–2019 IEEE International Conference on Acoustics, Speech and Signal Processing (ICASSP), pp. 1882–1886. IEEE (2019)

3. Han, D., Jiao, Z., Zhou, L., Ding, C., Wu, Y.: Geometric constraints based 3d reconstruction method of tomographic sar for buildings. Sci. China Inf. Sci. **66**(1), 1–13 (2023)
4. Han, K., Xiang, W., Wang, E., Huang, T.: A novel occlusion-aware vote cost for light field depth estimation. IEEE Trans. Pattern Anal. Mach. Intell., 1–1 (2021)
5. Honauer, K., Johannsen, O., Kondermann, D., Goldluecke, B.: A dataset and evaluation methodology for depth estimation on 4d light fields. In: Asian Conference on Computer Vision (2016)
6. Hua, S., Liu, Q., Yin, G., Guan, X., Jiang, N., Zhang, Y.: Research on 3d medical image surface reconstruction based on data mining and machine learning. Int. J. Intell. Syst. **37**(8), 4654–4669 (2022)
7. Kim, C., Zimmer, H., Pritch, Y., Sorkine-Hornung, A., Gross, M.: Scene reconstruction from high spatio-angular resolution light fields. ACM Trans. Graph. **32**(4), 1 (2013)
8. Peng, J., Xiong, Z., Zhang, Y., Liu, D., Wu, F.: Lf-fusion: dense and accurate 3d reconstruction from light field images. In: 2017 IEEE Visual Communications and Image Processing (VCIP), pp. 1–4. IEEE (2017)
9. Perra, C., Murgia, F., Giusto, D.: An analysis of 3d point cloud reconstruction from light field images. In: 2016 Sixth International Conference on Image Processing Theory, Tools and Applications (IPTA), pp. 1–6. IEEE (2016)
10. Raj, A.S., Lowney, M., Shah, R., Wetzstein, G.: Stanford light field archives (2016). http://lightfields.stanford.edu/
11. Ren, N., Levoy, M., Bredif, M., Duval, G., Hanrahan, P.: Light field photography with a hand-held plenoptic camera. Stanford University Cstr (2005)
12. Tomasi, C., Manduchi, R.: Bilateral filtering for gray and color images. In: International Conference on Computer Vision (2002)
13. Tsai, Y.J., Liu, Y.L., Ouhyoung, M., Chuang, Y.Y.: Attention-based view selection networks for light-field disparity estimation. In: Proceedings of the AAAI Conference on Artificial Intelligence, vol. 34, pp. 12095–12103 (2020)
14. Wang, T.C., Efros, A.A., Ramamoorthi, R.: Depth estimation with occlusion modeling using light-field cameras. IEEE Trans. Pattern Anal. Mach. Intell. **38**(11), 2170–2181 (2016)
15. Wang, Y., Wang, L., Liang, Z., Yang, J., An, W., Guo, Y.: Occlusion-aware cost constructor for light field depth estimation. In: Proceedings of the IEEE/CVF Conference on Computer Vision and Pattern Recognition (CVPR), pp. 19809–19818 (June 2022)
16. Wanner, S., Meister, S., Goldluecke, B.: Datasets and benchmarks for densely sampled 4d light fields. In: Vision, Modeling and Visualization, pp. 225–226 (2013)
17. Zhang, L., Liu, L., Chai, B., Xu, M., Song, Y.: Multi-resolution 3d reconstruction of cultural landscape heritage based on cloud computing and hd image data. J. Intell. Fuzzy Syst. **39**(4), 5097–5107 (2020)
18. Zhang, S., Sheng, H., Li, C., Zhang, J., Xiong, Z.: Robust depth estimation for light field via spinning parallelogram operator. Comput. Vis. Image Underst. **145**, 148–159 (2016)
19. Zhao, H., Liu, Y., Wei, L., Wang, Y.: Superpixel-based optimization for point cloud reconstruction from light field. In: 2022 IEEE International Conference on Multimedia and Expo Workshops (ICMEW), pp. 1–6. IEEE (2022)

TriAxial Low-Rank Transformer for Efficient Medical Image Segmentation

Jiang Shang[1] and Xi Fang[2(✉)]

[1] Department of Information Systems, City University of Hong Kong, Hong Kong, Kowloon Tong, China

[2] Department of Biomedical Engineering and Center for Biotechnology and Interdisciplinary Studies, Rensselaer Polytechnic Institute, Troy, USA

x.fang@ieee.org

Abstract. Transformer-CNN architectures have achieved state-of-the-art on 3D medical image segmentation due to their ability to capture both long-term dependencies and local information. However, directly using the existing transformers as encoders can be inefficient, particularly when dealing with high-resolution 3D medical images. This is due to the fact that self-attention computes pixel-to-pixel relationships, which is computationally expensive. Despite attempts to mitigate this through the use of local-window attention or axial-wise attention, these methods may result in the loss of interaction between certain local regions during the self-attention computation. Instead of using the sparsified attention, we aim to incorporate the relationships between all pixels while substantially reducing the computational demand. Inspired by the low-rank property of attention, we hypothesized that the pixel-to-pixel relationship can be approximated by the composition of the plane-to-plane relationship. We propose TriAxial Low-Rank Transformer Network (TALoRT-Net) for medical image segmentation. The core of this model lies in its attention module, which approximates pixel-to-pixel attention matrix using the low-rank representation of the product of plane-to-plane matrices and significantly reduces the computation complexity inherent in 3D self-attention. Moreover, we replaced the linear projection and vanilla Multi-Layer Perceptron (MLP) in Vision Transformer with a convolutional stem and depthwise convolution layer (DCL) to further reduce the number of model parameters. We evaluated the performance of the method on the public BTCV dataset, which significantly reduce the computational effort while maintaining uncompromised accuracy.

Keywords: Low-rank Representation · Efficient Self-attention · Vision Transformer · TriLoRa Attention · 3D Medical Image Segmentation

1 Introduction

Deep learning techniques have revolutionized medical image segmentation, an essential aspect of computer-aided medical systems. By accurately and automatically delineating important regions or tissues, these advanced algorithms

Q. Liu et al. (Eds.): PRCV 2023, LNCS 14426, pp. 91–102, 2024.
https://doi.org/10.1007/978-981-99-8432-9_8

greatly contribute to intelligent medical practices and offer invaluable support for various clinical applications. For example, convolutional neural network (CNN) based approaches have been proposed for automatic multi-organ segmentation in CT images [8,9,21], which significantly saves valuable time for clinicians and improves the accuracy of organ delineation, leading to significant advancements in downstream tasks, such as surgical planning, disease assessment, and personalized medicine [6,10,25].

CNN based architectures like U-Net [26], which utilizes skip connections to connect the encoder and decoder parts, have been widely adopted in the field of medical image segmentation. These networks excel in hierarchical feature extraction, enhancing the extraction of high-level features, which is crucial for segmentation task. However, the localized nature of the convolutional operation might limit its capacity to capture long-range dependencies and contextual information between pixels. This motivates the development of more effective approaches that models long-range dependencies within images. For example, Vision Transformers (ViTs) [7] divide the image into patches and leverage self-attention mechanisms to learn the interactions and dependencies between these patches. This technique has been successfully incorporated into medical image segmentation framework [7,22,27], resulting in notable improvements in performance and accuracy. State-of-the-art architectures for medical image segmentation have embraced the combination of ViT and CNNs [11,12]. By integrating the strengths of ViTs in capturing long-range context and CNNs in propagating high-level features, these hybrid models have attracted considerable attention and exhibited exceptional performance in various medical image segmentation tasks, establishing themselves as top-tier solutions in the field. However, traditional transformer models suffer from computational and memory inefficiency, especially when dealing with high-resolution 3D medical images. This is because when computing the global self-attention module, it is necessary to calculate the relationships between all pixel pairs in the input image.

A series of studies have aimed to improve the efficiency of self-attention matrix computation. These works can be generally grouped into two categories: (1) Replacing global self-attention with local self-attention [16,22,28]; (2) Computing self-attention within axial or cross-shaped window [15,30,31]. In the first category, a typical representative is Swin transformer [22], which uses local self-attention instead of global self-attention to limit the scope of the self-attention module in a local region. Although the interaction of information between local windows is done by shift operations, this local operation approach will restrict the receptive field and restrict the interaction between distant patches. On the other hand, some works aim to introduce axial or cross-shaped windows when computing self-attention. For example, Wang et al. [30] proposed to calculate self-attention separately in either horizontal or vertical directions. Moreover, some works proposed to compute self-attention within a cross-shaped window formed by multiple intersecting rows and columns to further enhance the ability to capture long-range dependencies [5,32]. Although these approaches can reduce computational complexity, they compromise the interaction between remaining pixels. Thus, we explore the following question: Could we formulate a self-attention mechanism that is capable of reducing computational complexity while still preserving all dependencies between pixels?

Inspired by the property of low-rank of self-attention mechanism [15,31] and redundancy of medical images [4], we introduce the TriAxial Low-Rank Transformer Network (TALoRT-Net). This network utilizes an attention mechanism that decomposes the self-attention matrix into low-rank representations across three views. By capturing relationships across planes along different views, the TALoRT-Net significantly reduces the computation complexity while maintaining its ability to understand and leverage spatial dependencies. Specifically, to compute the plane-wise relationships, we employ global average pooling (GAP) to compact the 3D features into three view-based 1D features. Each of these 1D features represents the global features of a unique plane within the 3D medical image. Subsequently, we calculate self-attention between planes individually for each view. These calculations exemplify the correlation between various planes across different views and serve as a factor of pixel-to-pixel self-attention. By multiplying the attention matrices across all three views, we construct a comprehensive self-attention map that indicates the pixel-to-pixel relationship.

Our proposed TALoRT Network was evaluated using the BTCV [20] multi-organ segmentation dataset. The results demonstrate that our proposed Triaxial Low-Rank Attention(TriLoRa attention) significantly reduces the computational complexity and parameter count in comparison to other efficient self-attention methods, such as local window attention and axial attention while maintaining uncompromised accuracy. The contribution can be summarized as:

- We design a novel efficient self-attention mechanism by approximation the conventional pixel-to-pixel relationship through a low-rank representation of the multiplication of the plane-to-plane relationship. This approach significantly reduces both the computational complexity and the number of parameters compared with another self-attention mechanism.
- We propose an efficient transformer-based encoder for medical image segmentation. Compared to previous transformers applied for medical image segmentation, our encoder requires less computational resources and has lower computational complexity.
- Experimental results on the BTCV dataset demonstrate that TALoRT-Net significantly reduces the network complexity in comparison to traditional transformers while maintaining uncompromised accuracy.

2 Related Work

2.1 Efficient Attention in Transformer

Considering the high-resolution characteristics of medical images, it's essential to develop efficient self-attention mechanisms. Efficient attention methods in the transformer model can be generally classified into two types: 1) adopting local self-attention as an alternative to global self-attention; 2) Computing self-attention within an axial or cross-shaped window. The first type of approaches employ local self-attention to replace the global self-attention [11,22]. These methods restrict the computation of self-attention layer within a designated local region. For example, Liu et al. [22] proposed Swin transformer, where self-attention was computed within a local window. They also introduced a shifted window mechanism to guarantee the interaction across different windows and expand the receptive field. Ali et al. [11] further adapted it into 3D context, and combined it with a CNN-based decoder for 3D medical image segmentation. However, this local operation approach may ignore long-distance interactions in the computation of self-attention. The second type of approach leverages axial attention, which computes attention along each axis independently [5,14,17,29,33]. Such kind of attention is often implemented in a 2D manner, performing self-attention across each row or column of the feature map. For example, Huang et al. [17] introduced a method that utilizes cross-shaped attention to extract contextual features after CNN. In this approach, attention is computed within a cross-shaped window region, formed by a single intersecting row and column. Dong et al. [5] proposed a general ViT backbone with cross-shaped window self-attention. This method performs self-attention in parallel across multiple intersecting rows and columns. Zhu et al. [33] further extended the axial attention to 3D version for pulmonary nodule classification in medical images. However, this method computes self-attention only for pixels along the axis, potentially overlooking pixels not positioned on the axis.

2.2 Low Rank Approximation of Self-attention

Many works have proposed low-rank decomposition or low-rank approximation techniques for model compression or model finetuning [15,19,30]. Jaderberg et al. [19] suggested two straightforward strategies to speed up the convolution operation. The first approach uses a set of basic filters to approximate the model filter via linear combinations. On the other hand, the filter with a spatial dimension of $d \times d$ can be factorized into the multiplication of two axial kernels in the shape of $d \times 1$ and $1 \times d$, respectively. Ultimately, both methods can be viewed as approximating the filter by using low-rank approximation matrix. Many works [15,31] indicated that the self-attention matrix may potentially be approximated through low-rank representation. For example, Hu et al. [15] suggested to decompose the original self-attention matrix using low-rank factorization as a means to effectively fine-tune transformer models for various tasks. Wang et al. [30] suggested to factorize 2D self-attention matrix into two 1D self-attention vectors along the height and width dimension, which reduce the computation complexity and expand the receptive field. Wang et al. [31] proposed to scale the first dimension of key and value by two linear projection matrices, to reduce the complexity associated with keys and values without changing their dimensions.

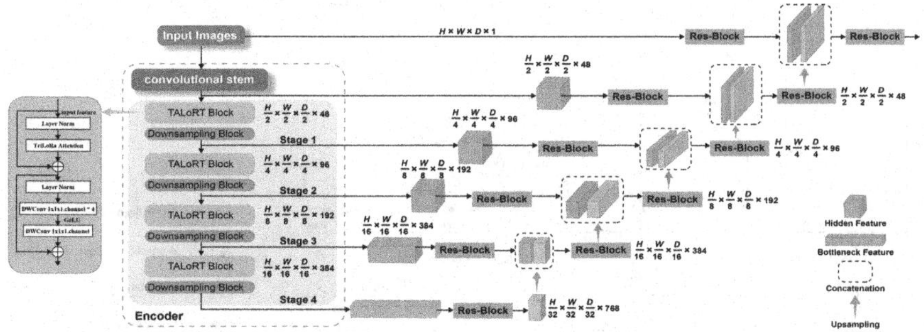

Fig. 1. Overview of the proposed TALoRT-Net.

3 Methodology

As shown in Fig. 1, our model is built with a transformer-based encoder and a CNN-based decoder. These components are designed to capture long-range dependencies and detailed spatial information, respectively. In this section, we begin by providing a detailed explanation of the TALoRT Transformer. Then, we discuss the integration with CNN for segmentation.

3.1 TALoRT Based Encoder

Overall Architecture. The TALoRT transformer is constructed of a convolutional stem succeeded by four stages of hierarchical feature extraction. For an input image with size of $H \times W \times D$, we've developed a convolutional stem to capture low-level features efficiently. This stem includes three convolution layers to obtain feature maps $\frac{H}{2} \times \frac{W}{2} \times \frac{D}{2}$. Each layer consists of a $3 \times 3 \times 3$ convolution, coupled with Batch Normalization [18], and GELU [13] activation. The strides for these layers are 2, 1, and 1, respectively.

Each stage of hierarchical feature extraction consists of two sequential TALoRT Transformer Blocks and a downsampling block. The TALoRT Transformer block has an overall similar topology as the vanilla multi-head self-attention but replaces the vanilla self-attention mechanism and MLP with our proposed TriLoRa Attention and a modified MLP module. The TriLoRa Attention will be described in the following subsection in detail. The vanilla MLP module is replaced by a DCL with LayerScale and a kernel size of $1 \times 1 \times 1$. Compared to the vanilla linear projection, DCL further reduces the number of parameters. Therefore, the output of the TALoRT Transformer block in layer l can be articulated as follows:

$$\begin{aligned} \hat{z}^l &= TriLoRa(LN(z^{l-1})) + z^{l-1}, \\ z^l &= DCL(LN(\hat{z}^l)) + \hat{z}^l, \end{aligned} \tag{1}$$

where $\hat{z}^l$ is the output from TriLoRa attention layer in different depth levels; LN denotes the Layer Normalization (LN) layer [1]. The downsampling block is

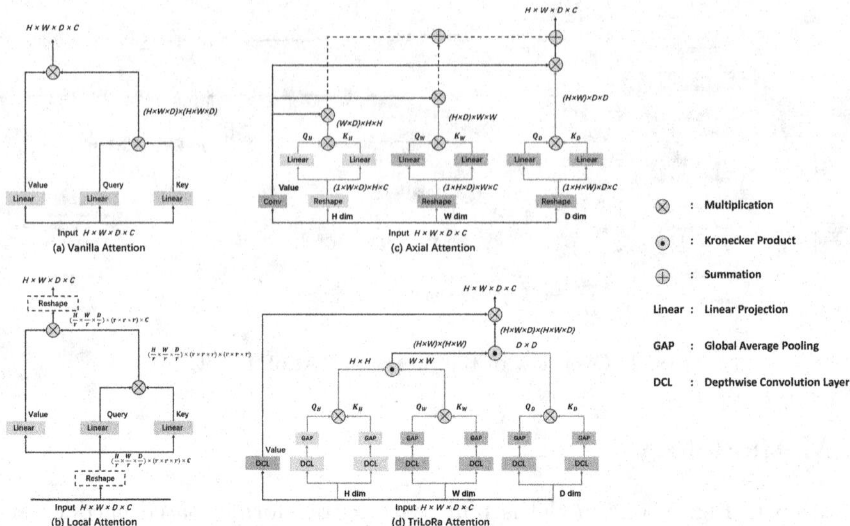

Fig. 2. Illustration of different self-attention approaches: (a) Vanilla attention, (b) Local attention, (c) Axial attention, and (d) our proposed TriLoRa attention, respectively.

comprised of a convolution layer (kernel size $2 \times 2 \times 2$, stride 2) and an LN layer. The size of feature map is hierarchically reduced by a factor of two after each stage, which is then used to fuse with the decoder in the subsequent stages to enhance the feature representation.

TriLoRa Attention. Despite the ability of the vanilla global self-attention mechanism to model long-range context, it comes with a computational complexity that increases in a quadratic manner with the size of feature map. In order to reduce the computational complexity of the global self-attention module, we've developed the TriLoRa attention, which is achieved by decomposing the pixel-to-pixel attention matrix into the low-rank approximation of plane-to-plane attention matrices across three views - axial, sagittal, and coronal. Figure 2 presents a detailed comparison between the conventional self-attention methods and our TriLoRa attention mechanism. Instead of using linear projection, we generated query, key and value by using DCL of $1 \times 1 \times 1$ convolution to reduce the number of parameters. For example, the value feature $\mathbf{V}$ can be written as:

$$\mathbf{V} = DCL_V(X), \tag{2}$$

Specifically, to calculate the plane-to-plane attention map for each view, we employed GAP along different dimensions to create the plane-wise query and key for each view, each represented as a vector. For example, the procedure of computing coronal view can be defined as:

$$\mathbf{Q}_H, \mathbf{K}_H = Sq(GAP_H(DCL_H(X))), Sq(GAP_H(DCL_H(X))), \tag{3}$$

Table 1. Comparison of different attention mechanism configurations in 3D Transformers

Method	Computational Complexity	Number of Parameters
ViT Attention [12]	$\mathcal{O}(n^2)$	2.4M
Swin ViT Attention [11]	$\mathcal{O}(n^2/r)$	62K
Axial Attention [33]	$\mathcal{O}(n\sqrt[3]{n})$	14.3K
TriLoRa Attention	$\mathcal{O}((\sqrt[3]{n})^2)$	1.3K

where $Sq(\cdot)$ means the operation of squeezing, and DCL_H and GAP_H denote the DCL and GAP along the axial view (H dim). To exploit the contextual information between neighboring planes, the kernel in the DCL for query feature extraction extends over multiple planes for each view. Given the input resolution of $H \times W \times D \times C$. we depicted how the queries and keys for the coronal, sagittal, and axial views are represented in Fig 2, where they are represented as $[\mathbf{Q}_H \in \mathcal{R}^{H\times C}, \mathbf{K}_H \in \mathcal{R}^{H\times C}]$, $[\mathbf{Q}_W \in \mathcal{R}^{W\times C}, \mathbf{K}_W \in \mathcal{R}^{W\times C}]$, and $[\mathbf{Q}_D \in \mathcal{R}^{D\times C}, \mathbf{K}_D \in \mathcal{R}^{D\times C}]$, respectively. Sequentially, we computed the plane-to-plane attention for the three views utilizing the derived queries and keys.

$$\mathbf{Attn}(\mathbf{Q}_M, \mathbf{K}_M) = Softmax(\frac{\mathbf{Q}_M \mathbf{K}_M^T}{\sqrt{C}}), M \in [H, W, D]. \quad (4)$$

Finally, the TriLoRa attention matrix is represented by the matrix direct product of the three plane-to-plane attentions:

$$\mathbf{Attn}_{Final}(\mathbf{Q}, \mathbf{K}) = \mathbf{Attn}(\mathbf{Q}_H, \mathbf{K}_H) \odot \mathbf{Attn}(\mathbf{Q}_W, \mathbf{K}_W) \odot \mathbf{Attn}(\mathbf{Q}_D, \mathbf{K}_D), \quad (5)$$

where $\odot$ refers to Kronecker product [2], $\mathbf{Attn}_{Final}(\mathbf{Q}, \mathbf{K})$ represents the final pixel-to-pixel attention matrix. Table 1 provides the comparison between our proposed TriLoRa attention method against three other commonly used self-attention modules, i.e., ViT attention, Swin ViT attention, and Axial-attention. Note, r and n represent the number of local windows and pixels, respectively. It shows that the TriLoRa attention significantly reduces the computational complexity and the number of parameters in comparison other approaches. Specifically, the TriLoRa attention module only requires about 1.3K parameters to compute the self-attention, approximately 1/11 of the parameters needed by Axial Attention, 1/60 of those used by Swin ViT Attention, and 1/1900 compared to ViT Attention's parameter count.

3.2 CNN Based Decoder

For the decoder, inspired by the CNN-based decoder from U-Net [26], we merge the multi-scale features from each stage in the encoder with the decoder. In each stage, we extract the output feature which will be fed into a residual block with two convolutional layers(kernel size $3 \times 3 \times 3$) by using instance normalization.

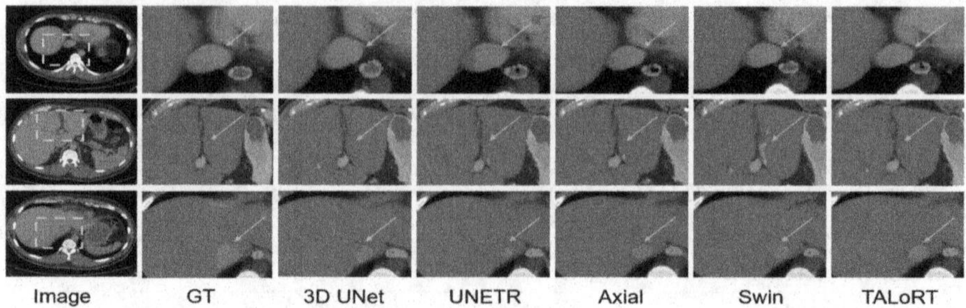

Fig. 3. Examples of segmentation comparisons of different models. Regions of evident improvements are highlighted with a rectangle in the original image. Each row from left to right shows the original image, the ground truth, as well as the segmentation results of 3D UNet, UNETR, AxialUNETR, SwinUNETR, and our proposed TALoRT-Net.

Subsequently, we employ deconvolutional layers to upsample the output feature from each stage by a factor of two, and the results will be concatenated with outputs from the encoder. The final segmentation outputs will use a residual block with convolutional layer(kernel size $1 \times 1 \times 1$) with a activation function(Softmax).

4 Experiments

4.1 Dataset

The performance of our method was evaluated on BTCV dataset [20] for 3D medical image semantic segmentation in CT imaging modality. The BTCV dataset consists of 30 abdominal CT scans, which have been randomly divided into 24 scans for training and 6 scans for validation. All 13 organs from BTCV dataset were annotated manually by radiologists at Vanderbilt University Medical Center. The segmented organs include the spleen, right kidney, left kidney, gallbladder, esophagus, liver, stomach, aorta, inferior vena cava, portal vein and splenic vein, pancreas, right adrenal gland, and left adrenal gland, which was from metastatic liver cancer patients or post-operative ventral hernia patients. Each CT scan has 80 to 225 slices with 512×512 pixels. We reported the average Dice score on 13 abdominal organs for evaluation.

4.2 Implementation Details

TALoRT-Net was implemented based PyTorch [24] and MONAI. The model was trained using a NVIDIA RTX A5000 server in a batch size of 4 for one GPU by using the AdamW optimizer [23] with original learning rate of 1e-4 for 30,000 iterations. We used layer normalization [1] in each downsampling block, and put it before TriLoRa Attention and DCL module in each stage. The number of blocks for each stage in the proposed transformer was $\{2, 2, 2, 2\}$. The channel number after each stage were $\{96, 192, 384, 768\}$. During the training phase, we

Table 2. Dice score (%) of multi-organ segmentation results of various methods (Spl: spleen, RKid: right kidney, LKid: left kidney, Gall: gallbladder, Eso: esophagus, Liv: liver, Sto: stomach, Aor: aorta, IVC: inferior vena cava, Veins: portal and splenic veins, Pan: pancreas, AG: left and right adrenal glands).

Methods	Spl	RKid	LKid	Gall	Eso	Liv	Sto	Aor	IVC	Veins	Pan	AG	Avg
3D U-Net [3]	94.09	91.22	91.52	59.01	**77.66**	96.08	83.91	90.11	85.57	69.26	78.02	64.68	80.45
3D UX-Net [21]	95.58	93.57	93.87	59.21	75.19	96.81	83.09	89.88	86.44	**74.99**	77.67	**69.68**	81.97
UNETR [12]	92.84	93.58	93.62	48.12	73.44	95.82	79.52	88.36	80.81	68.31	73.77	66.42	78.54
Swin UNETR [11]	96.05	94.23	94.31	63.92	73.22	96.91	81.48	**90.37**	84.55	74.59	**81.81**	68.45	82.18
Axial UNETR	94.54	94.24	94.07	59.62	74.16	96.75	82.18	89.79	85.44	73.31	76.15	69.12	81.42
TALoRT Net(Ours)	**96.28**	**94.73**	**94.52**	**66.28**	76.71	**96.94**	**85.20**	90.31	**86.98**	71.37	81.09	69.64	**83.05**

employed several data augmentation techniques: We performed random intensity shifts within the [-0.1, 0.1] range, with a probability of 0.5. Additionally, images were randomly flipped across axial, sagittal, and coronal views with a likelihood of 0.1. Moreover, we applied random rotations to images at angles of 90, 180, and 270°C. For the purpose of improving the contrast of abdominal organs, we restricted the Hounsfield Unit (HU) values within the $[-175, 250]$ range and normalize them to $[0, 1]$. Subsequently, these values were normalized into $[0, 1]$. We also randomly cropped the input images with volume sizes of $[96, 96, 96]$ and ensured that the ratio of foreground and background in the random patches was 1:1. For inference, we used a sliding window approach with an overlapping of 0.25 for neighboring patches.

4.3 Comparison with Others

Table 2 shows the comparison between our method with the state-of-the-art CNN and hybrid models for medical image segmentation. Specifically, we have replaced the Swin ViT in Swin UNETR with an axial attention-based transformer [33] for comparison, referred to as Axial UNETR in Table 2. The result indicated that hybrid models generally outperform CNN-only models. This may be because of their ability to effectively capture both global dependencies and local information [12]. Moreover, by comparing different hybrid models, we found that TALoRT-Net achieved superior performance with the least computational effort. Especially, when compared with the state-of-the-art SwinUNETR, the proposed SwinUNETR demonstrates improvements in 11 out of 13 organs. Notably, it only requires 3.3M parameters compared to the encoder of UNETR (88.3M), Axial (5.7M), and Swin (8.1M).

4.4 Ablation Study

Effectiveness of Different Components. We first conducted ablation studies to investigate the effectiveness of different components within TALoRT, i.e., the convolutional stem (ConvStem), the TriLoRa Attention module (TriLoRa), and the DCL. UNETR [12], which employed 3D ViT as its encoder, served as the

Table 3. Effect of different components

ConvStem	TriLoRa	DCL	Dice
✗	✗	✗	80.51
✗	✓	✓	83.16
✓	✗	✓	82.62
✓	✓	✗	84.26
✓	✓	✓	**84.30**

Table 4. Ablation study in TriLoRa Attention

(a)Effect of multi-plane and multi-head

Multi-plane	Multi-Head	Dice
✗	✗	82.32
✗	✓	84.30
✓	✗	83.96
✓	✓	**84.73**

(b)Hyper-parameter study for multi-plane

Kernel-Size	Dilation	Dice
1	1	84.30
3	1	84.41
3	2	**84.73**
5	1	84.23
5	2	84.58

baseline. Average dice score of the 14 classes (including background) was used for comparison. As shown in Table 3, the inclusion of all three proposed components elevated the Dice score from 80.51% to 84.30%. In cases where either ConvStem, TriLoRa, or DCL were not implemented, we observed a corresponding decrease in the Dice score by 1.14%, 1.68%, and 0.04%, respectively. Importantly, all three components utilized fewer computational resources than their counterparts in the baseline model while achieving improvement, further demonstrating their efficiency and effectiveness.

Effectiveness of the TriLoRa Attention. We also conducted ablation studies to investigate the hyper-parameters inside the TriloRa Attention, including the effect of multi-head, and hyper-parameters for multi-plane contextual query features in self-attention. DCL with different kernel sizes and dilation rates was used to investigate the effect of the multi-plane contextual query feature. As shown in Table 4(a), the inclusion of multi-plane and multi-head enhances the Dice score from 82.32% to 83.96% and 84.30%, respectively. Moreover, when both of them are utilized, the Dice score is further improved to 84.73%. As shown in Table 4(b), the DCL with a kernel size of 3 and a dilation of 2 achieves the best performance. It indicates the effectiveness of the contextual query feature for self-attention computation. However, it's important to note that an excessively large kernel size could potentially lead to the neglect of local information, especially for feature maps in low resolution.

5 Conclusion

In summary, we proposed the TALoRT-Net, a hybrid architecture for efficient medical image segmentation. Our method features an efficient transformer with TriLoRa attention, which is designed to significantly reduce the computational burden associated with self-attention calculations. TriLoRa Attention utilizes a low-rank representation, leveraging the multiplication of plane-to-plane attention matrices across various views to approximate conventional pixel-to-pixel attention. Additionally, we employ DCL layers to replace the linear projections and vanilla MLP layers to reduce the parameter count within the TALoRT Transformer Block. Extensive experimental results of our method on multi-organ segmentation have shown that our proposed model significantly reduces the computation effort while maintaining comparable performance with the state-of-the-art models. We believe that the efficient attention mechanism can be easily adapted for other vision tasks, thus potentially improving their computational efficiency.

References

1. Ba, J.L., Kiros, J.R., et al.: Layer normalization. arXiv preprint arXiv:1607.06450 (2016)
2. Broxson, B.J.: The kronecker product (2006)
3. Çiçek, Ö., Abdulkadir, A., Lienkamp, S.S., Brox, T., Ronneberger, O.: 3D U-Net: learning dense volumetric segmentation from sparse annotation. In: Ourselin, S., Joskowicz, L., Sabuncu, M.R., Unal, G., Wells, W. (eds.) MICCAI 2016. LNCS, vol. 9901, pp. 424–432. Springer, Cham (2016). https://doi.org/10.1007/978-3-319-46723-8_49
4. Devaraj, S.J.: Emerging paradigms in transform-based medical image compression for telemedicine environment. In: Telemedicine technologies, pp. 15–29. Elsevier (2019)
5. Dong, X., Bao, J., et al.: CSwin transformer: a general vision transformer backbone with cross-shaped windows. In: Proceedings of the IEEE/CVF Conference on Computer Vision and Pattern Recognition, pp. 12124–12134 (2022)
6. Dong, X., Lei, Y., et al.: Synthetic MRI-aided multi-organ segmentation on male pelvic CT using cycle consistent deep attention network. Radiother. Oncol. **141**, 192–199 (2019)
7. Dosovitskiy, A., Beyer, L., et al.: An image is worth 16x16 words: Transformers for image recognition at scale. arXiv preprint arXiv:2010.11929 (2020)
8. Fang, X., Yan, P.: Multi-organ segmentation over partially labeled datasets with multi-scale feature abstraction. IEEE Trans. Med. Imaging **39**(11), 3619–3629 (2020)
9. Gibson, E., Giganti, F., et al.: Automatic multi-organ segmentation on abdominal ct with dense V-networks. IEEE Trans. Med. Imaging **37**(8), 1822–1834 (2018)
10. Gou, S., Tong, N., et al.: Self-channel-and-spatial-attention neural network for automated multi-organ segmentation on head and neck ct images. Phys. Med. Biol. **65**(24), 245034 (2020)
11. Hatamizadeh, A., Nath, V., et al.: Swin UNETR: swin transformers for semantic segmentation of brain tumors in MRI images. In: International MICCAI Brainlesion Workshop, pp. 272–284. Springer (2021). https://doi.org/10.1007/978-3-031-08999-2_22
12. Hatamizadeh, A., Tang, Y., et al.: UNETR: transformers for 3d medical image segmentation. In: Proceedings of the IEEE/CVF Winter Conference on Applications of Computer Vision, pp. 574–584 (2022)
13. Hendrycks, D., Gimpel, K.: Gaussian error linear units (GELUs). arXiv preprint arXiv:1606.08415 (2016)
14. Ho, J., Kalchbrenner, N., et al.: Axial attention in multidimensional transformers. arXiv preprint arXiv:1912.12180 (2019)
15. Hu, E.J., Shen, Y., et al.: Lora: low-rank adaptation of large language models. arXiv preprint arXiv:2106.09685 (2021)
16. Huang, L., Yuan, Y., et al.: Interlaced sparse self-attention for semantic segmentation. arXiv preprint arXiv:1907.12273 (2019)

17. Huang, Z., Wang, X., et al.: CCNet: criss-cross attention for semantic segmentation. In: Proceedings of the IEEE/Cvf International Conference on Computer Vision, pp. 603–612 (2019)
18. Ioffe, S., Szegedy, C.: Batch normalization: Accelerating deep network training by reducing internal covariate shift. In: International Conference on Machine Learning, pp. 448–456. PMLR (2015)
19. Jaderberg, M., Vedaldi, A., et al.: Speeding up convolutional neural networks with low rank expansions. arXiv preprint arXiv:1405.3866 (2014)
20. Landman, B., Xu, Z., et al.: Miccai multi-atlas labeling beyond the cranial vault-workshop and challenge. In: Proceedings of MICCAI Multi-Atlas Labeling Beyond Cranial Vault-Workshop Challenge, vol. 5, p. 12 (2015)
21. Lee, H.H., Bao, S., et al.: 3d UX-Net: a large kernel volumetric convnet modernizing hierarchical transformer for medical image segmentation. arXiv preprint arXiv:2209.15076 (2022)
22. Liu, Z., Lin, Y., et al.: Swin transformer: Hierarchical vision transformer using shifted windows. In: Proceedings of the IEEE/CVF International Conference on Computer Vision, pp. 10012–10022 (2021)
23. Loshchilov, I., Hutter, F.: Decoupled weight decay regularization. arXiv preprint arXiv:1711.05101 (2017)
24. Paszke, A., Gross, S., et al.: Pytorch: an imperative style, high-performance deep learning library. In: Advances in Neural Information Processing Systems 32 (2019)
25. Peng, Z., Fang, X., et al.: A method of rapid quantification of patient-specific organ doses for ct using deep-learning-based multi-organ segmentation and gpu-accelerated monte carlo dose computing. Med. Phys. **47**(6), 2526–2536 (2020)
26. Ronneberger, O., Fischer, P., Brox, T.: U-Net: convolutional networks for biomedical image segmentation. In: Navab, N., Hornegger, J., Wells, W.M., Frangi, A.F. (eds.) MICCAI 2015. LNCS, vol. 9351, pp. 234–241. Springer, Cham (2015). https://doi.org/10.1007/978-3-319-24574-4_28
27. Touvron, H., Cord, M., et al.: Training data-efficient image transformers & distillation through attention. In: International Conference on Machine Learning, pp. 10347–10357. PMLR (2021)
28. Tu, Z., Talebi, H., et al.: Maxvit: multi-axis vision transformer. In: Computer Vision-ECCV 2022: 17th European Conference Proceedings, Part XXIV, pp. 459–479. Springer (2022). https://doi.org/10.1007/978-3-031-20053-3_27
29. Valanarasu, J.M.J., Oza, P., Hacihaliloglu, I., Patel, V.M.: Medical transformer: gated axial-attention for medical image segmentation. In: de Bruijne, M., Cattin, P.C., Cotin, S., Padoy, N., Speidel, S., Zheng, Y., Essert, C. (eds.) MICCAI 2021. LNCS, vol. 12901, pp. 36–46. Springer, Cham (2021). https://doi.org/10.1007/978-3-030-87193-2_4
30. Wang, H., Zhu, Y., Green, B., Adam, H., Yuille, A., Chen, L.-C.: Axial-deeplab: stand-alone axial-attention for panoptic segmentation. In: Vedaldi, A., Bischof, H., Brox, T., Frahm, J.-M. (eds.) ECCV 2020. LNCS, vol. 12349, pp. 108–126. Springer, Cham (2020). https://doi.org/10.1007/978-3-030-58548-8_7
31. Wang, S., Li, B.Z., et al.: Linformer: self-attention with linear complexity. arXiv preprint arXiv:2006.04768 (2020)
32. Wu, S., Wu, T., et al.: Pale transformer: a general vision transformer backbone with pale-shaped attention. In: Proceedings of the AAAI Conference on Artificial Intelligence, vol. 36, pp. 2731–2739 (2022)
33. Zhu, Q., Wang, Y., et al.: Multi-view coupled self-attention network for pulmonary nodules classification. In: Proceedings of the Asian Conference on Computer Vision, pp. 995–1009 (2022)

SACFormer: Unify Depth Estimation and Completion with Prompt

Shiyu Tang, Di Wu, Yifan Wang, and Lijun Wang(✉)

Dalian University of Technology, Dalian 116024, China
ljwang@dlut.edu.cn

Abstract. Monocular depth estimation and depth completion are closely correlated, yet they have long been approached as two distinct and separate tasks. In this paper, we propose a new Transformer architecture dubbed SACFormer to unify these two tasks with Spatially Aligned Cross-modality (SAC) attention modules. Unlike existing unified methods, SACFormer is able to take advantage of the correlation between two tasks using a single model and one set of network parameters without task- or modality-specific modules. To better identify their unique characteristics, we further introduce a window-based deep prompt learning scheme that enables SACFormer to seamlessly switch between the two tasks. By integrating the above two contributions, we are able to enforce the synergy between depth estimation and completion, while respecting their differences. As a result, our method using one model trained on multi-domain data can simultaneously handle the two related tasks, thereby significantly reducing memory footprint during deployment. The proposed method is extensively evaluated on popular benchmarks and performs favorably in both indoor and outdoor scenes.

Keywords: Monocular depth estimation · Depth completion · Unified model · Prompt learning

1 Introduction

Monocular depth estimation [1–3] and depth completion [4,5] are two fundamental ways for 3D scene geometry perception, which have wide applications in many real-world scenarios, including autonomous driving, robotics, augmented reality, *etc.* Among them, monocular depth estimation aims to predict the depth map from a single-view RGB image, while depth completion focuses on depth restoration from a sparse LiDAR depth map guided by its corresponding RGB image. Though they differ in terms of the input, the two tasks are intrinsically correlated to each other as both of them aims to infer the missing depth values through not only geometry- but also semantic-level interpretation of the input scenes.

With the development of deep learning techniques, recent years have witnessed significant progress in both fields, ranging from architectural design [6–8]

Q. Liu et al. (Eds.): PRCV 2023, LNCS 14426, pp. 103–114, 2024.
https://doi.org/10.1007/978-981-99-8432-9_9

to new training strategies [9]. Though the precision of depth prediction has been significantly improved, the two tasks are largely investigated as two isolated topics. For one thing, the correlations between the two tasks are unexplored, restricting cross-task mutual benefits as well as further performance improvement. For anther, one need to design and train an independent model for each task, causing higher cost for deployment.

Based on the above observations, Guizilini *et al.* [10] propose one of the first unified models to simultaneously address both monocular depth estimation and completion. It uses two separate encoders for feature extraction from input images and sparse depth maps, respectively. The extracted features from two domains are fused during depth completion. Otherwise, only image features are used for depth estimation. Although the idea of unified depth estimation and completion has been partially validated by [10], it is still an open question with many issues left unsolved. **First**, with task-specific encoders, the interactions between image and sparse depth domains are still limited in [10]. Besides, the cross-domain feature fusion are dynamically conducted according to the current task during training, which leads to sudden distribution shift of the input feature to the decoder, causing training instability. *It is therefore unclear whether depth estimation and completion can be better unified in a more appropriate manner without task-specific or modal-specific modules.* **Second**, although depth estimation and completion are highly correlated, they have different focuses (*ie.*, estimation *vs.* restoration) with heterogeneous input. *It is unknown how to leverage their correlation while respecting their heterogeneity.*

In this paper, we propose a new approach to deeply unify monocular depth estimation and completion. To answer the first question, we design the **S**patially **A**ligned **C**ross-modality (**SAC**) self-attention module, which can better promote RGB-D feature interaction in a more flexible manner. By stacking multiple SAC attention blocks into an encoder-decoder architecture, we obtain a new Transformer architecture named SACFormer. Compared with prior methods [10] using convolutional structures, SACFormer with sequential input-output permits a more natural way towards the unification of depth estimation and completion, which is able to accept heterogeneous input without task-specific modules or dynamic network architecture adjustment, giving rise to more stable network training. Besides, with the strong global modeling ability of SAC attention, our method facilitates cross-modality synergy, enables a shared feature representation learning across tasks, and thus yields significant mutual benefits.

To address the second question, we develop a window-based deep prompt learning scheme. Trainable continuous prompts are assigned to individual tasks as pre-fix input tokens, which are optimized in an end-to-end manner to characterize the task-specific properties. By incorporating these input prompts as task indicators into the SACFormer, it can better identify task differences between depth estimation and completion, allowing to take full advantages of their correlation.

It should be noted that recent research on generalist models [11] with prompt-based learning [12] has achieved significant breakthrough. Nevertheless, it is still very rare to explore these cutting-edge techniques to unify depth estimation and completion. We are motivated by the recent progress and make a step further towards this goal. Extensive experiments demonstrate that our method achieves mutual benefits and delivers state-of-the-art performance on both tasks with a single model.

2 Related Work

2.1 Monocular Depth Estimation

With the development of CNN networks, [13] extended these methods to monocular depth estimation tasks, resulting in a significant improvement in performance. Following that, monocular depth estimation was mainly divided into two approaches. The first approach involves extracting features from the image and directly regressing continuous depth values [6,14]. The second approach discretizes the depth space and transforms the depth estimation into a classification problem. DORN [1] introduced the concept of converting the regression problem of depth estimation into a classification problem, aiming to address issues such as slow convergence and unsatisfactory local solutions in depth estimation. Additionally, some methods also incorporate geometric constraints to assist in the depth estimation task [15].

Recently, with the rise of Transformers in the field of computer vision, an increasing number of methods have incorporated attention mechanisms into depth estimation. Adabins [2] utilized min-ViT to calculate bin width, while DPT [6] innovatively employed ViT as an encoder to tackle dense tasks with global perception. Due to the strong multi-scale feature extraction capability of Swin Transformer [16], which aligns well with depth estimation, more and more works have utilized it as an encoder for their tasks [17,18]. In this paper, we also adopt Swin Transformer as the backbone. However, we do not simply use it as is. Instead, we enhance it with the ability of cross-modality modeling through SAC attention. This attention mechanism allows us to flexibly accept either an RGB image or a combination of an image and sparse depth map as input, using a merge-attention-stitch approach.

2.2 Depth Completion

Since RGB images can reflect the details of color and texture changes, most recent depth completion work is based on the SPN [4] strategy to achieve end-to-end neural network training. SPN is based on the strategy of anisotropic diffusion, obtaining the final dense depth map through affinity with neighbors and refining iterative predictions. Representative works such as CSPN [5] and NLSPN [19] are based on the SPN strategy to improve the network and perform the next round of experiments. Packnet-SAN [10] proposed sparse auxiliary networks (SAN) to assist monocular depth estimation networks in performing both depth estimation and depth completion tasks. Coefficient point cloud information is added to the depth prediction network through jump connections to

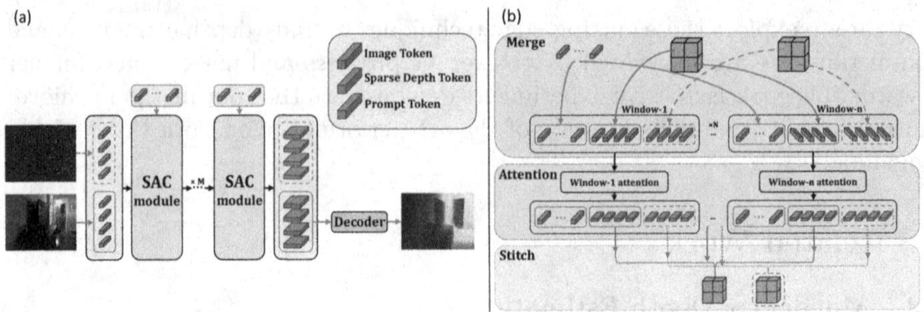

Fig. 1. (a) The proposed SACFormer overview. SACFormer takes RGB image tokens and an optional sparse depth tokens (indicated by the dashed box) as inputs, and outputs a depth map. (b) One iteration of the SAC attention module, operating in a merge-attention-stitch manner. The SAC attention module consists of two iterations with different window configurations. And the sparse depth feature is also optional.

enhance its features. However, this approach still uses a significant number of additional parameters and significantly modifies the monocular depth estimation network, making the network less unified.

3 Method

As shown in Fig. 1 (left panel), we implement the SACFormer with an encoder-decoder architecture. The encoder is mainly built upon the SAC self-attention modules. As for the decoder, we adopt a lightweight structure [14] for depth inference, which already works well. In addition, we introduce a deep prompt-based learning scheme to capture the task-specific information, serving as a task indicator.

Given an input RGB image $\mathbf{I}$ and (during depth completion) an optional sparse depth map[1] $\mathbf{S}$, we first split them into non-overlapping patches of 4×4 pixels, which are further embedded into feature tokens, serving as the input to SACFormer. Inspired by the remarkable performance of hierarchical Transformer architectures [20], SACFormer is designed in a similar philosophy. To this end, we follow [16] and divide the encoder into four stages. Each stage first spatially downsamples the input tokens by 2×2 times through patch merging and then processes the input tokens together with trainable prompts using SAC attention modules. The encoded features of the input image and sparse depth map are finally generated. Since the sparse depth map is absent during depth estimation, we select the image feature only as the input to the decoder to enforce feature consistency across tasks. Thanks to the strong non-local and cross-modality representational ability of SAC attention modules, the encoded image feature learns to capture both image and sparse depth information, providing essential cues for

[1] The input one-channel sparse depth map is duplicated into 3 channels to be consistent with the RGB image.

both tasks. In the following, we elaborate on the details of the SAC attention module and window-level prompt-based learning strategy.

3.1 Spatially Aligned Cross-Modality Self-attention

Swin Transformer improves computational efficiency by limiting self-attention computation within a local window. However, it is designed for RGB image input only. We empower it with the ability of cross-modality modeling through the SAC attention, which can flexibly accept as input either RGB image or the combination of image and sparse depth map in a merge-attention-stitch manner (*cf.*, Fig. 1 right panel).

Given the patch tokens of image and sparse depth map, we follow Swin Transformer by evenly partitioning them into non-overlapping windows respectively, each of which consists of $M \times M$ patch tokens. For the l-th SAC attention module, the merge step groups each window in the image modality with its spatially corresponding counterpart in the depth modality into a super-window, which is further translated into a token sequence $\mathbf{t}^{(l-1)} = \{\mathbf{t}_i^{(l-1)} | i = 1, 2, \ldots, 2M^2\}$. The attention module then performs multi-head self-attention (MSA) within the cross-modality super-window followed by a MLP:

$$\begin{aligned} \hat{\mathbf{t}}^{(l)} &= \mathrm{MSA}(\mathrm{LN}(\mathbf{t}^{(l-1)})) + \mathbf{t}^{(l-1)}, \\ \mathbf{t}^{(l)} &= \mathrm{MLP}(\mathrm{LN}(\hat{\mathbf{t}}^{(l)})) + \hat{\mathbf{t}}^{(l)}, \end{aligned} \tag{1}$$

where $\mathrm{LN}(\cdot)$ denotes layer normalization, and the input tokens from both domains share the same relative position embeddings. Finally, the stitch step put the output tokens by attention step back to their original location in either the image or depth map plane. Each SAC attention module is composed by two iterations of the above procedures. The first iteration performs a normal window partition, while the second one adopts a shifted window configuration [16] to obtain cross-window connection.

The SAC attention enables cross-modality feature interactions between spatially aligned image and depth tokens. Through the shifted window partition, cross-modality information is further propagated globally. Another appealing property of our SAC attention module is its great flexibility in handling heterogeneous input and strong compatibility with existing architectures. The above description explains how to perform feature encoding during depth completion. For monocular depth estimation, the SAC attention is equivalent conventional Swin attention blocks. This compatibility allows us to leverage Swin Transformers pre-trained on single-modality large-scale data for parameter initialization.

3.2 Window-Based Deep Prompt Learning

The SACFormer described above can already simultaneously perform monocular depth estimation and completion. However, by directly training SACFormer in a multi-task manner leads to unsatisfactory results. We conjecture the current design of SACFormer fails to recognize the discrepancy between the two tasks. To verify this, we introduce a window-based deep prompt learning strategy which may not only characterize the unique property of each tasks, but also serves as an indicator to guide the SACFormer to properly switch between different task modes.

Specifically, for a MSA layer, we initialize two trainable prompts for depth estimation and completion tasks, respectively, which have the same dimension as the input tokens. Considering the fact that SAC attention is performed within local windows and that task indicators should be spatially invariant, we determine to share one prompt for each window, and thus it is called window-based prompt. During forward propagation, the prompt is selected according to the current task and concatenated with the input image (and optionally depth) tokens to form the actual input to the MSA layer. Our preliminary experiments suggest that window-based prompt delivers similar or even better performance than its spatially unshared counterparts with a significantly reduced parameter number. Rather than performing prompt learning only at the input of SACFormer, we borrow the experience from recent works [21] in language processing by introducing window-based prompts at each MSA layers, giving rise to the window-based deep prompt learning.

3.3 Training Loss

We train SACFormer simultaneously on monocular depth estimation and completion in a multi-task manner using the scale-invariant loss [7]:

$$L = \alpha\sqrt{\frac{1}{T}\sum_i g_i^2 - \left(\frac{1}{T}\sum_i g_i\right)^2 + (1-\lambda)\left(\frac{1}{T}\sum_i g_i\right)^2}, \tag{2}$$

where $g_i = log(\tilde{d}_i) - log(d_i)$ with d_i and $\hat{d}_i$ being the ground truth and predicted depth, respectively; T denotes the number of valid pixels; λ and α are set to 0.85 and 10, respectively.

4 Experiments

4.1 Implementation Details

We train and evaluate our method on two popular datasets: NYUD-v2 [22] and KITTI [23]. NYUD-v2 contains 464 diverse indoor scenes, with 249 used for training and 215 for testing. We train on 24K images and test on 654 images. For KITTI dataset, we adopt Eigen split [13] with 23,488 training images, 888 validation images, and 697 testing images (of which only 652 contain accumulated ground-truth depth maps). We set the cropping size as 480 × 640 for NYUD-v2 and 352 × 704 for KITTI. Since the NYUD-v2 dataset has no sparse

Table 1. Ablation of SACFormer

Method	Params	DC			MDE		
		Abs.Rel	RMSE	$\delta < 1.25$	Abs.Rel	RMSE	$\delta < 1.25$
Separate embedd	+2423	**0.015**	0.103	**0.995**	0.098	0.335	0.921
Separate backbone	+86M	0.020	0.119	0.994	0.099	0.337	0.920
Sparse convolution	+4.8M	0.020	0.120	0.992	0.099	0.337	0.919
Concat RGBD	+72	0.019	0.106	0.994	0.100	0.338	0.919
SACFormer	89M	**0.015**	**0.099**	**0.995**	**0.096**	**0.334**	**0.923**

DC represents depth completion, and MDE represents monocular depth estimation.

Table 2. Ablation of the prompt token

Method	DC			MDE		
	Abs.Rel	RMSE	$\delta < 1.25$	Abs.Rel	RMSE	$\delta < 1.25$
w/o prompt	0.020	0.105	0.994	0.104	0.345	0.906
10 prompts	0.017	0.101	**0.995**	0.102	0.340	0.915
50 prompts	0.015	0.100	**0.995**	0.098	0.335	**0.923**
Only DC	**0.012**	**0.092**	**0.995**	-	-	-
Only MDE	-	-	-	0.099	0.336	0.921
25 prompts	0.015	0.099	**0.995**	**0.096**	**0.334**	**0.923**

depth, we follow prior methods [19] by randomly sampling 500 points from the ground truth to generate input sparse depth maps.

We implement our model in PyTorch [24] on 4 NVIDIA 3090 GPUs, using AdamW [25] as optimizer with $\beta_1 = 0.9$, $\beta_2 = 0.999$ and weight decay $= 10^{-4}$, learning is scheduled via polynomial decay from base learning rate 10^{-4} with power $p = 0.9$. The batch size per GPU is set to 4. We train the model for 40 epochs and randomly select one task in each iteration. We use Swin-L pre-trained on ImageNet [26] as the encoder and window size M is set to 12. The number of prompts is set to 25. We use root mean square error (RMSE), absolute relative difference (Abs.Rel), threshold accuracy (δ_1, δ_2 and δ_3) as the evaluation metric.

4.2 Ablation Study

We conduct ablation experiments to verify our contribution. The experiments are conducted on the joint depth estimation and depth completion tasks of the NYUD-v2 dataset. The experimental settings are consistent with those presented above, except for the use of a base-scale model as the backbone.

Ablation of SACFormer. We conduct an ablation study on the structure of the network to extract sparse depth features and compare several variations of

Table 3. Comparison With State-of-The-Art Methods on KITTI Eigen Split

Method	Input	Param(M)	Lower is better ↓		Higher is better ↑		
			Abs.Rel	RMSE	$\delta < 1.25$	$\delta < 1.25^2$	$\delta < 1.25^3$
VNL [15]	RGB	86	0.072	3.258	0.938	0.990	0.998
DPT-hybrid∗ [6]	RGB	117	0.062	2.573	0.959	0.995	0.999
GLP [14]	RGB	62	0.057	2.297	0.967	0.996	0.999
NewCRFs [18]	RGB	257	0.052	2.129	0.974	0.997	0.999
PixelFormer [8]	RGB	258	0.051	2.081	0.976	0.997	0.999
PackNet-SAN∗ [10]	RGB	106	0.052	2.233	0.970	0.996	0.999
	RGB+D		0.015	0.909	0.997	0.999	1.000
Ours	RGB	191	0.053	2.194	0.969	0.996	0.999
	RGB+D		0.015	0.885	0.996	0.999	1.000

∗ indicates that extra data is used for pretraining.

the network in Table 1. Common sense suggests that a larger number of parameters and modality-specific modules lead to better performance for the network. However, experimental result ('Separate embedding') indicates that using the same embedding for both the RGB and depth modalities produces better results. Additionally, we attempt to train a separate Swin Transformer ('Separate bakebone') specifically for sparse depth to extract features, but still obtain poor results. We believe that the sparse depth modality has too few effective points to provide sufficient training for its embedding or feature extractor, while the dense training of RGB can compensate for this. This idea is also reflected in other parts of the network: using the dense modality to assist in the training of the sparse modality. This more unified network with fewer parameters can achieve more remarkable results. To find a better network for extracting sparse depth features, we also try using sparse convolution ('Sparse convolution') to extract features and concatenate RGBD ('Concat RGBD') for feature extraction. However, experiments show that these commonly used methods, despite using a larger number of parameters compared to SACFormer, do not achieve satisfactory results in the joint task of depth completion.

Ablation of Prompt Token. We conduct ablation experiments on the number of prompts, as shown in Table 2. 'Only MDE' represents performing only depth estimation, and 'Only DC' represents performing only depth completion. The results indicate that prompts are necessary for the network to differentiate between two tasks, while the quantity of prompts has little impact. We select 25 prompts to balance performance and speed. Furthermore, the joint task does not have adverse effects on depth estimation and may even slightly improve the effectiveness of depth estimation.

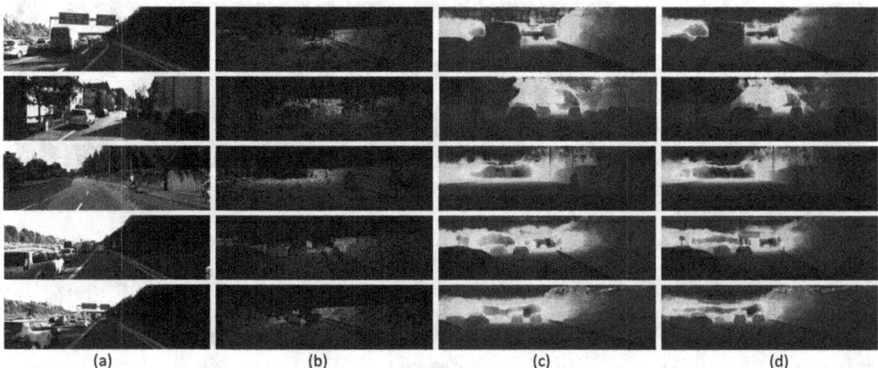

Fig. 2. Qualitative results of SACFormer on KITTI. (a) RGB input; (b) ground truth; (c) result of monocular depth estimation; (d) result of depth completion.

Table 4. Comparison With State-of-The-Art Methods on NYUD-v2

Method	Input	Param(M)	Lower is better ↓		Higher is better ↑		
			Abs.Rel	RMSE	$\delta < 1.25$	$\delta < 1.25^2$	$\delta < 1.25^3$
VNL [15]	RGB	86	0.108	0.416	0.875	0.976	0.994
GLP [14]	RGB	62	0.098	0.344	0.915	0.988	0.997
Depthformer [27]	RGB	261	0.100	0.345	0.913	0.988	0.997
NewCRFs [18]	RGB	257	0.095	0.334	0.922	0.992	0.998
PixelFormer [8]	RGB	258	0.090	0.322	0.929	0.991	0.998
CSPN [5]	RGB+D	17	0.016	0.117	0.992	0.999	1.000
NLSPN [19]	RGB+D	26	0.012	0.092	0.996	0.999	1.000
PackNet-SAN∗ [10]	RGB	106	0.106	0.393	0.892	0.979	0.995
	RGB+D		0.019	0.120	0.994	0.999	1.000
Ours(w/o prompt)	RGB	190	0.100	0.345	0.914	0.987	0.997
	RGB+D		0.023	0.111	0.996	0.999	1.000
Ours	RGB	191	0.094	0.326	0.921	0.989	0.998
	RGB+D		0.015	0.097	0.996	0.999	1.000

∗ indicates that extra data is used for pretraining.

4.3 Results on the KITTI Dataset

Table 3 presents the quantitative results of depth estimation and depth completion on KITTI. We follow PackNet-SAN [10] by using Eigen's split to jointly train two tasks, which is adopted by most monocular depth estimation methods but inconsistent with prior depth completion approaches. Therefore, our method cannot be directly compared with existing depth completion methods. It is also worth noticing that our comparison with PackNet-SAN on the depth completion task is also unfair since it uses additional training data. Compared to existing methods for monocular depth estimation, our approach achieves competitive performance. Our method demonstrates comprehensive improvements

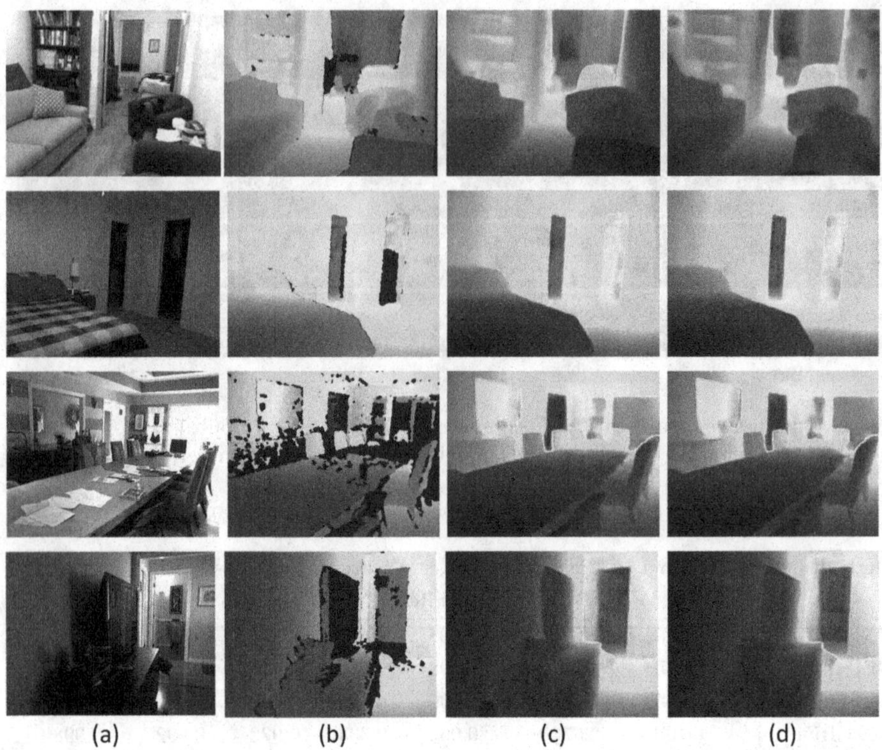

Fig. 3. Qualitative results of SACFormer on NYUD-v2. (a) RGB input; (b) ground truth; (c) result of monocular depth estimation; (d) result of depth completion.

over PackNet-SAN, despite the latter utilizing additional depth datasets for pretraining. In Fig. 2, we present the high-quality prediction results of our model. The experimental results demonstrate that SACFormer excels in both depth estimation and depth completion tasks on the outdoor dataset.

4.4 Results on the NYUD-V2 Dataset

In Table 4, we show the results on NYUD-v2, where similar trends to KITTI can be observed. Our proposed SACFormer achieves satisfactory results on both tasks. Our approach significantly outperforms PackNet-SAN in various tasks, although it should be noted that this is a unfair comparison since PackNet-SAN utilizes additional datasets for depth pretraining. We also showcase the experimental results without prompts, which clearly indicate the assistance prompts provide to the model. In Fig. 3, we present the prediction results of SACFormer. From both qualitative and quantitative perspectives, SACFormer performs exceptionally well on the indoor dataset while accomplishing both tasks.

5 Conclusion

In this paper, we present a unified model for monocular depth estimation and depth completion tasks. We propose SACFormer, a Transformer architecture with Spatially Aligned Cross-modality attention modules, which handles both tasks using a single model and reduces memory usage. The window-based deep prompt learning scheme allows seamless switching between tasks by enforcing synergy while respecting their differences. Our method achieves satisfactory results on NYUD-v2 and KITTI datasets, as demonstrated by extensive evaluations.

Acknowledgement. The paper is supported in part by the National Natural Science Foundation of China (62276045, 62006036), Dalian Science and Technology Talent Innovation Support Plan (2022RY17), and Fundamental Research Funds for Central Universities (DUT22LAB124, DUT21RC(3)025, DUT22QN228).

References

1. Fu, H., Gong, M., Wang, C., Batmanghelich, K., Tao, D.: Deep ordinal regression network for monocular depth estimation. In: Proceedings of the IEEE Conference on Computer Vision and Pattern Recognition, pp. 2002–2011 (2018)
2. Bhat, S.F., Alhashim, I., Wonka, P.: Adabins: Depth estimation using adaptive bins. In: Proceedings of the IEEE/CVF Conference on Computer Vision and Pattern Recognition, pp. 4009–4018 (2021)
3. Dosovitskiy, A., et al.: An image is worth 16x16 words: transformers for image recognition at scale. arXiv preprint arXiv:2010.11929 (2020)
4. Liu, S., De Mello, S., Gu, J., Zhong, G., Yang, M.H., Kautz, J.: Learning affinity via spatial propagation networks. In: Advances in Neural Information Processing Systems 30 (2017)
5. Cheng, X., Wang, P., Yang, R.: Depth estimation via affinity learned with convolutional spatial propagation network. In: Proceedings of the European Conference on Computer Vision (ECCV), pp. 103–119 (2018)
6. Ranftl, R., Bochkovskiy, A., Koltun, V.: Vision transformers for dense prediction. In: Proceedings of the IEEE/CVF International Conference on Computer Vision, pp. 12179–12188 (2021)
7. Lee, J.H., Han, M.K., Ko, D.W., Suh, I.H.: From big to small: multi-scale local planar guidance for monocular depth estimation. arXiv preprint arXiv:1907.10326 (2019)
8. Agarwal, A., Arora, C.: Attention attention everywhere: monocular depth prediction with skip attention. In: Proceedings of the IEEE/CVF Winter Conference on Applications of Computer Vision, pp. 5861–5870 (2023)
9. Long, X., et al.: Adaptive surface normal constraint for depth estimation. In: Proceedings of the IEEE/CVF International Conference on Computer Vision, pp. 12849–12858 (2021)
10. Guizilini, V., Ambrus, R., Burgard, W., Gaidon, A.: Sparse auxiliary networks for unified monocular depth prediction and completion. In: Proceedings of the IEEE/CVF Conference on Computer Vision and Pattern Recognition, pp. 11078–11088 (2021)

11. Chen, T., Saxena, S., Li, L., Fleet, D.J., Hinton, G.: Pix2seq: a language modeling framework for object detection. arXiv preprint arXiv:2109.10852 (2021)
12. Jia, M., et al.: Visual prompt tuning. In: Computer Vision-ECCV 2022: 17th European Conference, Tel Aviv, Israel, October 23–27, 2022, Proceedings, Part XXXIII, pp. 709–727. Springer (2022). https://doi.org/10.1007/978-3-031-19827-4_41
13. Eigen, D., Puhrsch, C., Fergus, R.: Depth map prediction from a single image using a multi-scale deep network. In: Advances in neural information processing systems 27 (2014)
14. Kim, D., Ga, W., Ahn, P., Joo, D., Chun, S., Kim, J.: Global-local path networks for monocular depth estimation with vertical cutdepth. arXiv preprint arXiv:2201.07436 (2022)
15. Yin, W., Liu, Y., Shen, C.: Virtual normal: enforcing geometric constraints for accurate and robust depth prediction. IEEE Trans. Pattern Anal. Mach. Intell. **44**(10), 7282–7295 (2021)
16. Liu, Z., et al.: Swin transformer: hierarchical vision transformer using shifted windows. In: Proceedings of the IEEE/CVF International Conference on Computer Vision, pp. 10012–10022 (2021)
17. Li, Z., Wang, X., Liu, X., Jiang, J.: Binsformer: revisiting adaptive bins for monocular depth estimation. arXiv preprint arXiv:2204.00987 (2022)
18. Yuan, W., Gu, X., Dai, Z., Zhu, S., Tan, P.: New CRFs: neural window fully-connected CRFs for monocular depth estimation. arXiv preprint arXiv:2203.01502 (2022)
19. Park, J., Joo, K., Hu, Z., Liu, C.-K., So Kweon, I.: Non-local spatial propagation network for depth completion. In: Vedaldi, A., Bischof, H., Brox, T., Frahm, J.-M. (eds.) ECCV 2020. LNCS, vol. 12358, pp. 120–136. Springer, Cham (2020). https://doi.org/10.1007/978-3-030-58601-0_8
20. Wang, W., et al.: Pyramid vision transformer: a versatile backbone for dense prediction without convolutions. In: Proceedings of the IEEE/CVF International Conference on Computer Vision, pp. 568–578 (2021)
21. Liu, X., et al.: P-tuning: prompt tuning can be comparable to fine-tuning across scales and tasks. In: Proceedings of the 60th Annual Meeting of the Association for Computational Linguistics (Volume 2: Short Papers), pp. 61–68 (2022)
22. Silberman, N., Hoiem, D., Kohli, P., Fergus, R.: Indoor segmentation and support inference from RGBD images. ECCV **5**(7576), 746–760 (2012)
23. Geiger, A., Lenz, P., Stiller, C., Urtasun, R.: Vision meets robotics: the kitti dataset. Int. J. Robot. Res. **32**(11), 1231–1237 (2013)
24. Paszke, A.,et al.: Pytorch: an imperative style, high-performance deep learning library. In: Advances in Neural Information Processing Systems 32 (2019)
25. Loshchilov, I., Hutter, F.: Decoupled weight decay regularization. arXiv preprint arXiv:1711.05101 (2017)
26. Deng, J., Dong, W., Socher, R., Li, L.J., Li, K., Fei-Fei, L.: Imagenet: a large-scale hierarchical image database. In: 2009 IEEE Conference on Computer Vision And Pattern Recognition, pp. 248–255. IEEE (2009)
27. Li, Z., Chen, Z., Liu, X., Jiang, J.: Depthformer: exploiting long-range correlation and local information for accurate monocular depth estimation. arXiv preprint arXiv:2203.14211 (2022)

Rotation-Invariant Completion Network

Yu Chen(✉) and Pengcheng Shi

School of Software Engineering, Xi'an Jiaotong University, Xi'an, China
{chenyu123,spcbruea}@stu.xjtu.edu.cn

Abstract. Real-world point clouds usually suffer from incompleteness and display different poses. While current point cloud completion methods excel in reproducing complete point clouds with consistent poses as seen in the training set, their performance tends to be unsatisfactory when handling point clouds with diverse poses. We propose a network named **R**otation-**I**nvariant **C**ompletion **Net**work (RICNet), which consists of two parts: a Dual Pipeline Completion Network (DPCNet) and an enhancing module. Firstly, DPCNet generates a coarse complete point cloud. The feature extraction module of DPCNet can extract consistent features, no matter if the input point cloud has undergone rotation or translation. Subsequently, the enhancing module refines the fine-grained details of the final generated point cloud. RICNet achieves better rotation invariance in feature extraction and incorporates structural relationships in man-made objects. To assess the performance of RICNet and existing methods on point clouds with various poses, we applied random transformations to the point clouds in the MVP dataset and conducted experiments on them. Our experiments demonstrate that RICNet exhibits superior completion performance compared to existing methods.

Keywords: Point cloud completion · Rotation invariance · 3D vision

1 Introduction

Point clouds are widely used for representing the 3D world in computer vision and robotics tasks, but they are often incomplete in real-world scenarios due to limitations of LiDAR scannings, such as object occlusion and range constraints, which result in information loss and constrain point cloud practicality.

Among the existing point cloud completion methods, [2–5] rely on deterministic partial-to-complete mappings to generate complete shapes, lacking the ability to conditionally generate depending on the incomplete point cloud. Additionally, they fail to capture important attributes such as geometric symmetries, regular arrangements, and surface smoothness. To address these issues, VRCNet [7] further incorporates relational structural attributes to enhance the recovery of fine details in point clouds. Although numerous point cloud completion methods proposed in recent years have shown impressive completion performance on various datasets, they only perform well on point clouds with a specific pose in the training dataset, while performance significantly deteriorates on rotated

Q. Liu et al. (Eds.): PRCV 2023, LNCS 14426, pp. 115–127, 2024.
https://doi.org/10.1007/978-981-99-8432-9_10

point clouds. However, incomplete point clouds in the real world exhibit various poses, and it is essential for point cloud completion networks to have good generalization performance for point clouds with different poses.

To address these challenges, we propose the **R**otation-**I**nvariant **C**ompletion **Net**work (RICNet), comprising a Dual Pipeline Completion Network (DPCNet) and an enhancing module. DPCNet generates a coarse complete point cloud, while the enhancing module enhances the detailed features and structural relationships of the coarse point cloud. DPCNet adopts an encoder-decoder architecture. We extract the rotation invariant features by considering the correlation not only between points and their neighbors, but also the internal relationships among the neighboring points. Our feature extraction module ensures consistent predictions during training and testing, regardless of whether the input point cloud has undergone rigid transformations like rotation and translation. Inspired by Pluralistic image completion [8], our first module DPCNet uses a dual-path architecture comprising a reconstruction path of the complete point cloud and a completion path of the incomplete point cloud. The reconstruction path adopts a VAE framework, with an encoder that embeds the point cloud into a feature space to obtain an intermediate latent distribution of point cloud features. The Decoder then reconstructs the complete shape from this code. In a similar vein, the completion path attempts to reconstruct a complete shape using the latent distributions and features of the partial input. The encoder and decoder weights are shared between both paths, with the exception of the distribution inference layers. During the training process, we regularize the consistency between the posterior distribution of the encoded incomplete point cloud and the prior distribution of the complete point cloud. Regarding the enhancing module, there exist several plausible options. Inspired by the effectiveness of fine-grained detail and structural relationship enhancement in the coarse point cloud completion of VRCNet [7], we adopt RENet to infer correlated structures from the incomplete point cloud observations and the generated rough framework after the completion network of the coarse point cloud stage.

We conducted experiments on the MVP dataset. To assess the completion performance and compare it with existing methods on point clouds with different poses, we randomly initialized rigid transformations for each point cloud and trained on them. Then we test the network on both the original and transformed point clouds. Numerous experiments show RICNet's exceptional completion outcomes on both original and transformed point clouds in the MVP dataset.

In conclusion, the following contributions are made by this paper: **1)** We propose a network architecture for rotation-invariant point cloud completion, consisting of a Dual Pipeline Completion Network (DPCNet) and an enhancing module. **2)** We have designed a rotation-invariant feature extraction module for learning features of rotated point clouds. This module can be transferred to other point cloud tasks to enhance the network's rotation invariance. **3)** We compared the completion rates of our approach with other methods on the MVP dataset point clouds after random rotations. The results show that our RICNet outperformed the existing methods.

2 Related Works

2.1 Rotation-Invariant Convolution

Previous models for point clouds like [9,21] often ignore the rotation invariance. PointNet [1] only used a T-net to learn the rotational features of point clouds, but its performance suffered greatly from the simple object rotations if there is no data augmentation. The disorderliness of the point cloud makes it difficult to capture the rotational features. For instance, if some rotation is applied to the point cloud in PointNet++ [9], the segmentation performance becomes particularly poor. [11] proposed a simpler approach using hand-crafted features based on Euclidean distance and angle. However, the resulting local features may not be sufficiently expressive, leading to reduced accuracy. [12] proposes a global context-aware convolution approach that utilizes anchor points and local reference frames (LRF). [13,14] further utilized LRF to learn rotation-invariant local descriptors to improve performance. [15] proposed an effective framework based on several features. However, it will be unstable in the presence of noise or outliers, limiting overall performance. To tackle these difficulties, RIConv++ [16] proposed a straightforward and powerful convolutional operator. This operator is specifically designed to capture robust rotation-invariant features from local regions. It enhances feature distinctiveness by considering the correlation not only between the points and their neighboring points, but also the internal relationships among the neighboring points. To address these challenges, RIConv++ [16] proposed a simple and effective convolutional operator, which is designed to extract robust rotation-invariant features from local regions and improve feature discrimination by considering the relationships between interest points and their neighbors, as well as the internal relationships among neighbors.

2.2 Point Cloud Completion

PCN [2] employs folding operations for upsampling to generate coarse completion by leveraging global features learned from incomplete point clouds. The decoder proposed by TopNet [6] can effectively predicts complete shapes. [4,5,11] utilize local features to enhance the quality of their completion results. The aforementioned approaches focus on the generation of overall shape frameworks while neglecting fine-grained details. Moreover, they largely learn deterministic mappings from partial to complete, ignoring structural relationships in man-made objects. Slice Sequential Network [20] utilizes a slice-based approach to process incomplete point clouds. By slicing the input point cloud, it can effectively capture the inter-slice information and generates the missing parts with high fidelity. VRCNet [7] proposed a variational framework consisting of PMNet and RENet. PMNet leverages the advantages of explicit reconstruction and generative modeling to generate coarse and complete point clouds. RENet further extracts relational point features through point self-attention and point selection modules to refine local shape details. The recently proposed methods have

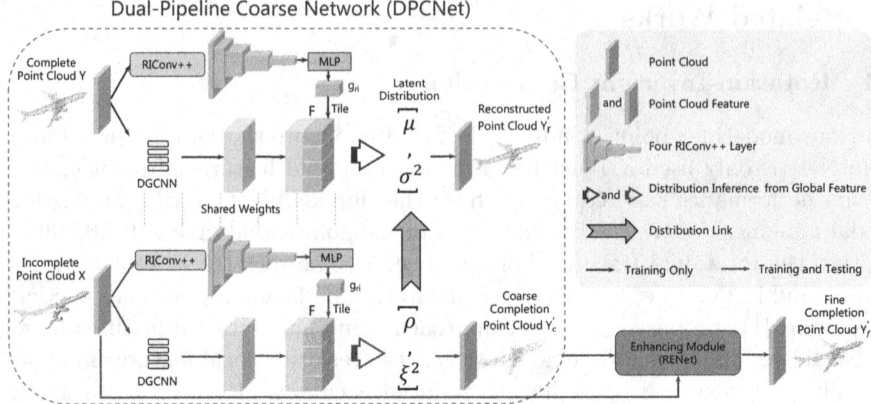

Fig. 1. The architecture of RICNet. DPCNet generates a coarse complete point cloud $Y_c{}'$, while the enhancing module refines the fine-grained details of the final output to produce the final completion $Y_f{}'$. DPCNet employs an encoder-decoder architecture, with the two parallel paths sharing weights. The distribution link ensures consistency between the posterior distribution of the encoded partial point cloud and the prior distribution of the complete point cloud.

achieved impressive completion results on various point cloud datasets. However, they lack rotation invariance and generalization ability for handling rotated point clouds, leading to unsatisfactory results for point clouds with diverse poses. The neglect of rotation invariance in current point cloud completion networks restricts their practical applicability in real-world scenarios.

3 Our Method

To address the issue of unsatisfactory results in point clouds with varied poses, we propose a network architecture for rotation-invariant point cloud completion, as shown in Fig. 1. We define X as a partial observation, while Y represents the ground truth. The objective of our network is to generate a predicted complete point cloud $Y_f{}'$ based on the provided input X.

3.1 Rotation-Invariant Embedding Module

We employ RIConv++ to explicitly extract global features, while simultaneously integrating DGCNN [17] to enhance the extraction of local features. The RIConv++ network, comprising four convolutional layers, processes the input point cloud to obtain a global feature g_{ri}. The definition of the convolutional layer can be found in Rotation-Invariant Convolution. Besides, DGCNN converts each point into a point feature vector f_i. We connect g_{ri} to each f_i to create a point feature matrix F where each row represents a concatenated feature vector $[g_{ri}, f_i]$. Extensive experiments demonstrate that our feature extraction module

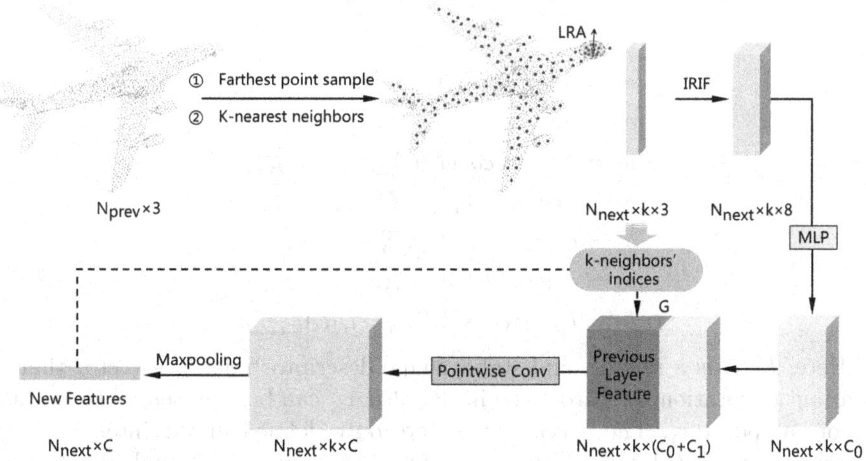

Fig. 2. A Rotation-Invariant Convolutional Layer. We perform farthest point sampling to select a set of reference points $\{r_j\}$ (red dots) from the point cloud. K-NN is utilized to obtain local point sets $X_j = \{x_n\}$ (blue dots) for each reference point r_j. IRIFs are then calculated and transformed into a high-dimensional space using an MLP. These features are combined with previous layer features (if any), followed by pointwise convolution and Maxpooling to generate the New Features.

inherits the rotation invariance and global feature extraction capabilities from RIConv++, while also effectively extracting local features with the assistance of DGCNN.

Rotation-Invariant Convolution. RIConv++ employs a sequence of four convolutional layers to explicitly extract rotation-invariant global features of point clouds, denoted as g_{ri}. Figure 2 illustrates the structure of a rotation-invariant convolutional layer. The reference point set obtained in a convolutional layer will serve as the input point cloud for the subsequent layer, and the New Features obtained in the current layer will be combined with the point cloud features of the next layer, generating the feature matrix G for the subsequent layer, as indicated by the dashed line in Fig. 2.

IRIF serves as a powerful feature representation. Figure 3 can facilitate our comprehension of IRIF, which is a transformation that maps each neighboring point x_n to a tuple, comprising seven attributes:

$$t_n = [s, \delta, a_1, a_2, a_3, b_1, b_2, b_3], \tag{1}$$

where t_n represents the IRIF feature tuples of x_n. s, a_1, a_2, and a_3 quantify the correlation between neighbor point x_n and the reference point r (radial direction). δ, b_1, b_2, and b_3 encode the correlation between x_n and its adjacent neighbor x_{n+1} :

$$
\begin{aligned}
s &= \|x_n - r\|, \\
a_1 &= arccos(LRA_r, \overrightarrow{x_n r}), \\
a_2 &= arccos(LRA_{x_n}, \overrightarrow{x_n r}), \\
a_3 &= k_a \cdot arccos(LRA_{x_n}, LRA_r). \\
\delta &= arccos(\overrightarrow{x_{n+1} r}, \overrightarrow{x_n r}), \\
b_1 &= arccos(LRA_{x_n}, \overrightarrow{x_n x_{n+1}}), \\
b_2 &= arccos(LRA_{x_{n+1}}, \overrightarrow{x_n x_{n+1}}), \\
b_3 &= k_b \cdot arccos(LRA_{x_n}, LRA_{x_{n+1}}).
\end{aligned} \tag{2}
$$

Here, LRA is a reliable and stable shape descriptor reference vector that is invariant to rotation, as introduced in [10]. LRA_{p_i} can be represented by normal vectors at point p_i. The $arccos$ values are only defined in the interval $[0, \pi]$, introducing a signed ambiguity. To capture angle and directional information between two vectors, we employ the utilization of signed angles as indicated in Eq. (3).

$$
k_a = \begin{cases} +1, if\ a_1 \leq a_2 \\ -1, otherwise \end{cases}, k_b = \begin{cases} +1, if\ b_1 \leq b_2 \\ -1, otherwise \end{cases}. \tag{3}
$$

After MLP transformation, we obtain a $N_{next} \times k \times C_0$ feature matrix. The

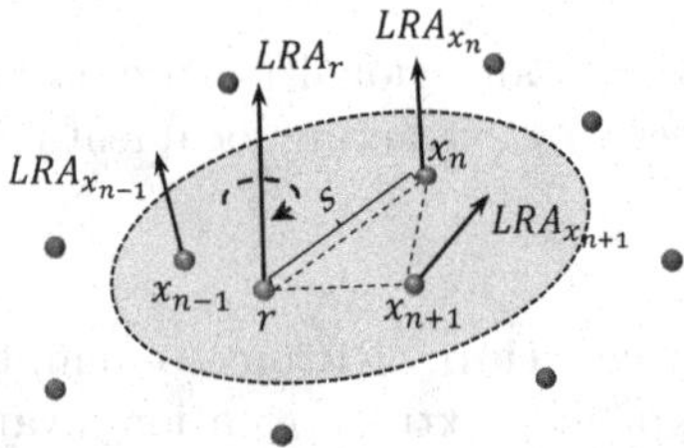

Fig. 3. IRIF Construction. Given a reference point r, the K-NN algorithm retrieves k nearest neighbors $X_j = \{x_n\}$ and orders the points clockwise. The IRIF of a neighbor point x_n is represented by (1).

previous convolutional layer's features represented as a $N_{prev} \times C_1$ matrix will be transmitted to this layer. For each reference point r_j, C_1-dimensional features are assigned to its k neighboring points based on the k-neighbors' indices. Therefore, we obtain a $N_{next} \times k \times C_1$ feature matrix, which will be combined with the features obtained in the current layer to form the feature matrix G. We order the points clockwise by projecting each point x_n onto the local tangent disk. The ordering is established by selecting a starting point (e.g. x_0) and setting its projection $\overrightarrow{x_0 p}$ as the reference point on the disk. Then, we compute the angles between the projected points $\overrightarrow{x_n p}$ and $\overrightarrow{x_0 p}$ and sort them in ascending order from 0 to 360°C. In our implementation, x_0 is chosen as the farthest point from the reference point p in the local neighborhood. Then we apply a 1D convolution to the feature tensor, followed by Maxpooling along the k-nearest neighbors dimension, to obtain the New Feature for this layer.

In more formulaic terms, we define a point set $\mho = \{p_i\}$ where p_i denotes the 3D coordinates of the point i, and t_i represents the IRIF feature of p_i. The convolution operation h for learning the features of $\mho$ is expressed as follows:

$$h(\mho) = \sigma(\kappa(MLP(t_i) : \forall i)), \tag{4}$$

Here, σ is an activation function, and κ is a Maxpooling function. It consists of a 1D convolutional kernel and an ordering function, ensuring rotational invariance is preserved. We denote the IRIF features after passing through the MLP as m_i.

$$\kappa(m_i) = Maxpool(1Dconv\,(order\,(m_i))), \tag{5}$$

Here, the function *order* arranges the points in clockwise order by projecting p_i onto the local tangent disk, and $1Dconv$ refers to a one-dimensional convolution.

3.2 DPCNet: Dual-Pipeline Completion Network

To construct the overall framework of the complete point cloud, we introduce a dual-path design. Figure 1 shows that the architecture consists of a reconstruction path dedicated to the ground truth Y and a completion path for the incomplete point cloud X. The reconstruction path utilizes a VAE framework, with an encoder (Rotation-Invariant Embedding Module) that embeds the point cloud into a feature space to obtain an intermediate latent distribution of point cloud features. The Decoder then reconstructs the complete shape from this code.

Specifically, the reconstruction path first extracts point cloud features to obtain a feature matrix F. MLP and Maxooling are then applied to derive the global feature v_r, from which we extract the latent distributions $\lambda(v_r|Y)$ for the complete shape Y, and then use a decoding distribution to recover a complete shape $Y_r{}'$. The reconstruction path is only used for training. The loss function of the reconstruction path is formulated as:

$$L_{rec} = -KL[\lambda(v_r|Y)\,||\,p(v_r)] + L_{CD}\left(Y_r{}', Y\right) \tag{6}$$

Here, KL represents the Kullback-Leibler divergence, $p(v_r) = N(0, I)$ is a predefined Gaussian conditional prior, and CD denotes the symmetric chamfer distance (CD) loss formulated as:

$$L_{CD}(P, Q) = \frac{1}{|P|}\sum_{x \in P} \min_{y \in Q} \|x - y\|^2 + \frac{1}{|Q|}\sum_{y \in Q} \min_{x \in P} \|x - y\|^2 . \tag{7}$$

Here, x and y stand for points that belong to point clouds P and Q.

The structure of the completion path is similar to that of the reconstruction path and shares its encoder and decoder weights, except for the distribution inference layer. Its main objective is to reconstruct the complete shape $Y_c{}'$ based on the global feature v_c and the latent distribution $\varphi(v_c|X)$ obtained from the input X. To effectively use the most significant features from the incomplete point cloud, a learned conditional distribution $\lambda(v_r|Y)$, which encodes its corresponding complete 3D shape Y, is employed to adjust the latent distribution φ during training. Thus, $\lambda(v_r|Y)$ forms the prior distribution, and $\varphi(v_c|X)$ serves as the posterior importance sampling function. The completion path is characterized by the following loss function formulation:

$$L_{com} = -KL[\lambda(v_r|Y)||\varphi(v_c|X)] + L_{CD}\left(Y_c{}', Y\right) \tag{8}$$

3.3 Enhancing Module

Inspired by the effectiveness of fine-grained detail and structural relationship enhancement in the coarse point cloud completion of VRCNet [7], we utilize the Relation Enhancement Network (RENet) as our enhancement module to improve structural relationships and recover the fine completion $Y_f{}'$ based on the initial completion $Y_c{}'$. The R-PSK module, comprising an MLP and PSK module, plays a crucial role in RENet. The PSK module enables the adaptive fusion of structure relations learned at various scales, allowing neurons to dynamically adjust their receptive field sizes to match the distinct scales of PSA's relation structures. PSA effectively aggregates point features in local neighborhoods by leveraging learned relations. The loss function for the enhancing module is formulated as:

$$L_{fine} = L_{CD}\left(Y_f{}', Y\right) \tag{9}$$

3.4 Loss Function

The joint training loss L of our RICNet is formulated as follows:

$$L = \omega_{rec} L_{rec} + \omega_{com} L_{com} + \omega_{fine} L_{fine} \tag{10}$$

Here, L_{rec}, L_{com} and L_{fine} are the losses defined in (6), (8), and (9). ω_{rec}, ω_{com}, and ω_{fine} are weighted parameters.

4 Experiments

We utilized PyTorch to implement our network and our model was trained on an NVIDIA TITAN Xp GPU. We employ the Adam optimizer [18], with an initial learning rate of $1e^{-4}$, decaying by 0.7 every 40 epochs. The Chamfer distance (7) is used to calculate the distance between the final predicted complete point cloud (Y') generated by the network and the ground truth point cloud (Y), while the F-score [19] is utilized to measure the distance between object surfaces.

To evaluate the proposed network's completion performance under different point cloud poses, we randomly apply a rigid transformation to each point cloud and train the network on these transformed point clouds, testing it on both the original and transformed point clouds.

Table 1. Completion results of the original cloud without transformation. Our RICNet outperforms all existing methods.

Method	$CD \downarrow$	$F1 \uparrow$
TopNet [6]	29.46	0.167
PCN [2]	26.13	0.173
Cascade [5]	21.48	0.309
MSN [3]	22.79	0.344
ECG [4]	13.48	0.424
VRCNet [7]	10.58	0.461
RICNet(Ours)	**9.14**	**0.470**

Table 2. Completion results of the transformed cloud. Our RICNet also outperforms all existing methods.

Method	$CD \downarrow$	$F1 \uparrow$
TopNet [6]	20.561	0.193
PCN [2]	15.68	0.236
Cascade [5]	16.46	0.318
MSN [3]	15.81	0.362
ECG [4]	9.44	0.440
VRCNet [7]	7.75	0.469
RICNet(Ours)	**7.57**	**0.477**

Table 3. Completion results on the transformed cloud of MVP. Here we adopt the CD loss (multiplied by 10^4) as the evaluation metric. Lower values indicate better performance. RICNet stands out as the top-performing model overall.

Method	Airplane	Cabinet	Car	Chair	Lamp	Sofa	Table	Watercraft	Bed	Bench	Bookshelf	Bus	Guitar	Motorbike	Pistol	Skateboard	**Avg.**
PCN [2]	6.556	22.618	11.042	19. 017	22.916	17.612	18.55	13.197	27.329	13.012	21.363	15.028	4.482	9.956	8.831	7.157	14.916
TopNet [6]	8.986	31.152	14.174	25.553	29.211	24.133	23.326	14.861	34.320	16.196	28.045	20.389	6.506	11.706	10.966	9.350	19.304
MSN [3]	5.678	25.246	13.853	18.138	21.559	19.218	16.064	13.624	25.646	10.483	20.125	17.017	4.546	10.122	9.210	8.391	14.932
Cascade [5]	6.435	29.065	11.847	19.728	20.418	19.819	20.836	11.205	26.301	12.92	24.034	17.094	4.453	8.942	8.364	6.659	15.507
ECG [4]	3.522	15.009	7.631	11.191	12.196	11.754	10.326	8.463	18.117	6.888	13.469	9.327	2.409	5.858	5.515	4.504	8.510
VRCNet [7]	2.679	13.904	**7.108**	**8.932**	8.031	9.644	**8.629**	6.572	14.939	5.287	**10.386**	7.451	1.816	**4.713**	4.004	**2.959**	7.315
RICNet(Ours)	**2.644**	**12.950**	7.607	9.096	**7.341**	**9.277**	8.774	**6.115**	**13.771**	**4.950**	11.582	**7.134**	**1.669**	4.859	**3.929**	2.975	**7.167**

Table 4. Completion results on the transformed cloud of MVP. Here we adopt the F-score [20] as the evaluation metric. Higher values indicate better performance. RICNet also stands out as the top-performing model overall.

Method	Airplane	Cabinet	Car	Chair	Lamp	Sofa	Table	Watercraft	Bed	Bench	Bookshelf	Bus	Guitar	Motorbike	Pistol	Skateboard	**Avg**
PCN [2]	0.509	0.102	0.153	0.143	0.211	0.116	0.192	0.267	0.108	0.278	0.120	0.197	0.519	0.243	0.346	0.470	0.248
TopNet [6]	0.409	0.078	0.131	0.120	0.170	0.098	0.170	0.243	0.092	0.245	0.097	0.150	0.464	0.226	0.309	0.400	0.212
MSN [3]	0.650	0.193	0.239	0.302	0.440	0.23	0.337	0.408	0.249	0.446	0.238	0.289	0.640	0.375	0.439	0.494	0.373
Cascade [5]	0.615	0.145	0.209	0.230	0.392	0.179	0.252	0.394	0.197	0.362	0.171	0.268	0 .639	0.326	0.436	0.538	0.294
ECG [4]	0.737	0.242	0.300	0.375	0.547	0.283	0.403	0.495	0.308	0.556	0.289	0.391	0.748	0.442	0.541	0.656	0.382
VRCNet [7]	**0.780**	**0.250**	0.308	0.403	0.619	0.302	0.430	0.530	0.333	0.606	0.315	0.405	**0.824**	0.474	0.579	0.700	0.491
RICNet(Ours)	0.772	0.246	**0.310**	**0.414**	**0.636**	**0.308**	**0.439**	**0.541**	**0.346**	**0.617**	**0.318**	**0.406**	0.814	**0.486**	**0.579**	**0.706**	**0.496**

Quantitative Evaluation. The completion results on the original point cloud are shown in Table 1, while the results on randomly rotated point clouds are presented in Table 2. Previous methods only consider point clouds with fixed poses, and notable performance degradation can be observed when applied to rotated point clouds. RICNet performs better than existing methods on both non-rotated and randomly rotated point clouds.

MVP dataset comprises 16 categories of point clouds. Tables 3 and 4 quantitatively the completion results on incomplete point clouds with different poses among all categories. Although RICNet may have slightly inferior performance in certain specific categories compared to VRCNet, it still stands out as the top-performing model overall.

Table 5. Qualitative completion results on the transformed cloud using different methods. The point clouds in the dataset consist of 2,048 points. RICNet surpasses VRCNet in capturing the overall structure in completion and exhibits significantly superior completion capability compared to other existing methods.

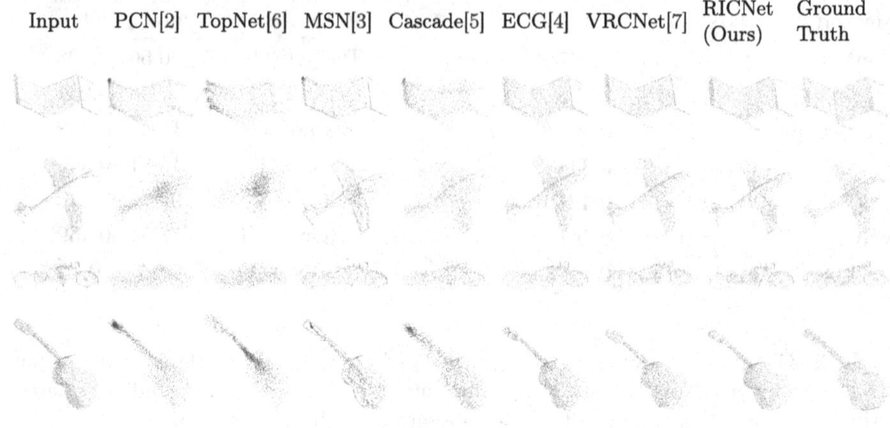

Qualitative Evaluation. Table 5 qualitatively shows the completion results on the transformed cloud by different methods. As we discussed in the previous section, [2–6] neglect the fine-grained local details. Furthermore, unlike the completion results on fixed-pose point clouds in previous studies, we can clearly observe that existing methods exhibit a significant deterioration in completion performance on rotated point clouds. For example, TopNet [6] almost lacks the ability to complete local features on the rotated point cloud, and MSN [3] only roughly recovers the outline, leaving many intermediate points missing. RICNet generates complete shapes with finer details compared to other alternatives, with clearly observable relational structures.

More specifically, compared with the results of VRCNet [7], RICNet achieves a more comprehensive completion of specific global characteristics. In the first row, RICNet produces chair completions with straighter legs. In the second row, VRCNet shows a higher occurrence of outliers in the completion of airplanes, while RICNet demonstrates a cleaner and more precise representation of the surface. In the third row, VRCNet neglects the completion of the missing motorcycle headlight, whereas RICNet exhibits remarkable proficiency in recovering the overall point cloud structure of motorcycles, highlighting its enhanced completion capabilities for local features. Lastly, in the fourth row, RICNet accurately recovers the smooth and rounded edge contours of guitars, preserving intricate details and ensuring faithful reconstruction. These achievements are attributed to our robust rotation-invariant point cloud completion network, which effectively handles the completion of randomly rotated point clouds.

Ablation Study. This section presents the results of the ablation study conducted on RICNet to evaluate the effectiveness of each component. The study focuses on three key components of our model: the rotation-invariant encoder, the dual-path architecture, and the enhancing module. We use PCN [2] to denote the model that does not incorporate any of the three aforementioned modules. Table 6 illustrates the results of the ablation study conducted on our proposed modules. The completion results are evaluated with 2048 points. The results of the ablation study convincingly highlight the crucial significance of the proposed module within our network.

Table 6. Ablation experiments are conducted on the proposed network modules. These experiments examine the effectiveness of rotation-invariant encoder, dual-path architecture, and the enhancing module.

Enhancing Module	Dual-Path Architecture	Rotation-Invariant Encoder	CD↓	F1↑
			15.68	0.236
√			8.51	0.468
√	√		7.75	0.469
√	√	√	**7.57**	**0.477**

5 Conclusion

In this paper, we propose RICNet, a rotation-invariant point cloud completion network. Our feature extraction module exhibits robustness to rigid transformations such as rotation and translation of the input point clouds. This module can be transferred to other point cloud tasks to enhance the network's rotation invariance. We evaluate the completion performance of existing methods and our network on rotated point clouds through comprehensive experiments. RICNet exhibits excellent generalization performance on incomplete point clouds with different poses, making it suitable for practical applications like robotics, autonomous driving, and 3D reconstruction. RICNet benefits downstream tasks like point cloud segmentation, classification, and registration. Our network modules are also available for future research on incomplete point clouds.

References

1. Qi, C.R., Su, H., Mo, K., et al.: PointNet: deep learning on point sets for 3D classification and segmentation. In: Proceedings of the IEEE Conference on Computer Vision and Pattern Recognition, pp. 652–660 (2017)
2. Yuan, W., Khot, T., Held, D., et al.: Pcn: point completion network. In: International Conference on 3D Vision (3DV), pp. 728–737 (2018)

3. Liu, M., Sheng, L., Yang, S., et al.: Morphing and sampling network for dense point cloud completion. In: Proceedings of the AAAI Conference on Artificial Intelligence, pp. 11596–11603 (2020)
4. Pan, L.: Ecg: edge-aware point cloud completion with graph convolution. In: IEEE Robotics and Automation Letters (2020)
5. Wang, X., Ang, M.H., Jr., Lee, G.H.: Cascaded refinement network for point cloud completion. In: Proceedings of the IEEE/CVF Conference on Computer Vision and Pattern Recognition, pp. 790–799 (2020)
6. Tchapmi, L.P., Kosaraju, V., Rezatofighi, H., et al.: Topnet: structural point cloud decoder. In: Proceedings of the IEEE Conference on Computer Vision and Pattern Recognition, pp. 383–392 (2019)
7. Pan, L., Chen, X., Cai, Z., et al.: Variational relational point completion network. In: Proceedings of the IEEE/CVF Conference on Computer Vision and Pattern Recognition, pp. 8524–8533 (2022)
8. Zheng, C., Cham, T.-J., Cai, J.: Pluralistic image completion. In: Proceedings of the IEEE Conference on Computer Vision and Pattern Recognition, pp. 1438–1447 (2019)
9. Qi, C.R., Yi, L., Su, H., et al.: Pointnet++: deep hierarchical feature learning on point sets in a metric space. In: Advances in Neural Information Processing Systems, pp. 5105–5114 (2017)
10. Tombari, F., Salti, S., Di Stefano, L.: Unique signatures of histograms for local surface description. In: Daniilidis, K., Maragos, P., Paragios, N. (eds.) ECCV 2010. LNCS, vol. 6313, pp. 356–369. Springer, Heidelberg (2010). https://doi.org/10.1007/978-3-642-15558-1_26
11. Zhang, Z., Hua, B..S, Rosen, D.W., et al.: Rotation invariant convolutions for 3d point clouds deep learning. In: International Conference on 3D Vision, pp. 204–213 (2019)
12. Zhang, Z., Hua, B.S., Chen, W., et al.: Global context aware convolutions for 3d point cloud understanding. In: International Conference on 3D Vision, pp. 210–219 (2020)
13. Kim, S., Park, J., et al.: Rotation-invariant local-to-global representation learning for 3d point cloud. In: Advances in Neural Information Processing Systems, pp. 8174–8185 (2020)
14. Thomas, H.: Rotation-invariant point convolution with multiple equivariant alignments. In: 2020 International Conference on 3D Vision (3DV), pp. 504–513 (2020)
15. Li, X., Li, et al.: A rotation-invariant framework for deep point cloud analysis. IEEE Trans. Visualizat. Comput. Graph. 4503–4514 (2021)
16. Zhang, Z., Hua, S.K.: RIConv++: effective rotation invariant convolutions for 3D point clouds deep learning. Inter. J. Comput. Vis. 1228–1243 (2022)
17. Wang, Y., Sun, Y., et al.: Dynamic graph cnn for learning on point clouds. ACM Trans. Graph. 1–12 (2019)
18. Knapitsch, A., Park, J., Zhou, Q., et al.: Tanks and temples: benchmarking large-scale scene reconstruction. ACM Trans. Graph. (ToG), 1–13 (2017)
19. Tatarchenko, M., Richter, S.R., Ranftl, R., et al.: What do single-view 3d reconstruction networks learn? In: Proceedings of the IEEE/CVF Conference on Computer Vision and Pattern Recognition, pp. 3405–3414 (2019)
20. Chen, B., Fan, J., Zhao, P., et al.: Slice sequential network: a lightweight unsupervised point cloud completion network. In: Pattern Recognition and Computer Vision: 4th Chinese Conference, PRCV 2021, pp. 103–114 (2021)

21. Tao, M., Zhao, C., Wang, J., et al.: Global patch cross-attention for point cloud analysis. In: Pattern Recognition and Computer Vision: 5th Chinese Conference, PRCV 2022, pp. 96–111 (2022)
22. Pan, L., Chen, X., Cai, Z., et al.: Variational relational point completion network for robust 3D classification. IEEE Trans. Pattern Anal. Mach. Intell. (2023)

Towards Balanced RGB-TSDF Fusion for Consistent Semantic Scene Completion by 3D RGB Feature Completion and a Classwise Entropy Loss Function

Laiyan Ding[1], Panwen Hu[1], Jie Li[2], and Rui Huang[1](✉)

[1] School of Science and Engineering, The Chinese University of Hong Kong (Shenzhen), Shenzhen, Guangdong, China
{laiyanding,panwenhu}@link.cuhk.edu.cn, ruihuang@cuhk.edu.cn

[2] School of Artificial Intelligence, Shenzhen Polytechnic University, Shenzhen, Guangdong, China
jieli1@szpt.edu.cn

Abstract. Semantic Scene Completion (SSC) aims to jointly infer semantics and occupancies of 3D scenes. Truncated Signed Distance Function (TSDF), a 3D encoding of depth, has been a common input for SSC. Furthermore, RGB-TSDF fusion, seems promising since these two modalities provide color and geometry information, respectively. Nevertheless, RGB-TSDF fusion has been considered nontrivial and commonly-used naive addition will result in inconsistent results. We argue that the inconsistency comes from the sparsity of RGB features upon projecting into 3D space, while TSDF features are dense, leading to imbalanced feature maps when summed up. To address this RGB-TSDF distribution difference, we propose a two-stage network with a 3D RGB feature completion module that completes RGB features with meaningful values for occluded areas. Moreover, we propose an effective classwise entropy loss function to punish inconsistency. Extensive experiments on public datasets verify that our method achieves state-of-the-art performance among methods that do not adopt extra data.

Keywords: Semantic Scene Completion · RGB-TSDF fusion · Entropy-based loss function

1 Introduction

Semantic scene completion is the task that reconstructs the entire scene and infers the voxel-wise semantics, including both visible surfaces and occluded

Supplementary Information The online version contains supplementary material available at https://doi.org/10.1007/978-981-99-8432-9_11.

Q. Liu et al. (Eds.): PRCV 2023, LNCS 14426, pp. 128–141, 2024.
https://doi.org/10.1007/978-981-99-8432-9_11

areas, given a single depth image or a pair of RGB and depth images [8,21]. This ability to infer 3D semantics and geometry from a single 2D observation can benefit diverse computer vision applications, e.g., indoor navigation of robots and autonomous driving.

In the task of 3D semantic scene completion, the pioneering work SSCNet [21] takes Truncated Signed Distance Function (TSDF), an encoding of depth in 3D space, as the sole input. Later works [8,9] propose naive ways to incorporate RGB information into the depth-only network and verify the benefit of including RGB as inputs. RGB provides rich texture and color information and depth describes the local geometry or shape information. These two modalities encode different but complementary information. Afterward, most existing methods take both RGB and depth as inputs and adopt various fusion methodologies [14,23].

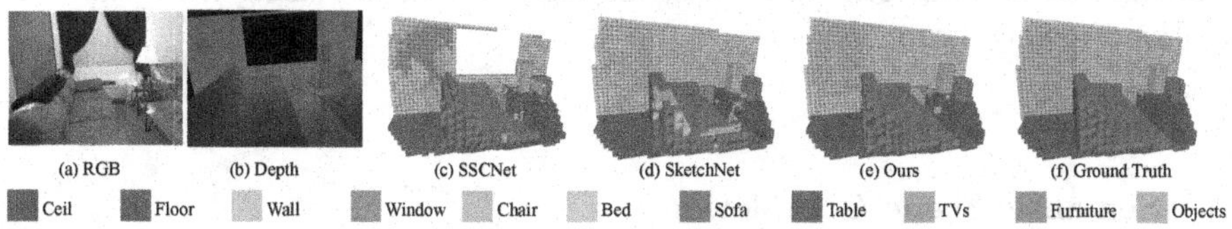

Fig. 1. Visualization of semantic scene completion results on NYUCAD dataset. From left to right: (a) RGB input, (b) depth map, (c) results of SSCNet [21], (d) results of SketchNet [2], (e) results of our proposed method, (f) ground truth. Our method can achieve better instance consistency on *sofa* and *wall*, which are in occluded areas, compared with SSCNet [21] and SketchNet [2]. Best viewed in color and zoomed in.

However, previous works mostly consider fusion in 2D space, ignoring that TSDF can provide a rich supervision signal [19]. Nevertheless, RGB-TSDF fusion is nontrivial. EdgeNet [5] argues that we cannot encode RGB in 3D in a TSDF-like way since RGB is not binary. Consequently, EdgeNet [5] projects 2D image edges into 3D space and applies TSDF-encoding to it making it dense. Yet, RGB information will be greatly reduced since only edge information is preserved. SketchNet [2] uses TSDF to predict a 3D sketch, a geometry prior, and adds it to projected RGB features. Nevertheless, the sketch locates on semantic edges, while RGB features upon projection only exist on visible surfaces. This distribution difference will still lead to inconsistent results shown in Fig. 1 (d). Similar to SketchNet [2], simply adding the dense TSDF and sparse RGB features will also result in imbalanced feature maps shown in Fig. 2. TSDF features will dominate the occluded areas leading to the failure of RGB-TSDF fusion.

To tackle the aforementioned distribution difference, we propose a two-stage network that takes RGB and TSDF as inputs. The 3D RGB feature completion stage will produce useful TSDF and completed RGB features with a 3D RGB Feature Completion Module (FCM), which assigns meaningful features for occluded areas based on visible surfaces. Then the refined semantic scene completion stage will produce refined results using the features from the first stage as inputs. To further reduce inconsistency, we propose an effective Classwise

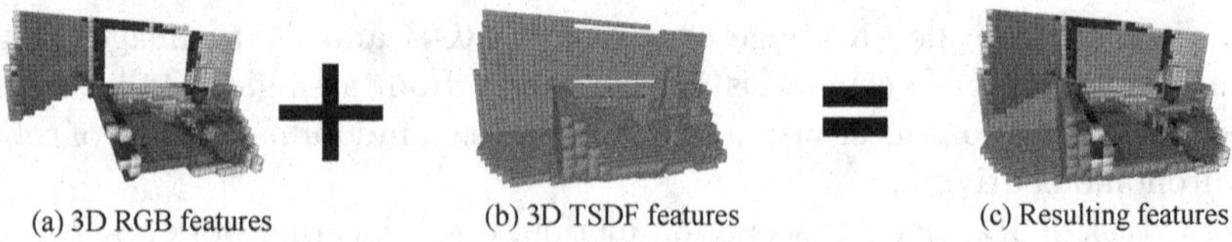

Fig. 2. Addition of 3D RGB and TSDF features. We visualize RGB and TSDF in (a) and (b), respectively, for better illustration. In the resulting features (c), we visualize RGB on the visible surfaces and TSDF in occluded areas.

Entropy Loss (CEL) which calculates the entropy of the mean probability vector for all classes, as additional supervisions aiming at punishing inconsistency throughout classes. Furthermore, the proposed method achieves state-of-the-art performance on NYUCAD [6] dataset among methods not using extra data or iterative learning strategy.

2 Related Work

2.1 Semantic Scene Completion

SSCNet [21] is the first work to introduce the task of Semantic Scene Completion (SSC). They use TSDF as the sole input and infer the complete scene structure and voxel-wise semantics. Though TSDF/Depth-only methods can achieve impressive results [3,22,28], it is evident that RGB can help identify objects with color and texture information. Some of later works develop RGB-Depth fusion methods in 2D [23] or 3D [14,15]. Nevertheless, they ignore the benefits that TSDF brings, i.e., its dense 3D representation and viewpoint invariance [21], which can be suitable for scene representation. Thus, we only consider methods involving RGB-TSDF fusions here.

RGB-TSDF methods conduct the fusion of RGB and TSDF features in 3D space and seem promising since they provide complementary information. Moreover, TSDF is a meaningful signal for the network to learn scene representation [21]. EdgeNet [5] reasons that the sparsity of RGB leads to the failure of fusion. They propose fusing TSDF-like encoding of 2D edges from the Canny detector and TSDF and achieve promising improvement. SketchNet [2] proposes embedding TSDF information as 3D sketch, which is resolution-invariant. MFFNet [7] first trains an 3D object detection network and uses the detection results to help refine coarse SSC predictions. However, these methods still suffer from the sparsity of 3D RGB features and inconsistent completion predictions. Unlike previous methods, we transform 3D RGB features from sparse to dense for balanced RGB-TSDF fusions. Furthermore, one possible way to mitigate the sparsity is by depth completion [25,26] to increase the number of points that can be projected.

2.2 Intra-class Consistency

Intra-class consistency is a general problem in many computer vision tasks, e.g., image segmentation [24,27], clustering [11,12]. Wang et al. [24] transfer robust

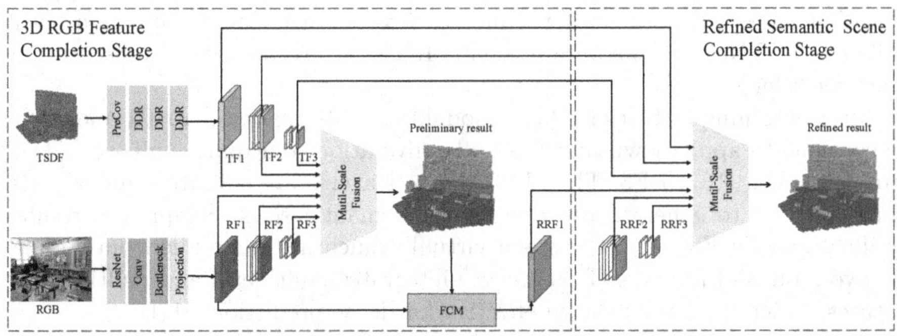

Fig. 3. The overview of the proposed network. In 3D RGB feature completion stage, we generate useful TSDF features (TF1, TF2, TF3) and completed 3D RGB features (RRF1) with the proposed 3D RGB Feature Completion Module (FCM). The refined semantic scene completion stage will utilize features from the previous stage to produce the refined result.

intra-class feature representations from teacher network to student network. For neural deep learning-based clustering, Hsu et al. [11] motivate using pairwise constraints to learn clustering in an end-to-end manner and achieve higher purity.

We propose a classwise entropy loss similar to the intra-class part of constraint in the work of Hsu et al. [11]. Since RGB features are not balanced between visible surfaces and occluded areas, there exists differences predicting semantic labels for them. Therefore, we hope to reduce the variance of predicted probabilities in each object. We impose such constraint by minimizing the entropy of the average probabilities in an object since entropy is a measurement of uncertainty in probabilistic theory [18].

3 Method

In this section, we will first present the structure of our network shown in Fig. 3, then the FCM and the classwise entropy loss for completion and segmentation consistency, and finally, the overall loss function to train our network.

3.1 Overall Network Architecture

3D RGB Feature Completion Stage. As illustrated in Fig. 3, the 3D RGB feature completion stage takes RGB images and TSDF volumes as inputs. TSDF is computed from a depth image and encodes its information in 3D voxel space where each voxel has a value d indicating the distance to its nearest surface, and the sign of d shows whether the voxel is in free or occluded areas.

In terms of feature extraction, similar to SketchNet [2], for TSDF feature extraction, we employ several convolutional layers to transform TSDF into high dimensional features (TF1) for later fusion with RGB features. For RGB feature extraction, a ResNet [10] is employed as the backbone which is pre-trained on

ImageNet [4] and freezed during training. Next, by utilizing a projection layer similar to SATNet [16], we obtain the 3D RGB features in voxel representation which we refer to as RF1.

After obtaining features for both modalities in 3D space, i.e., RF1 and TF1, we continue to apply downsampling and convolution to acquire multi-scale feature maps, i.e., RF2, RF3, TF2, TF3. These features are fed into a multi-scale fusion module to generate the preliminary semantic scene completion result. As illustrated in Figure 4, there is a virtual branch in the middle from adding the two groups of features. This module integrates multi-level and multi-modal features, which have been shown effective for dense predictions [17].

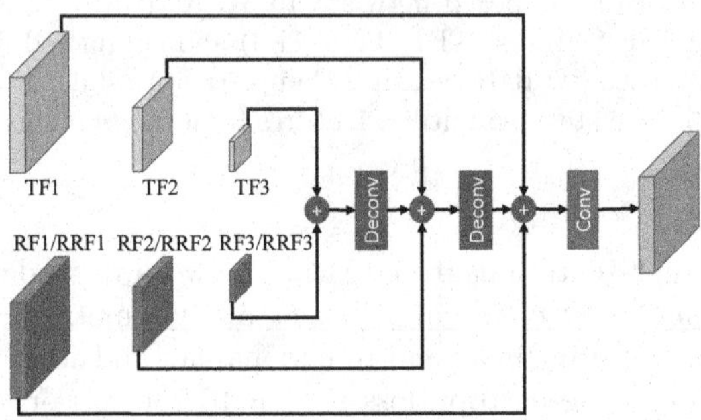

Fig. 4. Multi-scale fusion module. This module performs addition and deconvolution in a sequential way.

Lastly, the 3D RGB feature completion stage will utilize the preliminary results to transform sparse RF1 into dense RGB features (RRF1) by feeding them into the proposed 3D RGB feature completion module.

Refined Semantic Scene Completion. The inputs to the refined semantic scene completion stage will be TSDF features from the previous stage, i.e., TF1, TF2, TF3, and the completed RGB features RRF1. After we obtain RRF1, we feed it through successive convolutions similar to stage one. The multi-scale fusion module takes RGB features RRF1, RRF2, RRF3, and TSDF features from stage one and infers the refined result. Reusing TSDF features from the previous stage reduces computation cost and helps refine the result obtained in stage one. Notice we do not reuse RGB features from stage one since convolution layers will learn different weights for sparse and dense inputs.

3.2 3D RGB Feature Completion Module

Our motivation is that RGB features throughout an instance should be similar and similar features can lead to similar predictions. Consequently, for each instance, we assign RGB features to occluded areas based on the visible surfaces.

Nevertheless, we do not have instance supervision and thus, we conduct the RGB feature completion classwisely. We propose to assign the same feature vector for all the occluded areas using the mean feature vector of visible surfaces for each class. For the sake of simple notations, considering a specific class,

$$f_j = \frac{1}{|S_v|} \sum_{i \in S_v} f_i, \forall j \in S_o \tag{1}$$

where f_i and f_j are the feature vectors for voxel i and voxel j, S_o is the group of occluded voxels, S_v is the group of voxels on the surface.

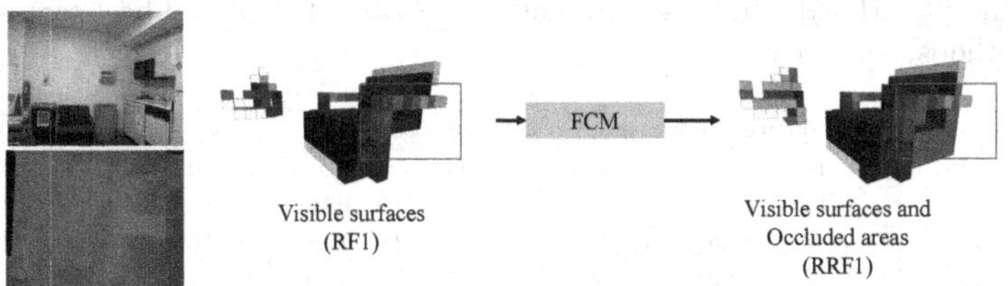

Fig. 5. Example of applying FCM on the class *chair*. The 3D RGB feature maps are transformed from sparse to dense.

The above operation is our 3D RGB FCM. An example on a single class is shown in Fig. 5, where we detach and transform RF1 to RRF1. By applying FCM, first, the RGB features for the scene excluding seen atmospheres are transformed from sparse to dense since both visible surfaces and occluded areas now contain meaningful features. Second, the RGB features throughout the instances are consistent in semantics and prominence since the features filled in are the average of features from visible surfaces. In this way, the later fusion with TSDF features will not be dominated by the TSDF. Thus, we can obtain consistent completion results.

Evidently, we cannot inject more information through FCM, since we know nothing about the occluded areas upon projection into 3D space. Nevertheless, this FCM enables easier learning for the RGB branch and, consequently the TSDF branch during the multi-modality fusion.

3.3 Classwise Entropy Loss

Imbalanced RGB and TSDF features can lead to inconsistent results, and our FCM can mitigate such imbalance. From another perspective, inconsistent predictions directly come from different predicted probabilities inside instances. Thus, to achieve more consistent predictions for each instance in the scene, we can enforce less variability and randomness in the predicted probabilities for each object.

Consequently, we impose such constraint by minimizing the entropy of the average probabilities in an object since entropy is a measurement of uncertainty

[18]. Since entropy is a measurement of randomness, reducing entropy reduces the variance, where the entropy is defined as

$$Entropy(p) = -\sum_i p_i log p_i \tag{2}$$

for any probability vector p and p_i is the i-th element of p.

Table 1. Semantic scene completion results on NYUCAD. ED means using extra data, i.e., 2D semantic labels, high resolution ground truth, or 3D instance labels. IL means iterative learning where multiple passes through the network are required. **Bold** numbers and underlined numbers represent the best and the second best scores among similar methods, respectively.

			Scene Completion			Semantic Scene Completion											
Methods	ED	IL	prec	recall	IoU	ceil	floor	wall	win	chair	bed	sofa	table	tvs	furn	objs	avg.
CCPNet [28]	✓		91.3	92.6	82.4	56.2	94.6	58.7	35.1	44.8	68.6	65.3	37.6	35.5	53.1	35.2	53.2
FFNet [23]	✓		94.8	90.3	85.5	62.7	94.9	67.9	35.2	52.0	74.8	69.9	47.9	27.9	62.7	35.1	57.4
MFFNet [7]	✓		88.7	92.5	84.8	54.5	94.8	63.3	29.3	50.9	73.6	70.9	56.4	31.7	61.3	42.0	57.2
IMENet [13]	✓	✓	84.8	92.3	79.1	-		-	-	-	-	-	-	-	-	-	47.5
SISNet [1]	✓	✓	94.2	91.3	86.5	63.4	94.4	67.2	52.4	59.2	77.9	71.1	51.8	46.2	65.8	48.8	63.5
SSCNet [21]			75.4	**96.3**	73.2	32.5	92.6	40.2	8.9	33.9	57.0	59.5	28.3	8.1	44.8	25.1	40.0
DDRNet [14]			88.7	88.5	79.4	54.1	91.5	56.4	14.9	37.0	55.7	51.0	28.8	9.2	44.1	27.8	42.8
SketchNet [2]			90.6	92.2	84.2	59.7	**94.3**	64.3	32.6	51.7	72.0	68.7	45.9	19.0	60.5	38.5	55.2
PVANet [22]			**95.1**	90.3	**86.3**	**71.5**	94.1	66.6	23.7	60.0	**78.5**	72.2	45.3	16.7	60.1	36.9	56.9
Ours			94.5	87.5	83.3	64.7	**94.3**	**68.0**	**35.3**	**62.7**	76.9	**73.6**	**49.4**	**20.9**	**61.7**	**41.7**	**59.0**

Same as our RGB feature completion method, we conduct the supervision at the class level since we do not have instance-level information. Thus we define our loss as the mean entropy of the mean probability vector for each class, excluding the empty class. Yet we only apply our loss when the class c indeed contains voxels in the ground truth, and the mean probability has the highest probability at position c, that is,

$$\begin{aligned} L_E &= \frac{1}{|S|}\sum_{c \in S} Entropy(\frac{1}{N_c}\sum_{i=1}^{N_c} p_{ci}) \\ &= \frac{1}{|S|}\sum_{c \in S} Entropy(\tilde{p}_c) \end{aligned} \tag{3}$$

where N_c is the number of voxels for class c in ground truth, p_{ci} is the predicted probability for voxel i in ground truth class c, i.e., $i \in S_c$ and S_c is the voxels that are labeled as class c in ground truth, $\tilde{p}_c$ is the mean probability vector for class c and $S = \{c : N_c > 0, argmax(\tilde{p}_c) = c, c > 0\}$. In this way, we hope the mean probability to be concentrated and as large as possible at the correct position.

3.4 Overall Loss Function

Given pairs of RGB-TSDF and 3D ground truth, the total loss function is calculated as:

$$L = L_{ce}^{(1)} + \lambda_1 L_E^{(1)} + L_{ce}^{(2)} + \lambda_2 L_E^{(2)} \tag{4}$$

where $L_{ce}^{(1)}$, $L_E^{(1)}$, and λ_1 are cross entropy loss, our classwise entropy loss and weight of the entropy loss for the 3D RGB feature completion stage, respectively. $L_{ce}^{(2)}$, $L_E^{(2)}$, and λ_2 are defined similarly but for the refined semantic scene completion stage.

4 Experiments

4.1 Implementation Details

We follow SATNet [16] to generate TSDF from depth images and downsample the high-resolution 3D ground truth. The overall loss function used is defined as Eq. 4 without extra data, e.g., dense 2D semantic labels and 3D instance information, that originally were not provided in semantic scene completion. Our network is trained with 2 GeForce GTX 2080 Ti GPUs and a batch size of 4 using Pytorch framework. We follow SketchNet [2] using mini-batch SGD with momentum 0.9 and weight decay 0.0005. The network is trained for 300 epochs. We use a poly learning rate policy and learning rate is updated by $(1 - \frac{iteration}{max\_iteration})^{0.9}$. We report the highest mean intersection over union (mIoU) on the validation set among models saved every 10 epochs.

4.2 Datasets and Metrics

Following SketchNet [2], we evaluate our method on NYUCAD dataset. NYUCAD is a scene dataset consisting of 1449 indoor scenes. Each sample is a pair of RGB and depth images, and low-resolution ground truth is obtained following SATNet [16]. NYUCAD provides depth images projected from 3D annotations to reduce misalignment and missing depth. There are 795 training samples and 654 test samples. Following SSCNet [21], we validate our methods on two tasks, scene completion (SC) and semantic scene completion (SSC). For SC, we evaluate the voxel-wise predictions, i.e., empty or non-empty on occluded areas. For SSC, we evaluate the intersection over union (IoU) for each class on both visible surfaces and occluded areas in the view frustum.

4.3 Comparisons with State-of-the-art Methods

Quantitative Comparison. We compare our method with other state-of-the-art methods. Table 1 shows the results on NYUCAD dataset. Our method achieves the best among methods without using extra data for supervision, e.g., 2D semantic labels or 3D instance information. Compared with the recent FFNet [23] and PVANet [22], we obtain an increase of 1.6% and 2.1% on SSC mIoU metric, respectively. Furthermore, we obtain supreme performance on classes that are difficult to maintain consistency during completion, i.e., *chair*, *sofa*, and *wall*. Performance comparisons and analyses on NYU [20] dataset are in the supplementary material.

Qualitative Comparison. We provide some visualization results on NYUCAD dataset in Fig. 6. First, we can observe that our preliminary result has been better and more consistent compared with SketchNet [2]. For example, in the first row, our preliminary result can recover the whiteboard more, and in the third row, we identify the small group of objects as the same class while SkecthNet will predict heterogeneous results. Second, our refined result is superior to the preliminary result. Take the second row, for example, where the predictions are presented in the rear view. Refined results predict more of the cabinet to be *furniture* instead of *wall* since with the proposed FCM, it is easier to predict more coherent results even in occluded areas. Furthermore, with our proposed classwise entropy loss function, the results in both stages are consistent throughout classes, for instance, the fourth row in Fig. 6.

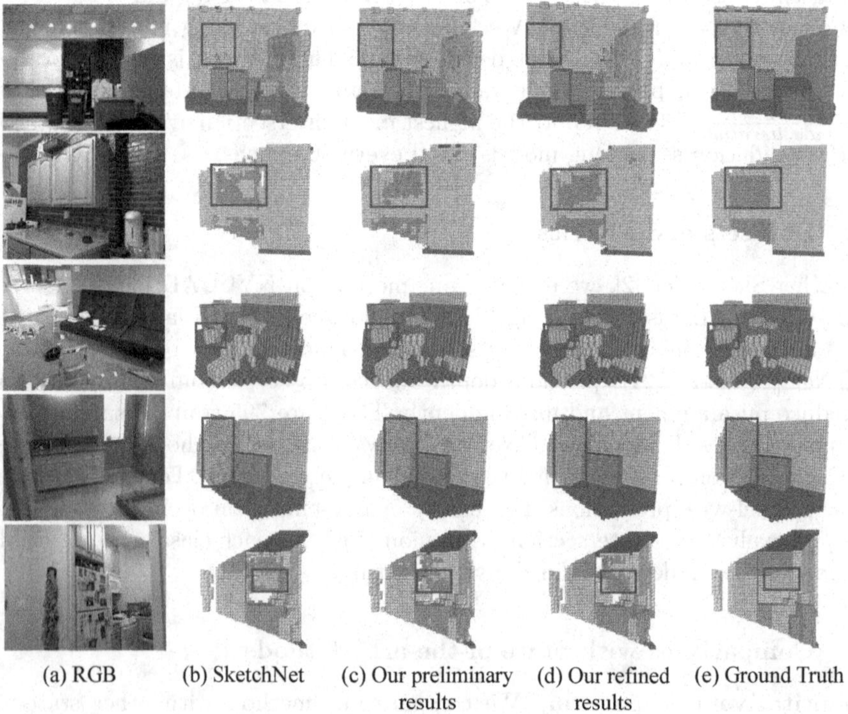

Fig. 6. Visualization results on NYUCAD dataset. Both FCM and the CEL can help produce consistent results. The second row is the rear view. The last row is a failure case.

4.4 Numerical Analysis of Class Consistency

Consider an instance that is partially predicted as *chair* and partially as *sofa*. It means that the confidence or the predicted probability is not high enough. This is shown in Fig. 7 where we visualize the predicted probability, which is obtained by applying a softmax function on the network output, for *chair* for one sample and the entire validation set. The baseline method would produce relatively more predictions of probabilities around 0.5, showing its uncertainty which would lead to inconsistent completion results. With the aid of our two-stage paradigm and classwise entropy loss function, the predicted probabilities are pushed to the extremes, i.e., 0,1. This validates that our methods can indeed produce consistent results no matter it is correct or wrong predictions.

4.5 Ablation Study

The Effectiveness of FCM and Reusing TSDF Features. We first design three models to validate the effectiveness of the initialization method and resuing TSDF features separately. Due to limited GPU memory, we design a relatively lightweight model A based on the 3D RGB feature completion stage shown in Fig. 3, where we remove the three DDR [14] blocks before generating TF1. We then design model B, where we take model A as the 3D RGB feature completion stage and do not reuse the TSDF features in refined semantic scene completion, i.e., we replicate the TSDF branch in the refined semantic scene completion stage as in the previous stage. Model C takes model A as the 3D RGB feature completion stage and reuses the TSDF features in the refined semantic scene completion stage. Results are shown in Table 2. By comparing model A and model B, we can see that, FCM can boost the performance by 1.4% on SSC. Moreover, it shows that the dense 3D RGB features can indeed help identify instances that were overwhelmed by dense TSDF features. Looking at model B and model C, our reusing mechanism can not only improve both the preliminary results and the refined results but also reduce memory usage considerably during training.

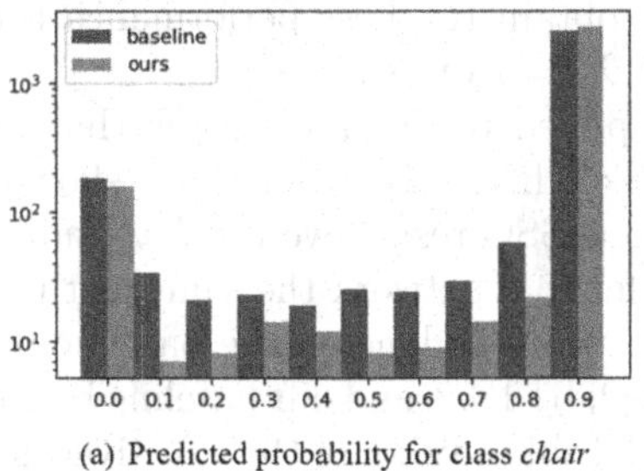

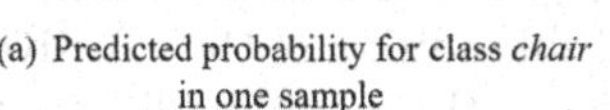
(a) Predicted probability for class *chair* in one sample

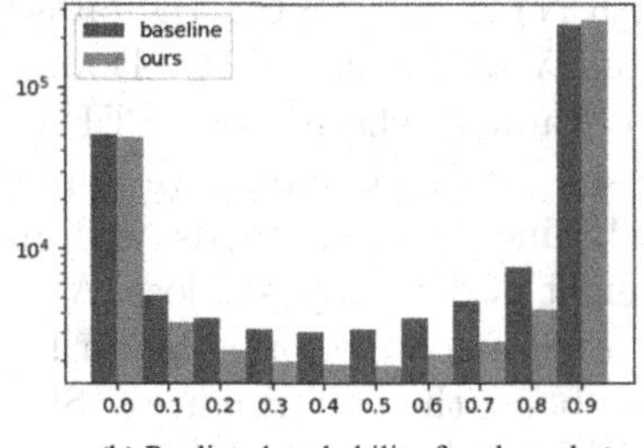

(b) Predicted probability for class *chair* in the entire validation set

Fig. 7. Histogram of predicted probabilities on NYUCAD dataset in log scale.

Table 2. Ablation study of effects of FCM and reusing TSDF features on NYUCAD dataset. Here pre. denotes preliminary results and ref. denotes refined results.

Method	FCM	reuse	pre. (SSC)	ref. (SSC)	memory
Model A			54.5	-	5266M
Model B	✓		54.3	55.9	9052M
Model C	✓	✓	55.7	56.7	7871M

Table 3. Ablation study of effects of the weight of our classwise entropy loss in two stages.

		NYUCAD(SSC)		NYU(SSC)	
λ_1	λ_2	pre	ref	pre	ref.
0	0	56.7	57.5	41.0	41.5
0.5	0.5	57.9	**59.0**	40.8	41.7
0.5	1	57.6	58.5	41.6	41.6
1	0.5	57.9	58.4	41.5	41.8
1	1	58.0	58.6	41.8	**42.3**

Table 4. Ablation study of effects of our proposed classwise entropy loss function on different network architectures.

Method	$\lambda = 0$	$\lambda = 0.5$	$\lambda = 1$
SketchNet [2]	55.2	56.1	56.3
Baseline	56.1	57.5	57.7

The Effectiveness and Generalizability of Our Classwise Entropy Loss to Architectures that Take RGB and TSDF as Inputs. For better illustration, we first provide results on our model as shown in Table 3. We experiment on the model in Fig. 3 and conduct ablation study on the choice of hyperparameters λ_1 and λ_2. With our loss applied, SSC performance can be boosted on both NYU and NYUCAD datasets. We obtain the best performance using $\lambda_1 = 0.5, \lambda_2 = 0.5$ on NYUCAD and $\lambda_1 = 1, \lambda_2 = 1$ on NYU.

To examine whether our CEL can be applied to other networks that take RGB and TSDF as input, we apply it to SketchNet [2]. Results are shown in Table 4. Since these methods only produce one SSC result, we use λ to indicate the weight of our proposed loss. Also, we refer to a network the same as the 3D RGB feature completion stage in Fig. 3 but trained alone as the baseline. Our loss can boost the performance of SketchNet [2] by 1.1% and our baseline by 1.6% on NYUCAD dataset. Results in Table 3 and Table 4 validate the effectiveness of the proposed loss on our model and other models if the modality distributions in 3D space are different. i.e., RGB-TSDF or RGB-Sketch. Nevertheless, λ should be carefully selected for better performance.

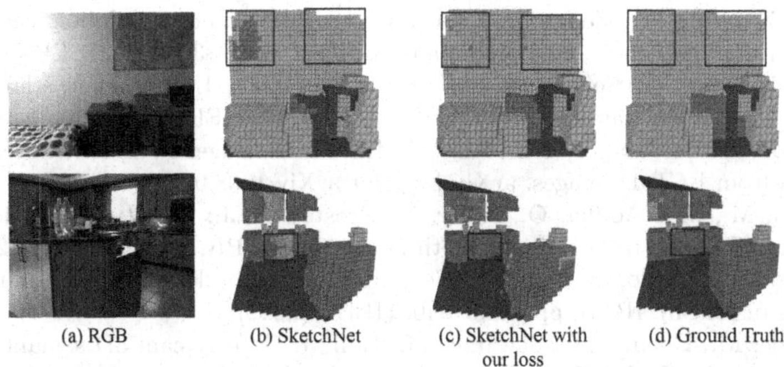

Fig. 8. Visualization results comparing applying our loss to SketchNet [2] or not. With our classwise entropy loss, SketchNet [2] can produce more consistent results.

We provide visualization results on NYUCAD dataset, comparing SketchNet [2] trained with our proposed loss or not in Fig. 8. For the first row, our loss can help produce more consistent results on visible walls. As for the second row, our loss helps achieve consistent completion results shown in red boxes. Unseen areas can be predicted extremely differently from their visible counterparts, yet our loss can help alleviate such inconsistency.

5 Conclusion

In this paper, we identify that different distributions between sprase RGB features and dense TSDF features in 3D space can lead to inconsistent predictions. To alleviate this inconsistency, we propose a two-stage network with FCM, which transforms 3D RGB features from sparse to dense. Besides, a novel classwise entropy loss function is introduced to punish inconsistency. Experiments demonstrate the effectiveness of our method compared with our baseline methods and other state-of-the-art methods on public benchmarks. Mainly, our methods can produce consistent and reasonable results.

Acknowledgment. This work was partially supported by Shenzhen Science and Technology Program (JCYJ20220818103006012, ZDSYS20211021111415025), Shenzhen Institute of Artificial Intelligence and Robotics for Society, and the Research Foundation of Shenzhen Polytechnic University (6023312007K).

References

1. Cai, Y., Chen, X., Zhang, C., Lin, K.Y., Wang, X., Li, H.: Semantic scene completion via integrating instances and scene in-the-loop. In: CVPR, pp. 324–333 (2021)
2. Chen, X., Lin, K.Y., Qian, C., Zeng, G., Li, H.: 3D sketch-aware semantic scene completion via semi-supervised structure prior. In: CVPR, pp. 4193–4202 (2020)

3. Chen, X., Xing, Y., Zeng, G.: Real-time semantic scene completion via feature aggregation and conditioned prediction. In: ICIP, pp. 2830–2834. IEEE (2020)
4. Deng, J., Dong, W., Socher, R., Li, L.J., Li, K., Fei-Fei, L.: Imagenet: a large-scale hierarchical image database. In: CVPR, pp. 248–255. IEEE (2009)
5. Dourado, A., de Campos, T.E., Kim, H., Hilton, A.: Edgenet: semantic scene completion from RGB-D images. arXiv preprint arXiv:1908.02893 1 (2019)
6. Firman, M., Mac Aodha, O., Julier, S., Brostow, G.J.: Structured prediction of unobserved voxels from a single depth image. In: CVPR, pp. 5431–5440 (2016)
7. Fu, R., Wu, H., Hao, M., Miao, Y.: Semantic scene completion through multi-level feature fusion. In: IROS, pp. 8399–8406. IEEE (2022)
8. Garbade, M., Chen, Y.T., Sawatzky, J., Gall, J.: Two stream 3D semantic scene completion. In: CVPRW (2019)
9. Guedes, A.B.S., de Campos, T.E., Hilton, A.: Semantic scene completion combining colour and depth: preliminary experiments. arXiv preprint arXiv:1802.04735 (2018)
10. He, K., Zhang, X., Ren, S., Sun, J.: Deep residual learning for image recognition. In: CVPR, pp. 770–778 (2016)
11. Hsu, Y.C., Kira, Z.: Neural network-based clustering using pairwise constraints. arXiv preprint arXiv:1511.06321 (2015)
12. Karim, M.R., et al.: Deep learning-based clustering approaches for bioinformatics. Brief. Bioinform. **22**(1), 393–415 (2021)
13. Li, J., Ding, L., Huang, R.: Imenet: Joint 3D semantic scene completion and 2d semantic segmentation through iterative mutual enhancement. In: IJCAI, pp. 793–799 (2021)
14. Li, J., et al.: RGBD based dimensional decomposition residual network for 3d semantic scene completion. In: CVPR, pp. 7693–7702 (2019)
15. Li, J., Song, Q., Yan, X., Chen, Y., Huang, R.: From front to rear: 3D semantic scene completion through planar convolution and attention-based network. IEEE Transactions on Multimedia (2023)
16. Liu, S., et al.: See and think: Disentangling semantic scene completion. In: NIPS 31 (2018)
17. Park, S.J., Hong, K.S., Lee, S.: Rdfnet: RGB-D multi-level residual feature fusion for indoor semantic segmentation. In: ICCV, pp. 4980–4989 (2017)
18. Robinson, D.W.: Entropy and uncertainty. Entropy **10**(4), 493–506 (2008)
19. Roldao, L., De Charette, R., Verroust-Blondet, A.: 3D semantic scene completion: a survey. In: IJCV, pp. 1–28 (2022)
20. Silberman, N., Hoiem, D., Kohli, P., Fergus, R.: Indoor segmentation and support inference from RGBD images. In: Fitzgibbon, A., Lazebnik, S., Perona, P., Sato, Y., Schmid, C. (eds.) ECCV 2012. LNCS, vol. 7576, pp. 746–760. Springer, Heidelberg (2012). https://doi.org/10.1007/978-3-642-33715-4_54
21. Song, S., Yu, F., Zeng, A., Chang, A.X., Savva, M., Funkhouser, T.: Semantic scene completion from a single depth image. In: CVPR, pp. 1746–1754 (2017)
22. Tang, J., Chen, X., Wang, J., Zeng, G.: Not all voxels are equal: semantic scene completion from the point-voxel perspective. In: AAAI, vol. 36, pp. 2352–2360 (2022)
23. Wang, X., Lin, D., Wan, L.: Ffnet: Frequency fusion network for semantic scene completion. In: AAAI. vol. 36, pp. 2550–2557 (2022)
24. Wang, Y., Zhou, W., Jiang, T., Bai, X., Xu, Y.: Intra-class feature variation distillation for semantic segmentation. In: Vedaldi, A., Bischof, H., Brox, T., Frahm, J.-M. (eds.) ECCV 2020. LNCS, vol. 12352, pp. 346–362. Springer, Cham (2020). https://doi.org/10.1007/978-3-030-58571-6_21

25. Yan, Z., Wang, K., Li, X., Zhang, Z., Li, J., Yang, J.: Rignet: repetitive image guided network for depth completion. In: ECCV, pp. 214–230. Springer (2022). https://doi.org/10.1007/978-3-031-19812-0_13
26. Yan, Z., Wang, K., Li, X., Zhang, Z., Li, J., Yang, J.: Desnet: decomposed scale-consistent network for unsupervised depth completion. In: AAAI, vol. 37, pp. 3109–3117 (2023)
27. Yu, C., Wang, J., Peng, C., Gao, C., Yu, G., Sang, N.: Learning a discriminative feature network for semantic segmentation. In: CVPR, pp. 1857–1866 (2018)
28. Zhang, P., Liu, W., Lei, Y., Lu, H., Yang, X.: Cascaded context pyramid for full-resolution 3D semantic scene completion. In: ICCV, pp. 7801–7810 (2019)

FPTNet: Full Point Transformer Network for Point Cloud Completion

Chunmao Wang[1], Xuejun Yan[2], and Jingjing Wang[2(✉)]

[1] Hangzhou Hikrobot Co., Ltd., Hangzhou, China
wangchunmao@hikrobotics.com
[2] Hikvision Research Institute, Hangzhou, China
{yanxuejun,wangjingjing9}@hikvision.com

Abstract. In this paper, we propose a novel point transformer network (FPTNet) for point cloud completion. Firstly, we exploit the local details as well as the long term relationships from incomplete point shapes via residual point transformer blocks. Secondly, we realize the deterministic mapping learning is a challenging task as point completion is a many-to-one problem. To address this, the shape memory layer is designed to store general shape features. The network infers complete point shapes from both incomplete clouds and shape memory features. Thirdly, the recurrent learning strategy is proposed to gradually refine the complete shape. Comprehensive experiments demonstrate that our method outperforms state-of-the-art methods on PCN and Completion3D benchmarks.

Keywords: Point cloud completion · Transformer · Recurrent learning

1 Introduction

Due to the limitations of the sensor resolution, self-occlusion and view angles, raw point clouds captured by 3D scanning devices are usually partial, sparse and noisy. Therefore, it is desirable and important to infer complete point clouds before being applied to various downstream tasks such as shape classification [11], point cloud registration [2,23] and 3D reconstruction [3,4].

As we know, the unordered and unstructured properties make the point cloud completion become challenging to deal with. Enlightened by the success of Pointnet [11], the pioneering work PCN [23] uses the shared multilayer perception (MLP) layers and symmetrical max-pooling operation to extract global features for shape completion. The following work TopNet [13] infers object shapes using a hierarchical rooted tree structured decoder. SA-Net [17] uses skip-attention mechanism to fuse local region features from encoder into the decoder features for the point cloud completion task. Recently, Point Selective Kernel Module is proposed to exploit and fuse multi-scale point features [10]. To generate more accurate shapes, several methods [10,16,19] learn the global features via siamese

C. Wang and X. Yan—Equal Contribution

Q. Liu et al. (Eds.): PRCV 2023, LNCS 14426, pp. 142–154, 2024.
https://doi.org/10.1007/978-981-99-8432-9_12

auto-encoder network which maps the partial and complete shapes into the shared latent space. The coarse-to-fine strategy is also a widely used technology for point completion. The first sub-network is applied to generate coarse shapes from partial point clouds and then a refinement network is used to get fine-grained results based on coarse shapes [10,15,19–21]. Although these works achieved impressive results on completing object shape, the point feature fusion and local detail restoration are still key issues for point completion task.

Enlightened by the success of transformers in natural language processing [14] and computer vision [18], we propose a Full Point Transformer Network for point completion (entitled as FPTNet), which utilizes point transformer blocks to fuse point cloud features and decode the complete shape, as show in Fig. 1. FPTNet consists of two consecutive sub-networks that serve as the coarse completion network and the refinement network, respectively. Specifically, the coarse net is a point transformer-based encoder-decoder network which extracts global features and generates coarse shapes. Both local and long-term point features are fused via self-attention blocks and the multi-scale aggregation strategy is applied cross these blocks. As a result, our encoder-decoder network is more efficient compared to the MLP based coarse nets [19,23]. With the help of generated global features and the coarse shape, the second sub-network strives to refine coarse result by learning the local details from the shape memory features and partial point cloud. Learning complete shape from partial input is an ill-posed problem. To alleviate this problem, we devise a shape memory block with learnable memory features. The memory block stores the general shape information and improve completion performance through the cross attention strategy with coarse shape features. Furthermore, drawn by the success of recurrent learning strategy in image super resolution [22], we take the point refinement stage as a gradual optimization process. The output of the refinement network is treated as a better coarse result, and inputted to the refinement network recurrently, which further improves the completion performance. Our key contributions are as follows:

- We propose a novel point transformer network (FPTNet) for point cloud completion, which exploits the local details as well as the long term relationships from incomplete point clouds via residual point transformer blocks.
- We design a shape memory block which stores general shape features in the refinement network. The complete point cloud is inferred from both the incomplete cloud and learnable memory features.
- Our recurrent learning strategy gradually refines the output of refinement network, which further improves the proposed method's performance.
- Extensive experiments demonstrates that FPTNet outperforms previous state-of-the-art methods on PCN dataset and Completion3D benchmark.

2 Approach

We firstly define the partial point cloud X of size $N \times 3$ as the input of completion net, and the complete point cloud Y of size $N_{gt} \times 3$ as the ground truth (GT). The overall network architecture of our FPTNet is shown in Fig, 1, which predicts

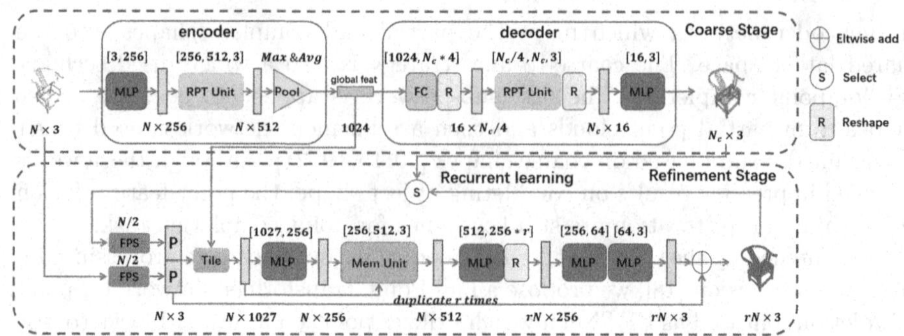

Fig. 1. The overall architecture of **Full Point Transformer Network for Point Cloud Completion**. The network consists of the coarse sub-network **Point transformer encoder-decoder network** (Top) and the **refinement network with shape memory blocks** (bottom). The **recurrent learning strategy** is applied in refinement stage.

the final shape Y_f conditioned on partial cloud X in a coarse-to-fine strategy. The point transformer encoder-decoder network predicts a coarse cloud Y_c of size $N_c \times 3$ firstly, and the following refinement net generates a high-quality completion Y_f via the shape memory layers and the recurrent learning strategy. The size of Y_f is $rN \times 3$ where $r = N_{gt}/N$.

2.1 Point Transformer Encoder-Decoder Network

The fruitful results in coarse stage is beneficial for generating fine-grained shapes in refinement stage. Previous networks [10,16] tend to use PCN-like network to predict coarse point cloud in coarse stage. PCN [23] learns global features with stacked MLPs and max pooling which processes every point independently and not exploits the relationship among points.

Inspired by previous transformer works in point processing [6,7], a Residual Point Transformer (RPT) block is proposed to adaptively aggregate point features via the self-attention mechanism, shown in Fig. 3 (a). The self-attention mechanism is ideally-suited for point feature learning due to ability of exploiting not only the local details but also the long-term relationships among points. Using the same terminology as in [14], let Q, K, V be the query, key and value features respectively, which are generated by MLP operation as follows:

$$\begin{aligned} V &= MLP(Feat_{in}, in\_c, in\_c) \\ K &= MLP(Feat_{in}, in\_c, in\_c/4) \\ Q &= MLP(Feat_{in}, in\_c, in\_c/4) \end{aligned} \tag{1}$$

where $V, Feat_{in} \in \mathbb{R}^{N \times in\_c}, Q, K \in \mathbb{R}^{N \times in\_c/4}$.

The Q, K features is used to calculate the attention map, and then a softmax is applied to generate normalized weights. Then, the self-attention features is calculated via matrix multiplication:

$$A = Softmax(Q \otimes K^T, dim = 0)$$
$$Feat_{sa} = A \otimes V \tag{2}$$

As point features in $Feat_{sa}$ are the weighted sums of V features cross all points. The self-attention mechanism can exploit both local and long-term relationships among points.

The RPT block takes $Feat_{sa}$ as residual features and the output of RPT is generated as follows:

$$Feat_{RPT} = MLP(Feat_{in} + Feat_{sa}, in\_c, ou\_c) \tag{3}$$

The overall architecture of point transformer encoder-decoder network is show in the top of Fig. 1. In the encoder sub-network, a RPT Unit with three blocks is applied to learn point features. The RPT Uint stacks several RPT blocks to extract multi-scale features, and these features are aggregated via concatenation operation. Then, MLPs are used to reduce dimension of aggregated features and generate final output, as shown in Fig. 2. Finally, we use both maximum and average pooling to pool RPT Unit features and concatenate them to generate global shape features.

The target of the decoder sub-network is to produce a coarse but complete shape Y_c. The decoder network firstly extends dimensions of global features through full connection layer and splits features to 16 groups via the Reshape layer. Next, the grouped features is further refined through another RPT Unit. Finally, per-point features is generated though the reshape operation and fed to MLPs to generate coarse cloud Y_c of size $N_c \times 3$.

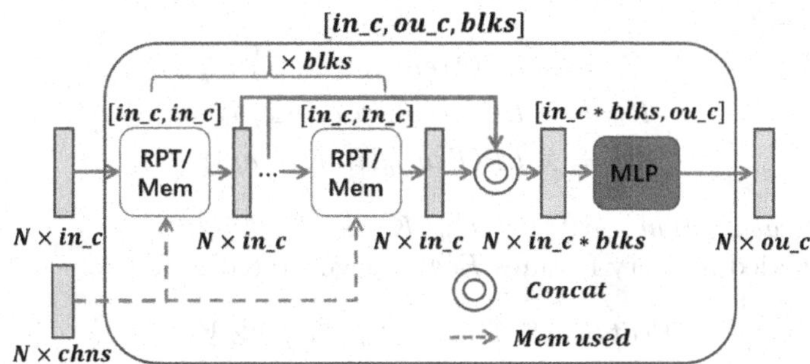

Fig. 2. The multi-scale aggregation strategy of the PRT and Mem units. The difference between two units is that the Mem unit take two inputs and RPT take only one input.

2.2 Refinement Network

The refinement network aims to enhancing local geometric details based on the coarse cloud Y_c generated at the coarse stage. As shown in the bottom of Fig. 1, we first downsample both Y_c and X to the same size $N/2 \times 3$ via farthest points sampling (FPS) algorithm and merge them to synthesize a novel point cloud of

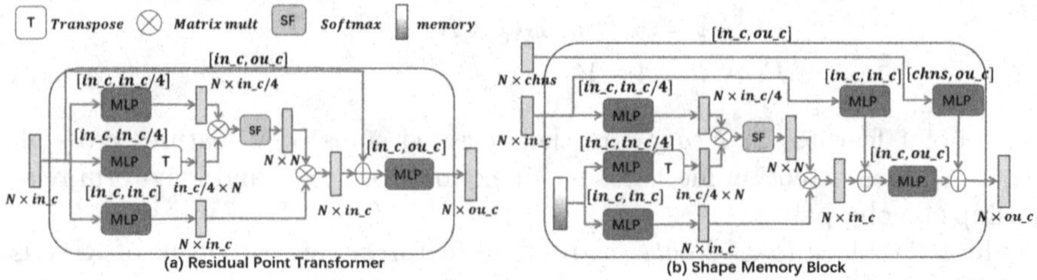

Fig. 3. The detailed structure of residual point transformer block and shape memory block

size $N \times 3$. Next, we tile global features with each point of synthetic point cloud. As a result, we get the fused feature IN as the input of the refinement network with size $N \times (3 + 1024)$.

Inferring complete shapes from partial inputs is an ill-posed problem as input only contains partial information of the target. Nevertheless, we hypothesize that there exists a general pattern set for 3D objects and the partial features can search useful patterns to complete unseen parts. Therefore, we propose the shape memory block with learnable memory features with size $N \times c$. The memory features store general patterns and provide good priors for the target which is learnt during training. As show in Fig. 3(b), the cross-attention strategy is applied to learn relationships between input features and memory features. Value and key features are created by memory features. Then, the query features generated through input features is used to query useful patterns in memory features as:

$$\begin{aligned} V_m &= MLP(Mem, in\_c, in\_c) \\ K_m &= MLP(Mem, in\_c, in\_c/4) \\ Q_m &= MLP(Feat_{in}, in\_c, in\_c/4) \end{aligned} \tag{4}$$

where $V_m, Feat_{in}, Mem \in \mathbb{R}^{N \times in\_c}, Q_m, K_m \in \mathbb{R}^{N \times in\_c/4}$.

Then, queried memory features $Feat_{ca}$ is generated as follows:

$$Feat_{ca} = Softmax(Q_m \otimes K_m^T) \otimes V_m \tag{5}$$

The output of the memory block is computed as follows:

$$Feat_{MEM} = MLP(MLP(Feat_{in}) + Feat_{ca}) + MLP(IN) \tag{6}$$

In Mem Unit, memory blocks are stacked similarly as RPT Unit, shown in Fig. 2.

The overall refinement net is shown in the bottom of Fig. 1. The objective of refinement net is to produce $rN \times 3$ offset of synthetic point cloud. The Mem Unit is used to extract point features, and point shuffle strategy [12] is applied to split one point features to r times point features. Then, these features is fed to MLPs to generate rN point offsets. Finally, we duplicate input point cloud r times and add offset to generate final results.

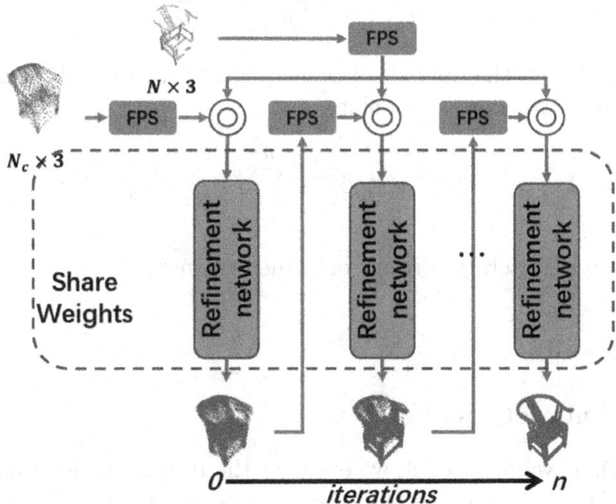

Fig. 4. The recurrent learning strategy for refinement network. Each iteration take a synthetic point cloud as input, which is sampled from partial cloud and the output of previous iteration. In first iteration, the merged cloud is created from the partial cloud and the coarse stage result.

2.3 Recurrent Learning Strategy

As described in previous subsections, the refinement network takes coarse shape generated by the coarse network and the partial shape as input. It is reasonable that the better coarse shape input will make completion task easier and help refinement network to generate fine-grained shape with more details. Motivate by the success of recurrent learning strategy in image super resolution [22], we propose the point cloud recurrent learning strategy as shown in Fig. 4. The shape memory refinement net is reused with shared weights cross iterations. In every iteration, it takes the synthetic point clouds based on the partial clouds and outputs of previous iteration as inputs. The output of the refinement network contains more local details compared to coarse cloud and can be used to synthesize more fruitful inputs for the next iteration.

2.4 Loss Function

Our FPTNet is trained in an end-to-end manner, with Chamfer Distance (CD) as the reconstruction loss. The definitions of CD distance and CD-P loss between two point clouds P, Q are as follows:

$$L_{P,Q} = \frac{1}{N_P} \sum_{p \in P} \min_{q \in Q} \|p - q\|^2$$

$$L_{Q,P} = \frac{1}{N_Q} \sum_{q \in Q} \min_{p \in P} \|q - p\|^2$$

$$L_{CD-P}(P,Q) = (\sqrt{L_{P,Q}} + \sqrt{L_{Q,P}})/2$$

where N_P and N_Q are the number of points in P, Q respectively. Consequently, the joint loss function can be formulated as:

$$L_{total} = L_{CD-P}(Y_c, Y) + \sum_{i=1}^{n} L_{CD-P}(Y_{f,i}, Y)$$

where $Y_{f,i}$ is the i-th iteration result of refinement net.

3 Experiments

3.1 Evaluation Metrics

Following previous works [10,20], we evaluate the reconstruction quality by calculating the CD-T between the last iteration shape $Y_{f,n}$ and the ground truth shape Y. The CD-T distance is defined as follows:

$$L_{CD-T}(P,Q) = L_{P,Q} + L_{Q,P}$$

3.2 Implementation Details

Our method is implemented using PyTorch and trained on NVIDIA V100 16G GPU using the Adam [8] optimizer with batch size 48. The initialization learning rate is set to 0.0001, and the learning rate is decayed by 0.7 every 16 epochs. The total epochs of training is set to 100. The default number of iterations for recurrent learning strategy is 2, and the resolution of Y_c is set to 2048.

For the coarse sub-network, the RPT Unit stacks 3 RPT blocks. Both input and output channels of RPT block in the encoder are set to 256, while the ones in the decoder are set to 512 and 2048, respectively, also shown in Fig. 1.

For the refinement sub-network, we firstly downsample both partial cloud and coarse cloud to the same size 1024×3 via FPS and then merge them to a new point cloud with size 2048×3. If recurrent learning strategy applied, we replace coarse complete cloud with the $i-1$-th iteration's output cloud to synthesize new point cloud for i-th iteration($i > 1$). Global features is tiled to each point of synthetic point cloud and input size of refinement net is 2048×1027. The input channel, output channel and block number of Mem unit are set to 256, 512 and 3, respectively.

3.3 Evaluation on PCN Dataset

PCN dataset [23] covers 30974 CAD models from 8 categories, which is created from the ShapeNet dataset [1]. The same PCN dataset partition is adopted as pervious work [23]. The resolution of incomplete point cloud is 2048 and the resolution of ground truth is 16384.

Table 1. Quantitative comparison on all 8 categories on PCN dataset (16384 points). Our FPTNet outperforms all methods in CD distance error ($\times 10^4$). Lower is better.

Methods	Airplane	Cabinet	Car	Chair	Lamp	Sofa	Table	Watercraft	Average
PCN [23]	1.55	4.01	2.55	4.29	5.71	4.80	3.44	4.08	3.82
TopNet [13]	1.50	4.71	2.95	5.25	4.78	4.71	4.35	3.46	4.11
MSN [9]	1.28	4.30	2.52	3.73	4.33	4.77	2.99	2.96	3.36
CRN [15]	1.23	3.79	2.38	3.79	4.14	4.29	2.93	2.72	3.18
VRCNet [10]	1.15	3.68	2.39	3.37	3.46	4.94	2.47	2.60	3.03
Ours(iter=1)	1.06	**3.54**	2.16	3.22	3.41	3.90	2.61	2.46	2.80
Ours(iter=2)	**0.94**	3.69	**2.12**	**3.09**	**3.25**	**3.86**	**2.40**	**2.30**	**2.71**

We retrain several state-of-the-art point cloud completion methods such as PCN [23],TopNet [13], MSN [9], CRN [15], VRCNet [10] and compare these methods with our method. The quantitatively results are shown in Table 1. FPTNet achieves the lowest CD errors in all eight categories. Compared with VRCNet [10], FPTnet improves the performance of averaged CD error with a margin of 10%, which demonstrates the superior performance of our method.

We also trained FPTNet without the recurrent learning strategy, which set number of iterations to 1 (iter=1). As show in Table 1, without recurrent learning strategy, FPTNet also gets competitive result compare to other methods. When the recurrent learning strategy applied (iter=2), our method further reduces the average CD error by 0.09. The visualization results are shown in Fig. 5, our method can not only get lower reconstruction errors but also recover fine-grained geometric details of the targets. Take the fifth row for example, our FPTNet can precisely recover the missing wire of the lamp, while PCN, CRN and VRCNet can only generate coarse lamp shape and fail to recover the fine-grained details.

3.4 Evaluation on Completion3D Benchmark

Completion3D benchmark is an online platform for evaluating shape completion methods, the resolution of both incomplete and complete point cloud is 2048. Following the benchmark instructions, we upload our best results to the platform. The quantitative comparison between our method and state-of-the-art point cloud completion approaches is shown in Table 2. The proposed FPTNet outperforms all SOTA methods and is ranked first on the Completion3D benchmark across all eight categories. Compared with the second-best method ASFM-Net [19], our method significantly improves the performance of averaged reconstruction error (chamfer distance) with a margin of 12%.

3.5 Evaluation on KITTI Dataset

We further evaluate our method on KITTI [5] dataset, which includes many real-scanned partial cars captured by a LiDAR. We directly test all models trained

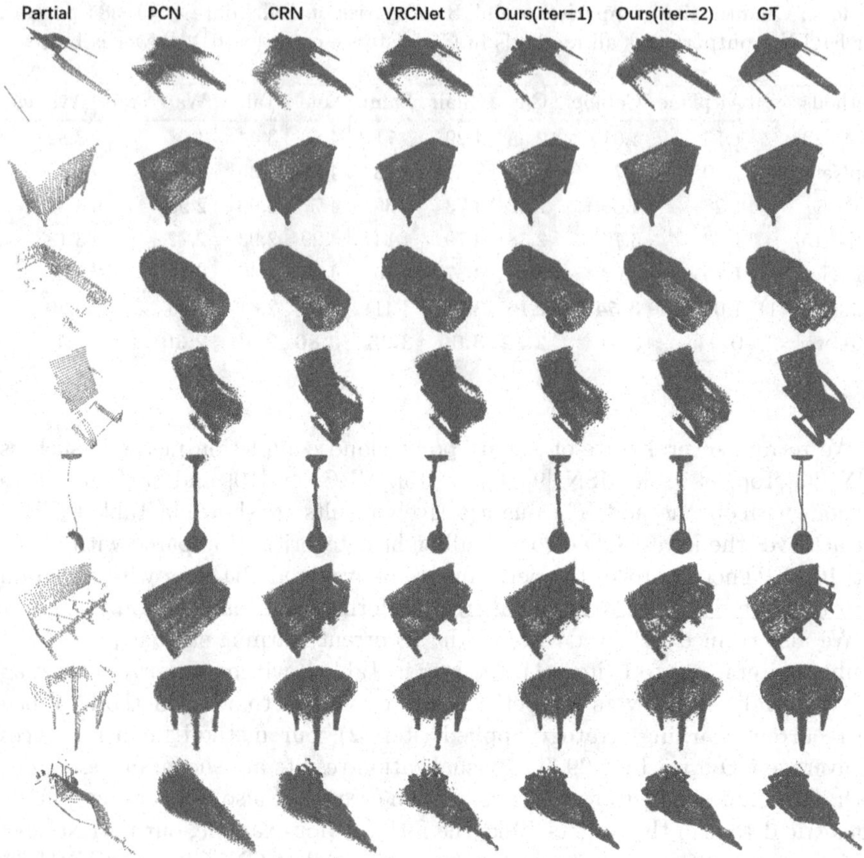

Partial PCN CRN VRCNet Ours(iter=1) Ours(iter=2) GT

Fig. 5. Qualitative comparison of different methods on the PCN dataset. Our FPTNet can generate better complete shapes with fine-grained details than other methods.

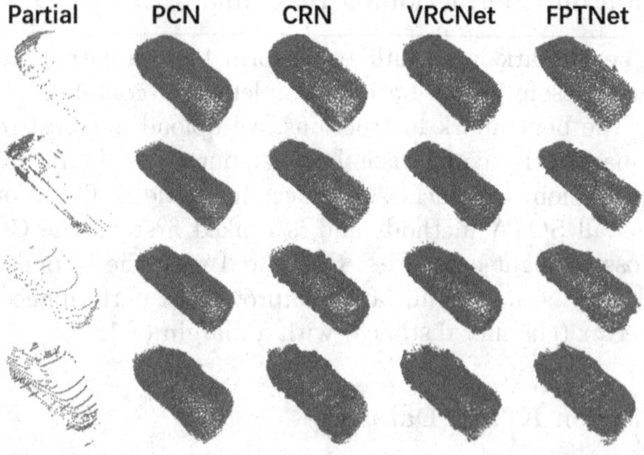

Fig. 6. Qualitative comparison on the KITTI dataset

Table 2. Quantitative comparison on the Completion3D benchmark (2048 points). Our FPTNet is ranked first on this benchmark across all eight categories.

Methods	Airplane	Cabinet	Car	Chair	Lamp	Sofa	Table	Watercraft	Average
PCN [23]	9.79	22.70	12.43	25.14	22.72	20.26	20.27	11.73	18.22
TopNet [13]	7.32	18.77	12.88	19.82	14.60	16.29	14.89	8.82	14.25
SANet [17]	5.27	14.45	7.78	13.67	13.53	14.22	11.75	8.84	11.22
GRNet [20]	6.13	16.90	8.27	12.23	10.22	14.93	10.08	5.86	10.64
CRN [15]	3.38	13.17	8.31	10.62	10.00	12.86	9.16	5.80	9.21
VRCNet [10]	3.94	10.93	6.44	9.32	8.32	11.35	8.60	5.78	8.12
ASFM-Net [19]	2.38	9.68	5.84	7.47	7.11	9.65	6.25	5.78	6.68
Ours	**2.20**	**8.91**	**5.44**	**6.53**	**6.25**	**8.14**	**5.26**	**3.99**	**5.87**

Table 3. Fidelity error (FD) comparison on the KITTI dataset

PCN [23]	CRN [15]	VRCNet [10]	FPTNet
3.011	1.951	1.946	**1.892**

on PCN [23] dataset without any fine-tuning or retraining operations as previous methods [19,23] did on shapeNet car dataset. As shown in Fig. 6, we can see that FPTNet can not only generate a general shape of car but also preserve the observed fine-grained details. The fidelity error evaluates the average distance between each point in input and the nearest point in completed shape. In Table 3, our method gets lower fidelity error compared to CRN [15] and VRCNet [10].

3.6 Resource Usages

The resource usages of PCN, CRN, VRCNet and our FPTNet on PCN dataset (16384 points) is shown in Table 4. All methods are implemented using Pytorch. For a fair comparison, we use the same batch size 32 and test these methods by using the NVIDIA Titan XP 12G GPU. As shown in Table 4, our FPT takes shorter inference time compared to VRCNet. Note that, kNN operation is applied frequently in VRCNet for aggregating neighboring point features, which incurs high data structuring cost. Our FPTNet gets superior performance in completion qualities with lower computational cost.

3.7 Ablation Study

We evaluate the effectiveness of each part of our method. All experiments are conducted on PCN dataset. We downsample the ground truth shapes from 16384 to 2048 via FPS and evaluate performance on the resolution of 2048. The CD-T and F-score metrics are adopted to evaluate completion performance.

To evaluate the effectiveness of each part in FPTNet, we design four different networks as follows. (1) The *w/o RPT blocks* network replaces self-attention

Table 4. Resource usages

Methods	Model Size (Mb)	Infer Time (ms)
PCN [23]	26	34.7
CRN [15]	20	23.8
VRCNet [10]	66	72.7
Ours(iter=2)	74	41.9

Table 5. Ablation studies of our method on the PCN dataset. Point resolutions for output and ground truth are 2048. For CD-T, lower is better. For F1-Score, higher is better

Methods	Average CD	F1-Score
w/o RPT blocks	6.02	0.4523
w/o Memory	5.83	0.4558
FPTNet(iter=1)	5.77	0.4582
FPTNet(iter=2)	**5.69**	**0.4656**
FPTNet(iter=3)	5.76	0.4654

operation in RPT block with a single MLP in coarse network. As a result, $Feat_{sa}$ calculation in Eq. 2 degrades to $Feat_{sa} = V$. (2) The *w/o Memory* network replaces cross-attention strategy with self-attention strategy in refinement network. The learnable shape memory features is discarded and V_m, K_m is calculated based on $Feat_{in}$ as same as Q_m. (3) The *FPTNet(iter=1)* network does not apply the recurrent learning strategy. (4) The *FPTNet(iter=3)* network iterates refinement network 3 times, and the latest 2 iterations of refinement network take partial cloud and previous fine-grained shape as input. It should be noted that both *w/o RPT blocks* and *w/o Memory* networks also do not apply recurrent learning strategy. The experiment results are shown in Table 5. By comparing *w/o RPT blocks* with *FPTNet(iter=1)*, we can find that the self-attention operation which exploits relationships among points in coarse shape learning achieves significant improvement. Without self-attention operation, the coarse network degrades to a PCN-like network which processes every point independently. By comparing *w/o Memory* with *FPTNet(iter=1)*, the better performance of *FPTNet(iter=1)* proves the effectiveness of proposed learnable memory features and cross-attention strategy. The comparison among *FPTNet(iter=1)*,*FPTNet(iter=2)* and *FPTNet(iter=3)* exhibits that 2 iterations is more suitable with lowest CD error and moderate computational efficiency.The recurrent learning strategy can improve the performance of our method.

4 Conclusion

We propose a novel point transformer network for point cloud completion, which outperforms state-of-the-art methods on PCN dataset and Completion3D benchmark. The experiments and discussions on these datasets prove the effectiveness of proposed residual point transformer block, shape memory block and recurrent learning strategy. These modules can be conveniently applied in other point cloud works and highly encourage researchers to use these modules on point cloud completion and other 3D perceptual tasks.

References

1. Chang, A.X., et al.: Shapenet: an information-rich 3D model repository. ArXiv:1512.03012 (2015)
2. Dai, W., Yan, X., Wang, J., Xie, D., Pu, S.: Mdr-mfi:multi-branch decoupled regression and multi-scale feature interaction for partial-to-partial cloud registration. In: IEEE International Conference on Acoustics, Speech and Signal Processing (ICASSP) (2023)
3. Du, H., Yan, X., Wang, J., Xie, D., Pu, S.: Point cloud upsampling via cascaded refinement network. In: Asian Conference on Computer Vision (2022)
4. Du, H., Yan, X., Wang, J., Xie, D., Pu, S.: Rethinking the approximation error in 3D surface fitting for point cloud normal estimation. In: Proceedings of the IEEE/CVF Conference on Computer Vision and Pattern Recognition (CVPR), pp. 9486–9495 (June 2023)
5. Geiger, A., Lenz, P., Urtasun, R.: Are we ready for autonomous driving? the kitti vision benchmark suite. In: CVPR, pp. 3354–3361 (2012)
6. Guo, M.H., Cai, J., Liu, Z.N., Mu, T.J., Martin, R.R., Hu, S.: PCT: point cloud transformer. Comput. Vis. Media **7**, 187–199 (2021)
7. Han, X.F., Kuang, Y., Xiao, G.Q.: Point cloud learning with transformer. ArXiv **2104**, 13636 (2021)
8. Kingma, D.P., Ba, J.: Adam: A method for stochastic optimization. In: NeurIPS (2016)
9. Liu, M., Sheng, L., Yang, S., Shao, J., Hu, S.: Morphing and sampling network for dense point cloud completion. In: AAAI (2020)
10. Pan, L., et al.: Variational relational point completion network. In: CVPR, pp. 8520–8529 (2021)
11. Qi, C., Su, H., Mo, K., Guibas, L.J.: Pointnet: deep learning on point sets for 3D classification and segmentation. In: VPR, pp. 77–85 (2017)
12. Qian, G., Abualshour, A., Li, G., Thabet, A.K., Ghanem, B.: PU-GCN: point cloud upsampling using graph convolutional networks. In: 2021 IEEE/CVF Conference on Computer Vision and Pattern Recognition (CVPR), pp. 11678–11687 (2021)
13. Tchapmi, L.P., Kosaraju, V., Rezatofighi, H., Reid, I.D., Savarese, S.: Topnet: structural point cloud decoder. In: CVPR, pp. 383–392 (2019)
14. Vaswani, A., et al.: Attention is all you need. ArXiv:1706.03762 (2017)
15. Wang, X., Ang, M.H., Lee, G.H.: Cascaded refinement network for point cloud completion. In: CVPR, pp. 787–796 (2020)
16. Wang, X., Ang, M.H., Lee, G.H.: Point cloud completion by learning shape priors. In: IROS, pp. 10719–10726 (2020)
17. Wen, X., Li, T., Han, Z., Liu, Y.S.: Point cloud completion by skip-attention network with hierarchical folding. In: CVPR, pp. 1936–1945 (2020)
18. Wu, B.,et al.: Visual transformers: token-based image representation and processing for computer vision. ArXiv:2006.03677 (2020)

19. Xia, Y., Xia, Y., Li, W., Song, R., Cao, K., Stilla, U.: Asfm-net: asymmetrical siamese feature matching network for point completion. In: ACM International Conference on Multimedia (2021)
20. Xie, H., Yao, H., Zhou, S., Mao, J., Zhang, S., Sun, W.: Grnet: gridding residual network for dense point cloud completion. In: ECCV (2020)
21. Yan, X., et al,: Fbnet: Feedback network for point cloud completion. In: European Conference on Computer Vision, pp. 676–693. Springer (2022). https://doi.org/10.1007/978-3-031-20086-1_39
22. Yang, X., et al.: Drfn: deep recurrent fusion network for single-image super-resolution with large factors. IEEE Trans. Multimedia **21**, 328–337 (2019)
23. Yuan, W., Khot, T., Held, D., Mertz, C., Hebert, M.: PCN: Point completion network. In: International Conference on 3D Vision (3DV), pp. 728–737 (2018)

Efficient Point-Based Single Scale 3D Object Detection from Traffic Scenes

Wenneng Tang[1], Yaochen Li[1(✉)], Yifan Li[1], and Bo Dong[2]

[1] School of Software Engineering, Xi'an Jiaotong University, Xi'an 710049, China
yaochenli@mail.xjtu.edu.cn

[2] School of Continuing Education, Xi'an Jiaotong University, Xi'an 710049, China

Abstract. In the field of 3D object detection, the point-based method faces significant limitations due to the need to process large-scale collections of irregular point clouds, resulting in reduced inference speed. To address this issue, we propose an efficient single-scale single-stage 3D object detection algorithm called SS-3DSSD. Our method eliminates the time-consuming multi-scale feature extraction module used in PointNet++ and adopts an efficient single-scale feature extraction method based on neighborhood-attention, significantly improving the model's inference speed. Additionally, we introduce a learning-based sampling method to overcome the limited receptive fields of single-scale methods and a multi-level context feature grouping module to meet varying feature requirements at different levels. On the KITTI test set, our method achieves an inference speed of 66.7 frames per second on the RTX 2080Ti, with an average precision of 81.35% for the moderate difficulty car category. This represents a better balance between inference speed and detection accuracy, offering promising implications for real-time 3D object detection applications.

Keywords: 3D Object Detection · Neighborhood Attention · Feature Extraction · Deep Learning

1 Introduction

3D object detection is becoming increasingly crucial in various fields such as autonomous driving, robot navigation and beyond. Notably, point cloud-based 3D object detection has garnered more attention from researchers as the cost of LiDAR sensors continues to decrease. However, in outdoor tasks like autonomous driving which need to use large-scale point data, it is difficult to meet the real-time requirements directly processing point clouds. Therefore, improving the efficiency of these method has become a popular research topic.

Inspired by the success of PointNet [12] and PointNet++ [13], researchers have proposed a series of point-based methods. Point-based methods directly

This work was supported by the Key Research and Development Program in Shaanxi Province of China (No. 2022GY-080).

Q. Liu et al. (Eds.): PRCV 2023, LNCS 14426, pp. 155–167, 2024.
https://doi.org/10.1007/978-981-99-8432-9_13

operate on the raw point cloud data, which helps to preserve more geometric information and often leads to better detection accuracy compared to grid-based methods. However, due to the need to handle massive irregular point clouds, point-based methods tend to have slower inference speeds and higher memory usage than grid-based methods.

The majority of existing point-based methods utilize PointNet++ as their backbone for extracting point-wise features. Set Abstraction(SA) is the main structure for extracting features in PointNet++, which contains three layers: sampling layer, grouping layer, and PointNet layer. The sampling layer is employed to downsample the input point cloud. The grouping layer is utilized to query the neighborhood points of the sampled points. The PointNet layer is responsible for aggregating the neighborhood features of the sampled points. Our experiments show that the PointNet layer is the main bottleneck for efficiency within the SA module. Specifically, when the SA module processes 4096 points, the PointNet layer consumes over 90% of the processing time. Thus, simplifying the structure of this layer can greatly improve the efficiency. In 3D object detection tasks, the SA module typically groups points in a multi-scale style. By using multiple PointNets to extract neighborhood features of various scales and then combining them, rich multi-scale context features can be obtained.

In order to study the effect of multi-scale structure, we try to replace the multi-scale structure of SA module in 3DSSD backbone with a single-scale one. The results of the experiment show that the single-scale method can significantly speed up the inference, but the accuracy is seriously degraded.

To address the slow detection speed of multi-scale methods and low detection accuracy of single-scale methods, we propose a precise and efficient single-scale method. We are the first to attempt single-scale feature extraction in 3D object detection.

The main contributions of our work are summarized as follows:

- We propose an efficient single-scale point feature encoder (S-PFE) which significantly improves the detection efficiency.
- We propose a learning-based Top-K sampling method that explicitly makes the model focus on the area around the object, and thus the small receptive field problem caused by single-scale feature extraction is mitigated.
- We propose a novel Neighborhood-Attention module that effectively aggregates neighborhood features channel-wise by leveraging the semantic and geometeric relationships among points.
- We propose a lightweight and efficient multi-level contextual feature grouping (M-CFG) module, thus satisfying the need for different features of the object detection task.

2 Related Work

In this section, we briefly introduce several works related to our method. We simply divide 3D object detection methods for point clouds into grid-based methods and point-based methods. Grid-based methods divide the point cloud into regular grids and then apply a convolutional neural network (CNN) to extract features. Point-based methods utilize networks such as PointNet to directly process raw point cloud data. As the point-based methods need to deal with the irregular point cloud data, they are memory-consuming and the inference speed is slower than the grid-based methods.

In PointPillar [7], Lang et al. divide the point cloud into 2D pillars, then use PointNet to extract the features of each pillar. And 2D CNN is applied to process the feature map. Unlike PointPillar, VoxelNet [26] divides the point cloud into 3D voxels and directly uses 3D CNN to process the voxels. In order to solve the performance problem brought by 3D convolution, SECOND [18] introduces a sparse convolution method to extract voxel features. Inspired by CenterNet, CenterPoint [21] proposes an anchor-free method to directly predict the center and size of the object.

PointNet [12,13] is one of the pioneers of point-based methods. They use the raw point cloud as input and then extract the features of the point cloud using a shared MLP and symmetric function. PointRCNN [15] first uses PointNet++ to segment the point cloud and then uses foreground points to extract features of RoI. 3DSSD [19] improves the detection speed of the model by discarding the time-consuming Feature Propogation (FP) stage. Inspired by the Hough voting method, VoteNet [11] attempts to learn the offset of the surface point to the center point. Point-based methods all use PointNet++ as the backbone network to extract multi-scale features. Unlike conventional methods, we employ a single-scale approach to greatly enhance the model's efficiency.

3 Method

The architecture of our method can be divided into four parts: the module for extracting single-scale features, the module for candidate point generation, the module for grouping multi-level contextual features and the detection head for generating detection results. The full pipeline of our method is shown in Fig. 1. The details about each part are introduced in the following.

3.1 Single-Scale Point Feature Encoder

To reduce the time consuming of multi-scale grouping and improve the model inference speed, we propose an efficient Single-scale Point Feature Encoder (S-PFE) whose principle is shown in the S-PFE subfigure in Fig. 1. And the subfigure(a) in Fig. 2 shows the difference between multi-scale grouping and single-scale grouping.

The module is mainly divided into three stages. Firstly, the input point cloud P_{input} is downsampled using the learning-based Top-K sampling method. Then the neighborhood points of the sampled points P_{sample} are aggregated using only a single-scale grouping stratergy. After that, the neighborhood point features fused with geometric and semantic information are aggregated using the

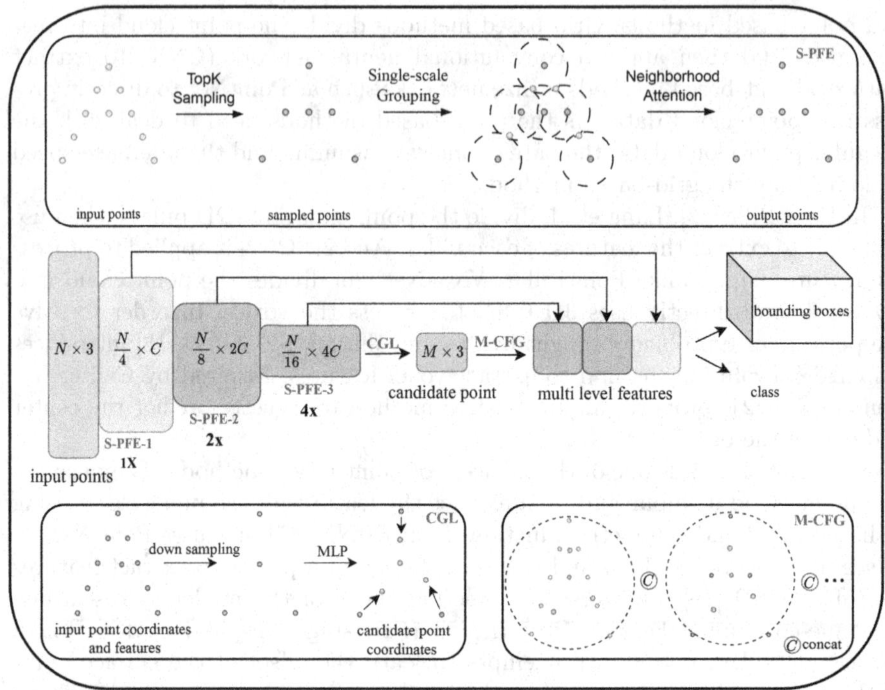

Fig. 1. The full pipeline of SS-3DSSD. The points of different colors correspond to the outputs of their respective color modules.

neighborhood attention module and the final output features are used as the features of each sampled point. Each sub-structure of S-PFE is described in detail below.

Top-K Sampling Method. Traditional downsampling methods such as FPS, F-FPS are difficult to apply to our single-scale method. The reason is that single-scale method usually has a limited receptive field, and it is difficult to extract meaningful semantic information from the background points that are far away from the object. Our experimental results in Table 2 also prove this point.

To address these problems, a learning-based sampling method called Top-K method is proposed. The purpose of this method is to make the model avoid sampling meaningless background points as much as possible, improve the learning efficiency of the model, and make the model explicitly focus on the local points around the object. The entire Top-K sampling layer consists of several shared MLPs that use the coordinates of P_{input} and the corresponding features as input to generate an score for each point. And the K points with the highest scores are selected as P_{sample}.

In Top-K sampling, how to assign positive samples is an issue worth investigating. It is a very natural idea to define points located in the ground truth

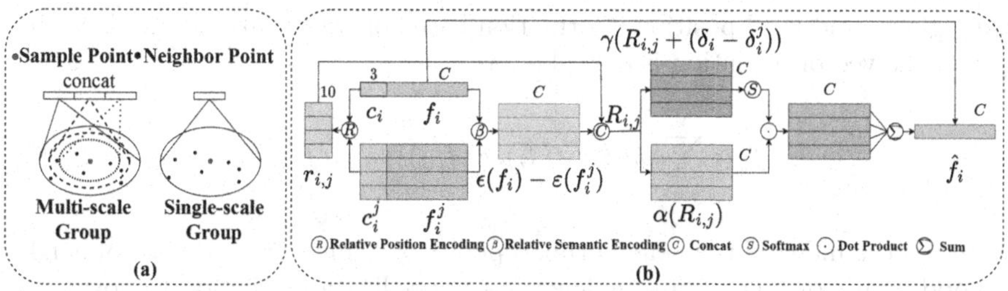

Fig. 2. (a) The difference between Multi-scale Grouping and Single-scale Grouping; (b) The principle of Neighborhood-Attention

as positive samples but this suffers from a serious class imbalance problem. For some objects that are close to the LiDAR sensor, they usually contain thousands of points, while for more distant and heavily occluded objects, they usually contain only a few points. Another approach is to select the N closest points to the object center as positive samples. Although this method alleviates the class imbalance problem, it is difficult to set a reasonable N for it. For objects with few foreground points, when N is large, most of the nearest N points are background points, which makes the training process unstable. When N is small, it is difficult to provide enough information to predict the object.

Inspired by the label assignment method in YOLOF [1], we propose a novel method that randomly selects N points located in the ground truth as positive samples. For objects with less than N number of foreground points, only the points located in the ground truth are selected as positive, and for objects with far more than N foreground points, N points in the ground truth are randomly selected to avoid the problem of class imbalance. And we use a simple binary cross-entropy loss to train the sampling layer by setting the labels of positive sample to 1 and the rest of points to 0. In the experiment section, we compare the effects of the above label assignment strategies on the model performance.

Neighborhood-Attention Aggregating. For grouping neighborhood points, we use the same settings as 3DSSD to search for points within a certain radius in P_{input} with the points in P_{sample} as the center. In the process of downsampling, the search radius is continuously expanded, and the same settings as 3DSSD are used in our experiment. The difference is that we only need to search the neighborhood points of a single scale, while 3DSSD needs to search the neighborhood points of several different scales. In order to ensure sufficient receptive field, our search radius is set to the maximum of multi scales of 3DSSD.

In terms of feature aggregation, unlike 3DSSD which uses simple max pooling to aggregate features of points in the neighborhood, we propose a more robust neighborhood feature aggregation method which is based on vector attention, namely the Neighborhood-Attention method. The traditional vector attention approach that only uses semantic features to model the relationship between neighborhood points, we further improve the performance of the method by fusing global and local position information based on vector attention. The principle of the vector attention is described as:

$$\widetilde{f_i} = \sum_{j=1}^{N_i} \rho(\gamma(\beta(\varphi(f_i), \psi(f_i^j)))) \odot \alpha(f_j), \tag{1}$$

where j is the index of the neighborhood points of point i, N_i is the set of neighborhood points of point i, and ρ denotes a normalization function to normalize the attention weights. γ is a function used to calculate attention weights which is implemented using a linear layer in our method. β denotes a relational function to calculate the relationship between features f_i and f_i^j, which is commonly done by adding, subtracting, and multiplying operations. φ, ψ and α are some linear variations of the input features, which is usually done using MLP. And $\odot$ denotes the dot product operation. Vector attention is very suitable for processing point cloud data to aggregate point cloud features efficiently, but it ignores the important relative position relationships between sampled points and neighboring points as well as global position relationships, which makes it difficult for the model to extract robust point cloud features efficiently.

To address this problem, we propose a novel method called neighborhood attention, which is based on the vector attention and incorporates the relative position relationship and the absolute global position relationship between the two types of points. The subfigure(b) in Fig. 2 depicts its underlying principle.

In terms of relative position, we use the spatial coordinates between the two types of points to calculate it, which is shown as:

$$r_{i,j} = C(D_{i,j}, c_i - c_i^j, c_i, c_i^j), \tag{2}$$

where $D_{i,j}$ denotes the Euclidean distance between point i and point j. c_i, c_i^j denote the 3D coordinates of point i and point j in the point cloud space, respectively. In terms of absolute position, inspired by the position embedding in self-attention, we use the spatial coordinates of points as the position embedding for representing the global spatial information of points. Also, in order to avoid problems such as gradient vanishing with the increasing of network depth, we introduce a short connection similar in ResNet [4]. Thus, the principle of neighborhood attention can be described as:

$$R_{i,j} = C((\epsilon(f_i) - \varepsilon(f_i^j)), r_{i,j}), \tag{3}$$

$$\hat{f_i} = (\sum_{j=1}^{N_i} \rho(\gamma(R_{i,j} + (\delta_i - \delta_i^j))) \odot (\alpha(R_{i,j}))) + f_i, \tag{4}$$

where $R_{i,j}$ denotes the relative relationship between points i and points j which contains two main aspects. The first is the relative relationship $\epsilon(f_i) - \varepsilon(f_i^j)$ of semantic features and the second is the relative position relationship $r_{i,j}$ of geometric features. The fusion of the two types of features is achieved by concatenation. In addition, the global position relations δ_i and δ_i^j generated by sinusoidal positional encoding are introduced during the calculation of attention weights to provide more comprehensive and fine spatial information.

Besides, for more accurate prediction as well as better aggregation of features, we introduce a Candidate point Generation Layer (CGL) to explicitly predict the offset of the sampled point to the corresponding object center which is same as 3DSSD. The principle of this module is shown in the CGL subfigure in Fig 1.

3.2 Multi-level Contextual Feature Grouping

Unlike the 3DSSD method that aggregates the output of only the last SA layer, we propose a lightweight Multi-level Contextual Feature Grouping (M-CFG) module to aggregate the output of different S-PFE layers to meet the requirements for different features of the object detection task. With the help of efficient S-PFE module, the detection speed of the model is not greatly affected. The principle of this module is shown in the M-CFG subfigure in Fig. 1.

The module takes the candidate points and the output of each S-PFE layer as input. By grouping and concatenating the output features of different S-PFEs with the candidate points as the center, it constitutes a multi-level feature. Then, the multi-level features are fed into the detection head which consists of two MLPs. One MLP predicts the class and the other predicts the location and object size. In the experiment section, we explore the effect of grouping different levels of features on performance.

3.3 Loss Function

The loss function of our method is composed of four parts: the classification loss L_{sample} required for the sampling task, the regression loss L_{cgl} required for the candidate point generation task, and the classification loss L_{cls} and the localization loss L_{box} of the object generation task. For L_{sample} in the sampling task, we use a simple cross-entropy loss. And we set different weights for different layers in the S-PFE. The L_{cgl} uses the Smooth-L1 loss which is same as 3DSSD. For classification and localization loss of objects, we use the same settings as 3DSSD [19]. Therefore, the overall loss function can be written as:

$$L = L_{sample} + L_{cgl} + L_{cls} + L_{box}. \tag{5}$$

4 Experiment

We verify the effectiveness of the proposed method on the KITTI dataset. Compared with other SOTA methods, our method achieve higher detection accuracy as well as faster detection speed. In addition, we conduct extensive ablation experiments to analyze the effectiveness of each of the model's design choices.

4.1 Dataset and Implementation Details

The KITTI dataset is widely used in 3D object detection. It contains 7481 training samples and 7518 testing samples. The training set only contains annotations of the 3D bounding boxes in the front view, and no labels are provided for the test set. All our experiments are based on the OpenPCDet toolbox[1]. We use the commonly used one-cycle learning rate strategy, the maximum learning rate is 0.005, the batch size is 16, and two RTX2080Ti are used for training. During training, GT-paste and other common data augmentation tricks such as rotation and scaling are applied. All results are calculated by recall 40 positions.

4.2 Results on KITTI Dataset

We submit the test results to the KITTI test server to obtain the test set accuracy. From the results in Table 1, we can draw the following conclusions.

In terms of efficiency, the inference speed of different methods is tested on a single 2080Ti with batch size set to 1. The inference speed of our method surpasses all existing methods, being able to process 66.7 frames per second and far exceeding the second place PointPillar's 42.4 frames. The accuracy improves by more than 5%, which is attributed to the efficient S-PFE module we designed.

In terms of accuracy, it can be seen that our method achieves the highest accuracy for point-based methods and, at the same time, exceeds most of the grid-based methods. In addition, our method achieves the highest detection accuracy among all one-stage methods, far exceeding the baseline network 3DSSD and significantly improving the inference speed. Compared to the two-stage grid-based method PV-RCNN, our method only differs from it by 0.08% at moderate difficulty, but our inference speed is improved by more than 50 FPS. From the experimental results, we can see that our method achieves a favorable balance of detection accuracy and inference speed.

In order to further explore the efficiency of our method, we compare the memory usage of different methods when a single 2080Ti is fully loaded. For a fair comparison, we also list the input scale of each method. As can be seen from Table 1, our method only costs 180 MB of GPU memory, and can parallel infer 56 frames on a single 2080Ti. The inference speed also exceeds all other methods. This shows that the efficiency of the model is greatly improved by extracting single-scale features.

In addition, we also show the points sampled by our Top-K sampling method and compare our detection results with those of 3DSSD. As can be seen from the Fig. 3, our sampling method covers the object area better, and the detection accuracy is also significantly better than that of the 3DSSD method.

4.3 Ablation Experiment

The validation set is obtained from the training set by dividing it in a 1:1 ratio. This section focuses on ablation experiments on the validation set of KITTI to verify the effectiveness of the proposed method. All results are for the Car class.

[1] https://github.com/openmmlab/OpenPCDet.

Table 1. Results on the KITTI test set. † is the best result of one-stage methods. * is the best result of point-based methods.

Method	Type	Car(IoU=0.7) R40			FPS	Memory	Batch	Input
		Easy	Mod.	Hard				
RGB+LiDAR								
3D-CVF [22]	2-stage	89.20	80.05	73.11	13.3			
Fast-CLOCs [10]	2-stage	89.10	80.35	**76.99**	<10			
CAT-Det [23]	2-stage	89.87	81.32	76.78	3.18			
Voxel Based								
SECOND [18]	1-stage	83.34	72.55	65.82	30.0	377	27	11-17K
PointPillar [7]	1-stage	82.58	74.31	68.99	42.4	354	28	2-9K
TANet [9]	2-stage	88.81	75.38	67.66	29.0	3000	3	<12K
Part-A^2 [16]	2-stage	87.81	78.49	73.51	12.5	702	13	11-17K
SA-SSD [3]	1-stage	88.75	79.79	74.16	25.9			
CIA-SSD [25]	1-stage	89.59†	80.28	72.87	32.5			
SVGA-Net [5]	1-stage	87.33	80.47	75.91	16.1			
Point-Voxel Based								
F-PointRCNN [2]	2-stage	84.28	75.73	67.39	15.4			
STD [20]	2-stage	87.95	79.71	75.09	12.5			
PV-RCNN [14]	2-stage	**90.25**	**81.43**	76.82	12.5	1223	8	11-17K
VIC-Net [6]	1-stage	88.25	80.61	75.83	16.9			
Point Based								
PointRCNN [15]	2-stage	86.96	75.64	70.70	10.0	560	18	11-17K
Point-GNN [17]	1-stage	88.33	79.47	72.29	1.6			
3DSSD [19]	1-stage	88.36	79.57	74.55	26.3	502	19	16384
3D IoU-Net [8]	1-stage	87.96	79.03	72.78	10.0			
PC-RGNN [24]	2-stage	89.13 *	79.90	75.54				
Ours	1-stage	87.98	81.35† *	76.43†*	**66.7**† *	**180**	**56**	16384

In this section, we study the difference in performance between the F-FPS method in 3DSSD and our Top-K method. For the Top-K method, the effects of different label assignment methods on the model performance are also compared. As can be seen from Table 2, the Top-K method significantly improves the detection accuracy of the model. In addition, for different label assignment strategies, the performance of the GT-In assignment strategy is lower than the other two. The reason is that it brings a serious class imbalance problem, which makes it difficult for the model to sample difficult samples. And for the hard object, GT-K-In method is better than the KNN since it does not assign the background points near these objects as positive samples.

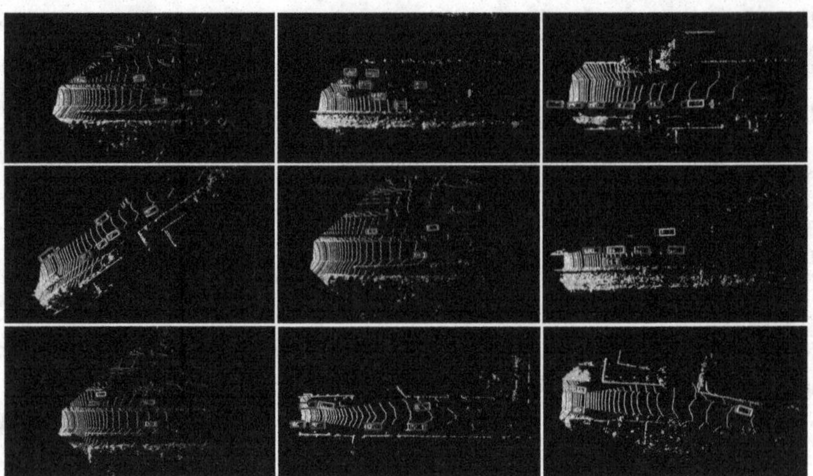

Fig. 3. Green Points are sampled by Top-K. Red boxes are the GT boxes, green .boxes are results of our method, and blue boxes are results of 3DSSD. (Color figure online)

Table 2. Effect of sampling method and label assignment method on performance

FPS&F-FPS	Top-K			Easy	Mod.	Hard
	GT-In	KNN	GT-K-In			
√				82.38	73.67	68.70
	√			91.31	83.46	81.36
		√		91.73	84.52	80.57
			√	**91.84**	**85.20**	**82.55**

Table 3. Effect of location information in Neighborhood-Attention. M.P. denotes the Max Pooling method. V.A. denotes the traditional vector attention method. G.Info. denotes the introduction of global position embedding, and L.Info. denotes the introduction of local relative position information.

M. P	V.A.	G.Info.	L.Info.	Easy	Mod.	Hard
√				88.37	82.61	78.56
	√			87.92	81.27	76.89
	√	√		90.37	83.76	81.91
	√		√	91.69	84.87	81.64
	√	√	√	**91.84**	**85.20**	**82.55**

The impact of different information in the Neighborhood-Attention is also studied and the traditional Max Pooling method is compared, too. As shown in Table 3, the detection accuracy of the vector attention method (V.A.) alone is lower than that of the maximum pooling method (M.P.) because the V.A.

Table 4. The effect of the M-CFG layer number. 0 layer means directly use features of candidate points. 1 layer means to group features only using the output of the last S-PFE layer and 2 means the last two layers, and so on.

M-CFG Layer Number	Easy	Mod.	Hard	FPS
0	91.92	82.66	80.06	**70.6**
1	91.77	84.64	82.36	68.2
2	91.80	85.12	82.47	67.5
3	**91.84**	**85.20**	**82.55**	66.7

lacks important spatial position information in modeling the neighborhood point features. After introducing global position information (G.Info.) and local relative information (L.Info.) respectively, the performance is improved in both cases.

We also explore how the M-CFG layer number influence the performance. From the experimental results of Table 4, we find that with the increase of the M-CFG layer number, the performance of the model has been improved.

5 Conclusion

We propose a single-scale single-stage 3D object detection method, which is fast and efficient. To address the issue of subpar detection performance caused by single-scale feature extraction, we design a learning-based sampling strategy and a robust Neighborhood-Attention feature aggregation module. And a multi-level contextual feature grouping module is followed to efficiently aggregate features at different level. With the support of these effective modules, our approach achieves a better balance of accuracy and speed. Our method provides a practical and reliable solution for the application of point-based 3D object detection methods to real-world autonomous driving scenarios. In addition, our proposed neighborhood attention module with fused position information is also inspiring for other 3D object detection algorithms.

References

1. Chen, Q., Wang, Y., Yang, T., Zhang, X., Cheng, J., Sun, J.: You only look one-level feature. In: 2021 IEEE/CVF Conference on Computer Vision and Pattern Recognition (CVPR), pp. 13034–13043 (2021)
2. Chen, Y., Liu, S., Shen, X., Jia, J.: Fast point r-cnn. In: Proceedings of the IEEE/CVF International Conference on Computer Vision, pp. 9775–9784 (2019)
3. He, C., Zeng, H., Huang, J., Hua, X.S., Zhang, L.: Structure aware single-stage 3D object detection from point cloud. In: Proceedings of the IEEE Conference on Computer Vision and Pattern Recognition (2020)
4. He, K., Zhang, X., Ren, S., Sun, J.: Deep residual learning for image recognition. In: Proceedings of the IEEE Conference on Computer Vision and Pattern Recognition, pp. 770–778 (2016)

5. He, Q., Wang, Z., Zeng, H., Zeng, Y., Liu, Y.: Svga-net: sparse voxel-graph attention network for 3D object detection from point clouds. In: Proceedings of the AAAI Conference on Artificial Intelligence. vol. 36, pp. 870–878 (2022)
6. Jiang, T., Song, N., Liu, H., Yin, R., Gong, Y., Yao, J.: Vic-net: voxelization information compensation network for point cloud 3d object detection. In: 2021 IEEE International Conference on Robotics and Automation (ICRA), pp. 13408–13414 (2021). https://doi.org/10.1109/ICRA48506.2021.9561597
7. Lang, A.H., Vora, S., Caesar, H., Zhou, L., Yang, J., Beijbom, O.: Pointpillars: fast encoders for object detection from point clouds. In: Proceedings of the IEEE/CVF Conference on Computer Vision and Pattern Recognition, pp. 12697–12705 (2019)
8. Li, J., et al.: 3D iou-net: iou guided 3d object detector for point clouds. arXiv preprint arXiv:2004.04962 (2020)
9. Liu, Z., Zhao, X., Huang, T., Hu, R., Zhou, Y., Bai, X.: Tanet: robust 3D object detection from point clouds with triple attention. In: Proceedings of the AAAI Conference on Artificial Intelligence. vol. 34, pp. 11677–11684 (2020)
10. Pang, S., Morris, D., Radha, H.: Fast-clocs: fast camera-lidar object candidates fusion for 3d object detection. In: Proceedings of the IEEE/CVF Winter Conference on Applications of Computer Vision (WACV), pp. 187–196 (January 2022)
11. Qi, C.R., Litany, O., He, K., Guibas, L.J.: Deep hough voting for 3d object detection in point clouds. In: proceedings of the IEEE/CVF International Conference on Computer Vision, pp. 9277–9286 (2019)
12. Qi, C.R., Su, H., Mo, K., Guibas, L.J.: Pointnet: deep learning on point sets for 3d classification and segmentation. In: Proceedings of the IEEE conference on computer vision and pattern recognition, pp. 652–660 (2017)
13. Qi, C.R., Yi, L., Su, H., Guibas, L.J.: Pointnet++: deep hierarchical feature learning on point sets in a metric space. arXiv preprint arXiv:1706.02413 (2017)
14. Shi, S., et al.: Pv-rcnn: point-voxel feature set abstraction for 3D object detection. In: Proceedings of the IEEE/CVF Conference on Computer Vision and Pattern Recognition, pp. 10529–10538 (2020)
15. Shi, S., Wang, X., Li, H.: Pointrcnn: 3D object proposal generation and detection from point cloud. In: Proceedings of the IEEE/CVF Conference on Computer Vision and Pattern Recognition, pp. 770–779 (2019)
16. Shi, S., Wang, Z., Shi, J., Wang, X., Li, H.: From points to parts: 3d object detection from point cloud with part-aware and part-aggregation network. IEEE Trans. Pattern Anal. Mach. Intell. **43**(8), 2647–2664 (2020)
17. Shi, W., Rajkumar, R.R.: Point-gnn: graph neural network for 3d object detection in a point cloud. In: The IEEE Conference on Computer Vision and Pattern Recognition (CVPR) (June 2020)
18. Yan, Y., Mao, Y., Li, B.: Second: sparsely embedded convolutional detection. Sensors **18**(10), 3337 (2018)
19. Yang, Z., Sun, Y., Liu, S., Jia, J.: 3dssd: point-based 3d single stage object detector. In: Proceedings of the IEEE/CVF Conference on Computer Vision and Pattern Recognition, pp. 11040–11048 (2020)
20. Yang, Z., Sun, Y., Liu, S., Shen, X., Jia, J.: Std: sparse-to-dense 3D object detector for point cloud. In: Proceedings of the IEEE/CVF International Conference on Computer Vision, pp. 1951–1960 (2019)
21. Yin, T., Zhou, X., Krähenbühl, P.: Center-based 3D object detection and tracking. arXiv preprint arXiv:2006.11275 (2020)
22. Yoo, J.H., Kim, Y., Kim, J., Choi, J.W.: 3D-CVF: generating joint camera and lidar features using cross-view spatial feature fusion for 3D object detection. In: European Conference on Computer Vision, pp. 720–736. Springer (2020)

23. Zhang, Y., Chen, J., Huang, D.: Cat-det: contrastively augmented transformer for multimodal 3d object detection. In: 2022 IEEE/CVF Conference on Computer Vision and Pattern Recognition (CVPR), pp. 898–907 (2022)
24. Zhang, Y., Huang, D., Wang, Y.: Pc-rgnn: point cloud completion and graph neural network for 3D object detection. ArXiv:2012.10412 (2020)
25. Zheng, W., Tang, W., Chen, S., Jiang, L., Fu, C.W.: Cia-ssd: confident iou-aware single-stage object detector from point cloud. In: Proceedings of the AAAI Conference on Artificial Intelligence.,vol. 35, pp. 3555–3562 (2021)
26. Zhou, Y., Tuzel, O.: Voxelnet: end-to-end learning for point cloud based 3D object detection. In: Proceedings of the IEEE Conference on Computer Vision and Pattern Recognition, pp. 4490–4499 (2018)

Matching-to-Detecting: Establishing Dense and Reliable Correspondences Between Images

Haobo Xu[1,2], Jun Zhou[1,2], Renjie Pan[1,2], Hua Yang[1,2(✉)], Cunyan Li[1,2], and Xiangyu Zhao[1,2]

[1] Institute of Image Communication and Network Engineering, Shanghai Jiao Tong University, Shanghai 200240, China
hyang@sjtu.edu.cn

[2] Shanghai Key Lab of Digital Media Processing and Transmission, Shanghai 200240, China

Abstract. We present a novel view for local image feature matching, which is inspired by the difference between existing methods. Detector-based methods detect predefined keypoints in local regions, so that the stability and reliability of established matches are ensured. In contrast, detector-free methods usually directly match dense features and refine the filtered results, which can help generate more matches. In order to combine their advantages, we propose a novel Matching-to-Detecting (M2D) process for feature matching, in which we first perform global reasoning for patch-level matching and subsequently identify discriminative matches within local areas to obtain pixel-level matches. At the patch-level, dense matching provides our pipeline with the ability to find plenty of matches even in low-texture areas, while at the pixel-level, our method can be viewed as detecting from a matching perspective, so that the established matches have higher stability and reliability and are remarkable in local regions. Experimental results demonstrate that our method outperforms state-of-the-art methods by a significant margin in terms of matching accuracy and the number of matches. Moreover, the computational complexity of our model is quite low, making it more suitable for real-world applications.

Keywords: Local feature matching · Detector-based matching · Detector-free matching

1 Introduction

Finding correspondences between images is a fundamental task in computer vision, playing a significant role in various applications such as 3D reconstruction [9,32], visual localization [28], pose estimation [11], and Structure-from-Motion (SfM) [24]. Despite significant research efforts invested in feature matching pipelines, existing methods still face numerous challenges due to the complexity of real-world scenes, including variations in viewpoint, illumination, low-texture, and motion blur.

Q. Liu et al. (Eds.): PRCV 2023, LNCS 14426, pp. 168–179, 2024.
https://doi.org/10.1007/978-981-99-8432-9_14

Detector-based methods [2,6,17,18,22,23] have been proven effective in feature matching by matching a set of keypoints with corresponding descriptors, in which repeatable and reliable keypoints ensure the stability and accuracy of matches. However, pre-filtering the matching space can exclude many potential matches, leading to poor performance in low-texture areas where keypoints are scarce [27]. To address this concern, detector-free methods [4,12,27,31,33] directly match dense features and learn matches from matching scores, so they can generate matches even in low texture areas. However, the saliency of matches in local areas is less considered in these methods, so that more stable and higher-accuracy matches may be ignored, leading to a waste of matching space. Besides, detector-free methods also face the problem of high computational complexity, which is brought about by the matching of dense features and the utilization of context information.

To tackle these concerns, we propose a novel detector-free matching method called Matching-to-Detecting (M2D). Inspired by the difference between detector-based and detector-free methods, our matching process consists of two stages: At the patch-level, M2D directly matches between dense patch-level features, so that the number of matches can be easily ensured. At the pixel-level, M2D detects from a matching perspective by searching for pixel-level matches that are discriminative between local regions. Overall, M2D can be seen as turning over the detecting-to-matching process of detector-based methods. By finding more matches first and then detecting, this pipeline guarantees both the number and the accuracy of matches. Our **contributions** can be summarized as follows:

a) We present a novel view for feature matching. By patch-level reasoning and pixel-level detectively matching, our method can combine the advantages of both detector-based and detector-free methods.
b) Through carefully designed feature extraction network and matching process, our model has much lower computational complexity compared to current detector-free methods. As an illustration, our model's runtime is merely 36.5% and 31.0% of that demanded by LoFTR [27] and ASpanFormer [4], respectively.
c) Experimental results show that our method outperforms the state-of-art methods a lot in terms of matching number and accuracy, and leads to better performance on the homography estimation task. Specifically, our approach achieves a Mean Matching Accuracy (MMA) of 66.5% for the 1-pixel threshold on the HPatches dataset [1], outperforming ASpanFormer [4] by 11.6%.

2 Related Work

2.1 Detector-Based Matching

Detector-based methods have played a significant role in feature matching tasks over decades, some hand-crafted methods such as SIFT [17], SURF [2], ORB [22] are still widely used today. Among them, some detect predefined structures in the image, such as SuperPoint [5] and some hand-crafted methods [2,17,22].

Some methods make assumptions about the positions of keypoints. For example, D2Net assumes that the extrema in descriptor space make good features [6], while R2D2 [18] defines the repeatability and reliability of the keypoints. However, these methods filter out many potential matches, and their assumptions may not be consistent with the matching conditions.

For detector-based methods, after keypoints are detected and described, image features can be matched using simple methods like nearest neighbor (NN) search. However, simple matching methods are often vulnerable to matching noise. To address this issue, researchers have proposed context-based matching methods, such as SuperGlue [23] and some of its variants [3,25]. These methods use additional contextual information to improve the accuracy and robustness of feature matching, making them more effective in real-world scenarios.

2.2 Detector-Free Matching

Detector-free methods are a type of feature matching approach that does not rely on feature detectors. This idea can be traced back to the SIFT Flow [15]. To get better results, detector-free methods usually learn in an end-to-end manner and establish a differentiable matching layer. For example, NCNet [20] uses 4D convolutions to build and filter matches from the correlation tensor of dense feature maps, which is also used in [12,33]. Different kinds of methods are also applied to obtain more precise matches [4,19,27,31,33].

However, since the detector-free matching directly matches between dense features, it usually leads to a much larger search space and higher computation complexity. To address this issue, advanced methods often match between low-resolution feature maps and refine the results. For example, DRC-Net [12] extracts feature maps with different resolutions and uses coarse confidences to weight the fine confidences. Patch2Pix [33] constructs a network that uses stepwise regression refinement and jointly rejects outlier matches with confidence scores. LoFTR [27] uses a transformer architecture [10] for global inference and employs a correlation-based approach to obtain fine matches, which has also been used in other transformer-based methods [4,31].

3 Method

For image feature matching, we need more accurate matches with lower computational cost. Different approaches aim to achieve this through the matching network and method. An overview of our proposed network and matching process is shown in Fig. 1.

3.1 Feature Extraction Network

For the subsequent matching process, our network needs to extract patch-level features at 1/8 of the original image dimension and pixel-level features at full

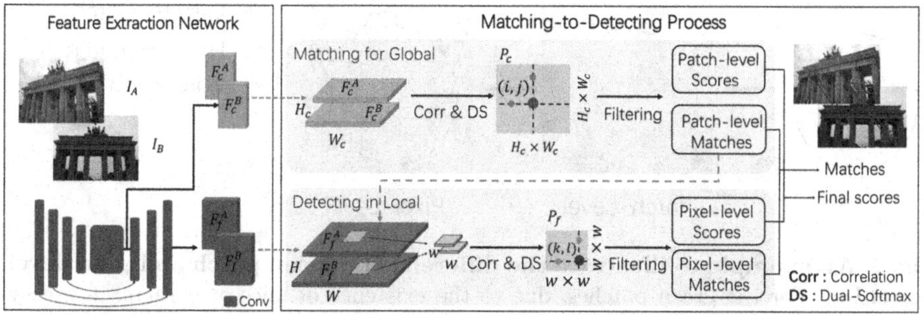

Fig. 1. Overview of the proposed method. The pipeline consists of two parts, in which the network (left part) extracts features for the matching process (right part): **1.** Feature Extraction Network. A VGG-style [26] encoder extracts patch-level features F_c^A and F_c^B from image pairs I^A and I^B. Pixel-level (fine) features F_f^A and F_f^B are obtained by adding coarse and underlying features through upsampling and convolutions (Sect. 3.1). **2.** Matching-to-Detecting (M2D) process. In this part, patch-level features F_c^A and F_c^B are first correlated, and a dual-softmax operator is applied to obtain the coarse confidence matrix $\mathcal{P}_c$. A matching threshold is used to filter patch-level matches. For each corresponding patch pair, pixel-level features on F_f^A and F_f^B are correlated, and the dual-softmax operator is applied to obtain the local confidence matrix $\mathcal{P}_f$. Pixel-level matches with high discriminability between local windows are selected. Matching scores are calculated as the product of the patch-level and pixel-level scores (Sect. 3.2).

resolution. Each patch-level point corresponds to an 8×8 window in the pixel-level feature map. We use a VGG-style [26] encoder as a backbone to extract coarse features, similar to SuperPoint [5]. The feature map size is reduced to 1/8 of the original image size through convolutional operations and downsampling. To expand the receptive field, we attach 6 residual blocks [8]. Fine features are obtained by fusing the coarse features with the underlying features in the CNNs through interpolation and convolution layers, akin to [14,21].

3.2 Matching-to-Detecting Process

The two stages of matching, patch-level and pixel-level matching, have different focuses: patch-level matching allows for global inference and enables the matching of relatively flat regions, while pixel-level matching relies on distinguishing adjacent pixels. Low confidence and large errors may arise at the pixel-level if the pixels cannot be effectively distinguished, as illustrated in Fig. 2.

Matching Globally for Patches. The existence of dense matching allows for large-scale reasoning from a global perspective at the patch-level. As a detector-free method, we directly match dense feature maps without selection. To establish a differentiable matching pipeline, we first calculate the correlation matrix between the coarse features F_c^A and F_c^B:

$$\mathcal{S}_c(i,j) = \langle F_c^A(i), F_c^B(j) \rangle, \tag{1}$$

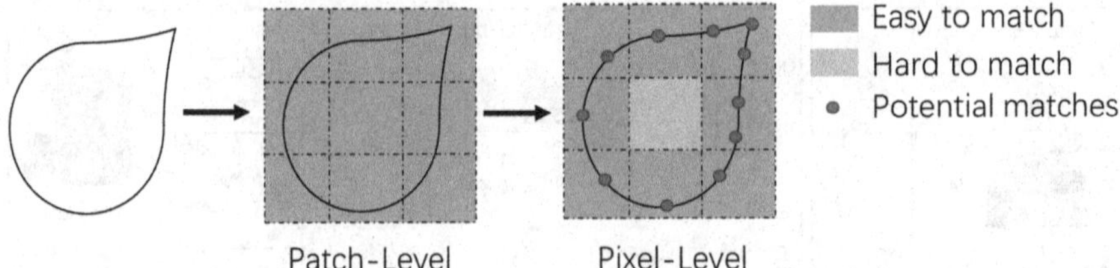

Fig. 2. An example to illustrate the difference between patch and pixel level matching. For the green patches, due to the existence of lines or corners, it is easy to match at both patch and pixel level. For the orange patch, since it is relatively flat, although it can be matched through context information, it's hard to establish accurate pixel-level matches. (Color figure online)

where $\langle\cdot\rangle$ denotes the correlation process. Inspired by [27,29], we employ a single dual-softmax operator to obtain the confidence of the matches:

$$\mathcal{P}_c(i,\ j) = \text{softmax}\left(\mathcal{S}_c(i,\cdot)\right)_j \cdot \text{softmax}\left(\mathcal{S}_c(\cdot,j)\right)_i. \tag{2}$$

The confidence score in the matrix can be considered as the overall confidence level of patch-level matching. To mitigate the computational complexity in the subsequent steps, a preselection of a set of patch-level matches is performed:

$$\mathcal{M}_c = \{(i,j)|\forall(i,j) \in \mathcal{P}_c, \mathcal{P}_c(i,j) \geq \theta_c\}, \tag{3}$$

where θ_c denotes the patch-level threshold. It is worth noting that the requirement for matches to be mutual nearest neighbors is not imposed due to the presence of scaling and viewpoint changes between images. Consequently, a single patch in the query image may correspond to multiple patches in the queried image, and this also holds true for the queried image.

Detecting Locally for Pixels. Detector-based methods typically locate good pixel-level matches in local regions by detecting predefined or pretrained keypoints. However, this approach remains indirect, and the definition of keypoints may not always match the actual situation. We found that points detected by the detector are usually salient in local regions and exhibit considerable uniqueness, which is a key factor in ensuring their accuracy. This inspired us to approach the problem from a matching point of view and detect significant matches in local areas to ensure confidence in the matching process. In our method, each coarse match corresponds to a pair of patches on the fine feature maps, which can be viewed as a small image pair. We directly correlate them to obtain $\mathcal{S}_f$ and then apply the dual-softmax operator to the resulting confidence matrix:

$$\mathcal{P}_f(i,\ j,m) = \text{softmax}\left(\mathcal{S}_f(i,\cdot,m)\right)_j \cdot \text{softmax}\left(\mathcal{S}_f(\cdot,j,m)\right)_i, \tag{4}$$

in which m denotes the $m'th$ patch pair. Due to the softmax operator, the confidence will be very low when matches cannot be effectively distinguished from there surrounding pixels, even if the similarity between the pixel-level descriptors is high. High confidence means the corresponding matches have high uniqueness, which often occurs in corners and line endpoints. Therefore, our method can adaptively detect locally salient matching points without pre-definition. Specifically, for the $m'th$ patch pair, the matches are:

$$\mathcal{M}_f^m = \{(k,l,m)|\forall(k,l,m) \in \mathcal{P}_f, \mathcal{P}_f(i,j,m) \geq \theta_f\}, \tag{5}$$

where θ_f denotes the pixel-level threshold. To reduce the computational burden, we often select the top-k matches within the patch-to-patch matching space and subsequently filter them using a predefined threshold. Given that the matching score reflects the quality of pixel-level matching, it can be considered local confidence.

Final Matching Proposals and Confidences. Assuming the patch size is $r \times r$, let $(i_c, j_c) \rightarrow (i'_c, j'_c)$ represents a patch-level match with confidence p_c, and let $(i_f, j_f) \rightarrow (i'_f, j'_f)$ denotes the related pixel-level match with confidence p_f. The final matching proposal $(i,j) \rightarrow (i',j')$ and its corresponding confidence p can be denoted as:

$$\begin{aligned}(i,j,i',j') &= (i_c r + i_f, j_c r + j_f, i'_c r + i'_f, j'_c r + j'_f),\\ p &= p_c \cdot p_f.\end{aligned} \tag{6}$$

The final confidence level can be viewed as a composite measure of both local and global confidence, thereby ensuring both the overall correctness and the local discrimination of the matches.

Number of Matches and Searching Space Analysis. For image matching problems, the number of extracted matches and the size of the search are factors worth analyzing. An image pair of dimensions $M \times N$ leads to a large matching space of M^2N^2, and can potentially result in up to MN pixel-to-pixel matches. As to our method, at the patch-level, the search space is $(M/r)^2(N/r)^2 = M^2N^2/r^4$, where r denotes the patch size. Furthermore, at the pixel-level, if a patch is associated with at most k patches, the largest search space becomes $k(M/r)(N/r)r^4 = kMNr^2$. As a result, the total search space is reduced to $M^2N^2/r^4 + kMNr^2$. Fortunately, since matching is neighbourhood consistent, our method can extract multiple matches between a single patch pair. Ideally, all patch-level matches and their corresponding pixel-level matches can be successfully identified, allowing for matching of all pixels between the images. For most methods, the matches that can be extracted are much smaller than this scale [4,23,27,31,33].

3.3 Supervision

To train our matching pipeline, we need a set of matching correspondences between images, which can be obtained by processing real-world datasets [13] or artificially distorting images [23]. For given pixel-to-pixel ground-truth correspondences, they are decomposed into patch-level and pixel-level (within the windows) correspondences. During training, we employ negative log-likelihood loss to supervise both patch-level and pixel-level matching:

$$\mathcal{L}_c = -\frac{1}{\left|\mathcal{M}_c^{gt}\right|} \sum_{(i,j)\in\mathcal{M}_c^{gt}} \log P_c(i,j),$$
$$\mathcal{L}_f = -\frac{1}{\left|\mathcal{M}_f^{gt}\right|} \sum_{(k,l,m)\in\mathcal{M}_f^{gt}} \log P_f(k,l,m), \tag{7}$$

where $\mathcal{M}_c^{gt}$ and $\mathcal{M}_f^{gt}$ denote patch-level and pixel-level ground-truth correspondences, respectively. Total loss is the sum of the two:

$$\mathcal{L} = \mathcal{L}_c + \mathcal{L}_f. \tag{8}$$

Note that our patch-level and pixel-level matching share a consistent matching process and training loss. Since patch-level and pixel-level features share the same underlying network architecture, their training can promote each other.

3.4 Implementation Details

Training Data. We train our model on the large-scale outdoor dataset MegaDepth [13]. To generate pixel-to-pixel ground-truth matches, we process it in two ways: (a) Following [6], for each image pair, we cut the image so that they have a size of 840 × 840, and then project all points of the second image with depth information into the first image. A depth-check with respect to the depth map of the first image was run to remove occluded pixels. Finally, we get about 100k image pairs with pixel level ground-truths. (b) Following [27], we resize the image to make its longer size 840, and padding the image to adjust the size to 840 × 840. The process of obtaining ground-truths is similar to (a) and 60k image pairs are generated. We do not perform any additional data augmentation.

Training Parameters. For feature extraction, the dimension of patch-level and pixel-level features are set to 256 and 64, respectively. Due to the large training dataset, we randomly select 20,000 image pairs for training in each round. During training, batch size is set to 4, initial learning rate is set to 10^{-4} with a decay rate of 90% for each round. The AdamW [16] is used to optimize the model with weight decay set to 10^{-4}. The model converges after 1 day of training on 2 V100 GPUs.

4 Experiments

4.1 Image Matching

We first test our method on widely used HPatches dataset [1]. It contains 116 sequences, in which 57 sequences have larger illumination changes and others have larger viewpoint changes. We adopt the setup proposed in D2Net [6]. For the image pairs, we extracted matches for each of them and used the provided ground-truth $\mathcal{H}$ to calculate the mean matching accuracy (MMA) of our results. Our model used a patch-level threshold θ_c of 0.15, with 7 matches selected between each patch pair and filtered by a pixel-level threshold θ_f of 0.1.

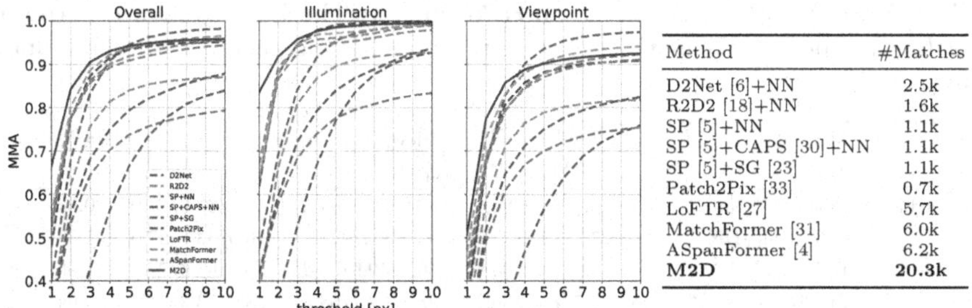

Method	#Matches
D2Net [6]+NN	2.5k
R2D2 [18]+NN	1.6k
SP [5]+NN	1.1k
SP [5]+CAPS [30]+NN	1.1k
SP [5]+SG [23]	1.1k
Patch2Pix [33]	0.7k
LoFTR [27]	5.7k
MatchFormer [31]	6.0k
ASpanFormer [4]	6.2k
M2D	**20.3k**

Fig. 3. Image Matching on HPatches [1]. The MMA for each method is presented as a function of the matching threshold (in pixels). We denote our methods with solid lines and other with dashed lines to make the results more clear.

Results. Based on the results presented in Fig. 3, it can be concluded that our proposed model M2D, outperforms state-of-the-art methods by a significant margin in low error thresholds. It improves the 1-pixel accuracy of ASpanFormer [4] from 54.9% to 66.5%, and performs well under both viewpoint and illumination changes, particularly at low error thresholds. Furthermore, our approach establishes significantly more matches than existing methods, partly because of the dense matching and the selection of multiple matches between patch pairs.

4.2 Homography Estimation

Estimating the geometric relationships between two images is one of the important downstream tasks of image feature matching. We evaluate our model also on HPatches [1] dataset for homography estimation. Inspired by [5], we adjust the image size to 640×480 to facilitate interpretation of the results (through cropping and resizing). For our method, we set the patch-level threshold θ_c to 0.15 and selected 1 pixel-level matches per patch pair without threshold. We use OpenCV with RANSAC [7] as the robust estimator to estimate the homography matrix $\hat{\mathcal{H}}$, and computed the corner error by warping corners with both $\hat{\mathcal{H}}$ and ground-truth $\mathcal{H}$. The RANSAC parameters were set to default. Consistent with previous work [23,27], we report the area under the cumulative curve (AUC) of the corner errors, with threshold values set to 3, 5, and 10 pixels.

Results. From Table 1, we can clearly see that our method achieves the best in overall evaluation, and we are the second best under viewpoint evaluation. It is interesting since our model doesn't use complex attention layers. It can be attributed to our model's ability for global interference and find high-quality matches between local regions, which makes it more stable under various changes. Under viewpoint changes, our model is a little weaker than ASpanFormer [4], but is comparable to the results of SuperGlue [23], LoFTR [27], which confirms the effectiveness of our matching framework.

Table 1. Homography estimation on HPatches [1]. The AUC of the corner error in percentage is reported. We mark the best results in **bold**, and underline the suboptimal results.

Category	Method	Illumination	Viewpoint	Overall	#Matches
		AUC (%, @3, 5, 10px)			
Detector-based	SP [5] + NN	60.4/73.9/86.0	37.0/52.5/68.6	48.5/62.9/77.0	0.5k
	D2Net [6]+NN	27.8/49.8/73.2	6.5/18.4/42.5	16.9/33.7/57.0	1.3k
	R2D2 [18]+NN	64.9/78.0/88.6	36.8/49.9/66.2	50.5/63.6/77.2	1.6k
	SP [5]+CAPS [30]+NN	59.7/74.2/86.7	38.2/52.7/68.5	48.7/63.2/77.4	0.6k
	SP [5]+SG [23]	61.6/75.7/87.7	47.2/60.7/<u>75.1</u>	54.2/68.0/81.2	0.5k
Detector-free	Patch2Pix [33]	72.8/82.1/90.0	35.4/47.7/63.4	53.7/64.5/76.4	0.4k
	LoFTR [27]	79.7/87.3/93.4	49.1/61.6/74.6	64.0/74.1/83.8	2.5k
	MatchFormer [31]	<u>82.6</u>/<u>88.9</u>/<u>94.2</u>	45.7/58.8/72.5	63.8/73.5/83.1	2.8k
	ASpanFormer [4]	79.7/87.2/93.5	**50.8/63.5/76.7**	<u>64.9</u>/<u>75.1</u>/<u>84.8</u>	2.7k
	M2D	**86.5/91.4/95.6**	<u>49.5</u>/<u>62.1</u>/<u>75.1</u>	**67.6/76.4/85.1**	2.9k

4.3 Understanding M2D

Runtime and Parameter Number Analysis. Besides performing well on various tasks, our method also shows superiority in efficiency. We report the number of parameters (M), and runtime (ms, on 1 V100 GPU) in Table 2. The image size is cut and resized to 640 × 480. We don't report GFLOPs because the computational cost of feature matching is related to not only input image size but also image contents. M2D performs the best in all metrics and it is even faster than the detector-based method SP [5]+SG [23] and can achieve better performance. As a detector-free method, M2D also greatly reduces runtime compared to LoFTR [27], MatchFormer [31], and ASpanFormer [4], which also match dense features, indicating a more efficient architecture. Note that, for fair comparison, we use the large-LA version of the MatchFormer [31] model, which is also used in image matching and homography estimation task.

Discussion of M2D Settings. M2D extracts and filters multiple matches within a patch, ensuring both the quality and quantity of matches. In this section, we investigate the impact of related two key parameters, namely the "Detecting" threshold θ_f and the number n of samples per patch pair. We employ the HPatches dataset, with images cropped and scaled to a resolution of 640 × 480, and report mean matching accuracy (MMA), homography estimation AUC, and the number of matches. Results in Table 3 show that increasing θ_f improves MMA when patch sampling matches are constant, and increasing matches per patch leads to more matches without significantly affecting MMA when θ_f is held constant. Our findings demonstrate the effectiveness of pixel-level match filtering and the ability of our pipeline to generate many matches with suitable parameters.

Table 2. Runtime and parameter number analysis. Image size is 640 × 480.

Method	Params (M)	Runtime (ms)
SP [5]+SG [23]	13	41
LoFTR [27]	11	96
MatchFormer [31]	19	185
ASpanFormer [4]	15	113
M2D	**9**	**35**

Table 3. Discussion of M2D settings. Image size is 640 × 480.

n	θ_f	MMA @1px	Hom. est AUC @3px	#Matches
1	0.0	68.6	**67.6**	2.95k
1	0.1	76.7	67.2	1.81k
1	0.2	**77.2**	64.8	1.17k
3	0.1	76.9	66.6	4.82k
5	0.1	77.0	66.5	7.38k
7	0.1	77.1	66.2	**9.64k**

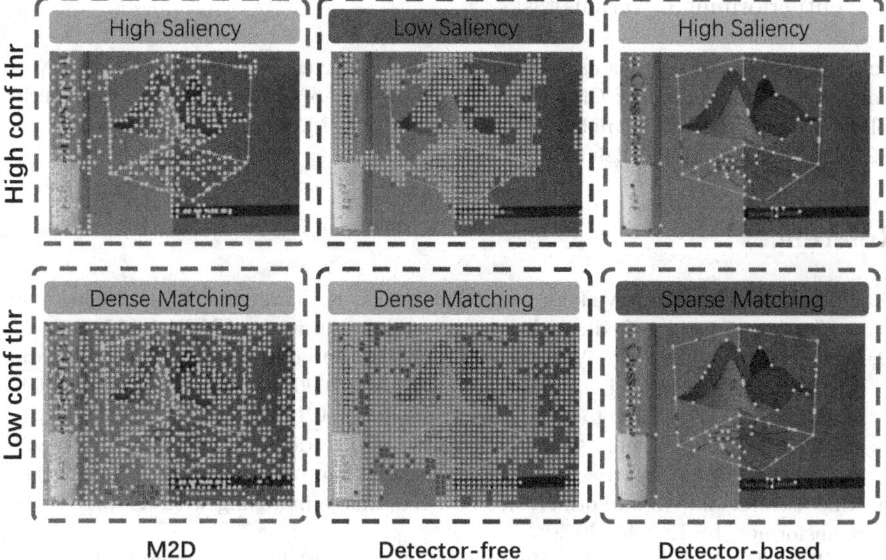

Fig. 4. Matching Distribution Comparison: The distribution of matches between M2D (our pipeline), ASpanFormer [4] (detector-free), and SuperPoint [5]+SuperGlue [23] (detector-based) are compared. (Color figure online)

Visual Analysis. To better illustrate the effectiveness of the M2D matching process, we selected an image pair from the HPatches dataset for matching and isolated a portion of the matching area. The baselines employed in our evaluation are the detector-free method ASpanFormer [4] and the detector-based method SP [5] + SG [23]. The query image is plotted, and matches with ≤1 pixel error are denoted in green, while others are marked in red. As demonstrated in the bottom figures obtained under low confidence thresholds, our approach leverages dense matching to produce numerous accurate matches, even in flat regions. On the other hand, at high thresholds (top figures), our matching performance is similar to that of detector-based methods, exhibiting high saliency in local regions (Fig. 4).

5 Conclusion

We propose a novel image feature matching method, called Matching-to-Detecting (M2D), which can establish dense and accurate correspondences between images. The process, in which first establishes patch-level matches from a global perspective and then finds discriminative pixel-level matches, ensures both the accuracy and the number of matches. Our method can be seen as turning over the detecting-to-matching process of detector-based methods and combining the advantages of both detector-based and detector-free methods. Through carefully designed feature extraction network and matching process, the overall pipeline can extract more accurate matches with lower computational cost, outperforming in various tasks.

Acknowledgements. This research was partly supported by grants of National Natural Science Foundation of China (NSFC, Grant No. 62171281), Science and Technology Commission of Shanghai Municipality (STCSM, Grant Nos. 20DZ1200203, 2021SHZDZX0102, 22DZ2229005).

References

1. Balntas, V., Lenc, K., Vedaldi, A., Mikolajczyk, K.: HPatches: a benchmark and evaluation of handcrafted and learned local descriptors. In: CVPR (2017)
2. Bay, H., Tuytelaars, T., Van Gool, L.: SURF: speeded up robust features. In: Leonardis, A., Bischof, H., Pinz, A. (eds.) ECCV 2006. LNCS, vol. 3951, pp. 404–417. Springer, Heidelberg (2006). https://doi.org/10.1007/11744023_32
3. Chen, H., et al.: Learning to match features with seeded graph matching network. In: Proceedings of International Conference on Computer Vision (ICCV) (2021)
4. Chen, H., et al.: ASpanFormer: detector-free image matching with adaptive span transformer. In: ECCV (2022)
5. DeTone, D., Malisiewicz, T., Rabinovich, A.: SuperPoint: self-supervised interest point detection and description. In: CVPR (2018)
6. Dusmanu, M., et al.: D2-Net: a trainable CNN for joint description and detection of local features. In: CVPR (2019)
7. Fischler, M.A., Bolles, R.C.: Random sample consensus: a paradigm for model fitting with applications to image analysis and automated cartography. Commun. ACM **24**, 381–395 (1981)
8. He, K., Zhang, X., Ren, S., Sun, J.: Deep residual learning for image recognition. In: CVPR (2016)
9. Heinly, J., Schonberger, J.L., Dunn, E., Frahm, J.M.: Reconstructing the world* in six days *(as captured by the yahoo 100 million image dataset). In: CVPR (2015)
10. Katharopoulos, A., Vyas, A., Pappas, N., Fleuret, F.: Transformers are RNNs: fast autoregressive transformers with linear attention. In: ICML (2020)
11. Kundu, J.N., Rahul, M.V., Ganeshan, A., Babu, R.V.: Object pose estimation from monocular image using multi-view keypoint correspondence. In: Leal-Taixé, L., Roth, S. (eds.) ECCV 2018. LNCS, vol. 11131, pp. 298–313. Springer, Cham (2019). https://doi.org/10.1007/978-3-030-11015-4_23
12. Li, X., Han, K., Li, S., Prisacariu, V.: Dual-resolution correspondence networks. In: NeurIPS (2020)

13. Li, Z., Snavely, N.: MegaDepth: learning single-view depth prediction from internet photos. In: CVPR (2018)
14. Lin, T.Y., Dollár, P., Girshick, R., He, K., Hariharan, B., Belongie, S.: Feature pyramid networks for object detection. In: CVPR (2017)
15. Liu, C., Yuen, J., Torralba, A.: SIFT flow: dense correspondence across scenes and its applications. IEEE Trans. Pattern Anal. Mach. Intell. **33**(5), 978–994 (2010)
16. Loshchilov, I., Hutter, F.: Decoupled weight decay regularization. arXiv preprint arXiv:1711.05101 (2017)
17. Lowe, D.G.: Distinctive image features from scale-invariant keypoints. IJCV **60**, 91–110 (2004)
18. Revaud, J., et al.: R2D2: repeatable and reliable detector and descriptor. arXiv preprint arXiv:1906.06195 (2019)
19. Rocco, I., Arandjelović, R., Sivic, J.: Efficient neighbourhood consensus networks via submanifold sparse convolutions. In: Vedaldi, A., Bischof, H., Brox, T., Frahm, J.-M. (eds.) ECCV 2020. LNCS, vol. 12354, pp. 605–621. Springer, Cham (2020). https://doi.org/10.1007/978-3-030-58545-7_35
20. Rocco, I., Cimpoi, M., Arandjelović, R., Torii, A., Pajdla, T., Sivic, J.: Neighbourhood consensus networks. In: NeurIPS (2018)
21. Ronneberger, O., Fischer, P., Brox, T.: U-Net: convolutional networks for biomedical image segmentation. In: Navab, N., Hornegger, J., Wells, W.M., Frangi, A.F. (eds.) MICCAI 2015. LNCS, vol. 9351, pp. 234–241. Springer, Cham (2015). https://doi.org/10.1007/978-3-319-24574-4_28
22. Rublee, E., Rabaud, V., Konolige, K., Bradski, G.: ORB: an efficient alternative to SIFT or SURF. In: ICCV (2011)
23. Sarlin, P.E., DeTone, D., Malisiewicz, T., Rabinovich, A.: SuperGlue: learning feature matching with graph neural networks. In: CVPR (2020)
24. Schonberger, J.L., Frahm, J.M.: Structure-from-motion revisited. In: CVPR (2016)
25. Shi, Y., Cai, J.X., Shavit, Y., Mu, T.J., Feng, W., Zhang, K.: ClusterGNN: cluster-based coarse-to-fine graph neural network for efficient feature matching. In: CVPR (2022)
26. Simonyan, K., Zisserman, A.: Very deep convolutional networks for large-scale image recognition. arXiv preprint arXiv:1409.1556 (2014)
27. Sun, J., Shen, Z., Wang, Y., Bao, H., Zhou, X.: LoFTR: detector-free local feature matching with transformers. In: CVPR (2021)
28. Svärm, L., Enqvist, O., Kahl, F., Oskarsson, M.: City-scale localization for cameras with known vertical direction. IEEE Trans. Pattern Anal. Mach. Intell. **39**(7), 1455–1461 (2016)
29. Tyszkiewicz, M., Fua, P., Trulls, E.: DISK: learning local features with policy gradient. In: NeurIPS (2020)
30. Wang, Q., Zhou, X., Hariharan, B., Snavely, N.: Learning feature descriptors using camera pose supervision. In: Vedaldi, A., Bischof, H., Brox, T., Frahm, J.-M. (eds.) ECCV 2020. LNCS, vol. 12346, pp. 757–774. Springer, Cham (2020). https://doi.org/10.1007/978-3-030-58452-8_44
31. Wang, Q., Zhang, J., Yang, K., Peng, K., Stiefelhagen, R.: MatchFormer: interleaving attention in transformers for feature matching. arXiv preprint arXiv:2203.09645 (2022)
32. Zhang, R., Zhu, S., Fang, T., Quan, L.: Distributed very large scale bundle adjustment by global camera consensus. In: ICCV (2017)
33. Zhou, Q., Sattler, T., Leal-Taixe, L.: Patch2Pix: epipolar-guided pixel-level correspondences. In: CVPR (2021)

Solving Generalized Pose Problem of Central and Non-central Cameras

Bin Li[1,2], Yang Shang[1,2(✉)], Banglei Guan[1,2(✉)], Shunkun Liang[1,2], and Qifeng Yu[1,2]

[1] College of Aerospace Science and Engineering, National University of Defense Technology, Changsha, China
{shangyang1977,guanbanglei12}@nudt.edu.cn

[2] Hunan Provincial Key Laboratory of Image Measurement and Vision Navigation, Changsha, China

Abstract. Recent applications in robotics, augmented reality, autonomous navigation, and self-driving involve cameras beyond the pinhole model, like fisheye cameras, multi-camera rigs, and other non-central cameras. We propose a unified method of solving the generalized pose problem of the central and non-central cameras. We first show the possibility of solving the generalized pose problem with the classical PnP solver initially designed for the central camera. We derive the closed formula of the translation in the generalized pose problem and transfer the classical PnP solver to the generalized absolute pose solver by computing a new coefficient matrix. In such a way, a category of classical PnP solvers can work with the non-central camera. Our generalized solvers inherit properties of the original PnP solvers, such as accuracy, robustness, numerical stability, and computational efficiency. Experiments on both synthetic data and publicly available real datasets show that our generalized solvers offer state-of-the-art performance.

Keywords: Non-central Cameras · PnP · Generalized Absolute Pose

1 Introduction

The Perspective-n-Point (PnP) algorithm addresses the problem of determining the absolute camera pose, which is a fundamental problem in photogrammetry and computer vision [1–3]. Given enough 3D world points and their corresponding 2D image points, the orientation and position of a central camera can be obtained. Over the last decades, vast amounts of PnP solvers have been proposed. In the minimal case, i.e., the P3P problem, the solvers use three 2D-3D point correspondences, build high-order polynomials of intermediate variables, and return at most four solutions [4, 5]. In the specific case with a limited number of points, i.e., the P4P, P5P problem, the methods solve the nonlinear geometric constraint equations, solutions of which have limited precision [6, 7]. Lepetit et al. [8] introduced four virtual control points to represent the 3D world points and first

Supplementary Information The online version contains supplementary material available at https://doi.org/10.1007/978-981-99-8432-9_15.

Q. Liu et al. (Eds.): PRCV 2023, LNCS 14426, pp. 180–192, 2024.
https://doi.org/10.1007/978-981-99-8432-9_15

reduced computation to linear complexity in cases with more than three points. However, its accuracy must be revised in slightly redundant and planar cases. Hesch et al. [9] proposed a direct least square (DLS) algorithm to handle all point configurations. The algorithm formulated the PnP problem into a minimization problem, in which the translation is a linear function of the rotation parametrized by Cayley representation. Zheng et al. [10] used non-unit quaternion representation to parameterize rotation and obtain state-of-art performance. Kneip et al. [11] presented a generalized PnP solver under the DLS mechanism using unit quaternion representation. Nakano et al. [12] derived a new optimality condition of DLS and solved it with Cayley parameters. Kun et al. [13] used Zheng's strategy to solve the minimization problem of DLS. Wang et al. [14] defined an intermediate object frame where the rotation axis Z is known and found the optimal solution by solving a seventh-order and a fourth-order univariate polynomial. Lourakis et al. [15] solved the optimality condition of Nakano's work with the rotation parameterized by modified Rodrigues parameters. Yu et al. [16] applied fixed pre-rotations to handle singularities of Cayley parameters. These methods yield a complicated system of multivariate polynomials, which can be solved with Gröbner basis solvers generated by automatic generators [17, 18]. Unlike the methods above, solving polynomial systems of the optimality condition of the minimization problem, Schweighofer et al. [19] have approached the PnP problem as a quadratically constrained quadratic program. Terzakis et al. [20] found a globally optimal solution with a sequential quadratic programming scheme. Zhou et al. [21] and Urban et al. [22] calculated the null space vector to retrieve the initial estimation of the pose.

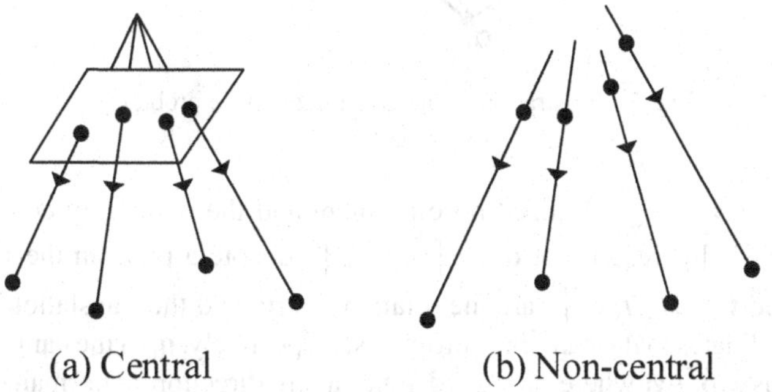

Fig. 1. Cameral model: the central (a) and non-central (b)

Recent applications have been equipped with cameras beyond the pinhole model, like fisheye cameras, multi-camera rigs, etc. [23–26]. Figure 1 depicts the central and non-central models. However, much work focuses on the central camera [4–16, 19–22], while little work handles the non-central case [11, 19]. The largest class of PnP solvers in the literature comprises methods that formulate the PnP problem into a functional minimization problem. Interestingly, such PnP solvers decouple the rotation from the translation and express the translation as a linear function of the rotation. Motivated by this, we derive generalized solvers from classical PnP solvers. Our contribution is the theoretical insight into the generalized pose problem. We show that the framework of the classical PnP solver can solve the generalized pose problem. The non-central can share the solver of the central. The above provides the following advantages:

- Our method enables a category of classical PnP solvers to work with generalized cases.
- The generalized pose solver derived by our method inherits properties of the original PnP solver and applies to both central and non-central cases.
- Our solvers outperform existing generalized solvers in experiments on synthetic data and real datasets.

2 Notation and Preliminaries

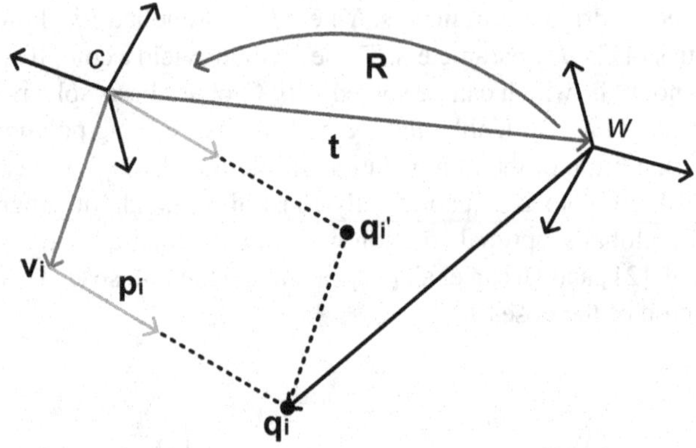

Fig. 2. Illustration of the generalized pose problem.

We first review the generalized pose problem and the optimality condition of the central case [9–13] (Fig. 2). Let $\mathbf{q_i} = \left[x_i, y_i, z_i\right]^T$ denote a point in the world frame. $\mathbf{R} \in \mathrm{SO}(3)$ and $\mathbf{t} = \left[t_x, t_y, t_z\right]^T$ are the rotation matrix and the translation vector from the world coordinates to the camera coordinates, respectively. $\mathbf{q_i}$ in the camera frame can be expressed as $\lambda_i \mathbf{p_i} + \mathbf{v_i}$ where $\mathbf{p_i} \in \mathbb{R}^3$ denotes a unit direction or normalized direction of the image ray, $\mathbf{v_i} \in \mathbb{R}^3$ is the ray origin, and λ_i is the depth factor. The generalized projection equation reads [11]:

$$\lambda_i \mathbf{p_i} + \mathbf{v_i} = \mathbf{R}\mathbf{q_i} + \mathbf{t}. \tag{1}$$

For the central camera, i.e., $\mathbf{v_i} = \mathbf{0}$, classical PnP solvers aim to retrieve $\mathbf{R}$ and $\mathbf{t}$ such that the world point falls on the image ray, i.e., $\lambda_i \mathbf{p_i} = \mathbf{R}\mathbf{q_i} + \mathbf{t}$. Due to noise, the constraint could not be perfectly satisfied. Many well-designed error functions have been proposed to measure the residuals, forming either algebraic or reprojection errors $\mathbf{e}_i$. Then, the PnP problem can be formulated into a functional minimization problem

$$\min_{\mathbf{R},\mathbf{t}} C(\mathbf{e}_1, ..., \mathbf{e}_i, ..., \mathbf{e}_n), \tag{2}$$

where the cost function C is the sum of the squared error $\mathbf{e}_i$ of each 2D-3D corresponding, i.e., $C = \sum_{i=1}^{n} \|\mathbf{e}_i\|^2$.

By zeroing the derivative of C with respect to $\mathbf{t}$, the translation can be expressed as a linear function of the rotation $\mathbf{t}(\mathbf{R})$. Substituting $\mathbf{t}(\mathbf{R})$ in Eq. (2), the minimization problem can be rewritten in terms of the rotation only

$$\begin{aligned} &\min_{\mathbf{R}} \mathbf{r}^T \mathbf{M} \mathbf{r} \\ &\text{s.t. } \mathbf{R} = \mathrm{f}(\mathbf{r}), \end{aligned} \tag{3}$$

where $\mathbf{M}$ is a symmetric coefficient matrix computed from $\mathbf{p_i}$ and $\mathbf{q_i}$, $\mathbf{r}$ is a vector composed of monomials determined by the rotation parameters, and $\mathrm{f}(\mathbf{r}) \in \mathrm{SO}(3)$ denotes the function from the chosen rotation parameters to the rotation. Thus, $\mathbf{R}$ can be obtained by solving the polynomial equations of the first-order optimality condition of Eq. (3), and $\mathbf{t}$ is then computed from the corresponding function of the rotation. In the next section, we prove that the translation is still a linear function of the rotation in the generalized pose problem, ending up in the same optimality condition as above, only with minor changes in the computation of the coefficient matrix.

3 Proposed Method

3.1 Classical PnP Solver

We want to show the possibility of transferring the classical PnP solver to the generalized pose solver besides presenting new ones. Thus, we chose the original solvers of two recognized baseline methods, OPnP [10] and DLS [9]. Since OPnP is proven highly accurate in many papers and DLS is the origin of many other solvers. We first review the classical PnP solvers OPnP and DLS. Zheng's OPnP advocates non-unit quaternion of four variables $\begin{pmatrix} a & b & c & d \end{pmatrix}$ to parametrize the rotation

$$\mathbf{R} = \frac{1}{s}\begin{bmatrix} \mathbf{r}_1^T \\ \mathbf{r}_2^T \\ \mathbf{r}_3^T \end{bmatrix} = \frac{1}{a^2+b^2+c^2+d^2}\begin{bmatrix} a^2+b^2-c^2-d^2 & 2bc-2ad & 2bd+2ac \\ 2bc+2ad & a^2-b^2+c^2-d^2 & 2cd-2ab \\ 2bd-2ac & 2cd+2ab & a^2-b^2-c^2+d^2 \end{bmatrix}, \tag{4}$$

where a, b, c, d are the four unknowns, s is the reciprocal of the average depth i.e., $s = 1 \Big/ \frac{1}{n}\sum_{i=1}^{n} \lambda_i$, and $\mathbf{r}_i^T$ is the corresponding row.

The pose estimation problem is formulated as a nonlinear optimization

$$\min_{a,b,c,d,\hat{t}_x,\hat{t}_y} \sum_{i=1}^{n}\left[\left(1+\mathbf{r}_3^T\tilde{\mathbf{q}}_i\right)u_i - \mathbf{r}_1^T\mathbf{q}_i - \hat{t}_x\right]^2 + \sum_{i=1}^{n}\left[\left(1+\mathbf{r}_3^T\tilde{\mathbf{q}}_i\right)v_i - \mathbf{r}_2^T\mathbf{q}_i - \hat{t}_y\right]^2, \tag{5}$$

where $\hat{t}_x = st_x, \hat{t}_y = st_y, [u_i, v_i]^T$ is the normalized image coordinates, $\tilde{\mathbf{q}}_i$ is the i-th world point after centralization, i.e., $\tilde{\mathbf{q}}_i = \mathbf{q}_i - \overline{\mathbf{q}}$, and $\overline{\mathbf{q}}$ is the centroid of the world points. The cost function in Eq. (5) is based on minimizing the sum of the squared algebraic errors.

The components $\hat{t}_x$ and $\hat{t}_y$ of the normalized translation vector are given by the function of the rotation

$$\begin{aligned} \hat{t}_x &= \overline{u} + \mathbf{r}_3^T \left(\frac{1}{n} \sum_{i=1}^{n} u_i \tilde{\mathbf{q}}_i \right) - \mathbf{r}_1^T \overline{\mathbf{q}} \\ \hat{t}_y &= \overline{v} + \mathbf{r}_3^T \left(\frac{1}{n} \sum_{i=1}^{n} v_i \tilde{\mathbf{q}}_i \right) - \mathbf{r}_2^T \overline{\mathbf{q}}, \end{aligned} \tag{6}$$

where $\left[\overline{u}\ \overline{v} \right]$ is the centroid of the normalized image coordinates, i.e., $\overline{u} = \sum_{i=1}^{n} u_i$, $\overline{v} = \sum_{i=1}^{n} v_i$. Eliminating $\hat{t}_x$ and $\hat{t}_y$ in Eq. (5), the minimization problem can be formulated into a matrix form

$$\min_{a,b,c,d} \mathbf{r}^T \mathbf{M} \mathbf{r} \tag{7}$$

where $\mathbf{r} = \left[1, a^2, ab, ac, ad, b^2, bc, bd, c^2, cd, d^2\right]^T$.

Moreover, Hesch's DLS employed the Cayley representation[1] of three variables $\left(b\ c\ d \right)$ to parametrize the rotation

$$\begin{aligned} \mathbf{R} &= \begin{bmatrix} r_{11} & r_{12} & r_{13} \\ r_{21} & r_{22} & r_{23} \\ r_{31} & r_{32} & r_{33} \end{bmatrix} \\ &= \frac{1}{1 + b^2 + c^2 + d^2} \begin{bmatrix} 1 + b^2 - c^2 - d^2 & 2bc - 2d & 2bd + 2c \\ 2bc + 2d & 1 - b^2 + c^2 - d^2 & 2cd - 2b \\ 2bd - 2c & 2cd + 2b & 1 - b^2 - c^2 + d^2 \end{bmatrix} \end{aligned} \tag{8}$$

The pose problem is formulated as an unconstrained least-squares minimization problem

$$\min_{b,c,d,\mathbf{t}} \sum_{i=1}^{n} \left\| \left(\mathbf{p_i} \mathbf{p_i}^T - \mathbf{I} \right) \left(\mathbf{R} \mathbf{q_i} + \mathbf{t} \right) \right\|^2 \tag{9}$$

where $\mathbf{p_i}$ is the unit direction of the image ray, and $\mathbf{I}$ is the 3×3 identity matrix. The cost function in Eq. (9) is based on minimizing the sum of the squared object-space errors.

The translation vector is computed as a function of the rotation

$$\mathbf{t} = \mathcal{H} \sum_{i=1}^{n} \left(\mathbf{p}_i \mathbf{p}_i^T - \mathbf{I} \right) \mathbf{R} \mathbf{q}_i \tag{10}$$

[1] The Cayley representation is equivalent to replacing the first element a in the non-unit quaternion by 1, i.e., $\left(1\ b\ c\ d \right)$.

where $\mathcal{H} = \left(n\mathbf{I} - \sum_{i=1}^{n} \mathbf{p}_i\mathbf{p}_i^T\right)^{-1}$. Plugging Eq. (10) into Eq. (9), the pose problem can be rewritten as

$$\min_{b,c,d} \mathbf{r}^T\mathbf{M}\mathbf{r} \tag{11}$$

where $\mathbf{r} = [r_{11}, r_{12}, ..., r_{32}, r_{33}]$ is the vector formed by stacking the rows of $\mathbf{R}$.

3.2 Generalized Pose Solver

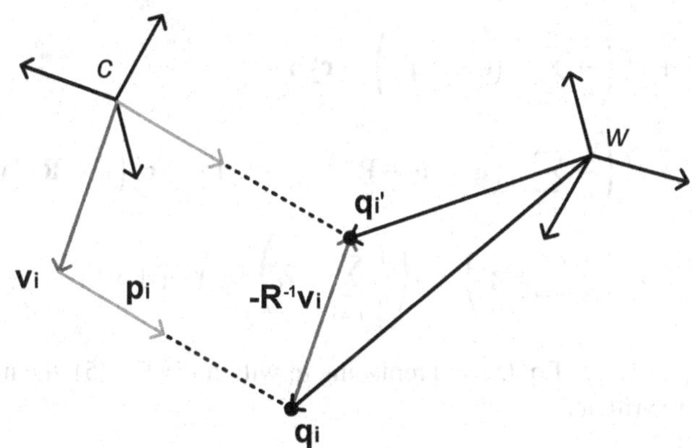

Fig. 3. The virtual world point $\mathbf{q'_i}$.

We begin to derive our generalized solvers. As shown in Fig. 3, we move the image ray from $\mathbf{v_i}$ to the camera frame origin and obtain a new virtual world point $\mathbf{q'_i}$. It is straightforward to recognize that $\mathbf{q'_i} = \mathbf{q_i} - \mathbf{R}^{-1}\mathbf{v}_i$.

Replacing $\mathbf{q_i}$ with $\mathbf{q'_i}$ in Eq. (6) of OPnP, we have

$$\begin{aligned} \hat{t}_x &= \overline{u} + \mathbf{r}_3^T\left(\frac{1}{n}\sum_{i=1}^{n} u_i\tilde{\mathbf{q}}_i\right) - s\left(\frac{1}{n}\sum_{i=1}^{n} u_i\tilde{\mathbf{v}}_i\right)_3 - \mathbf{r}_1^T\overline{\mathbf{q}} + s(\overline{\mathbf{v}})_1 \\ \hat{t}_y &= \overline{v} + \mathbf{r}_3^T\left(\frac{1}{n}\sum_{i=1}^{n} u_i\tilde{\mathbf{q}}_i\right) - s\left(\frac{1}{n}\sum_{i=1}^{n} v_i\tilde{\mathbf{v}}_i\right)_3 - \mathbf{r}_2^T\overline{\mathbf{q}} + s(\overline{\mathbf{v}})_2 \end{aligned} \tag{12}$$

where $\overline{\mathbf{v}}$ is the centroid of the ray origins, i.e., $\overline{\mathbf{v}} = \frac{1}{n}\sum_{i=1}^{n} \mathbf{v}_i$, $\tilde{\mathbf{v}}_i$ is the i-th image ray's origin after centralization, i.e., $\tilde{\mathbf{v}}_i = \mathbf{v}_i - \overline{\mathbf{v}}$, and $(\cdot)_i$ denotes the i-th value of the vector. More details are available in supplemental materials, and the simple derivation is as follows, where we have exploited the fact that $\mathbf{r}_i^T\mathbf{r}_j = 0$, $\mathbf{r}_i^T\mathbf{r}_i = s^2$.

$$\hat{t}_1 = \overline{u} + \mathbf{r}_3^T\left(\frac{1}{n}\sum_{i=1}^{n} u_i\mathbf{q}'_i\right) - \mathbf{r}_1^T\overline{\mathbf{q}}'$$

$$
\begin{aligned}
&= \bar{u} + \mathbf{r}_3^T\left(\frac{1}{n}\sum_{i=1}^{n} u_i(\mathbf{q}'_i - \bar{\mathbf{q}}')\right) - \mathbf{r}_1^T\bar{\mathbf{q}}' \\
&= \bar{u} + \mathbf{r}_3^T\left(\frac{1}{n}\sum_{i=1}^{n} u_i\left(\mathbf{q}_i - \bar{\mathbf{q}} - \mathbf{R}^{-1}(\mathbf{v}_i - \bar{\mathbf{v}})\right)\right) - \mathbf{r}_1^T\left(\bar{\mathbf{q}} - \mathbf{R}^{-1}\bar{\mathbf{v}}\right) \\
&= \bar{u} + \mathbf{r}_3^T\left(\frac{1}{n}\sum_{i=1}^{n} u_i\tilde{\mathbf{q}}_i\right) - s\left(\frac{1}{n}\sum_{i=1}^{n} u_i\tilde{\mathbf{v}}_i\right)_3 - \mathbf{r}_1^T\bar{\mathbf{q}} + s(\bar{\mathbf{v}})_1
\end{aligned} \tag{13}
$$

$$
\begin{aligned}
\hat{t}_2 &= \bar{v} + \mathbf{r}_3^T\left(\frac{1}{n}\sum_{i=1}^{n} v_i\mathbf{Q}_i\right) - \mathbf{r}_2^T\bar{\mathbf{q}}' \\
&= \bar{v} + \mathbf{r}_3^T\left(\frac{1}{n}\sum_{i=1}^{n} v_i(\mathbf{q}'_i - \bar{\mathbf{q}}')\right) - \mathbf{r}_2^T\bar{\mathbf{q}}' \\
&= \bar{v} + \mathbf{r}_3^T\left(\frac{1}{n}\sum_{i=1}^{n} v_i\left(\mathbf{q}_i - \bar{\mathbf{q}} - \mathbf{R}^{-1}(\mathbf{v}_i - \bar{\mathbf{v}})\right)\right) - \mathbf{r}_2^T\left(\bar{\mathbf{q}} - \mathbf{R}^{-1}\bar{\mathbf{v}}\right) \\
&= \bar{v} + \mathbf{r}_3^T\left(\frac{1}{n}\sum_{i=1}^{n} v_i\tilde{\mathbf{q}}_i\right) - s\left(\frac{1}{n}\sum_{i=1}^{n} u_i\tilde{\mathbf{v}}_i\right)_3 - \mathbf{r}_2^T\bar{\mathbf{q}} + s(\bar{\mathbf{v}})_2
\end{aligned} \tag{14}
$$

Plugging Eq. (12) into Eq. (5) and replacing $\mathbf{q_i}$ with $\mathbf{q'_i}$ in Eq. (5), the minimization problem can be rewritten as

$$
\min_{a,b,c,d} \mathbf{r}^T\mathbf{M}'\mathbf{r} \tag{15}
$$

where $\mathbf{r} = \left[1, a^2, ab, ac, ad, b^2, bc, bd, c^2, cd, d^2\right]^T$, $\mathbf{M}'$ is a new coefficient matrix computed from known parameters $\mathbf{p_i}$, $\mathbf{q_i}$, and $\mathbf{v_i}$.

Replacing $\mathbf{q_i}$ of DLS with $\mathbf{q'_i}$ in Eq. (10), we have

$$
\begin{aligned}
\mathbf{t} &= \mathcal{H}\sum_{i=1}^{n}\left(\mathbf{p}_i\mathbf{p}_i^T - \mathbf{I}\right)\mathbf{R}\mathbf{q}'_i \\
&= \mathcal{H}\sum_{i=1}^{n}\left(\mathbf{p}_i\mathbf{p}_i^T - \mathbf{I}\right)\mathbf{R}\left(\mathbf{q_i} - \mathbf{R}^{-1}\mathbf{v}_i\right) \\
&= \mathcal{H}\sum_{i=1}^{n}\left(\mathbf{p}_i\mathbf{p}_i^T - \mathbf{I}\right)\left(\mathbf{R}\mathbf{q}_i - \mathbf{v}_i\right)
\end{aligned} \tag{16}
$$

Plugging Eq. (16) into Eq. (9) and replacing $\mathbf{q_i}$ with $\mathbf{q'_i}$ in Eq. (9), the minimization problem can be rewritten as

$$
\min_{b,c,d} \mathbf{r}^T\mathbf{M}'\mathbf{r} \tag{17}
$$

where $\mathbf{r} = \left[1, b, c, d, b^2, bc, bd, c^2, cd, d^2\right]^T$.

Thus, we can use the same strategy as the original solvers to solve Eqs. (15) and (17), then obtain the optimal rotation and compute the translation from Eqs. (12) and (16), which yield two new generalized absolute pose solvers. Henceforth, we referred to the generalized solver originating from OPnP as GO and the generalized solver from DLS as GD.

4 Experimental Evaluation

4.1 Synthetic Data

We compare our generalized solvers to the original PnP solvers and the existing generalized absolute pose solvers using the Matlab PnP toolbox [10, 14]. We implement our solvers in Matlab. We use publicly available source codes of DLS [9], OPnP [10], SDP [19], and the mex-file implementation of UPnP [11]. Though UPnP has been presented for years, it is the state-of-the-art solution to the generalized pose problem. It has been used as the baseline in recent papers. In the central case, our generalized solvers are identical to the original PnP solvers. Hence, we perform simulations in which GO, GD, SDP, and UPnP handle the generalized case while OPnP and DLS handle the central case. The experiments show that our generalized pose solver inherits properties of the original PnP solvers, offering state-of-the-art performance in accuracy and robustness.

Numerical Stability
We test the numerical stability of our generalized solvers over 50,000 trials. Ray origins are randomly placed in the cube $[-1, 1] \times [-1, 1] \times [-1, 1]$. World points are randomly distributed in the volume $[-2, 2] \times [-2, 2] \times [4, 8]$. The origins and points determine ray directions. Random rotations and translations are applied. For each trial, we sample six 2D-3D correspondences and compute solutions. We compute the rotation errors and plot the results in Fig. 4(a). The errors are stable and comparable with the original solvers. The rotation error between $\mathbf{R}_{true}$ and the estimated $\mathbf{R}$ is defined as $e_{rot} = \|\mathbf{R} - \mathbf{R}_{true}\|_F / \|\mathbf{R}_{true}\|_F$. The horizontal axis shows the log 10 value of the error, while the vertical axis shows the counts over 50,000 independent runs.

Computational Efficiency
We vary the number of points from 4 to 100. We conduct 1000 independent tests for each trial and compute the average running time in milliseconds (ms). As shown in Fig. 4(b), the computational time of our generalized solvers is at the same level as the original ones. GO is the $O(1)$ method due to the vectorization implementation in OPnP [10], while GD is the $O(\mathrm{n})$ method like DLS [9].

Simulations with Varying Point Numbers and Increasing Image Noise
We perform two experiments to investigate the performance of our generalized solvers as the number of correspondences increases and as the amount of image noise increases. Since the performance is closely related to the point configuration, for both experiments, we generate uniformly random point configurations placing world points in three different cases:

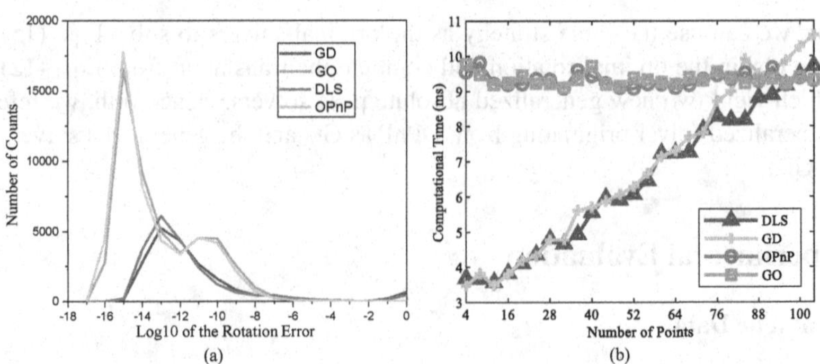

Fig. 4. Experimental results: (a) Numerical stability with the number of points fixed at six. (b) Computational time with the number of points varying from 4 to 100.

Ordinary Case. The world points are randomly distributed in the x-, y-, and z-range of $[-2, 2] \times [-2, 2] \times [4, 8]$.

Planar Case. The world points are randomly distributed in the plane of $[-2, 2] \times [-2, 2] \times [0, 0]$.

Quasi-Singular Case. The world points are randomly distributed in the range of $[1, 2] \times [1, 2] \times [4, 8]$.

The rotation and translation errors were measured by

$$\begin{aligned} e_{rot}(\text{deg}) &= \arccos\left(\frac{trace(\mathbf{R}\mathbf{R}_{true}) - 1}{2}\right) \\ e_t(\%) &= \frac{\|\mathbf{t} - \mathbf{t}_{true}\|}{\|\mathbf{t}\|} \times 100 \end{aligned} \tag{18}$$

When testing OPnP and DLS, the ray origin is at [0, 0, 0]. When testing GD, GO, UPnP, and SDP, the ray origins are randomly placed in the cube $[-1, 1] \times [-1, 1] \times [-1, 1]$.

The first experiment analyzes the performance of all methods with the number of points varying from 4 to 20. The image noise is zero-mean Gaussian noise with a fixed deviation of 2 pixels. As shown in Fig. 5, all methods perform well in the ordinary case. While in the planar and quasi-singular case, the generalized pose solvers outperform the central PnP solvers since the non-central camera conquers the point configuration problem of the central camera. The second simulation checks the accuracy of all methods with increasing image noise from 0.5 to 5 pixels. The number of the points was fixed with $n = 10$. Figure 6 shows similar results to the previous experiment. From Fig. 5 and Fig. 6, we can observe that UPnP is not accurate enough. The major reason lies in its shrunk solution space. SDP is accurate on the average. However, it is sensitive to the point configuration in the central case and not fast enough for real-time applications [10]. While GO and GD maintain state-of-art performance in the generalized case, outperforming UPnP and SDP. In the experiments, the generalized solvers show modest improvement compared with the PnP solvers, and this is due to point configurations. As is known, more scattered points yield better results in pose estimation. Though a multi-camera rig

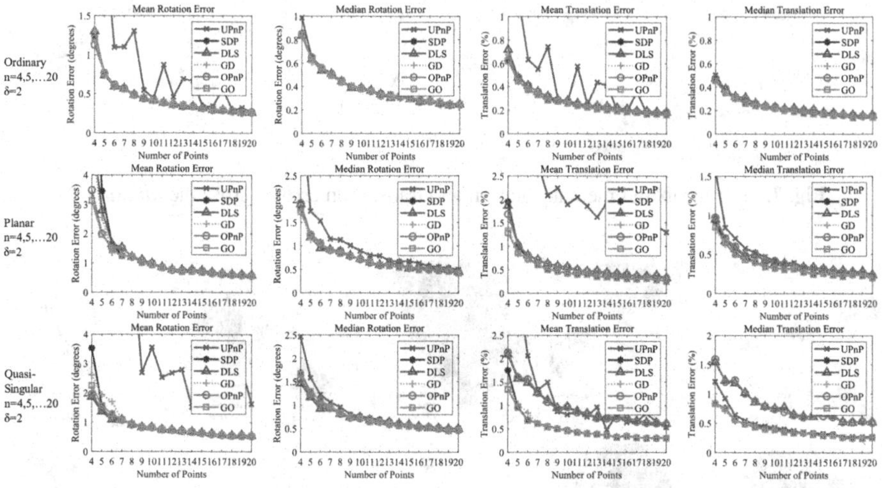

Fig. 5. The mean and median rotation and translation errors with the varying point numbers.

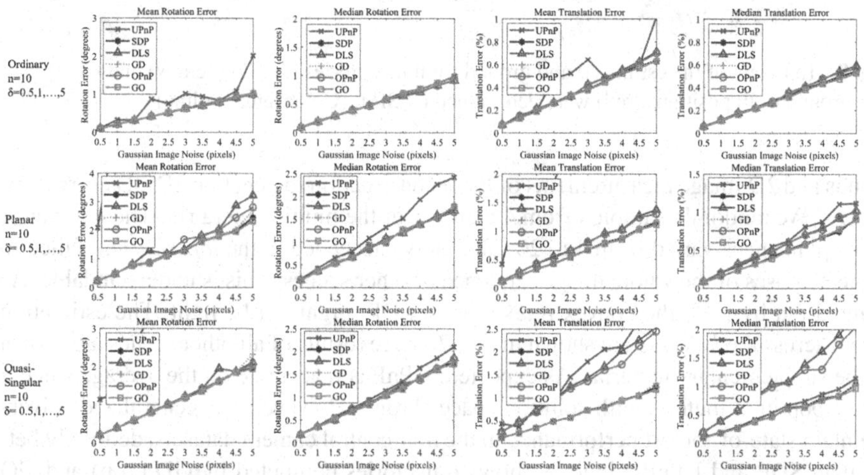

Fig. 6. The mean and median rotation and translation errors with the varying noise levels

can have a larger view than a single camera. To fairly evaluate the solvers, we set the camera catch the same view.

4.2 Real Data

We tested all the generalized solvers for recovering a non-central camera pose. In the tests, we used many-view datasets from ETH3D [27], in which a multi-camera rig of 4 cameras was adopted. The datasets contain ten scenes, and each containing camera poses collected by the SfM scene reconstruction using COLMAP [1]. We used 3D point

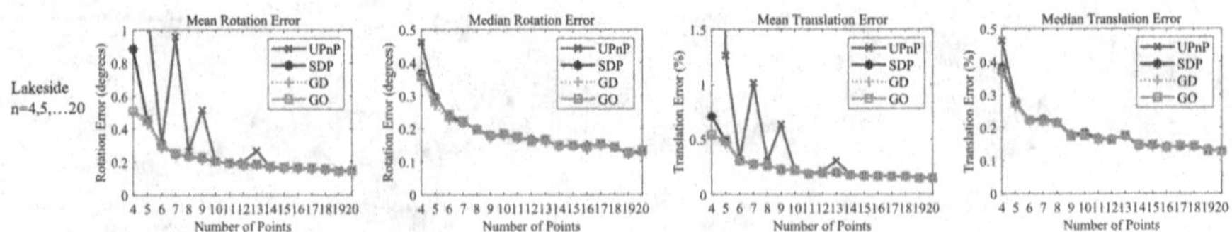

Fig. 7. The mean and median rotation and translation errors across the *lakeside.*

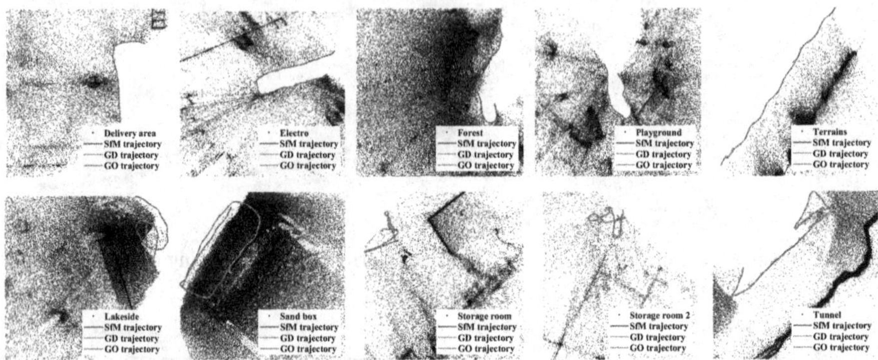

Fig. 8. Top view of the estimated and ground truth trajectories for GD (yellow) and GO (blue). The ground truth positions (red) were determined by SfM. (Color figure online)

clouds and 2D image measurements of the reconstruction to generate 2D-3D correspondences. We randomly sampled from 4 cameras of the multi-camera rig. The number of correspondences varied from 4 to 20. SDP only succeeded in the *lakeside* and failed to estimate poses of the whole image sequence of other scenes. This is understandable. As pointed out by [10], the convex relaxation in SDP might not be tight. The estimation errors across the *lakeside* are shown in Fig. 7. More results of all other scenes are shown in the supplementary material. As expected, UPnP is inaccurate on the average since it is a suboptimal method with solution space shrunk. However, our generalized solvers maintain state-of-the-art performance in the non-central camera datasets, definitely better than SDP and UPnP. Figure 8 shows trajectories computed by GD (10p) and GO (10p) for all scenes. Our generalized solvers offer visually pleasing results.

5 Conclusion

This work provides the possibility of solving the generalized pose problem by the framework of the classical PnP solver for the first time. We propose to transfer the classical PnP solver to the generalized pose solver. The critical process of our method is to derive the translation expression in the generalized case and compute the new form of the coefficient matrix. In such a way, we derive two generalized absolute pose solvers. Our generalized solvers inherit state-of-art performance of the original solvers. The experiment results show their superiority in accuracy when compared with existing generalized PnP solvers.

Acknowledgments. This work was supported in part by the Hunan Provincial Natural Science Foundation for Excellent Young Scholars under Grant 2023JJ20045 and in part by the Science Foundation under Grant KY0505072204 and Grant GJSD22006.

References

1. Schonberger, J.L., Frahm, J.M.: Structure-from-motion revisited. In: IEEE Conference on Computer Vision and Pattern Recognition, pp. 4104–4113 (2016)
2. Fu, Y., Shang, Y., Hu, W., et al.: Non-contact optical dynamic measurements at different ranges: a review. Acta. Mech. Sin. **37**(4), 537–553 (2021)
3. Campos, C., Elvira, R.J., Rodríguez, J.G., et al.: ORB-SLAM3: an accurate open-source library for visual, visual-inertial, and multimap SLAM. IEEE Trans. Rob. **37**(6), 1874–1890 (2021)
4. Ke, T., Roumeliotis, S.I.: An efficient algebraic solution to the perspective-three-point problem. In: IEEE Conference on Computer Vision and Pattern Recognition, pp. 4618–4626 (2017)
5. Yu, Q., Xu, G., Shi, J.: An efficient and reasonably simple solution to the perspective-three-point problem. In: International Conference on Pattern Recognition, pp. 46–59 (2021)
6. Bujnak, M., Kukelova, Z., Pajdla, T.: New efficient solution to the absolute pose problem for camera with unknown focal length and radial distortion. In: Asian Conference on Computer Vision, pp. 11–24 (2010)
7. Zhang, Z., Sun, C., Wang, P.: Two-step pose estimation method based on five reference points. Chin. Opt. Lett. **10**(7), 52–56 (2012)
8. Lepetit, V., Moreno-Noguer, F., Fua, P.: EPnP: an accurate O(n) solution to the PnP problem. Int. J. Comput. Vis. **81**, 155–166 (2009)
9. Hesch, J.A., Roumeliotis, S.I.: A direct least-squares (DLS) method for PnP. In: IEEE International Conference on Computer Vision, pp. 383–390 (2011)
10. Zheng, Y., Kuang, Y., Sugimoto, S., et al.: Revisiting the PnP problem: a fast, general and optimal solution. In: IEEE International Conference on Computer Vision, pp. 2344–2351 (2013)
11. Kneip, L., Li, H., Seo, Y.: UPnP: an optimal O(n) solution to the absolute pose problem with universal applicability. In: Fleet, D., Pajdla, T., Schiele, B., Tuytelaars, T. (eds.) ECCV 2014. LNCS, vol. 8689, pp. 127–142. Springer, Cham (2014). https://doi.org/10.1007/978-3-319-10590-1_9
12. Nakano, G.: Globally optimal DLS method for PnP problem with Cayley parameterization. In: British Machine Vision Conference, pp. 78.1–78.11 (2015)
13. Yan, K., Zhao, R., Tian, H., et al.: A high accuracy method for pose estimation based on rotation parameters. Measurement **122**, 392–401 (2018)
14. Wang, P., Xu, G., Cheng, Y., et al.: A simple, robust and fast method for the Perspective-n-Point Problem. Pattern Recogn. Lett. **108**(1), 31–37 (2018)
15. Lourakis, M., Terzakis, G.: A globally optimal method for the PnP problem with MRP rotation parameterization. In: International Conference on Pattern Recognition, pp. 3058–3063 (2020)
16. Yu, Q., Xu, G., Zhang, L., et al.: A consistently fast and accurate algorithm for estimating camera pose from point correspondences. Measurement **172**, 108914 (2021)
17. Kukelova, Z., Bujnak, M., Pajdla, T.: Automatic generator of minimal problem solvers. In: Forsyth, D., Torr, P., Zisserman, A. (eds.) ECCV 2008. LNCS, vol. 5304, pp. 302–315. Springer, Heidelberg (2008). https://doi.org/10.1007/978-3-540-88690-7_23

18. Larsson, V., Åström, K., Oskarsson, M.: Efficient solvers for minimal problems by syzygy-based reduction. In: IEEE Conference on Computer Vision and Pattern Recognition, pp. 2383–2392 (2017)
19. Schweighofer, G., Pinz, A.: Globally optimal O(n) solution to the PnP problem for general camera models. In: British Machine Vision Conference, pp. 1–10 (2008)
20. Terzakis, G., Lourakis, M.: A consistently fast and globally optimal solution to the Perspective-n-Point problem. In: Vedaldi, A., Bischof, H., Brox, T., Frahm, J.-M. (eds.) ECCV 2020. LNCS, vol. 12346, pp. 478–494. Springer, Cham (2020). https://doi.org/10.1007/978-3-030-58452-8_28
21. Zhou, L., Kaess, M.: An efficient and accurate algorithm for the Perspective-n-Point problem. In: IEEE/RSJ International Conference on Intelligent Robots and Systems, pp. 6245–6252 (2019)
22. Urban, S., Leitloff, J., Hinz, S.: MLPNP – a real-time maximum likelihood solution to the Perspective-n-Point problem. ISPRS Ann. Photogramm. Remote Sens. Spatial Inf. Sci. **3**, 131–138 (2016)
23. Ramalingam, S., Sturm, P.: A unifying model for camera calibration. IEEE Trans. Pattern Anal. Mach. Intell. **39**(7), 1309–1319 (2017)
24. Schöps, T., Larsson, V., Pollefeys, M., Sattler, T.: Why having 10,000 parameters in your camera model is better than twelve. In: IEEE/CVF Conference on Computer Vision and Pattern Recognition, pp. 2532–2541 (2020)
25. Campos, J., Cardoso, J.R., Miraldo, P.: POSEAMM: a unified framework for solving pose problems using an alternating minimization method. In: International Conference on Robotics and Automation, pp. 3493–3499 (2019)
26. Guan, B., Zhao, J., Barath, D., et al.: Minimal cases for computing the generalized relative pose using affine correspondences. In: IEEE International Conference on Computer Vision, pp. 6068–6077 (2021)
27. Schöps, T., Schönberger, J.L., Galliani, S., et al.: A multi-view stereo benchmark with high-resolution images and multi-camera videos. In: Conference on Computer Vision and Pattern Recognition, pp. 2538–2547 (2017)

RICH: Robust Implicit Clothed Humans Reconstruction from Multi-scale Spatial Cues

Yukang Lin[1], Ronghui Li[1], Kedi Lyu[2], Yachao Zhang[1(✉)], and Xiu Li[1(✉)]

[1] Tsinghua Shenzhen International Graduate School, Tsinghua University, Shenzhen, China
yachaozhang@stu.xmu.edu.cn, li.xiu@sz.tsinghua.edu.cn

[2] College of Computer Science and Technology, Jilin University, Changchun, China

Abstract. The pixel-aligned implicit functions (IFs) enable the reconstruction of 3D human with complete and detailed clothing from a single RGB image. To enhance robustness for poses, existing work introduce the parametric body model as prior, but this limits the recovery of the geometry details and makes it challenging to handle loose clothing. Our goal is to reconstruct both clothing and pose that highly align with the input image, even in cases of peculiar poses and complex clothing. To achieve this, we propose a multi-scale features-based implicit method, called RICH, which combines the flexibility of implicit function and the powerful prior of parametric body model. RICH introduces a 3D human body model as prior knowledge and adopts local feature to constrain human body generation. Furthermore, RICH employs a pretrained image encoder to extract global pixel-aligned feature, which contributes to high-precision and complete reconstruction of clothing geometry and of the external appearance such as hair and accessories. Besides, by establishing connections with the joints of the body model, RICH utilizes an attention mechanism to construct relative spatial feature, thereby increasing the robustness for poses. Finally, RICH takes as input local, relative, and global feature to IF to query occupancy and the clothed human is represented by the 0.5 iso-surface of the 3D occupancy field. Quantitative and qualitative evaluation on the THuman2.0 and CAPE datasets shows that RICH outperforms the state-of-the-art methods. In particular, RICH demonstrates strong generalization ability on in-the-wild images, even under the scenarios of challenging poses and complex clothing. The code and supplementary material will be available at https://github.com/lyk412/RICH.

Keywords: 3D clothed human · implicit reconstruction · loose clothing and challenging poses

Supplementary Information The online version contains supplementary material available at https://doi.org/10.1007/978-981-99-8432-9_16.

Q. Liu et al. (Eds.): PRCV 2023, LNCS 14426, pp. 193–206, 2024.
https://doi.org/10.1007/978-981-99-8432-9_16

1 Introduction

Realistic 3D humans play a crucial role in augmented reality and virtual reality. With the rise of sensor systems and immersive technologies, the demand for generating virtual 3D humans on a large scale is steadily increasing. Traditional methods [5,8] for reconstructing clothed human are tedious and rely on expensive 3D scanning devices. Nowadays, researchers focus on reconstructing clothed human directly from images [1–4,28]. The "body + clothing offset" methods [1–3,28] add offsets on the vertices of human body model such as SMPL [10] and SMPL-X [17], but the accuracy is limited by the number of vertices and cannot handle loose clothing. To enhance the representation of clothing geometry, the "body + clothing template" methods [12,14,23] specify clothing types and employs corresponding template to represent the clothing. However, these approaches are limited to a predefined set of clothing types and cannot generalize to arbitrary clothing.

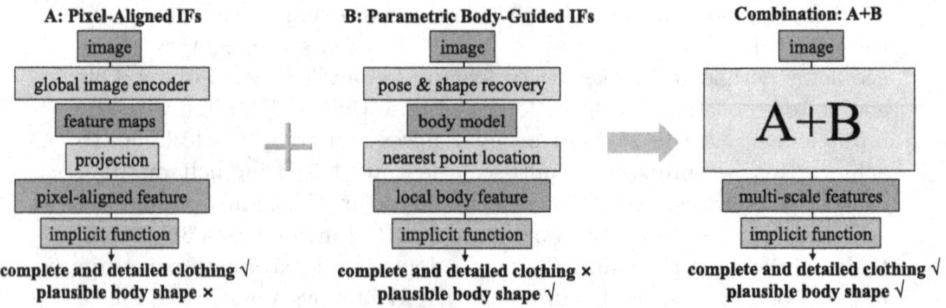

Fig. 1. An overview of implicit functions for clothed humans reconstruction.

To address the issues mentioned above, we base our method on implicit functions (IFs). Compared to explicit surface representations [10,17,25,26], IFs can capture complex shapes with arbitrary topology such as clothing surfaces and other external appearances, while being free from resolution limitations. However, IFs-based methods struggle with pose reconstructing, leading to unrealistic body shapes and broken limbs. Recently, the parametric body-guided implicit methods [21,27] introduce the human body models to constrain the reconstruction of body shape, but this approach weakens the power of IFs for clothing, especially loose clothing. To achieve robustness for both complex clothing and challenging poses, we propose a multi-scale features-based implicit method called RICH, capable of reconstructing 3D clothed human with complete and detailed clothing and plausible body shape from a single color image. Our motivation is illustrated by Fig. 1.

Quantitative and qualitative results on two challenging test sets, namely THuman2.0 [22] and CAPE [13], as well as in-the-wild images, demonstrate that RICH outperforms the SOTA methods. In particular, RICH exhibits strong robustness even in cases of complex clothing and peculiar poses. The main contributions of this work are summarized as follows:

RICH. We propose a multi-scale features-based implicit method, named RICH, that combines the strong prior of parametric body model with the high flexibility of implicit function. RICH achieves robust reconstruction of clothed humans with details on the challenging images, even under the scenarios of peculiar poses and extreme loose clothing.

Pretrained Image Encoder. We first pretrain the image encoder without the constraint of SMPL body and freeze its parameters when training RICH. This scheme ensures consistent global features with the input image and guarantees complete and detailed reconstruction of clothing.

Enhanced Local Feature. We enhance the local feature by correcting the nearest neighbor distance metric, thereby enhancing the reconstruction precision.

Overall, RICH takes into account both the clothing feature and pose feature in the image. Even in the presence of complex garments and unusual poses, RICH is able to robustly reconstruct complete clothing and physically plausible bodies, with high geometry quality of the reconstructed surface.

2 Related Work

2.1 3D Clothed Human Reconstruction

"body mesh vertex+offset". With the introduction of parametric body models [10,17], a lot of work [1–3,28] predicts 3D offsets on the body vertices to represent clothing geometry. This approach does not change the topology of the body mesh, thus the reconstructed clothed human retains all the attributes of the body model, such as animation, which is compatible with existing graphics pipelines. However, due to the fixed number of mesh vertices, this method has limited modeling capabilities for complex shape such as loose clothing.

"body mesh surface+offset". To overcome the limitation of the fixed number of body vertices, some recent methods [12,14,23] sample the points continuously on the surface of the mesh, allowing the number of points to go beyond that of mesh vertices. By decoupling the pose-induced offset from the clothing-specific offset, these methods present a more realistic deformation of clothing under different poses. However, this approach could only reconstruct predefined types of clothing and fail to generalize to arbitrary clothing.

"body mesh+clothing template". Recently, some work [4,9,20] designs clothing templates instead of representing clothing as an extension of the body mesh. Given images, these methods first estimate the body shape and identify the clothing type in the image. Then, the corresponding clothing template mesh is attached to the body mesh. Although this approach can reliably reconstruct accurate clothing, it does not scale up to a large variety of clothing types.

2.2 Clothed Human from Implicit Function

Implicit functions (IFs) [15,16] play a crucial role in the task of single-image-based implicit reconstruction of 3D clothed human. Once trained, IFs can be used to query the occupancy of any point in the 3D space. And the reconstructed surface is represented by the 0.5 iso-surface of the 3D occupancy field and is obtained using Marching Cubes [11]. The implicit reconstruction not only focuses on the accuracy of the clothed human, but also emphasizes the robustness and generalization ability to clothing and pose shown in the image.

To express the diversity of clothing and the complex geometry of garments, a lot of methods [6,18,19] utilize pixel-aligned IFs. Despite robust reconstruction for a wide range of clothing types, IFs show poor robustness such as incomplete limb and non-human shape, when it comes to challenging poses due to that the global image feature is sensitive to pose. To improve the generalization ability under the scenarios of challenging poses, some work [21,27] combines parametric body models with shape-free IFs to regulate the implicit reconstruction process. The parametric body-guided methods often optimize the body model during the inference stage to mitigate the impact of inaccurate initial body.

Our method, RICH, integrates IF with the body model SMPL to achieve robust reconstruction of a clothed human's 3D shape from a single RGB image. As a prior, SMPL regularizes the free-form IF using the semantic body features. In contrast to previous methods that solely focus on local body features or global image features, RICH considers both of them simultaneously, demonstrating robustness to clothing and poses unseen during training.

3 Method

Given a color image, RICH aims to infer a robust 3D clothed human while preserving the details present in the image. Specifically, RICH takes as input an image $\mathcal{I}$ with a segmented clothed human and the 3D SMPL body $\mathcal{M}$ estimated by PyMAF [24], and extracts multi-scale spatial features MSF. Then an implicit function $\mathcal{IF}$ represented by multi-layer perceptrons (MLPs) is employed to obtain the 3D occupancy field $\mathcal{O}$. The training process can be formulated as:

$$\mathcal{O} = \mathcal{IF}\left(\mathcal{G}_{MS}(\mathcal{I}, \mathcal{M})\right) : \mathcal{O} \in [0,1] \tag{1}$$

where $\mathcal{G}_{MS}$ is the network architecture of RICH. By using the Marching Cubes [11], RICH infers the 3D shape of the clothed human, where the implicit surface $\mathcal{S}$ is defined as the iso-surface with an occupancy value of 0.5, namely $\mathcal{S} : \mathcal{O} = 0.5$. Figure 2 illustrates the reconstruction pipeline. RICH contains two main modules: (1) 2.5D normal maps prediction, and (2) multi-scale features-based implicit surface reconstruction.

3.1 2.5D Normal Maps Prediction

The normal map provides the surface's normal vectors at each point, explicitly indicating the orientation and shape of the surface, which is beneficial for

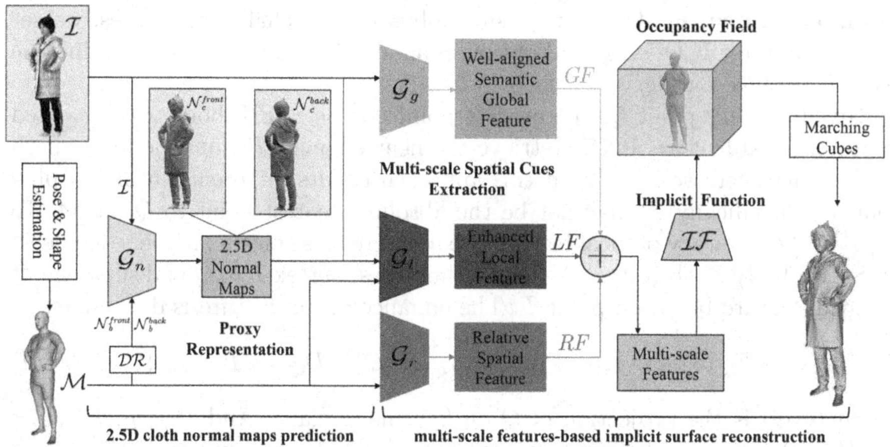

Fig. 2. Overview. RICH's architecture contains two main modules: (1) 2.5D cloth normal maps prediction, (2) multi-scale features-based implicit surface reconstruction.

improving the reconstruction accuracy. In addition, in our method, the normal map serves as an intermediate proxy representation between the input 2D image and the target 3D shape, so it is referred to as the 2.5D feature map.

To capture the geometric details of the front and back of the human body, we predict the normal maps for both sides, following the approach introduced by ICON [21]. Specifically, we utilize the differentiable renderer $\mathcal{DR}$ to render the estimated SMPL body $\mathcal{M}$ from the front and back views, obtaining the body normal maps $\mathcal{N}_b = \left\{\mathcal{N}_b^{front}, \mathcal{N}_b^{back}\right\}$. Given the input image $\mathcal{I}$ and $\mathcal{N}_b$, the network $\mathcal{G}_n = \left\{\mathcal{G}_n^{front}, \mathcal{G}_n^{back}\right\}$ predicts the cloth normal maps $\mathcal{N}_c = \left\{\mathcal{N}_c^{front}, \mathcal{N}_c^{back}\right\}$. It is worth noting that both networks share the same structure but have separate parameter updates. The process of predicting cloth normal maps can be defined as follows:

$$\mathcal{DR}\left(\mathcal{M}\right) \rightarrow \mathcal{N}_b \tag{2}$$

$$\mathcal{G}_n\left(\mathcal{I}, \mathcal{N}_b\right) \rightarrow \mathcal{N}_c \tag{3}$$

3.2 Multi-scale Features-Based Implicit Surface Reconstruction

To enhance the robustness to clothing and poses, suitable and comprehensive features need to be selected for implicit function $\mathcal{IF}$. We introduce multi-scale features, including enhanced local feature, well-aligned semantic global feature and attention-based relative spatial feature. The combined use of these features helps $\mathcal{IF}$ better model the clothing and body shape. In the following sections, we will discuss them separately.

Enhanced Local Feature. To ensure robustness to challenging poses, we get local feature like ICON [21] and obtain enhanced local feature by rectifying the nearest distance metric.

Given a query point Q, the nearest point on the SMPL body $\mathcal{M}$ is located to obtain local feature. ICON retrieve the nearest point P from the body mesh surface. However, we found that this approach results in "pseudo-nearest" phenomena. Specifically, P may not be the absolute nearest point to Q on $\mathcal{M}$. To improve the accuracy of local feature, we add a nearest metric, namely the nearest SMPL body mesh vertex. We locate the closest vertex T to Q and supplement the local feature based on point T. The enhanced local feature is defined as:

$$LF_Q = \oplus\{D_{Q\to P}, P_{normal}, \mathcal{N}_c\left(\Pi\left(Q\right)\right), D_{Q\to T}, T_{normal}\} \tag{4}$$

where $\Pi\left(Q\right)$ is the projection of Q on 2D image plane. And $D_{Q\to P}$, P_{normal}, $\mathcal{N}_c\left(\Pi\left(Q\right)\right)$ represent the distance from Q to P, the normal vector of P, and the pixel-aligned normal feature of Q on $\mathcal{N}_c$ respectively. The last two components represent the enhanced features, namely the distance from Q to T and the normal vector of T. It is worth noting that our enhanced local feature is also independent of the global pose.

Well-Aligned Semantic Global Feature. The pixel-aligned implicit methods train a global image encoder to generate semantic feature, and the SMPL body-guided ones regulate the implicit reconstruction process leveraging the pose and shape of the estimated body. However, neither of them can simultaneously capture complete clothing with details and accurate body shape (see Fig. 1).

To effectively integrate pixel-aligned features with the SMPL body in the reconstruction pipeline, we adopt a pretraining strategy. Specifically, we pretrain an image encoder $\mathcal{G}_g$ to extract semantic global feature map F_{map} in the absence of SMPL body constraints. Then we freeze its parameters when training $\mathcal{G}_{MS}$ that involves the SMPL body. This strategy ensures that the global feature is well-aligned with the input image all the time, and semantic enough to enable complete and detailed reconstruction of the clothing.

Given the input image $\mathcal{I}$ and the cloth normal maps $\mathcal{N}_c$, the pretrained image encoder $\mathcal{G}_g$ generates global feature map F_{map}. The well-aligned semantic global feature of query point Q can be defined as:

$$F_{map} = \mathcal{G}_g\left(\mathcal{I}, \mathcal{N}_c\right) \tag{5}$$

$$GF_Q = F_{map}\left(\Pi\left(Q\right)\right) \tag{6}$$

Attention-Based Relative Spatial Feature. In the field of human reconstruction, we are the first to utilize attention mechanism to construct relative spatial features. Inspired by RandLA-Net [7], a scene reconstruction method that utilize relative spatial encoding to model local structure, we establish a connection between the query points and the SMPL body joints to better learn the pose features from the image, which is illustrated in Fig. 3.

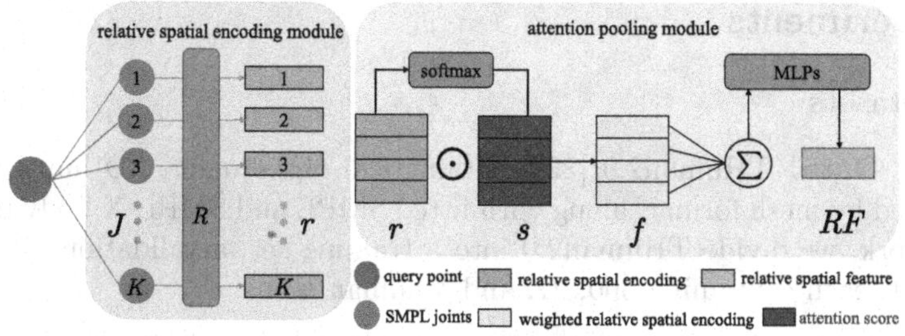

Fig. 3. The network architecture of relative spatial feature extraction module.

The relative spatial encoding for the query point Q with respect to the k_{th} joint can be obtained by operation R, which can be formulated as:

$$r_Q^k = \text{MLP}\left(\oplus\{q, p^k, q - p^k, \left\| q - p^k \right\|\}\right) \tag{7}$$

where q and p^k denote the 3D coordinates of the query point and the k_{th} joint respectively. $q-p^k$ is the relative position difference and $\left\| q - p^k \right\|$ is the euclidean distance. The attention score, denoted as s_Q^k, is obtained by applying a softmax operation to r_Q^k. The weighted relative spatial encoding can be defined as:

$$f_Q^k = s_Q^k \cdot r_Q^k \tag{8}$$

Finally, the relative spatial feature RF_Q is obtained by summing and mapping, formulated as Eq. (8). RF_Q provides a more intuitive and concrete representation of pose feature, which helps improve the robustness to poses.

$$RF_Q = \text{MLP}\left(\sum_{k=1}^{K} f_Q^k\right) \tag{9}$$

Now we can define the multi-scale features of query point Q as follows:

$$MSF_Q = \oplus\{LF_Q, GF_Q, RF_Q\} \tag{10}$$

$\mathcal{IF}$ maps MSF_Q to a 3D occupancy field $\mathcal{O}$, whose 0.5 level set is considered as the implicit surface of the clothed human. During inference, we employ $\mathcal{IF}$ to obtain the occupancy of each spatial point and utilize the Marching Cubes [11] to extract the iso-surface with an occupancy value of 0.5. Finally, we obtain a detailed human mesh that accurately represents the clothing and pose present in the input image.

We train the multi-scale features-based network $\mathcal{G}_{MS}$ with the following loss:

$$\mathcal{L}_{MS} = \frac{1}{n}\sum_{i=1}^{n}\left(\mathcal{IF}\left(MSF\left(X\right)\right) - f^*\left(X\right)\right)^2 \tag{11}$$

where X is the query point in the 3D space and n is the number of the query points. $f^*\left(\cdot\right)$ returns the ground truth occupancy of a query point. For training loss and implementation, see supplementary material for more details.

4 Experiments

4.1 Datasets

Training Data. THuman2.0 [22] contains 526 high-quality 3D human scans represented by mesh format, along with fitted SMPL and SMPL-X body models. In this work, we divide THuman2.0 into a training set, a validation set, and a testing set, each containing 500, 21, and 5 human scans.

Testing Data. The testing data consists of two parts: 5 scans with complex clothing from the THuman2.0, and the CAPE [13] dataset consisting of 50 3D scans with fashion pose and 100 3D scans with challenging pose. The results on THuman2.0 test set serve as a good evaluation of generalization ability to complex clothing, and the results on CAPE can reflect the generalization ability to challenging poses. See the supplementary material for some test samples.

Generating Synthetic Data. We employ the OpenGL scripts provided by ICON [21] to generate data samples, each consisting of a 3D clothed-human scan, its SMPL-X fit, an RGB image, camera parameters, 2D normal maps for the clothed human scan and the SMPL-X body (captured from two opposing views), as well as visibility for the SMPL-X mesh with respect to the camera. Please refer to the supplementary material for more details.

4.2 Evaluation

Metrics. We use point-to-surface (P2S) distance, chamfer distance (Chamfer), and normal maps difference (Normals) as quantitative evaluation metrics. We report the P2S distance from reconstructed mesh points to the closest ground-truth scan points. In addition, we compute average bi-directional P2S distances to measure how well the reconstructed mesh align with the ground-truth scan. These two metrics capture large geometric differences. To capture high-frequency differences, we render normal images for reconstructed and ground-truth surfaces from 4 viewpoints and calculate the L2 error between them.

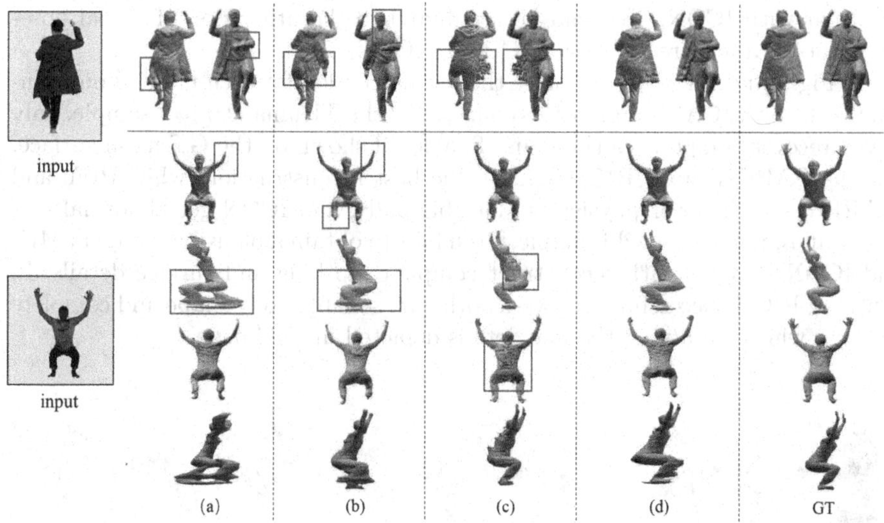

Fig. 4. Qualitative comparison of reconstruction: (a) by PIFu, (b) by PaMIR, (c) by ICON, (d) by RICH. The upper input image is from THuman2.0 test set and the other from CAPE.

Table 1. Quantitative evaluation (cm).

Methods	THuman2.0			CAPE			Preference
	Normals↓	Chamfer↓	P2S↓	Normals↓	Chamfer↓	P2S↓	
PIFu	0.123	1.728	1.725	0.162	2.726	1.693	20%
PaMIR	0.107	1.319	1.506	0.115	1.580	1.356	14.5%
ICON	0.117	1.583	2.099	**0.082**	1.224	1.107	12%
RICH	**0.090**	**1.125**	**1.169**	**0.082**	**1.193**	**0.967**	

Comparison. We compared RICH with the SOTA methods, including PIFu [18], PaMIR [27], and ICON [21], where PIFu demonstrates strong robustness in capturing clothing details, while PaMIR and ICON exhibit strong robustness to body poses. To ensure fairness, we use re-implementations of PIFu and PaMIR from ICON [21] so all four methods have the same architecture and input data. Besides, we retrained them on THuman2.0 training set under the same setting. To evaluate the performance for unseen clothing and poses, comprehensive comparisons are conducted on the THuman2.0 test set, CAPE test set, and in-the-wild images. See the supplementary for implementation details.

Table 1 shows the quantitative comparison results. Ours achieves the best results in all metrics on both THuman2.0 and CAPE test set. On the images containing out-of-distribution outfits (THuman2.0), RICH outperforms others by a large margin. In terms of out-of-distribution poses (CAPE), RICH achieves significantly smaller reconstruction errors compared to PIFu and PaMIR, and is on par with ICON. This can be attributed to the utilization of global pose-invariant local features in both RICH and ICON.

In Fig. 4 and Fig. 5, we evaluate the reconstruction for each method on THuman2.0 test set, CAPE and in-the-images. For the THuman2.0 test sample, only RICH recover complete clothing and fine detail shown on the GT mesh surface. For the CAPE sample, RICH achieve the best reconstruction, while PIFu and PaMIR cannot recover physically plausible body, and ICON get abnormal surface shapes. For in-the-wild sample, PIFu fails to obtain a plausible body. PaMIR and ICON struggle with reconstruct complete clothing and surface details. In contrast, RICH successfully recovers both a reasonable body shape and complete clothing while preserving the fine details depicted in the image.

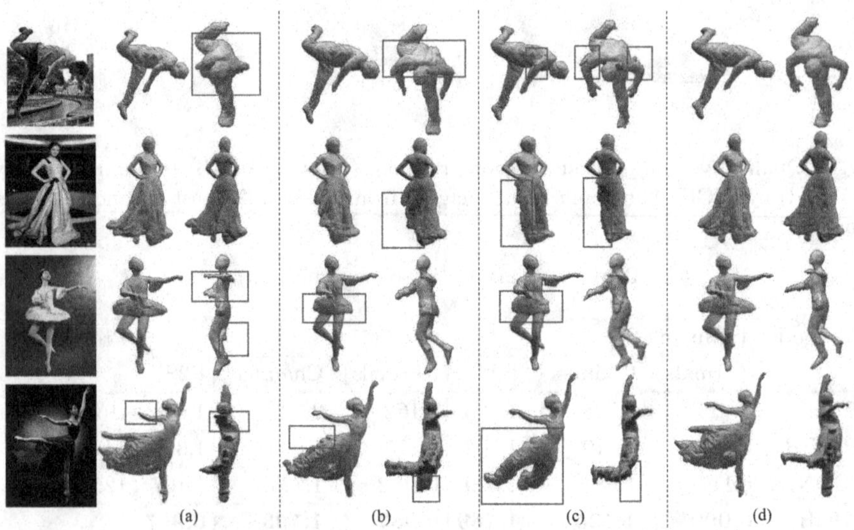

Fig. 5. Qualitative comparison of reconstruction on in-the-wild images: (a) by PIFu, (b) by PaMIR, (c) by ICON, (d) by RICH.

Table 2. Quantitative results of ablation.

Methods	THuman2.0			CAPE		
	Normals↓	Chamfer↓	P2S↓	Normals↓	Chamfer↓	P2S↓
RICH w/o Refine	0.091	1.153	1.218	0.086	1.191	0.983
RICH w/o Global	0.124	1.620	2.157	0.087	1.328	1.192
RICH w/o Relative	0.121	1.348	1.389	0.087	**1.166**	1.024
RICH	**0.090**	**1.125**	**1.169**	**0.082**	1.193	**0.967**

4.3 Ablation Study

In order to investigate the impact of the main components of RICH, certain modules are removed from RICH. "RICH w/o Global" refers to the base network without the global feature extraction module. "RICH w/o Refine" refers to the base network without the enhanced local feature module. "RICH w/o Relative" refers to the base network without the relative spatial feature extraction module. We present the quantitative results in Table 2 as well as the qualitative results in Fig. 6.

Fig. 6. Qualitative ablation of reconstruction: (a, b) by RICH, (c, d) by RICH w/o Global, (e, f) by RICH w/o Relative, (g, h) by RICH w/o Refine.

5 Application

Reconstruction of Robust and Detailed 3D Humans from In-the-Wild Images. We collect 50 in-the-wild images from the Internet that show people performing sports, dance and kung fu. etc. We show some qualitative results of RICH in Fig. 5 (d). Due to the lack of ground-truth geometry, we compare RICH to PIFu, PaMIR, and ICON in a perceptual study. Participants were asked to choose the best result that represents the clothing and body shape of the human in the image. We compute the chances that participants prefer the SOTA methods over RICH in Table 1. See supplementary for more details.

6 Conclusion

We propose a implicit function, which leverages enhanced local features, well-aligned semantic global features, and relative spatial features. Thanks to the design of multi-scale features, our method RICH can reconstruct robust implicit clothed humans. Unlike previous approaches, RICH exhibits robustness to both clothing and pose variations. Our experiment demonstrates the ability of RICH to reconstruct complete garments and plausible bodies even in cases of complex clothing and challenging poses, while preserving high-frequency details present in the image. Ablation study validates the design of multi-scale features.

Limitations. Compared to previous methods, RICH has a more complex network structure and utilizes more semantic features, resulting in higher computational cost. Besides, the accuracy of SMPL body estimation significantly affects the reconstruction performance of RICH. In the future, our goal is to devise more efficient methods to reduce reliance on the SMPL body model and accelerate the inference.

Acknowledgments. This work was supported in part by the Shenzhen Key Laboratory of next generation interactive media innovative technology (No. ZDSYS20210623092001004), in part by the China Postdoctoral Science Foundation (No. 2023M731957), in part by the National Natural Science Foundation of China under Grant 62306165.

References

1. Alldieck, T., Magnor, M., Bhatnagar, B.L., Theobalt, C., Pons-Moll, G.: Learning to reconstruct people in clothing from a single RGB camera. In: Proceedings of the IEEE/CVF Conference on Computer Vision and Pattern Recognition, pp. 1175–1186 (2019)
2. Alldieck, T., Pons-Moll, G., Theobalt, C., Magnor, M.: Tex2Shape: detailed full human body geometry from a single image. In: Proceedings of the IEEE/CVF International Conference on Computer Vision, pp. 2293–2303 (2019)
3. Alldieck, T., Zanfir, M., Sminchisescu, C.: Photorealistic monocular 3D reconstruction of humans wearing clothing. In: Proceedings of the IEEE/CVF Conference on Computer Vision and Pattern Recognition, pp. 1506–1515 (2022)
4. Bhatnagar, B.L., Tiwari, G., Theobalt, C., Pons-Moll, G.: Multi-Garment Net: learning to dress 3D people from images. In: Proceedings of the IEEE/CVF International Conference on Computer Vision, pp. 5420–5430 (2019)
5. Cui, Y., Chang, W., Nöll, T., Stricker, D.: KinectAvatar: fully automatic body capture using a single kinect. In: Park, J.I., Kim, J. (eds.) Computer Vision-ACCV 2012 Workshops: ACCV 2012 International Workshops, Daejeon, Korea, 5–6 November 2012, Revised Selected Papers, Part II 11. LNCS, vol. 7729, pp. 133–147. Springer, Cham (2013). https://doi.org/10.1007/978-3-642-37484-5_12
6. He, T., Collomosse, J., Jin, H., Soatto, S.: Geo-PIFu: geometry and pixel aligned implicit functions for single-view human reconstruction. Adv. Neural. Inf. Process. Syst. **33**, 9276–9287 (2020)
7. Hu, Q., et al.: RandLA-Net: efficient semantic segmentation of large-scale point clouds. In: Proceedings of the IEEE/CVF Conference on Computer Vision and Pattern Recognition, pp. 11108–11117 (2020)
8. Joo, H., et al.: Panoptic studio: a massively multiview system for social motion capture. In: Proceedings of the IEEE International Conference on Computer Vision, pp. 3334–3342 (2015)
9. Lin, S., Zhang, H., Zheng, Z., Shao, R., Liu, Y.: Learning implicit templates for point-based clothed human modeling. In: Avidan, S., Brostow, G., Cissé, M., Farinella, G.M., Hassner, T. (eds.) Computer Vision-ECCV 2022: 17th European Conference, Tel Aviv, Israel, 23–27 October 2022, Proceedings, Part III, pp. 210–228. Springer, Cham (2022). https://doi.org/10.1007/978-3-031-20062-5_13
10. Loper, M., Mahmood, N., Romero, J., Pons-Moll, G., Black, M.J.: SMPL: a skinned multi-person linear model. ACM Trans. Graph. (TOG) **34**(6), 1–16 (2015)

11. Lorensen, W.E., Cline, H.E.: Marching cubes: a high resolution 3D surface construction algorithm. In: Seminal Graphics: Pioneering Efforts that Shaped the Field, pp. 347–353 (1998)
12. Ma, Q., Saito, S., Yang, J., Tang, S., Black, M.J.: SCALE: modeling clothed humans with a surface codec of articulated local elements. In: Proceedings of the IEEE/CVF Conference on Computer Vision and Pattern Recognition, pp. 16082–16093 (2021)
13. Ma, Q., et al.: Learning to dress 3D people in generative clothing. In: Proceedings of the IEEE/CVF Conference on Computer Vision and Pattern Recognition, pp. 6469–6478 (2020)
14. Ma, Q., Yang, J., Tang, S., Black, M.J.: The power of points for modeling humans in clothing. In: Proceedings of the IEEE/CVF International Conference on Computer Vision, pp. 10974–10984 (2021)
15. Mescheder, L., Oechsle, M., Niemeyer, M., Nowozin, S., Geiger, A.: Occupancy networks: learning 3D reconstruction in function space. In: Proceedings of the IEEE/CVF Conference on Computer Vision and Pattern Recognition, pp. 4460–4470 (2019)
16. Park, J.J., Florence, P., Straub, J., Newcombe, R., Lovegrove, S.: DeepSDF: learning continuous signed distance functions for shape representation. In: Proceedings of the IEEE/CVF Conference on Computer Vision and Pattern Recognition, pp. 165–174 (2019)
17. Pavlakos, G., et al.: Expressive body capture: 3D hands, face, and body from a single image. In: Proceedings of the IEEE/CVF Conference on Computer Vision and Pattern Recognition, pp. 10975–10985 (2019)
18. Saito, S., Huang, Z., Natsume, R., Morishima, S., Kanazawa, A., Li, H.: PIFu: pixel-aligned implicit function for high-resolution clothed human digitization. In: Proceedings of the IEEE/CVF International Conference on Computer Vision, pp. 2304–2314 (2019)
19. Saito, S., Simon, T., Saragih, J., Joo, H.: PIFuHD: multi-level pixel-aligned implicit function for high-resolution 3D human digitization. In: Proceedings of the IEEE/CVF Conference on Computer Vision and Pattern Recognition, pp. 84–93 (2020)
20. Su, Z., Yu, T., Wang, Y., Liu, Y.: DeepCloth: neural garment representation for shape and style editing. IEEE Trans. Pattern Anal. Mach. Intell. **45**(2), 1581–1593 (2022)
21. Xiu, Y., Yang, J., Tzionas, D., Black, M.J.: ICON: implicit clothed humans obtained from normals. In: 2022 IEEE/CVF Conference on Computer Vision and Pattern Recognition (CVPR), pp. 13286–13296. IEEE (2022)
22. Yu, T., Zheng, Z., Guo, K., Liu, P., Dai, Q., Liu, Y.: Function4D: real-time human volumetric capture from very sparse consumer RGBD sensors. In: Proceedings of the IEEE/CVF Conference on Computer Vision and Pattern Recognition, pp. 5746–5756 (2021)
23. Zhang, H., et al.: CloSET: modeling clothed humans on continuous surface with explicit template decomposition. In: Proceedings of the IEEE/CVF Conference on Computer Vision and Pattern Recognition, pp. 501–511 (2023)
24. Zhang, H., et al.: PyMAF: 3D human pose and shape regression with pyramidal mesh alignment feedback loop. In: Proceedings of the IEEE/CVF International Conference on Computer Vision, pp. 11446–11456 (2021)

25. Zhang, Y., Qu, Y., Xie, Y., Li, Z., Zheng, S., Li, C.: Perturbed self-distillation: weakly supervised large-scale point cloud semantic segmentation. In: Proceedings of the IEEE/CVF International Conference on Computer Vision, pp. 15520–15528 (2021)
26. Zhang, Y., Xie, Y., Li, C., Wu, Z., Qu, Y.: Learning all-in collaborative multiview binary representation for clustering. IEEE Trans. Neural Networks Learn. Syst. 1–14 (2022). https://doi.org/10.1109/TNNLS.2022.3202102
27. Zheng, Z., Yu, T., Liu, Y., Dai, Q.: PaMIR: parametric model-conditioned implicit representation for image-based human reconstruction. IEEE Trans. Pattern Anal. Mach. Intell. **44**(6), 3170–3184 (2021)
28. Zhu, H., Zuo, X., Wang, S., Cao, X., Yang, R.: Detailed human shape estimation from a single image by hierarchical mesh deformation. In: Proceedings of the IEEE/CVF Conference on Computer Vision and Pattern Recognition, pp. 4491–4500 (2019)

An Efficient and Consistent Solution to the PnP Problem

Xiaoyan Zhou[1], Zhengfeng Xie[1], Qida Yu[1(✉)], Yuan Zong[2], and Yiru Wang[1]

[1] Nanjing University of Information Science and Technology, Nanjing 210044, China
003550@nuist.edu.cn

[2] Southeast University, Nanjing 211189, China

Abstract. In this paper, we present a novel non-iterative algorithm for solving the pose estimation problem from a set of 3D-to-2D point correspondences, known as the Perspective-n-Point (PnP) problem. The presented algorithm is capable of achieving both geometrical and statistical optimality by exploring the geometrical constraints of the PnP problem through a nonlinear least-squares fashion, as well as accounting for observation uncertainty in the solution process. In addition, to further improve the accuracy of the presented algorithm, we introduce a method that is able to eliminate the bias of solution caused by the propagation of uncertainty, resulting in a consistent estimate. Experimental tests on synthetic data and real images (i.e., TempleRing dataset) show that the presented algorithm can well adapt to different levels of noise, and outperform state-of-the-art (SOTA) PnP algorithms in terms of accuracy and computational cost. This makes the presented algorithm eminently suitable for a wide range of application scenarios.

Keywords: PnP Problem · Statistically Optimal · Pose Estimation

1 Introduction

Given a sparse set of point correspondences between 3D scene and 2D image, inferring the pose (a compact representation of orientation and position) of a calibrated camera is referred to as the Perspective-n-Point (PnP) problem. It is a fundamental problem in computer vision, and plays an indispensable role in many location-aware applications, such as large-scale Structure-from-Motion (SfM) [1], robot perception [2,3], novel view synthesis [4,5], and human interaction understanding [6]. Due to the importance of the PnP problem, extensive

This work is supported in part by the Startup Foundation for Introducing Talent of NUIST under Grant 2022r078.

Supplementary Information The online version contains supplementary material available at https://doi.org/10.1007/978-981-99-8432-9_17.

Q. Liu et al. (Eds.): PRCV 2023, LNCS 14426, pp. 207–220, 2024.
https://doi.org/10.1007/978-981-99-8432-9_17

efforts have been devoted to its study over the past few decades, with increasingly better algorithms for solving this problem continuing to be developed.

Most existing works set out from ideal geometric relationships to build least-squares solutions that are geometrically optimal. Particularly, these works assume the features to be affected by a specific model of noise (zero mean Gaussian noise), and treat feature location as equally accurate. Therefore, they do not take observation and propagation uncertainties into account in their problem formulations. However, the presence of uncertainty in the data inevitably arises in lots of real-world applications, where the above-mentioned methods may yield suboptimal results. Besides, some recent works have shown that incorporating the uncertainty associated to each feature point into solution process could indeed improve the accuracy of PnP methods. Thus, it is promising to revisit the PnP problem from the perspective of statistics.

In this paper, a novel formulation of a consistent and uncertainty-weighted solution to the PnP problem is presented, called CuPnP. For measurement noise of projection points, we model the uncertainty from scale and direction, and use covariance matrix to describe the uncertainty. Based on the covariance matrix, we construct an uncertainty weighting matrix, which fuses the uncertainty with the process of pose estimation. For the propagation problem of uncertainty, in refer to [9], estimate a uniform variance, and use deviation compensation to correct the least-squares solution to eliminate uncertainty in the quadratic solution. Therefore, our algorithm is geometrically optimal and statistically optimal. We compare the proposed solver with mainstream solvers using synthetic data and the TempleRing dataset[1]. The results indicate that our solver has the best estimation accuracy and the fastest time.

In summary, the main contributions of this paper have two aspects: 1) We consider the problem of uncertainty distribution from scale and direction, integrating uncertainty into pose estimation, making the algorithm more applicable; 2) We consider the problem of uncertainty propagation and eliminate bias in the quadratic solution.

2 Related Work

The minimal PnP problem (i.e., P3P) [7] has been proposed for a century and systematically studied by scholars. Closed-form solutions have been presented [8]. Recent years have witnessed significant progress on the study of the PnP problem ($n > 3$). Existing solvers can be roughly categorized into iterative solvers and non-iterative solvers.

Iterative solvers generally establish objective functions based on geometric or algebraic errors, directly minimizing the function through iterative methods. LHM [14] utilizes the inverse projection idea of monocular space to obtain the camera pose by minimizing the collinearity error in object space. EPPnP [15] uses iterative methods to improve EPnP [10], using redundant points as additional information to improve the stability of the results.

[1] http://grail.cs.washington.edu/projects/mview.

The early non-iterative solvers have high time complexity and inaccurate results, which are not convenient for practical application. EPnP [10] is the first solver whose time complexity is $O(n)$ and can accurately obtain the camera pose. It represents all world reference points as four control points and solves a polynomial equation system using linear techniques to obtain a closed-form solution for the pose. DLS [12] solve the PnP problem as a problem of minimizing the non-linear least-squares cost function. In addition, most advanced PnP solvers use the method of solving polynomial equations, such as RPnP [11], OPnP [13].

However, existing PnP solvers rarely model the observation errors of feature points. The solvers that took projection noise into account includes CEEPnP [16], MLPnP [17], EPnPU [18], and CPnP [19]. CEEPnP [16] uses an iterative method, which requires high accuracy of initial values and its solution is not stable. MLPnP [17] additionally uses the GN algorithm to optimize the solution, although the results are accurate, they sacrifice time costs and cannot balance accuracy and efficiency. EPnPU [18] approximates the uncertainty to isotropy, and needs to know the uncertainty covariance matrix of the initial value calculation feature points in advance. Although CPnP [19] takes into account the statistical characteristics of errors and eliminates deviations in the solution, it only considers the uncertainty of scale, and its effectiveness is not good in practical applications.

Our solver takes into account uncertainty from scale and direction, which is in line with practical application scenarios. Due to the spread of uncertainty, the least-squares solution is biased, we give a consistent estimate of the variance of projection noises, based on which the asymptotic bias is eliminated, yielding a consistent estimate of the camera pose. Moreover, our solver achieves both geometrically and statistically optimal. The accuracy and computational time of the results are better than the current best solvers.

3 Method

Notations: We use bold letters to represent matrices and vectors, while scalars are represented in italics, e.g., $\mathbf{R}$, $\mathbf{q}$, x. $\mathbf{1}_{i\times j}$ and $\mathbf{0}_{i\times j}$ represents a $i \times j$ matrix whose all elements are all 1 and 0. $\mathbf{I}_n$ is a $n \times n$ identity matrix.

3.1 Problem Formulation

Assuming that we have a set of 3D points and their corresponding 2D points. According to different reference frames, 3D points are represented as $\mathbf{p}_i^W = (x_i^W, y_i^W, z_i^W)^T$ and $\mathbf{p}_i^C = (x_i^C, y_i^C, z_i^C)^T$ in the world frame $\{W\}$ and camera frame $\{C\}$ respectively. Let $\mathbf{q}_i = (u_i, v_i)^T$ denote 2D points in the pixel plane, as shown in Fig. 1. Using the rigid body transformation formula and projection equation to align the three coordinate frames:

$$d_i \begin{bmatrix} \mathbf{q}_i \\ 1 \end{bmatrix} = \mathbf{K}\mathbf{p}_i^C = \mathbf{K}(\mathbf{R}\mathbf{p}_i^W + \mathbf{t}), \mathbf{K} = \begin{bmatrix} f_x & & u_0 \\ & f_y & v_0 \\ & & 1 \end{bmatrix} \tag{1}$$

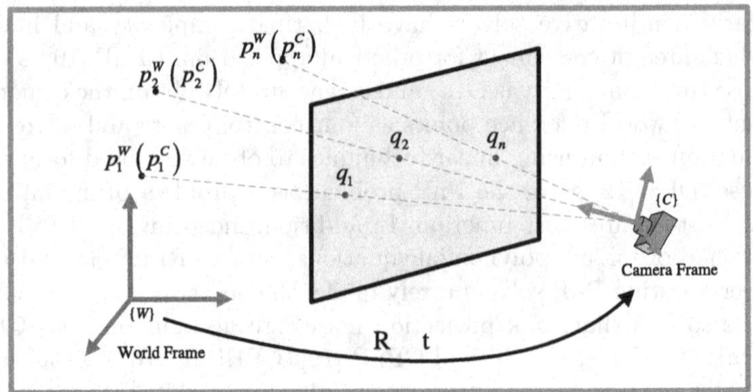

Fig. 1. Schematic representation of the PnP problem.

where $\mathbf{R}$ represents the rotation matrix and $\mathbf{t}$ represents the translation vector between the world frame and the camera frame. d_i is the depth information. $\mathbf{K}$ is the internal parameter matrix of the camera, composed of the pixel focal length f_x in the x-direction, the pixel focal length f_y in the y-direction, and the center coordinate (u_0, v_0) of the imaging plane. In reality, the internal parameter matrix of the camera is known.

By dividing the depth into both sides, Eq. (1) can be rewritten as:

$$\mathbf{q'}_i = \mathbf{q}_i - \left[u_0 \; v_0\right]^{\mathrm{T}} == \frac{\mathbf{FE}\left(\mathbf{Rp}_i^W + \mathbf{t}\right)}{d_i} \tag{2}$$

in which

$$\mathbf{F} = \begin{bmatrix} f_x & \\ & f_y \end{bmatrix}, \mathbf{E} = \begin{bmatrix} 1\,0\,0 \\ 0\,1\,0 \end{bmatrix} \tag{3}$$

We introduce $\mathbf{q'}_i = (u'_i, v'_i)^T$ to omit $(u_0, v_0)^T$. In reality, 2D points are corrupted with noise. We will analyze the uncertainty of 2D points in the next section. Assuming that the noise is $\boldsymbol{\eta}_i \in \mathbb{R}^{2\times 1}$, Eq. (2) can be rewritten as:

$$\boldsymbol{\eta}_i = \mathbf{q'}_i - \frac{\mathbf{FE}\left(\mathbf{Rp}_i^W + \mathbf{t}\right)}{d_i} \tag{4}$$

Then, the PnP problem is transformed into a closed-form least-squares problem as:

$$\arg\min \boldsymbol{\eta}_i^2, \text{s.t. } \mathbf{R} \in SO(3) \tag{5}$$

where $SO(3)$ is a constraint condition for $\mathbf{R}$, indicating that $\mathbf{R}$ should satisfy $\mathbf{R}^T\mathbf{R} = \mathbf{I}_3, \det \mathbf{R} = 1$.

3.2 Uncertainty Weighted

The local feature structure of an image introduces significant variations in the uncertainty of projection points. As show in Fig. 2(a), when the projection point

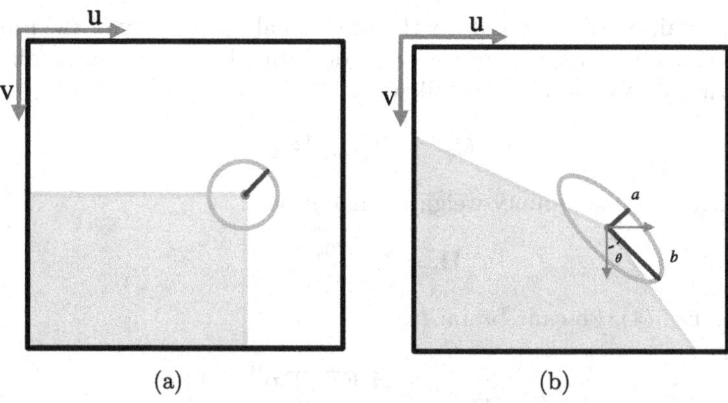

Fig. 2. Schematic diagram of 2D point uncertainty. (a) Uncertainty of right angle points, the uncertainty is scaled with no directional differences. (b) Uncertainty of non-right angle points, the uncertainty is directional with obvious directionality, where a and b represent the major and minor axes of the ellipse, θ is the angle at which the ellipse deviates

is a right-angled corner, the local features of the image do not cause any prominent directional uncertainty. Conversely, as indicated in Fig. 2(b), when the projection point is a non-right angle points, the local features of the image lead to substantial directionality, and the uncertainty of projection points is directional. Hence, in refer to [20,21], we formulate the uncertainty of projection point by modeling the image features surrounding it, as follows:

$$\mathbf{C}_i = \sum_{(u_i,v_i)\in\Omega_i} w_i(u_i,v_i) \begin{bmatrix} \nabla_u\nabla_u & \nabla_v\nabla_u \\ \nabla_u\nabla_v & \nabla_v\nabla_v \end{bmatrix} \tag{6}$$

where $\mathbf{C}_i$ is the covariance matrix of the error of the feature points, Ω_i is the ellipse area centered on the feature points $q_i = (u_i, v_i)^T$, and $w_i(u_i, v_i)$ is the sum of the pixel grayscale of the ellipse area centered (u_i, v_i), ∇_u and ∇_v are the gradients of the image in the u and v directions respectively. The noise $\boldsymbol{\eta}_i$ in Eq. (4) should be $\boldsymbol{\eta}_i \sim N(0, \mathbf{C}_i)$. Due to the different uncertainties of each feature point, their constraints on the solution are also different. Therefore, it is necessary to use a weighted method to consider the contribution of the uncertainty of each feature point to the solution. The covariance matrix $\mathbf{C}_i$ is a semi positive-definite matrix, and its SVD can obtain:

$$\mathbf{C}_i = \mathbf{U}_i \sum\nolimits_i \mathbf{U}_i^T \tag{7}$$

with

$$\sum\nolimits_i = \begin{bmatrix} \sigma_1^2 & \\ & \sigma_2^2 \end{bmatrix}, \mathbf{U}_i^T = \begin{bmatrix} \cos\theta & \sin\theta \\ -\sin\theta & \cos\theta \end{bmatrix} \tag{8}$$

As shown in Fig. 2(b), σ_1^2 and σ_2^2 is the variance of uncertainty along the maximum and minimum directions, namely the minor axis a and major axis b

of the uncertainty ellipse. θ reflects the directional uncertainty of the projection point, specifically manifested as the angle deviating from the u and v directions.

Performing SVD on $\mathbf{C}_i^{-1}$, as follows:

$$\mathbf{C}_i^{-1} = \mathbf{U}_i^T \mathbf{\Sigma}_i^{-1} \mathbf{U}_i \tag{9}$$

We can define an uncertainty-weighted matrix $\mathbf{H}_i$:

$$\mathbf{H}_i = \mathbf{\Sigma}_i^{-1/2} \mathbf{U}_i^T \tag{10}$$

Weighting Eq. (4), we can obtain:

$$\begin{aligned} \boldsymbol{\eta}'_i &= \mathbf{H}_i \mathbf{q}'_i - \frac{\mathbf{H}_i \mathbf{F} \mathbf{E} \left(\mathbf{R} \mathbf{p}_i^W + \mathbf{t} \right)}{d_i} \\ &= \hat{\mathbf{q}}_i - \frac{\mathbf{H}_i \mathbf{F} \mathbf{E} \left(\mathbf{R} \mathbf{p}_i^W + \mathbf{t} \right)}{d_i} \end{aligned} \tag{11}$$

with $\hat{\mathbf{q}}_i = \mathbf{H}_i \mathbf{q}'_i = (\hat{u}_i, \hat{v}_i)^T$. $\hat{\mathbf{q}}_i = (\hat{u}_i, \hat{v}_i)^T$ is the weighted imaging plane feature points. After weighting, the error η_i can be considered as an isotropic and independent distributed gaussian noise. Currently, we have used a weighted method to change the uncertainty of each feature point from direction to scale, also convert feature points from geometric space to statistical space. It is worth noting that $\hat{\mathbf{q}}_i = (\hat{u}_i, \hat{v}_i)^T$ is erroneous, which results in biased closed-form solutions. In the following section, we will introduce a method for estimating the global consistent noise variance and use bias compensation to correct the least-squares solution.

3.3 Modified Objective Function

To obtain a closed form solution, we rewrite Eq. (11). Multiplying d_i on both sides of Eq. (11), as follows:

$$\mathbf{H}_i \mathbf{F} \mathbf{E} \left(\mathbf{R} \mathbf{p}_i^W + \mathbf{t} \right) - d_i \hat{\mathbf{q}}_i + d_i \boldsymbol{\eta}'_i = 0 \tag{12}$$

Let $\mathbf{L}_i = [\mathbf{p}_i^{WT} \ 1] \otimes \mathbf{I}_{3\times3}$, and rewriting Eq. (12) into matrix form:

$$\mathbf{M}\mathbf{x} + \boldsymbol{\zeta} = 0 \tag{13}$$

with $\mathbf{M} = [(\mathbf{H}_1\mathbf{F}\mathbf{E} - \hat{\mathbf{q}}_1\mathbf{e})\mathbf{L}_1{}^T, ..., \mathbf{H}_n\mathbf{F}\mathbf{E} - \hat{\mathbf{q}}_n\mathbf{e})\mathbf{L}_n{}^T]^T$, $\boldsymbol{\zeta} = [d_1\boldsymbol{\eta}'_1{}^T, ..., d_n\boldsymbol{\eta}'_n{}^T]^T$. $\mathbf{x}$ is a column vector composed of unknowns $\mathbf{R}$ and $\mathbf{t}$. However, due to scale ambiguity, $\mathbf{M}$ is not a column full rank matrix, and because of the regression variable is $\mathbf{0}$, the closed-form solution cannot be obtained from Eq. (13). Therefore, we introduce the following constraints to eliminate scale ambiguity:

$$\alpha \sum_{i=1}^{n} d_i = n, \ \bar{\mathbf{p}}^W = \frac{1}{n} \sum_{i=1}^{n} \mathbf{p}_i^W \tag{14}$$

where α is a scaling factor, $\bar{\mathbf{p}}^W$ is the centroid of all points in the world coordinate. In order to derive the formula, let $\mathbf{R} = [\mathbf{r}_1, \mathbf{r}_2, \mathbf{r}_3]^T$, $\mathbf{r}_i$ is a 3×1 column vector, and $\mathbf{t} = [t_1, t_2, t_3]^T$. By combining two constraints Eq. (14), we obtain:

$$\alpha t_3 = 1 - \alpha \mathbf{r_3} \bar{\mathbf{p}}^W \tag{15}$$

Substituting Eq. (15) into Eq. (12) leads to:

$$\alpha \mathbf{H}_i \mathbf{F} \mathbf{E} \left(\mathbf{R} \mathbf{p}_i^W + \mathbf{t} \right) - \alpha (\mathbf{p}_i^W - \bar{\mathbf{p}}^W)^T \mathbf{r_3} \hat{\mathbf{q}}_i + [1 + \alpha (\mathbf{p}_i^W - \bar{\mathbf{p}}^W)^T \mathbf{r_3}] \boldsymbol{\eta}'_i = \hat{\mathbf{q}}_i \tag{16}$$

Equation (16) can be transformed into matrix form:

$$\mathbf{M}'\mathbf{x}' + \boldsymbol{\zeta}' = \mathbf{b} \tag{17}$$

Specifically, $\mathbf{b} = [\hat{\mathbf{q}}_1, ..., \hat{\mathbf{q}}_n]^T$, $\mathbf{x}'$ includes unknowns α, $\mathbf{R}$, t_1, t_2. The forms of $\boldsymbol{\zeta}'$, $\mathbf{x}'$ and $\mathbf{M}'$ are:

$$\boldsymbol{\zeta}' = \left[[1 + \alpha (\mathbf{p}_i^W - \bar{\mathbf{p}}^W)^T \mathbf{r_3}] \boldsymbol{\eta}'_1, ..., [1 + \alpha (\mathbf{p}_i^W - \bar{\mathbf{p}}^W)^T \mathbf{r_3}] \boldsymbol{\eta}'_n \right]^T \tag{18}$$

$$\mathbf{x}' = \alpha \left[\mathbf{r}_3^T \; \mathbf{r}_1^T \; t_1 \; \mathbf{r}_2^T \; t_2 \right]^T \tag{19}$$

$$\mathbf{M}' = \begin{bmatrix} -\hat{u}_1 (\mathbf{p_1^W} - \bar{\mathbf{p}}^\mathbf{W})^T & H_1^{11} f_x \mathbf{p}_1^{W^T} & H_1^{11} f_x & H_1^{21} f_y \mathbf{p}_1^{W^T} & H_1^{21} f_y \\ -\hat{v}_1 (\mathbf{p_1^W} - \bar{\mathbf{p}}^\mathbf{W})^T & H_1^{12} f_x \mathbf{p}_1^{W^T} & H_1^{12} f_x & H_1^{22} f_y \mathbf{p}_1^{W^T} & H_1^{22} f_y \\ \vdots & \vdots & \vdots & \vdots & \vdots \\ -\hat{u}_n (\mathbf{p_n^W} - \bar{\mathbf{p}}^\mathbf{W})^T & H_n^{11} f_x \mathbf{p}_n^{W^T} & H_n^{11} f_x & H_n^{21} f_y \mathbf{p}_n^{W^T} & H_n^{21} f_y \\ -\hat{v}_n (\mathbf{p_n^W} - \bar{\mathbf{p}}^\mathbf{W})^T & H_n^{12} f_x \mathbf{p}_n^{W^T} & H_n^{12} f_x & H_n^{22} f_y \mathbf{p}_n^{W^T} & H_n^{22} f_y \end{bmatrix} \tag{20}$$

with H_i^{mn} is the element in the m-th row and n-th column of $\mathbf{H}_i$. Obviously, $\mathbf{M}'$ is full rank. Therefore, the closed-form solution can be obtained according to Eq. (17), as follows:

$$\mathbf{x}' = \left(\mathbf{M}'^T \mathbf{M}' \right)^{-1} \mathbf{M}'^T \mathbf{b} \tag{21}$$

As previously mentioned, although we have removed the directional uncertainty of 2D points, scale uncertainty remains a challenge. In reference to Eq. (17), $\hat{\mathbf{q}}_i = (\hat{u}_i, \hat{v}_i)^T$ in $\mathbf{M}'$ and $\mathbf{b}$ still includes errors, ie. $\boldsymbol{\eta}'_i \sim N(0, \sigma_i^2 \mathbf{I})$, which causes bias in the closed-form solution $\mathbf{x}'$. Accounting for the influence of scale uncertainty (i.e. σ_i^2) of each point $\hat{\mathbf{q}}_i = (\hat{u}_i, \hat{v}_i)^T$ is complicated and challenging, and requires significant computing consumption. To resolve this issue, we estimate a consistent variance $\hat{\sigma}^2$ (assuming that $\boldsymbol{\eta}'_i \sim N(0, \hat{\sigma}^2 \mathbf{I})$) and utilize bias compensation to counteract the impact of errors on $\mathbf{x}'$.

3.4 Bias Compensation and Consistent Variance

Recalling Eq. (16), we assume that $\hat{\mathbf{q}}_i = (\hat{u}_i, \hat{v}_i)^T$ in $\mathbf{M}'$ and $\mathbf{b}$ is free-noise, denoted as $\mathbf{q}_i^o = (u_i^o, v_i^o)^T$:

$$\hat{\mathbf{q}}_i = \mathbf{q}_i^o + \boldsymbol{\eta}'_i \tag{22}$$

Substituting Eq. (22) into Eq. (16) leads to:

$$\alpha \mathbf{H}_i \mathbf{F}\mathbf{E}\left(\mathbf{R}\mathbf{p}_i^W + \mathbf{t}\right) - \alpha(\mathbf{p}_i^W - \bar{\mathbf{p}}^W)^T \mathbf{r}_3 \mathbf{q}_i^o = \mathbf{q}_i^o \tag{23}$$

Equation (23) can be transformed into matrix form:

$$\mathbf{M'}_1 \mathbf{x'}_1 = \mathbf{b}_1 \tag{24}$$

with $\mathbf{b}_1 = [q_1^o, ..., q_1^o]^T$. $\mathbf{x'}_1$ is as follows:

$$\mathbf{x'}_1 = \left(\mathbf{M'}_1^T \mathbf{M'}_1\right)^{-1} \mathbf{M'}_1^T \mathbf{b}_1 \tag{25}$$

$\mathbf{x'}_1$ does not include the influence of noise. But $\mathbf{q}_i^o = (u_i^o, v_i^o)^T$ is unknown, so we cannot directly obtain $\mathbf{x'}_1$. Assuming that 2D points in $\mathbf{M'}_1$ are still noise-free, while 2D points in $\mathbf{b}_1$ are corrupted by noise, Eq. (23) can be rewritten as follows:

$$\alpha \mathbf{H}_i \mathbf{F}\mathbf{E}\left(\mathbf{R}\mathbf{p}_i^W + \mathbf{t}\right) - \alpha(\mathbf{p}_i^W - \bar{\mathbf{p}}^W)^T \mathbf{r}_3 \mathbf{q}_i^o + \boldsymbol{\eta}'_i = \hat{\mathbf{q}}_i \tag{26}$$

Equation (26) can be transformed into matrix form:

$$\mathbf{M'}_1 \mathbf{x'}_2 + \boldsymbol{\xi} = \mathbf{b} \tag{27}$$

with $\boldsymbol{\xi} = [\boldsymbol{\eta}'_1, ..., \boldsymbol{\eta}'_n]^T$. The solution $\mathbf{x'}_2$ is as follows:

$$\mathbf{x'}_2 = \left(\mathbf{M'}_1^T \mathbf{M'}_1\right)^{-1} \mathbf{M'}_1^T \mathbf{b} \tag{28}$$

Lemma 1 [Theorem A.1 in [9]]: Let $\{X_k\}$ be a random sequence with $\mathbf{E}[X_k] = 0$ and $\mathbf{E}[X_k^2] < \infty$ for all k. Then, there holds $\sum_{k=1}^n X_k/\sqrt{n} = O_p(1)$.

By using Lemma 1, we obtain:

$$\begin{aligned}
\mathbf{x'}_2 &= \left(\frac{1}{n}\mathbf{M'}_1{}^T \mathbf{M'}_1\right)^{-1} \left(\frac{1}{n}\mathbf{M'}_1{}^T \mathbf{b}\right) \\
&= \left(\frac{1}{n}\mathbf{M'}_1{}^T \mathbf{M'}_1\right)^{-1} \left(\frac{1}{n}\mathbf{M'}_1{}^T \mathbf{M'}_1 \mathbf{x'}_1 + \frac{1}{n}\mathbf{M'}_1{}^T \eta'\right) \\
&= \mathbf{x'}_1 + O_p(1/\sqrt{n})
\end{aligned} \tag{29}$$

We conclude that $\mathbf{x'}_2 - \mathbf{x'}_1 = O_p(1/\sqrt{n})$. However, since $\mathbf{x'}_1$ is unknown, we cannot directly obtain the closed-form solution $\mathbf{x'}_2$. $\mathbf{M'}$ and $\mathbf{M'}_1$ exist bias as well. We define a bias matrix $\mathbf{Z}$ and a noise-free matrix $\mathbf{Z}_1$, as follows:

$$\mathbf{Z} = \mathbf{M'} - \mathbf{M'}_1 = \begin{bmatrix} -\eta'_{11}(\mathbf{p}_1^W - \bar{\mathbf{p}}^W)^{\mathrm{T}} & \mathbf{0}_{1\times 8} \\ -\eta'_{12}(\mathbf{p}_1^W - \bar{\mathbf{p}}^W)^{\mathrm{T}} & \mathbf{0}_{1\times 8} \\ \vdots & \vdots \\ -\eta'_{n1}(\mathbf{p}_n^W - \bar{\mathbf{p}}^W)^{\mathrm{T}} & \mathbf{0}_{1\times 8} \\ -\eta'_{n2}(\mathbf{p}_n^W - \bar{\mathbf{p}}^W)^{\mathrm{T}} & \mathbf{0}_{1\times 8} \end{bmatrix}, \mathbf{Z}_1 = \begin{bmatrix} -(\mathbf{p}_1^W - \bar{\mathbf{p}}^W)^{\mathrm{T}} & \mathbf{0}_{1\times 8} \\ -(\mathbf{p}_1^W - \bar{\mathbf{p}}^W)^{\mathrm{T}} & \mathbf{0}_{1\times 8} \\ \vdots & \vdots \\ -(\mathbf{p}_n^W - \bar{\mathbf{p}}^W)^{\mathrm{T}} & \mathbf{0}_{1\times 8} \\ -(\mathbf{p}_n^W - \bar{\mathbf{p}}^W)^{\mathrm{T}} & \mathbf{0}_{1\times 8} \end{bmatrix} \tag{30}$$

where $\boldsymbol{\eta}'_{\mathrm{ij}}$ is the j-th element of $\boldsymbol{\eta}'_{\mathrm{i}}$. Note that:

$$\begin{aligned}&\frac{1}{n}\mathbf{M}'^T\mathbf{M}' - \frac{1}{n}\mathbf{M}'_1{}^T\mathbf{M}'_1\\&= \frac{1}{n}\left(\mathbf{M}'_1{}^T\mathbf{Z}+\mathbf{Z}^T\mathbf{M}'_1 + \mathbf{Z}^T\mathbf{Z}\right)\\&= \frac{1}{n}\mathbf{Z}^T\mathbf{Z} + O_p(1/\sqrt{n})\\&= \hat{\sigma}^2\frac{1}{n}\mathbf{Z}_1{}^T\mathbf{Z}_1 + O_p(1/\sqrt{n})\end{aligned} \tag{31}$$

and

$$\begin{aligned}\frac{1}{n}\mathbf{M}'^T\mathbf{b} - \frac{1}{n}\mathbf{M}'_1{}^T\mathbf{b} &= \frac{1}{n}\mathbf{Z}^T\mathbf{b}\\&= \hat{\sigma}^2\frac{1}{n}\mathbf{Z}_1{}^T\mathbf{I}_{2n\times 1} + O_p(\frac{1}{\sqrt{n}})\end{aligned} \tag{32}$$

According to Eq. (31) and Eq. (32), we can obtain a corrected least-squares solution $\mathbf{x}'_3$:

$$\mathbf{x}'_3 = (\mathbf{M}'^T\mathbf{M}' - \hat{\sigma}^2\mathbf{Z}_1{}^T\mathbf{Z}_1)^{-1}(\mathbf{M}'^T\mathbf{b} - \hat{\sigma}^2\mathbf{Z}_1{}^T\mathbf{I}_{2n\times 1}) \tag{33}$$

$\mathbf{x}'_2$ satisfies $\mathbf{x}'_2 - \mathbf{x}'_1 = O_p(1/\sqrt{n})$, so $\mathbf{x}'_3 - \mathbf{x}'_1 = O_p(1/\sqrt{n})$. Therefore, the key to obtaining the closed-form solution $\mathbf{x}'_3$ is to estimate the variance $\hat{\sigma}^2$.

We reference the method [9] to obtain noise variance $\hat{\sigma}^2$ by solving generalized eigenvalues. Based on the existing information matrix, we can define $\boldsymbol{\Delta}$ and $\boldsymbol{\Phi}$, as follows:

$$\boldsymbol{\Delta} = \frac{1}{n}\begin{bmatrix}\mathbf{Z}^T\mathbf{Z} & \mathbf{Z}^T\mathbf{1}_{2n\times 1}\\ \mathbf{1}_{2n\times 1}^T\mathbf{Z} & 2n\end{bmatrix} \tag{34}$$

$$\boldsymbol{\Phi} = \frac{1}{n}\begin{bmatrix}\mathbf{M}'^T\mathbf{M}' & \mathbf{M}'^T\mathbf{x}'\\ \mathbf{M}'\mathbf{x}'^T & \mathbf{x}'^T\mathbf{x}'\end{bmatrix} \tag{35}$$

Based on $\boldsymbol{\Delta}$ and $\boldsymbol{\Phi}$, we define an equation $\mathbf{G}(\lambda) = \boldsymbol{\Phi} - \lambda\boldsymbol{\Delta}$ about variables λ. $\mathbf{G}(\lambda)$ is a polynomial of power 4 on λ. Solving equation $\mathbf{G}(\lambda) = 0$ yields four solutions $\hat{\sigma}^2$, and the noise variance $\hat{\sigma}^2$ is the minimum value of λ, as follows:

$$\hat{\sigma}^2 = \min\{\lambda_1, \lambda_2, \lambda_3, \lambda_4\} \tag{36}$$

3.5 Recovering Camera Pose

The corrected least-squares solution is:

$$\mathbf{x} = (\mathbf{M}'^T\mathbf{M}' - \hat{\sigma}^2\mathbf{M}'^T\mathbf{M}')^{-1}(\mathbf{M}'^T\mathbf{b} - \hat{\sigma}^2\mathbf{M}'^T\mathbf{1}_{2n\times 1}) \tag{37}$$

Recover α, $\mathbf{R}$ and $\mathbf{t}$ from Eq. (37):

$$\alpha = (\det([\,[\mathbf{x}]_{4:6}\,[\mathbf{x}]_{8:10}\,[\mathbf{x}]_{1:3}]^T))^{1/3} \tag{38}$$

$$\hat{\mathbf{R}} = \left[[\mathbf{x}]_{4:6}\ [\mathbf{x}]_{8:10}\ [\mathbf{x}]_{1:3} \right]^T / \alpha \tag{39}$$

$$\mathbf{t} = \left[[\mathbf{x}]_7\ [\mathbf{x}]_{11}\ 1 - \bar{\mathbf{p}}^{WT} [\mathbf{x}]_{1:3} \right]^T / \alpha \tag{40}$$

where $\hat{\mathbf{R}}$ does not necessarily satisfy constraint $\mathbf{R} \in SO(3)$. We can use SVD decomposition to satisfy the constraint:

$$\hat{\mathbf{R}} = \mathbf{U}_{\hat{\mathbf{R}}} \boldsymbol{\Sigma}_{\hat{\mathbf{R}}} \mathbf{V}_{\hat{\mathbf{R}}}^T \tag{41}$$

Then, the optimal rotation matrix satisfying the constraint is:

$$\mathbf{R} = \mathbf{U}_{\hat{\mathbf{R}}} \mathrm{diag}\left(\left[1\ 1\ \det(\mathbf{U}_{\hat{\mathbf{R}}} \mathbf{V}_{\hat{\mathbf{R}}}^T) \right] \right) \mathbf{V}_{\hat{\mathbf{R}}}^T \tag{42}$$

4 Experiments

In this section, we compare our solver with other solvers, using synthetic and real data. We evaluate the performance of solvers from both estimated accuracy and time cost. All the experiments are ran on a laptop equipped with Core i7-8750H 2.20 GHz and 8 Gb RAM. We use absolute rotation error $e_{\text{rot}} = \left| \mathrm{acos}\left(0.5 \left(\mathrm{trace}\left(\mathbf{R}_{\text{true}}^T \mathbf{R}\right) - 1\right)\right) \right|$ and relative translation error $e_{\text{trans}} = \|\mathbf{t}_{\text{true}} - \mathbf{t}\| / \|\mathbf{t}_{\text{true}}\| \times 100\%$ to evaluate the performance of the solve.

4.1 Synthetic Data Experiments

In the synthesis experiment, the 2D point $p_i^C = (x_i^C, y_i^C, z_i^C)^T$ is randomly generated in the cube $[-2, 2] \times [-2, 2] \times [4, 8]$ in the camera coordinate. We assume that the focal length of the virtual camera is $f_x = f_y = 800$, and the pixel image size is 640×480. We randomly generate rotation matrix $\mathbf{R}$ and translation vector $\mathbf{t}$. We obtain $p_i^W = (x_i^W, y_i^W, z_i^W)^T$ by transforming $p_i^C = (x_i^C, y_i^C, z_i^C)^T$ into the world coordinate system through $\mathbf{R}$ and $\mathbf{t}$. By using camera internal reference, we project $p_i^C = (x_i^C, y_i^C, z_i^C)^T$ onto the imaging plane to obtain $q_i = (u_i, v_i)^T$.

We add various directional uncertainty to the projection points $q_i = (u_i, v_i)^T$ by dividing 330 points into ten subsets with an equal number of points, increasing the standard deviation σ_1 from 1 to 10, and adding them to different point sets in sequence. The standard deviation σ_2 is fixed at 1, 5, 10, and randomly add to ten subsets from 1 to 10. Meanwhile, we give random directionality. We run each parameter condition 400 times and then take the average value as the experimental result.

Figure 3 presents a summary of the experimental results, wherein we simulate various noise situations for the projection points. Irrespective of the noise intensity, our solver consistently achieves the highest accuracy as the number of points increases. However, when the number of points is less than 60, the accuracy of our solver slightly decreases compared to other solvers. When the noise interference is severe, our solver still performs better. As the number of points increases, the estimated noise variance becomes more consistent. Notably, the

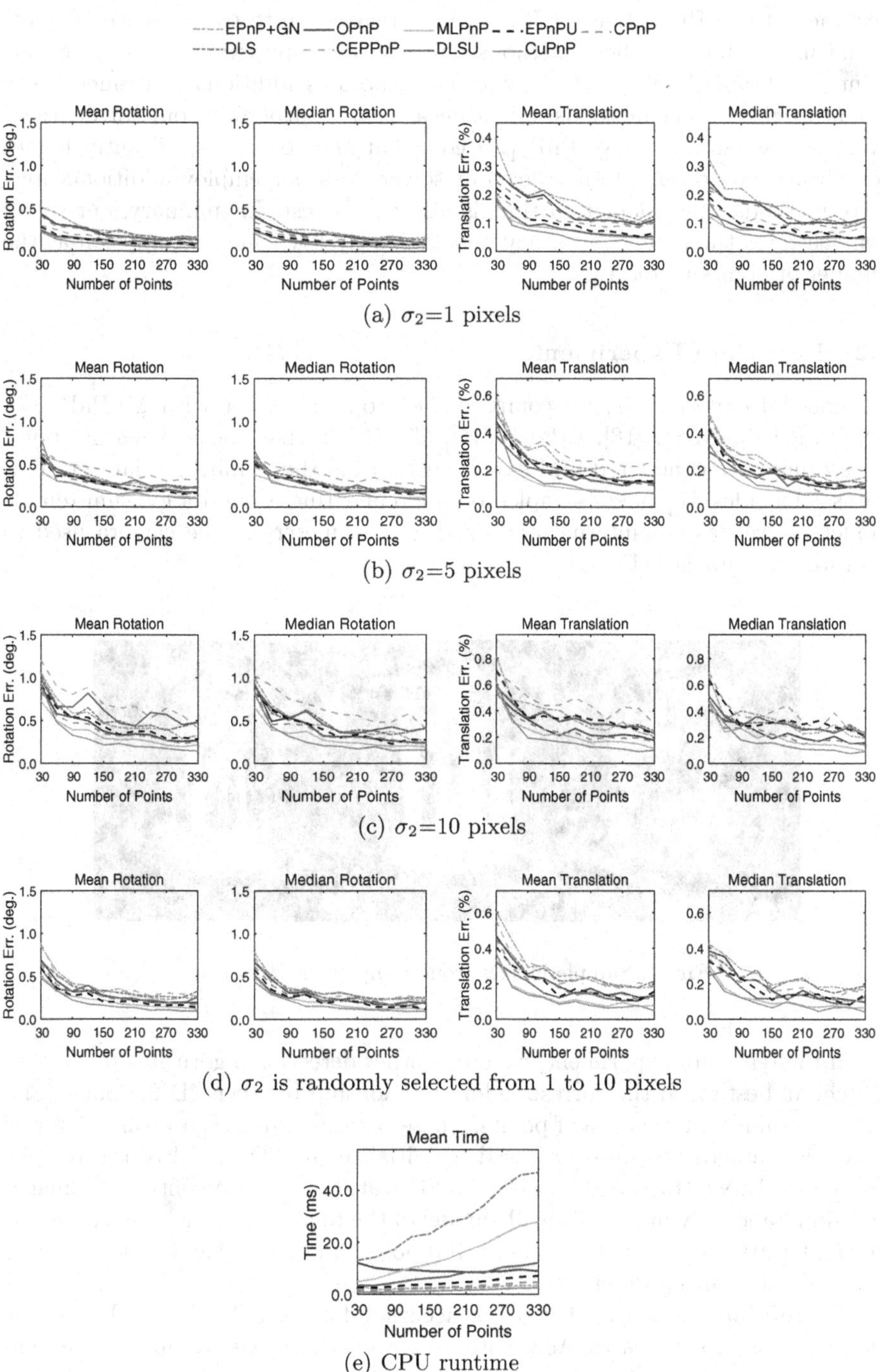

(a) σ_2=1 pixels

(b) σ_2=5 pixels

(c) σ_2=10 pixels

(d) σ_2 is randomly selected from 1 to 10 pixels

(e) CPU runtime

Fig. 3. Pose errors in synthetic experiments and CPU time comparison

accuracy of EPnPU [18] and CPnP [19] decreases due to the presence of directional uncertainty, as these methods only consider approximately scale uncertainty. Additionally, MLPnP [17], which incorporates additional refinement steps not utilized by other methods, might occasionally outperform our solver. However, this advantage of MLPnP [17] comes at the cost of significantly higher computational consumption. Since our solver does not employ additional iterative methods, the calculation time is also the fastest. In summary, our solver demonstrates the highest accuracy and extremely low time consumption as the number of points increases.

4.2 Real Data Experiments

In practical experiments, we compare the proposed solver with MLPnP [17], CEEPnP [16], DLSU [18], EPnPU [18], CPnP [19] that consider feature point uncertainty. We conducted SfM experiments using the TempRing dataset. This dataset includes 47 images sampled on a circular ring around a gypsum temple model, as well as the internal and external parameters of the camera used to capture these images (Fig. 4).

Fig. 4. Sample images from TempleRing data sets

In the real data experiment, we used corner detection algorithms to extract, match, and establish the corresponding relationship between 2D-2D points and the uncertainty information of points. These corresponding relationships, as well as known camera postures, are used to initialize the 3D model of the temple. Subsequently, we tracked these selected 2D feature points in continuous images and simultaneously updated the 3D model of the temple. For each image, we use different PnP algorithms to calculate 100 poses, and then calculate the average rotation and translation errors.

Figure 5 illustrates the changes in accuracy for each PnP algorithm as the number of images increases. As we utilize the estimated pose to update the temple model, errors tend to accumulate with a larger number of images used. The results obtained from CEEPnP [16] are heavily skewed due to the accumulation of errors. On the other hand, MLPnP [17] and our algorithm handle error

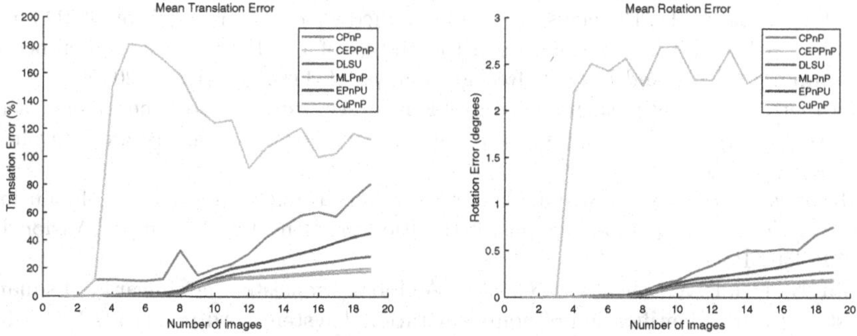

Fig. 5. Pose errors in real data experiments

accumulation remarkably well. In terms of both rotation error and translation error, our algorithm has demonstrated outstanding results. Due to space limitations, we have included the results of the SfM experiment in the supplementary materials.

5 Conclusion

In this paper, we present a novel approach to address the PnP problem by considering the uncertainty of feature points. To tackle the directional uncertainty caused by images, we employ covariance matrix to represent the uncertainty. In addition, we formulate a weighted objective function. To overcome the bias interference of errors on the closed-form solution, we estimate the noise variance and employ bias compensation to obtain a corrected least-squares solution, eliminating the noise bias. To validate the proposed method, a comprehensive evaluation is conducted, comparing it against the state-of-the-art PnP methods. Synthetic and real data experiments have shown that our algorithm outperforms other algorithms in terms of accuracy and efficiency.

References

1. Zhu, S., Zhang, R., Zhou, L., et al.: Very large-scale global SFM by distributed motion averaging. In: Proceedings of the IEEE Conference on Computer Vision and Pattern Recognition, pp. 4568–4577. IEEE (2018)
2. Liu, Y., Chen, G., Knoll, A.: Absolute pose estimation with a known direction by motion decoupling. IEEE Trans. Circ. Syst. Video Technol. (2023)
3. Du, G., Wang, K., Lian, S., et al.: Vision-based robotic grasping from object localization, object pose estimation to grasp estimation for parallel grippers: a review. Artif. Intell. Rev. **54**(3), 1677–1734 (2021)
4. Mildenhall, B., Srinivasan, P.P., Tancik, M., et al.: NERF: representing scenes as neural radiance fields for view synthesis. Commun. ACM **65**(1), 99–106 (2021)
5. Min, Z., Zhuang, B., Schulter, S., et al.: NeurOCS: neural NOCS supervision for monocular 3D object localization. In: Proceedings of the IEEE/CVF Conference on Computer Vision and Pattern Recognition, pp. 21404–21414. IEEE (2023)

6. Tekin, B., Bogo, F., Pollefeys, M.: H+O: unified egocentric recognition of 3D hand-object poses and interactions. In: Proceedings of the IEEE/CVF Conference on Computer Vision and Pattern Recognition, pp. 4511–4520. IEEE (2019)
7. Grunert, J.A.: Das pothenotische problem in erweiterter gestalt nebst bber seine anwen-dungen in der geodasie. Grunerts Archiv fur Mathematik und Physik **1**, 238–248 (1841)
8. Haralick, B.M., Lee, C.N., Ottenberg, K., et al.: Review and analysis of solutions of the three point perspective pose estimation problem. Int. J. Comput. Vision **13**, 331–356 (1994)
9. Mu, B., Bai, E.W., Zheng, W.X., et al.: A globally consistent nonlinear least squares estimator for identification of nonlinear rational systems. Automatica **77**, 322–335 (2017)
10. Lepetit, V., Moreno-Noguer, F., Fua, P.: EPnP: an accurate O(n) solution to the PnP problem. Int. J. Comput. Vision **81**, 155–166 (2009)
11. Li, S., Xu, C., Xie, M.: A robust O(n) solution to the perspective-n-point problem. IEEE Trans. Pattern Anal. Mach. Intell. **34**(7), 1444–1450 (2012)
12. Hesch, J.A., Roumeliotis, S.I.: A Direct Least-squares (DLS) method for PnP. In: 2011 International Conference on Computer Vision, pp. 383–390. IEEE (2011)
13. Zheng, Y., Kuang, Y., Sugimoto, S., et al.: Revisiting the PnP problem: a fast, general and optimal solution. In: Proceedings of the IEEE International Conference on Computer Vision, pp. 2344–2351. IEEE (2013)
14. Lu, C.P., Hager, G.D., Mjolsness, E.: Fast and globally convergent pose estimation from video images. IEEE Trans. Pattern Anal. Mach. Intell. **22**(6), 610–622 (2000)
15. Ferraz, L., Binefa, X., Moreno-Noguer, F.: Very fast solution to the PnP problem with algebraic outlier rejection. In: Proceedings of the IEEE Conference on Computer Vision and Pattern Recognition, pp. 501–508. IEEE (2014)
16. Ferraz Colomina, L., Binefa, X., Moreno-Noguer, F.: Leveraging feature uncertainty in the PnP problem. In: Proceedings of the BMVC 2014 British Machine Vision Conference, pp. 1–13. IEEE (2014)
17. Urban, S., Leitloff, J., Hinz, S.: MLPnP-a real-time maximum likelihood solution to the Perspective-n-Point problem. arXiv preprint arXiv:1607.08112 (2016)
18. Vakhitov, A., Ferraz, L., Agudo, A., et al.: Uncertainty-aware camera pose estimation from points and lines. In: Proceedings of the IEEE/CVF Conference on Computer Vision and Pattern Recognition, pp. 4659–4668. IEEE (2013)
19. Zeng, G., Chen, S., Mu, B., et al.: CPnP: consistent pose estimator for Perspective-n-Point problem with bias elimination. arXiv preprint arXiv:2209.05824 (2022)
20. Irani, M., Anandan, P.: Factorization with uncertainty. Int. J. Comput. Vision **49**, 101–116 (2002)
21. Steele, R.M., Jaynes, C.: Feature uncertainty arising from covariant image noise. In: 2005 IEEE Computer Society Conference on Computer Vision and Pattern Recognition, pp. 1063–1070. IEEE (2005)

Autoencoder and Masked Image Encoding-Based Attentional Pose Network

Longhua Hu[1], Xiaoliang Ma[1], Cheng He[1], Lei Wang[2(✉)], and Jun Cheng[2]

[1] Shenzhen University, Shenzhen 518060, China
{hulonghua2021,2110276237}@email.szu.edu.cn, maxiaoliang@szu.edu.cn
[2] Shenzhen Institute of Advanced Technology, Chinese Academy of Sciences, Beijing, China
{lei.wang1,jun.cheng}@siat.ac.cn
https://github.com/hulonghua-devin

Abstract. Despite recent advances in single-image-based 3D human pose and shape estimation, partial occlusion remains a major challenge for many methods, leading to significant prediction errors. Some existing methods fail to provide satisfactory performance for 3D human body reconstruction in occluded outdoor environments. To address these issues, we propose an autoencoder for feature extraction that integrates image masking methods to improve training stability. Our approach utilizes an attention mechanism to effectively capture the features of partially visible body parts, addressing partial occlusion. We further employ a partial attention mechanism to obtain the final features and use a regressor to estimate human model parameters. Experimental results on outdoor 3D poses in benchmark datasets demonstrate that our method outperforms state-of-the-art image-based methods in terms of robustness and efficiency. Qualitative evaluation shows that our method achieves more accurate and robust reconstruction results than existing methods, not only in occluded scenarios but also on standard benchmarks. Our approach exhibits excellent model robustness and training stability.

Keywords: 3D human pose and shape estimation · Autoencoder · Masked image encoding · Partial occlusion

1 Introduction

Direct regression of 3D human pose and shape (HPS) from RGB images is an important task in computer vision with numerous potential applications in areas such as virtual reality, robotics, and motion analysis. Such regression, achieved by using single images [13,17,25] or video sequences [14,15,21] to regress parameters of human models such as SMPL [19] as output, represents a more cost-effective and convenient alternative to methods that rely on depth cameras, IMUs, or multi-view inputs. Rapid progress in this area has been facilitated by the integration of deep convolutional neural networks (CNNs) [13,15,17,25].

Q. Liu et al. (Eds.): PRCV 2023, LNCS 14426, pp. 221–233, 2024.
https://doi.org/10.1007/978-981-99-8432-9_18

We have observed model instability and poor robustness in direct regression methods for 3D human pose and shape estimation. During training, Mean-Per-Joint-Position-Error (MPJPE) often exceed the normal range, which severely affects training efficiency. Furthermore, exaggerated poses may be reconstructed during testing. Therefore, the robustness of the model is also a critical issue to consider when reconstructing 3D human pose. Despite the solutions proposed in the literature [2,3,8,11,26,27,31,32,36], there are still limitations to the methods, such as a lack of robustness and stability in reconstructing occluded poses in real-world images, as well as suboptimal tradeoffs between reconstruction accuracy and model stability.

We designed an autoencoder to learn representations. Autoencoding is a classical method for learning representations. It consists of an encoder that maps an input to a latent representation and a decoder that reconstructs the input. One of the earliest examples of autoencoders is PCA and k-means [10]. In addition to autoencoding, we also developed an encoding method for masking images. This approach learns representations from images corrupted by masking. Furthermore, we are inspired by the effectiveness of attention-based approaches in other tasks [5,9,20,33], we leverage an attention mechanism to dynamically emphasize regions of the image that are most pertinent to the task of 3D pose estimation. This strategy enables the network to effectively utilize all relevant information from the body and surrounding pixels, while unsupervisedly learning to rely on the visible body parts to improve robustness to occlusion and overall 3D pose estimation performance.

In this work, we present AutoMaskPoseNet, a novel neural network designed for regressing 3D human body estimation from a single RGB image. AutoMaskPoseNet integrates the autoencoder and masked image encoding techniques to extract image features, resulting in reduced computational costs and improved model robustness. Additionally, our model incorporates an attention mechanism that dynamically focuses on relevant image regions, enabling effective utilization of pertinent information from both the body and surrounding pixels. By autonomously learning to rely on visible body parts, AutoMaskPoseNet enhances robustness to occlusion and enhances 3D pose estimation performance. Our evaluation on the 3DPW benchmark dataset demonstrates substantial improvements in both robustness and stability, leading to state-of-the-art performance in accurately estimating the 3D position of occluded objects. These findings provide empirical evidence supporting the effectiveness of our proposed method in addressing the challenges associated with robust 3D pose estimation.

In summary, our contributions are threefold:

- We present a novel model, called the e Masked Image Encoding Autoencoder (MIEA), which leverages both autoencoder and masked image encoding methods for extracting image features.
- We propose a novel neural network named Autoencoder and Masked Image Encoding-based Attentional Pose Network, which aims to estimate 3D human body poses from a single RGB image.

- We leverage pixel features in the vicinity of occluded joints to predict their positions, thereby achieving robustness against occlusion.
- We achieve state-of-the-art performance on benchmarks occluded as well as standard benchmarks for 3D pose estimation.

2 Related Work

The present study focuses on the task of estimating 3D human shape and pose from RGB images, with an emphasis on evaluating the strategies used by previous approaches to deal with occlusion in different scenarios. Specifically, this investigation examines a number of occlusion types, including self-occlusion, occlusion by the camera frame, and occlusion by scene objects, as well as the solutions proposed by previous studies to address these occlusion issues.

2.1 Methods for Estimating 3D Human Pose and Shape Based on Single Images

To estimate 3D human pose and shape from single images, current methods can be broadly classified into model-based and learning-based approaches.

Model-based approaches employ a pre-defined human model to estimate human pose and shape by manipulating the model's skeleton to match the observed body shape and infer the pose. While early attempts relied on keypoints and silhouettes to output 3D human body models, they exhibited limitations in terms of sensitivity to manual input, reliance on additional data, and poor generalization to real-world images. Recent advancements include using deep neural networks to directly regress SMPL parameters from pixels, and addressing the lack of 3D ground-truth data through weak supervision techniques such as 2D keypoint reprojection loss, body part segmentation, or sparse keypoints. In a different approach, Kolotouros et al. integrated HMR and SMPLify within a training loop to enhance network monitoring.

2.2 Handling Occlusions in 3D Human Shape and Pose Estimation

The ideal outcome for regressing 3D body shape should remain consistent regardless of occlusion. However, current state-of-the-art methods for pose and shape estimation such as those presented in [13,17] directly encode the entire input region as a single CNN feature after global average pooling, followed by body model parameter regression. Due to the lack of pixel-aligned structure, it is difficult for these networks to explicitly infer the positions and visibility of body parts. To achieve robustness against occlusion in these frameworks, data augmentation techniques are commonly employed. For example, frame occlusion can be simulated through image cropping as seen in [1,27], while object occlusion can be approximated by superimposing object patches on the image as in [7]. In contrast, Cheng et al. [2] propose augmentations to heatmaps containing richer

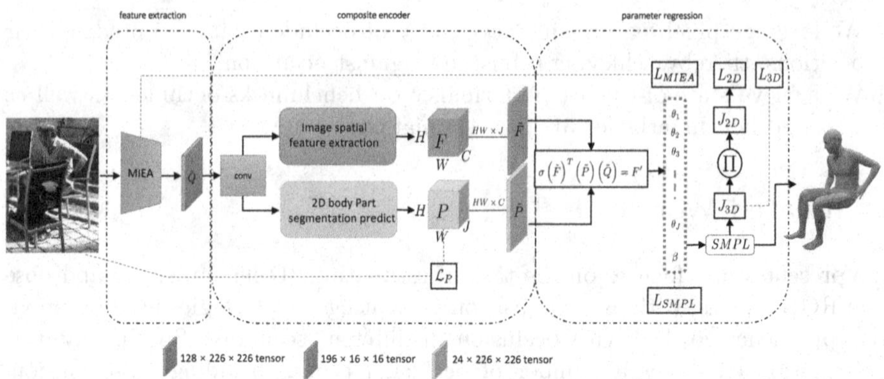

Fig. 1. Proposed Method Overview. An image is subjected to feature extraction utilizing the MIEA backbone. The derived features are then fed into both a 2D part and a 3 body branches, resulting in the generation of features P, F and Q. Subsequently, a regressor is applied to these features to obtain the SMPL parameters that represent the entire image.

semantic information, which provides a more intelligent way to simulate occlusions. However, these synthetic occlusions do not fully replicate the complexity of occlusions present in realistic images, nor do they provide insight into how to improve the network architecture to be inherently more robust to occlusion.

3 Method

The overall architecture of our method is shown in Fig. 1, which consists of three components: feature extraction module, composite encoder and a parameter regression module. One single image is passed through the feature extraction module to extract its features. These features are then fed into the composite encoder, which integrates information from both the 2D part and the 3D body branches to derive a novel 3D representation of the human body. This representation is used by a parameter regression module to estimate the 3D human pose and shape.

3.1 Feature Extraction

We use a Masked Image Encoding Autoencoder (MIEA) to extract image features. Our feature extraction network is an autoencoder method that learns image features by reconstructing the original signal under local observation conditions. The method consists of an encoder and a decoder. The encoder maps the input signal to a latent representation, from which the decoder reconstructs the original signal. Unlike the classical DAE [30] that use a symmetric structure, our approach adopts an asymmetric design. This allows the encoder to mask out a portion of the input signal and retain only a portion of it. The decoder then

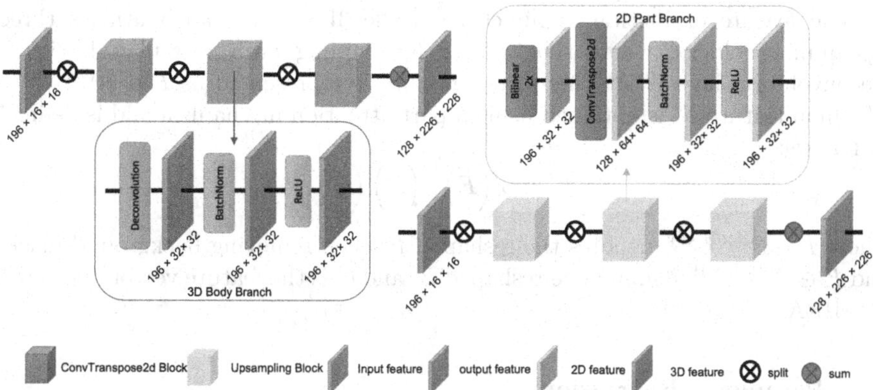

Fig. 2. The network architecture of the Composite Encoder involves the input of 3D body and 2D part branches, which are obtained by the image feature extraction process using MIEA (shown in pink). The 2D part branch employs an upsampling block to derive the feature maps (shown in yellow). Meanwhile, the 3D body branch (shown in blue) utilizes transposed convolution layers (Deconvolution Block) to further enhance the image feature extraction while performing the upsampling process. (Color figure online)

maps the latent representation of this partial signal to reconstruct the entire original signal, allowing the encoder to learn accurate image features.

After passing through the 2D body part and 3D body branches, the image feature vector is extracted into two pixel-level features and fused by partial attention to obtain the final feature for SMPL body regression, as depicted in Fig. 2.

3.2 Composite Encoder

The feature of our model consists of two segments. The first segment is obtained from the 2D part branch, with its features primarily employed for the calculation of attention weights. The second segment, sourced from the 3D body branch, is utilized to perform regression of SMPL [19] parameters.

To obtain the volume features of the 2D part and the 3D body branches, we designed two branches of feature encoder architecture. The feature extraction architecture for the 2D part branch consists of three Upsampling blocks and one attention module. The Upsampling blocks are composed of bilinear upsampling layers followed by batch normalization [12] and ReLU [24] activation after convolution. These blocks are used to implement 2x upsampling of the feature map size. The 2D part branch is represented as $P \in \mathbb{R}^{(J+1)\times H\times W}$, simulating J parts of attention and I background mask. Here, H and W represent the height and width of the feature volume, and each pixel $(h,\, w)$ stores the possibility of belonging to body part j.

The feature extraction architecture of the 3D body branch utilizes three Upsampling blocks, where the bilinear upsampling layer is replaced with a Deconvolution layer [35]. The output feature is represented as $F \in \mathbb{R}^{C \times H \times W}$. The final feature F' is obtained using a part attention mechanism and is defined as follows:

$$F' = \sigma\left(\tilde{F}\right)^T \left(\tilde{P}\right)\left(\tilde{Q}\right)$$

where $\tilde{P} \in \mathbb{R}^{J \times H \times W}$ denotes the reshaped P when removing background mask, and $\tilde{F} \in \mathbb{R}^{C \times H \times W}$ denotes the reshaped F, and $\tilde{Q}$ is the feature vector extracted by MIEA.

3.3 Parameter Regression

SMPL[19] defines human body shape and pose through the parameterization of movable joints. It uses two parameters, pose β and shape θ, to control body motion and pose changes. The shape parameter β describes the shape of a person, while the pose parameter θ describes pose through 24 joint rotations.

We use a weak-perspective camera model with scale and translation parameters $[s, t]$, $t \in \mathbb{R}^2$, following the approach in [15,17]. We use the tensor $\tilde{F}$ to predict the body shape β, and use the full feature tensor F' per row to predict the rotation angles of each body part through MLPs. This allows us to represent the rotation angles of each body part as a 6D vector parameter, in accordance with the VIBE [15] and SPIN [17] models.

3.4 Loss Function

When training the proposed network, we utilize loss functions the following. Of particular significance, we introduce the loss L_{MIEA}, to facilitate the adjustment of the MIEA model, the loss L_{MIEA} refers to the joint loss function encompassing parameters, vertices, and joints. These loss functions encompass the 3D joint loss L_{3D}, 2D keypoint loss L_{2D}, SMPL parameters loss L_{SMPL}, and the body part segmentation loss L_P. The overall loss is designated as:

$$L = \lambda_0 L_{2D} + \lambda_1 L_{3D} + \lambda_2 L_{SMPL} + \lambda_3 L_P + \lambda_4 L_{MIEA}$$

where each term is calculated as:

$$L_{2D} = \left\| X_{2D} - \hat{X}_{2D} \right\|_F^2,$$

$$L_{3D} = \left\| X_{3D} - \hat{X}_{3D} \right\|_F^2,$$

$$L_{MIEA} = \left\| X_{MIEA} - \hat{X}_{MIEA} \right\|_F^2,$$

$$L_{SMPL} = \left(\left\| \beta - \tilde{\beta} \right\|_2 + \left\| \theta - \tilde{\theta} \right\|_2 \right),$$

Table 1. Evaluating state-of-the-art methods on 3DPW, our method surpasses all but PARE in accuracy.

	Method	3DPW		
		MPJPE ↓	PA-MPJPE ↓	PVE ↓
temporal	HMMR [13]	116.5	72.6	-
	Doersch et al. [6]	-	74.7	-
	Sun et al. [29]	-	69.5	-
	VIBE [15]	93.5	56 5	113.4
	MEVA [21]	86.9	54.7	-
multi-stage	Pose2Mesh [4]	89.2	58.9	-
	Zanfir et al. [34]	90.0	57.1	-
	I2L-MeshNet [23]	93.2	58.6	-
	LearnedGD [28]		56.4	-
single-stage	HMR [13]	130.0	76.7	-
	CMR [18]	-	70.2	-
	SPIN [17]	96.9	59.2	135.1
	HMR-EFT [1]	-	54.2	-
	PARE [16]	82.9	52.3	99.7
	PARE [16] w. 3DPW	74.5	46.5	88.6
	Ours	85.6	53.7	102.3
	Ours w. 3DPW	**71.3**	**43.0**	**84.9**

Table 2. Our method achieves superior accuracy to HMMR and SPIN on MPI-INF-3DHP.

Method	MPI-INF-3DHP	
	MPJPE↓	PA-MPJPE↓
HMMR [13]	124.2	89.8
SPIN [17]	105.2	67.5
Ours	**92.5**	**55.8**

Table 3. Evaluation on 3DPW-OCC occlusion dataset shows that our method is superior to state-of-the-art PARE. All methods except SPIN are trained on COCO, Human3.6M, and 3DOH for consistency.

Method	3DPW-OCC		
	MPJPE↓	PA-MPJPE↓	PVE↓
Zhang et al. [36]	-	72.2	-
SPIN [17]	95.6	60.8	121.6
HMR-EFT [1]	94.4	60.9	111.3
PARE [16]	**90.5**	56.6	107.9
Ours	91.3	**56.1**	**105.6**

$$L_P = \frac{1}{HW} \sum_{h,w} \text{CrossEntropy} \left(\sigma \left(P_{h,w} \right), \hat{P}_{h,w} \right),$$

where X_{3D} refers to 3D joint locations, defined as $\mathcal{JM}(\theta, \beta)$, and is represented by J, which stands for a joint regression matrix in the SMPL [19] model. X_{2D} denotes 2D keypoint locations that are obtained by projecting the 3D joints X_{3D} onto the image space using camera parameters. $\hat{P}_{h,w}$ represents the ground-truth part label at location (h, w). The part segmentation loss L_P is calculated as the average cross-entropy loss between $P_{h,w}$ after applying softmax and the ground-truth label $\hat{P}_{h,w}$ over all $H \times W$ elements (Table 2, 3 and 4).

Table 4. All methods are trained on COCO-EFT. We compare our method with existing methods on 3DPW and 3DPW-OCC.

Method	3DPW		3DPW-OCC	
	MPJPE↓	PA-MPJPE↓	MPJPE↓	PA-MPJPE↓
HMMR [13]	99.0	59.9	97.9	64.7
PARE [13]	**93.4**	57.1	**93.9**	61.6
Ours	93.5	**56.8**	95.1	**59.9**

4 Experiments

We present experimental results based on the metrics of joint position accuracy, including the Procrustes-aligned Mean Per Joint Position Error (PAMPJPE) and Mean Per Joint Position Error (MPJPE). Additionally, for the 3DPW dataset, we provide information on the Per Vertex Error (PVE). We have performed the evaluation of on our model using the 3DPW-Test [22] and 3DPW-OCC [22, 36] datasets. Moreover, in order to thoroughly evaluate the performance of our model, we evaluate its robustness and convergence rate.

4.1 Comparison with State-of-the-Art Methods on Error Metric

As demonstrated in Table 1, our method outperforms all state-of-the-art temporal and multi-stage methods. For single-stage methods, our model is second only to PARE. Specifically, our method reduces the PA-MPJPE by 4.8% compared to the state-of-the-art multi-stage method LearnedGD [28] and by 1.8% compared to the state-of-the-art temporal method MEVA [21]. Our method also reduces MPJPE by 4% compared to the state-of-the-art multi-stage method Pose2Mesh [4] and by 1.5% compared to the state-of-the-art temporal method MEVA [21].

The results show that our method significantly outperforms both multi-stage and temporal methods. Compared to single-stage methods, our approach shows an improvement of 11.7% in MPJPE over SPIN [17] and almost an improvement of 1% over HMR-EFT [1] in terms of PA-MPJPE. Our method achieved competitive performance compared to the state-of-the-art method PARE in terms of PA-MPJPE, MPJPE, and PVE, while exhibiting better robustness and convergence, which will be detailed in the next section.

We evaluate our proposed method and conduct a comparative analysis with state-of-the-art approaches using the occlusion dataset. Except for SPIN [17], all methods are trained on a composite training set comprising COCO, Human3.6M, and 3DOH to ensure a fair evaluation. Our method and PARE [16] demonstrate superior performance, as evidenced by the MPJPE, PA-MPJPE, and PVE metrics on the masked dataset, surpassing other methods. Furthermore, we compare our method with PARE and HMMR on a single training set, where our approach achieves better PA-MPJPE results than PARE and comparable MPJPE performance.

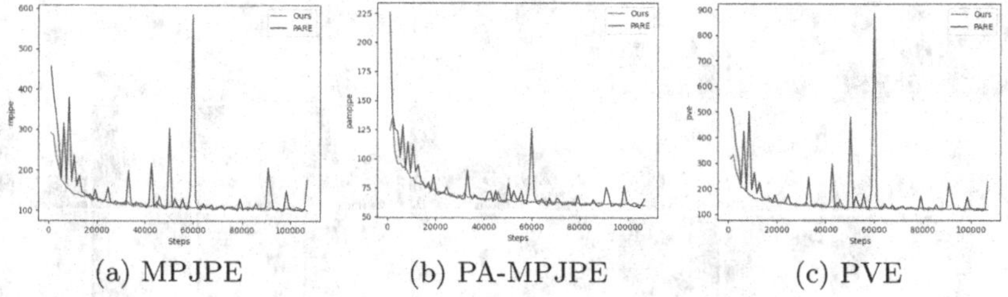

(a) MPJPE (b) PA-MPJPE (c) PVE

Fig. 3. Comparing the error variation during the training of our method (red curve) and PARE (blue curve), we use COCO as the training set and 3DPW as the validation set. (Color figure online)

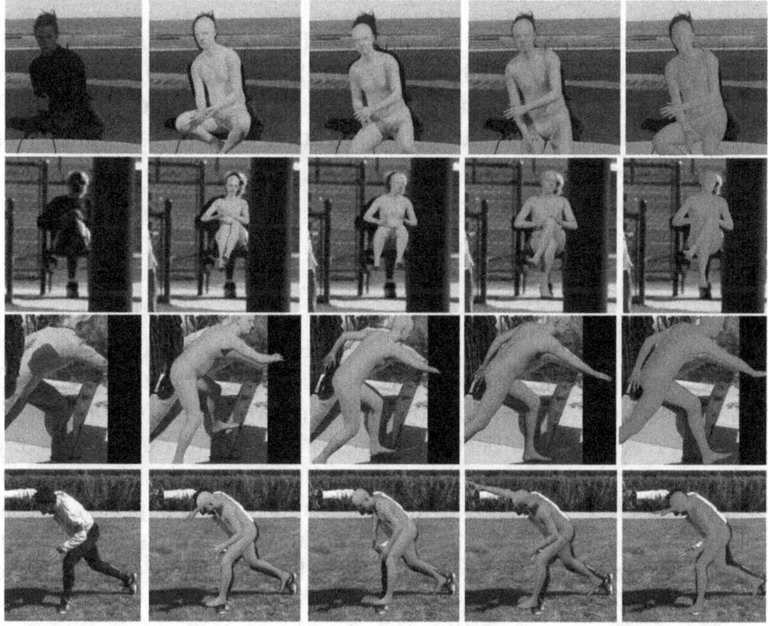

Fig. 4. present the 3D human body reconstruction results using various methods, shown from left to right in the following order: Input image, SPIN [17], HMR-EFT [1], PARE [16], and Our

4.2 Comparison with State-of-the-Art Methods on Robustness Metrics

We evaluate the performance of our proposed method and PARE on the COCO training dataset for 110k steps, with a validation set selected from 3DPW. The convergence and robustness of the models are evaluated by plotting the change in error during the training process in Fig. 3. It shows the decrease in MPJPE, PAMPJPE, and PVE, respectively, with the horizontal axis indicating the number of training steps and the vertical axis indicating the error in millimeters.

Fig. 5. Qualitative results on COCO and 3DPW datasets. From left to right: Input image, SPIN [17], HMR-EFT [1], PARE [16], and Our.

Results indicate that our method exhibits significantly less variable error difference compared to PARE, especially in the MPJPE and PVE metrics, where the error difference values show an almost linear trend. Meanwhile, the PA-MPJPE metric shows a moderate fluctuation, highlighting the greater stability of our method compared to PARE.

5 Visualization

In Figs. 4 and 5, we conducted a comparative evaluation of the 3D human body reconstruction performance of SPIN [17], HMR-EFT [1], PARE [16], and our proposed method in real-world scenarios with self-occlusion and environmental occlusion, respectively. The evaluation included various poses, such as single-person motion, reaching for objects, and natural sitting, as well as human poses ranging from natural standing, tilting to sitting, and occlusion areas ranging from local to major. Our proposed method demonstrated superior reconstruction performance compared to the other methods in both self-occlusion and environmental occlusion scenarios, indicating the robustness of our method in handling different occlusion scenarios.

6 Conclusions

We present a novel network designed to reconstruct a 3D human mesh from a single RGB image. Our approach is based on a partial attention mechanism and employs MIEA as the feature extraction technique for the image. Our method has been extensively evaluated and compared to previous approaches, demonstrating remarkable improvements in robustness and stability while still comparable results in terms of reconstruction quality to the state-of-the-art method. Overall, our proposed network represents a significant advancement in the field and holds great promise for further advancements in human mesh reconstruction from single RGB images.

Acknowledgment. This work was supported in part by the Shenzhen Technology Project (JCYJ20220531095810023), National Natural Science Foundation of China (61976143, U21A20487), Guangdong-Hong Kong-Macao Joint Laboratory of Human-Machine Intelligence-Synergy Systems (2019B121205007).

References

1. Joo, H., Neverova, N., Vedaldi, A.: Exemplar fine-tuning for 3D human model fitting towards in-the-wild 3D human pose estimation. In: 2021 International Conference on 3D Vision (3DV), pp. 42–52 (2021). https://doi.org/10.1109/3DV53792.2021.00015
2. Cheng, Y., Yang, B., Wang, B., Tan, R.T.: 3D human pose estimation using spatio-temporal networks with explicit occlusion training. Proc. AAAI Conf. Artif. Intell. **34**(07), 10631–10638 (2020)
3. Cheng, Y., Yang, B., Wang, B., Wending, Y., Tan, R.: Occlusion-aware networks for 3D human pose estimation in video. In: 2019 IEEE/CVF International Conference on Computer Vision (ICCV), pp. 723–732 (2019). https://doi.org/10.1109/ICCV.2019.00081
4. Choi, H., Moon, G., Lee, K.M.: Pose2Mesh: graph convolutional network for 3D human pose and mesh recovery from a 2D human pose. arXiv arXiv:2008.09047 (2020)
5. Choutas, V., Pavlakos, G., Bolkart, T., Tzionas, D., Black, M.J.: Monocular expressive body regression through body-driven attention. arXiv arXiv:2008.09062 (2020)
6. Doersch, C., Zisserman, A.: Sim2real transfer learning for 3D human pose estimation: motion to the rescue. arXiv arXiv:1907.02499 (2019)
7. Georgakis, G., Li, R., Karanam, S., Chen, T., Kosecka, J., Wu, Z.: Hierarchical kinematic human mesh recovery. arXiv arXiv:2003.04232 (2020)
8. Ghiasi, G., Yang, Y., Ramanan, D., Fowlkes, C.C.: Parsing occluded people. In: 2014 IEEE Conference on Computer Vision and Pattern Recognition, pp. 2401–2408 (2014). https://doi.org/10.1109/CVPR.2014.308
9. He, Y., Yan, R., Fragkiadaki, K., Yu, S.I.: Epipolar transformers. In: 2020 IEEE/CVF Conference on Computer Vision and Pattern Recognition (CVPR), pp. 7776–7785 (2020). https://doi.org/10.1109/CVPR42600.2020.00780
10. Hinton, G.E., Zemel, R.S.: Autoencoders, minimum description length and Helmholtz free energy. In: Neural Information Processing Systems (1993)
11. Huang, J.-B., Yang, M.-H.: Estimating human pose from occluded images. In: Zha, H., Taniguchi, R., Maybank, S. (eds.) ACCV 2009. LNCS, vol. 5994, pp. 48–60. Springer, Heidelberg (2010). https://doi.org/10.1007/978-3-642-12307-8_5
12. Ioffe, S., Szegedy, C.: Batch normalization: accelerating deep network training by reducing internal covariate shift. arXiv arXiv:1502.03167 (2015)
13. Kanazawa, A., Black, M.J., Jacobs, D.W., Malik, J.: End-to-end recovery of human shape and pose. In: 2018 IEEE/CVF Conference on Computer Vision and Pattern Recognition, pp. 7122–7131 (2018). https://doi.org/10.1109/CVPR.2018.00744
14. Kanazawa, A., Zhang, J.Y., Felsen, P., Malik, J.: Learning 3D human dynamics from video. In: 2019 IEEE/CVF Conference on Computer Vision and Pattern Recognition (CVPR), pp. 5607–5616 (2019). https://doi.org/10.1109/CVPR.2019.00576

15. Kocabas, M., Athanasiou, N., Black, M.J.: VIBE: video inference for human body pose and shape estimation. In: 2020 IEEE/CVF Conference on Computer Vision and Pattern Recognition (CVPR), pp. 5252–5262 (2020). https://doi.org/10.1109/CVPR42600.2020.00530
16. Kocabas, M., Huang, C.H.P., Hilliges, O., Black, M.J.: PARE: part attention regressor for 3D human body estimation. In: 2021 IEEE/CVF International Conference on Computer Vision (ICCV), pp. 11107–11117 (2021). https://doi.org/10.1109/ICCV48922.2021.01094
17. Kolotouros, N., Pavlakos, G., Black, M., Daniilidis, K.: Learning to reconstruct 3D human pose and shape via model-fitting in the loop. In: 2019 IEEE/CVF International Conference on Computer Vision (ICCV), pp. 2252–2261 (2019). https://doi.org/10.1109/ICCV.2019.00234
18. Kolotouros, N., Pavlakos, G., Daniilidis, K.: Convolutional mesh regression for single-image human shape reconstruction. In: 2019 IEEE/CVF Conference on Computer Vision and Pattern Recognition (CVPR), pp. 4496–4505 (2019). https://doi.org/10.1109/CVPR.2019.00463
19. Loper, M., Mahmood, N., Romero, J., Pons-Moll, G., Black, M.J.: SMPL: a skinned multi-person linear model. ACM Trans. Graph. (TOG) **34**(6), 1–16 (2015)
20. Lu, X., Wang, W., Ma, C., Shen, J., Shao, L., Porikli, F.: See more, know more: unsupervised video object segmentation with co-attention siamese networks. In: 2019 IEEE/CVF Conference on Computer Vision and Pattern Recognition (CVPR), pp. 3618–3627 (2019). https://doi.org/10.1109/CVPR.2019.00374
21. Luo, Z., Golestaneh, S.A., Kitani, K.M.: 3D human motion estimation via motion compression and refinement. arXiv arXiv:2008.03789 (2020)
22. von Marcard, T., Henschel, R., Black, M.J., Rosenhahn, B., Pons-Moll, G.: Recovering accurate 3D human pose in the wild using IMUs and a moving camera. In: Ferrari, V., Hebert, M., Sminchisescu, C., Weiss, Y. (eds.) ECCV 2018. LNCS, vol. 11214, pp. 614–631. Springer, Cham (2018). https://doi.org/10.1007/978-3-030-01249-6_37
23. Moon, G., Lee, K.M.: I2L-MeshNet: image-to-lixel prediction network for accurate 3D human pose and mesh estimation from a single RGB image. arXiv arXiv:2008.03713 (2020)
24. Nair, V., Hinton, G.E.: Rectified linear units improve restricted Boltzmann machines. In: International Conference on Machine Learning (2010)
25. Pavlakos, G., Zhu, L., Zhou, X., Daniilidis, K.: Learning to estimate 3D human pose and shape from a single color image. In: 2018 IEEE/CVF Conference on Computer Vision and Pattern Recognition, pp. 459–468 (2018). https://doi.org/10.1109/CVPR.2018.00055
26. Rafi, U., Gall, J., Leibe, B.: A semantic occlusion model for human pose estimation from a single depth image. In: 2015 IEEE Conference on Computer Vision and Pattern Recognition Workshops (CVPRW), pp. 67–74 (2015). https://doi.org/10.1109/CVPRW.2015.7301338
27. Rockwell, C., Fouhey, D.F.: Full-body awareness from partial observations. arXiv arXiv:2008.06046 (2020)
28. Song, J., Chen, X., Hilliges, O.: Human body model fitting by learned gradient descent. In: Vedaldi, A., Bischof, H., Brox, T., Frahm, J.-M. (eds.) ECCV 2020. LNCS, vol. 12365, pp. 744–760. Springer, Cham (2020). https://doi.org/10.1007/978-3-030-58565-5_44

29. Sun, Y., Ye, Y., Liu, W., Gao, W., Fu, Y., Mei, T.: Human mesh recovery from monocular images via a skeleton-disentangled representation. In: 2019 IEEE/CVF International Conference on Computer Vision (ICCV), pp. 5348–5357 (2019). https://doi.org/10.1109/ICCV.2019.00545
30. Vincent, P., Larochelle, H., Bengio, Y., Manzagol, P.A.: Extracting and composing robust features with denoising autoencoders. In: Proceedings of the 25th International Conference on Machine Learning, pp. 1096–1103 (2008)
31. Vosoughi, S., Amer, M.A.: Deep 3D human pose estimation under partial body presence. In: International Conference on Image Processing (2018)
32. Wang, J., Xu, E., Xue, K., Kidzinski, L.: 3D pose detection in videos: focusing on occlusion. arXiv arXiv:2006.13517 (2020)
33. Wang, X., Girshick, R., Gupta, A., He, K.: Non-local neural networks. arXiv arXiv:1711.07971 (2017)
34. Zanfir, A., Bazavan, E.G., Xu, H., Freeman, W.T., Sukthankar, R., Sminchisescu, C.: Weakly supervised 3D human pose and shape reconstruction with normalizing flows. In: Vedaldi, A., Bischof, H., Brox, T., Frahm, J.-M. (eds.) ECCV 2020. LNCS, vol. 12351, pp. 465–481. Springer, Cham (2020). https://doi.org/10.1007/978-3-030-58539-6_28
35. Zeiler, M.D., Krishnan, D., Taylor, G.W., Fergus, R.: Deconvolutional networks. In: 2010 IEEE Computer Society Conference on Computer Vision and Pattern Recognition, pp. 2528–2535 (2010). https://doi.org/10.1109/CVPR.2010.5539957
36. Zhang, T., Huang, B., Wang, Y.: Object-occluded human shape and pose estimation from a single color image. In: 2020 IEEE/CVF Conference on Computer Vision and Pattern Recognition (CVPR), pp. 7374–7383 (2020). https://doi.org/10.1109/CVPR42600.2020.00740

A Voxel-Based Multiview Point Cloud Refinement Method via Factor Graph Optimization

Hao Wu[1], Li Yan[1,2], Hong Xie[1,3(✉)], Pengcheng Wei[1], Jicheng Dai[1], Zhao Gao[2], and Rongling Zhang[1]

[1] School of Geodesy and Geomatics, Wuhan University, Wuhan 430079, China
{haowu2021,wei.pc,daijicheng,2022202140067}@whu.edu.cn

[2] School of Computer Science, Wuhan University, Wuhan 430072, China
lyan@sgg.whu.edu.cn, gaozzz@whu.edu.cn

[3] Hubei Luojia Laboratory, Wuhan University, Wuhan 430079, China
hxie@sgg.whu.edu.cn

Abstract. lidar enables fast reconstruction of the real world using high-precision point cloud maps. It usually requires the pose information (also called trajectory) of point clouds obtained by lidar at different times so that all scans are unified in the global coordinate system. Pairwise registration is one of the commonly used technology to estimate the poses of the lidar by measuring the relative pose of the scans and incrementally registering scans into the global map. However, the pose information obtained by pairwise registration largely suffers from low accuracy, inefficiency, and drift in trajectory. In this paper, we present a voxel-based multiview point cloud refinement method, which can effectively maintain the global consistency of point cloud maps. The proposed method store multiple point clouds into a voxel feature map indexed by the hash table, and then construct the factor graph using both point-to-point and point-to-plane factors to optimize the sensor's poses. A multi-scale voxel-grid optimization strategy is also presented to enhance the robustness as well as the accuracy of our algorithm. We conduct extensive experiments on the proposed method to demonstrate its accuracy and efficiency. It shows that the proposed method achieves higher accuracy than the state-of-the-art methods with a reasonable time cost.

Keywords: Point cloud registration · Multiple overlaps · factor graph optimization

---

Supplementary Information. The supplementary material available at https://github.com/WuHao-WHU/EVPA.

This work was supported in part by the Open Fund of Hubei Luojia Laboratory under Grant 220100053 and in part by the Science and Technology Major Project of Hubei Province under Grant 2021AAA010.

Q. Liu et al. (Eds.): PRCV 2023, LNCS 14426, pp. 234–245, 2024.
https://doi.org/10.1007/978-981-99-8432-9_19

1 Introduction

Light detection and ranging (lidar) has been widely used in many applications, such as robotic navigation [10], autonomous driving [17], mapping, etc. Owing to its immediate, compact, and dynamic collection of point clouds from the neighboring surroundings, lidar is capable of generating real-time three-dimensional information of the surrounding objects, thereby providing vital info to support state estimation for varied carriers (e.g., drones, and vehicles). Aligning two lidar scans measured at different states is a critical task in point cloud registration. This task forms the basis of an incremental registration pipeline that has applications in lidar-based odometry systems, multi-lidar calibration, etc. However, incrementally registering multiple scans will accumulate drift or registration error, leading to unstable state estimation or inconsistent global map for the subsequent task. Hence, we propose an efficient and consistent multiview point cloud refinement framework to address the accumulated drift caused by indirect registration among scans.

The main contributions of the proposed framework are as follows:

1. A multi-scale voxel-grid optimization strategy is proposed to register point clouds in a coarse-to-fine manner avoiding getting stuck into the local minima. The proposed strategy is also capable of efficiently constructing the adjacency matrix in each voxel patch to achieve context-aware optimization.
2. A quaternion-based factor graph optimization using both point-to-plane factor and point-to-point factor. The former captures the robust constraint in the structured context and the latter utilizes the non-structured surroundings to provide more constraint during the optimization process.
3. An open-source implementation of the proposed method is released on GitHub[1] and extensive evaluations of the proposed method are conducted to show the improved accuracy.

2 Related Work

2.1 Pairwise Registration

Pairwise registration is a fundamental research in 3D computer vision as its capability of recovering sensor's poses and reconstructing complete scenes [14]. The registration methods mentioned in this paper are rigid-body transformation. The ICP algorithm [3] is the most commonly used method in pairwise registration for its straightforward error metric and high usability. However, ICP is sensitive to the initial pose information, and its registration accuracy varies greatly under different point cloud density, noise level, and overlap ratio. To improve the performance of ICP, several ICP-based variants have been proposed using different error metric, such as point-to-plane [7], plane-to-plane [22]. The proposed error metrics perform well in those structured surroundings like urban areas, but still

[1] https://github.com/WuHao-WHU/EVPA.

suffer from noise and limited overlap. [6] introduced the sparse ICP where a new registration formulation using sparsity inducing norms is proposed to address the outliers and limited overlap.

The NDT [4] algorithm and its variants [8] are an alternative approach to register point cloud in a probability way. The referenced point clouds are represented by a set of Gaussian distributions and the cost functions are determined by the probability density functions. The main limitation of the NDT as well as ICP algorithm is highly sensitive to the initial pose, and may not converge to a satisfying optimum.

2.2 Multiview Registration

Pairwise registration only considers the overlap of two successive scans while ignoring the fact that the features can be shared by multiple scans, which will cause sudden drift in incremental registration. Multiview registration aims to maintain the global consistency of the point cloud map by exploring the shared overlap of all scans. Many researches [2,15,21] have been conducted to address the task. [20] presented the LUM approach to solve the drift in 2D scan matching, which is extended by [5] to the 3D scan matching. The proposed methods minimize the Mahalanobis distance of all related scans to achieve global consistency. [11] exploited the information redundancy in a set of 3D scans by using the averaging of relative motions.

Recently, many researchers have focused on addressing the multiview registration problem within the bundle adjustment framework. Since the points in the lidar scans are sparse, the strict point-to-point correspondences can not be found and shared through all scans. [13,16] segment each scan and estimate the local planes in advance and minimize the difference between the plane in the current scan and the plane predicted by scan poses and the plane parameters. However, segmentation is time-consuming and less reliable when facing the sparse lidar point clouds. [24] proposed to minimize the distance of each point to the predicted plane, resulting in a faster, more accurate, and more robust optimization of the sensor's poses. The proposed method also introduced a reduction technique to overcome the efficiency problems caused by the enumeration of each point.

[9] solves the planar bundle adjustment differently by minimizing the eigenvalue of the covariance matrix calculated from the local plane feature. But it was proven to be equivalent to the point-to-plane distance metric later on by [19]. [19] further analyzed plane features in detail. It is shown that all plane parameters can be solved with close-form solutions in advance. [18] also introduced a novel concept called point cluster to greatly improve the computational efficiency.

However, planar bundle adjustment framework mentioned above can only benefit from structured surroundings while ignoring potential information from the non-structured object.

3 Methodology

Given the initial poses of the sequential lidar scans, Our task is to further refine the poses to alleviate accumulated error and form a consistent point cloud map. First, points in all lidar scans with the initial poses are divided into different voxels for fast accessing and searching (Sect. 3.1). Then, we introduce the factor graph optimization using both point-to-plane and point-to-point factors (Sect. 3.2). Based on the established factor graph, the multi-scale voxel-grid strategy for coarse-to-fine optimization is illustrated in detail (Sect. 3.3). Figure 1 shows the overview of the proposed framework.

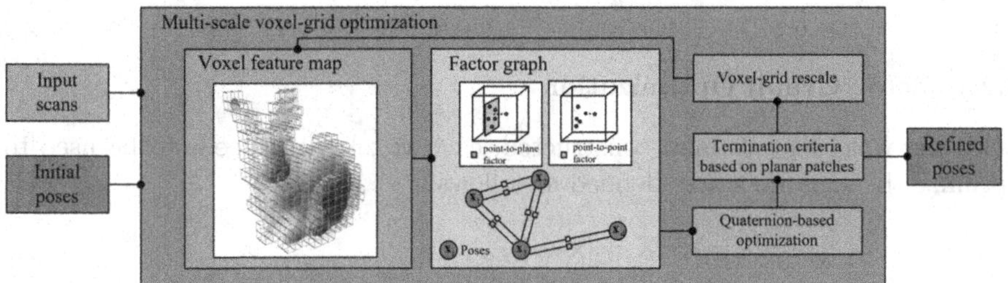

Fig. 1. Overview of the proposed framework.

3.1 Voxel-Grid Feature Map

The fundamental issue for multi-view registration to be addressed is how to quickly build an adjacency matrix and achieve context-aware optimization. Voxelization is capable of fast locating the overlapping patches and constructing registration factors within the analyzed voxel. Therefore, the voxel map is utilized for fast perception on adjacent scans in each voxel, which is shown in Fig. 2. Concretely, we use a set of voxels indexed by the hash table to store point clouds from different scans, where the scans are uniquely numbered. By traversing the numbers of the scans in each voxel, the adjacency matrix $\mathbf{A}$ can be constructed to indicate the shared overlap of adjacent scans.

The plane feature in point clouds was proven to be more robust in registration [12], and thus, the raw points of all shared scans in each voxel are analyzed using Principal Components Analysis (PCA) to find strong constraints in those structured environments.

So far, the voxel feature map has enabled us to quickly access the overlapping patches with different constraints attached, and add different factors for more accurate and robust optimization on the sensor's states.

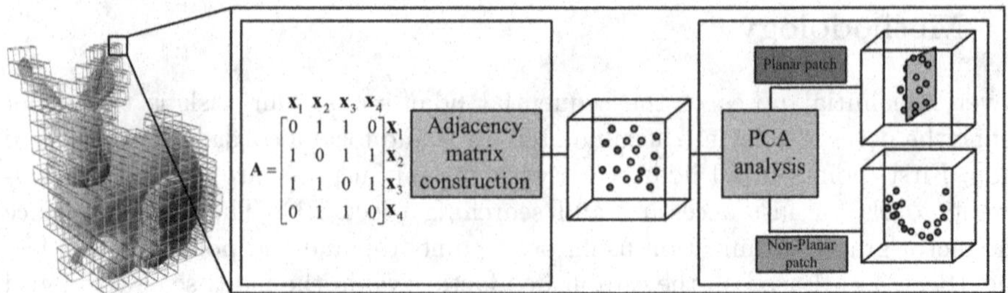

Fig. 2. All voxels are indexed by the hash table. Each voxel will use PCA analysis to determine whether it is a planar or non-planar patch. Points from different scans are colored by the index of the scan.

3.2 Factor Graph Optimization

With the voxel feature map constructed, each voxel patch needs to be used to optimize the sensor's states defined as follows:

$$\begin{aligned} \mathbf{X} &= [\mathbf{x}_0, \mathbf{x}_1, ..., \mathbf{x}_n] \\ \mathbf{x}_k &= [\mathbf{t}_k^w, \mathbf{q}_k^w], k \in [0, n] \end{aligned} \tag{1}$$

where $\mathbf{x}_k$ is the sensor's state at the kth timestamp, it contains the position and orientation of the sensor with respect to world frame.

We decouple the optimization into two stages: the first stage solves the relative poses of the adjacent scans as the previous adjacency matrix has revealed by two factors; the second stage merges the factors built in each voxel patch through a pose graph optimization.

The first stage optimization problem to be solved is as follows:

$$\begin{aligned} \sum_{\mathbf{t},\mathbf{q}\in\mathbf{A}} \Bigg\{ \min_{\mathbf{t}_k^{k+l}, \mathbf{q}_k^{k+l}} \Bigg\{ \sum \underbrace{\left\| \mathbf{r}_{plane}(n_i, {}^{k+l}p_j, {}^{k}p_i, \mathbf{t}_k^{k+l}, \mathbf{q}_k^{k+l}) \right\|^2}_{\textbf{point-to-plane factor}} \\ + \sum \underbrace{\left\| \mathbf{r}_{point}({}^{k+l}p_j, {}^{k}p_i, \mathbf{t}_k^{k+l}, \mathbf{q}_k^{k+l}) \right\|^2}_{\textbf{point-to-point factor}} \Bigg\} \Bigg\} \end{aligned} \tag{2}$$

where t_k^{k+l}, q_k^{k+l} are the translation and orientation from k frame to $k+l$ frame.

To construct the point-to-plane factor $\mathbf{r}_{plane}$, all planar patches in the voxel feature map are utilized. We start with a point ${}^{k}p_i$ in kth scan and search for its nearest five points ${}^{k+l}p = \{{}^{k+l}p_j\}_{j=1}^{5}$ in $k+l$th scans. The normal n_i of the plane feature can be calculated using least square, then the validity of the plane fitting can be assessed by evaluating the distance between each point and the plane, and thus we can obtain the following residual and its first order approximation made at initial poses:

$$\mathbf{r}_{plane} = f_{plane}(\mathbf{t}_k^{k+l}, \mathbf{q}_k^{k+l}) = n_i^T(\boldsymbol{R}\{\mathbf{q}_k^{k+l}\}\,{}^k p_i + \mathbf{t}_k^{k+l} - {}^{k+l}p_j)$$
$$\simeq f(\mathbf{t_0}_k^{k+l}, \mathbf{q_0}_k^{k+l}) + \mathbf{J_k^{k+1}} \begin{bmatrix} \delta \mathbf{t}_k^{k+l} \\ \delta \boldsymbol{\theta}_k^{k+l} \end{bmatrix} \quad (3)$$

where $\boldsymbol{R}\{\mathbf{q}_k^{k+l}\}$ denotes the rotation matrix represented by the quaternion $\mathbf{q}_k^{k+l}$. $\mathbf{J}_k^{\mathbf{k+1}}$ is the Jacobian matrix of $f(\mathbf{t}_k^{k+l}, \mathbf{q}_k^{k+l})$ with respect to $(\mathbf{t}_k^{k+l}, \boldsymbol{\theta}_k^{k+l})$. Note that, we optimize the rotation quaternion from vector space to SO(3) which leads to the small rotation $\delta\boldsymbol{\theta}_k^{k+l}$ in tangent space.

Similar to point-to-plane factor, point-to-point factor also starts with a point ${}^k p_i$ in kth scan and search for its nearest point within the distance threshold. Reciprocal point-to-point correspondences are used to guarantee the validity of point-to-point constraints. The point-to-point residual and its first order approximation made at initial poses are as follows:

$$\mathbf{r}_{point} = f_{point}(\mathbf{t}_k^{k+l}, \mathbf{q}_k^{k+l}) = (\boldsymbol{R}\{\mathbf{q}_k^{k+l}\}\,{}^k p_i + \mathbf{t}_k^{k+l} - {}^{k+l}p_j)$$
$$\simeq f(\mathbf{t_0}_k^{k+l}, \mathbf{q_0}_k^{k+l}) + \mathbf{J_k^{k+1}} \begin{bmatrix} \delta \mathbf{t}_k^{k+l} \\ \delta \boldsymbol{\theta}_k^{k+l} \end{bmatrix} \quad (4)$$

After solving the relative poses in the adjacency matrix, the second stage needs to fuse these factors measured by different pairwise scans and at different precision levels. So, the optimization for this stage comes with:

$$\min_{\mathbf{X}} \left\{ \sum_{\mathbf{t}_k^{k+l}, \mathbf{q}_k^{k+l} \in A} \left\| r_{pose}(\mathbf{t}_k^{k+l}, \mathbf{q}_k^{k+l}, \mathbf{X}) \right\|^2 \right\} \quad (5)$$

where the pose graph residual and its first order approximation is given by:

$$r_{pose} = F_{pose}(\mathbf{X}) = \begin{bmatrix} \boldsymbol{R}\{\mathbf{q}_w^{k+l}\}(\mathbf{t}_k^w - \mathbf{t}_{k+l}^w) - \mathbf{t}_k^{k+l} \\ 2\left[(\mathbf{q}_k^{k+l})^{-1} \otimes (\mathbf{q}_{k+l}^w)^{-1} \otimes \mathbf{q}_k^w\right]_{xyz} \end{bmatrix}$$
$$\simeq F(\mathbf{X}_0) + \mathbf{J}\delta\mathbf{X} \quad (6)$$

The $\otimes$ operation represents the quaternion product, and $[\cdot]_{xyz}$ extracts the vector part of a quaternion for the error-state representation. All the derivations of the Jacobian matrix mentioned above are given in supplementary materials. The ceres solver [1] is used for solving this nonlinear optimization problem.

3.3 Multi-scale Voxel-Grid Optimization

The refinement of the sensor's states is achieved by extracting factors from voxel patches in the feature map. However, this refinement method is highly sensitive to the size of the voxel grid and the initial pose of the sensor. Larger voxel grids tend to extract a coarse feature voxel map that lack the necessary detail to depict local details of point clouds. As a result, their convergence speed is slow, and they may not achieve higher accuracy, but they have a high tolerance

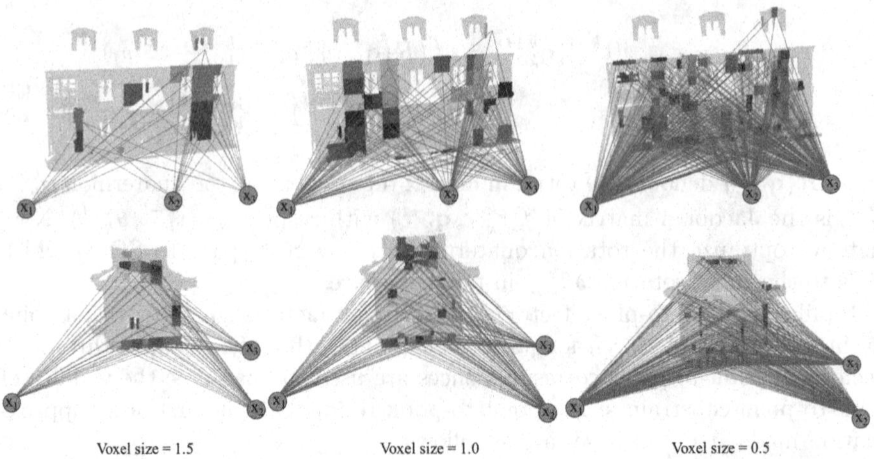

Fig. 3. The planar patches extracted at different voxel size using a coarse-to-fine multi-scale optimization strategy, different colors represent different planar patches and their intervisibility is shown by the line connected with pose nodes

for the initial values of sensor pose. Smaller voxel grids tend to provide more accurate feature voxel maps, which in turn offer more detailed constraints for optimization problems in Sect. 3.2, but relatively, it often requires a more precise initial pose. Therefore, a multi-scale voxel-grid optimization strategy is proposed to accelerate convergence speed, improve the accuracy of pose optimization, and reduce sensitivity to the initial pose of sensors.

As Fig. 3 shown, the proposed strategy starts with a coarse voxel-grid size to roughly associate planes or points. The number of the corresponding planar patches observed by all scans is relatively small at a coarse level, the constraints provided by the patches are rough but reliable even in the poor initial poses. The voxel size will gradually decrease before reaching the minimum voxel size, while the planar patches are also becoming more detailed and stable during the registration process, which allows the optimization to carefully home in on a better solution in a coarse-to-fine manner. We keep track of the number of planar patches to trace the registration quality as the shared overlap of finely-tuned scans should yield more planar patches than the coarse ones. If the number of corresponding planar patches found no longer increases or remains near a certain number during the optimization process, the entire iteration process will terminate. The details of the proposed refinement method is given in the supplementary materials using the pseudocode.

4 Experiment and Analysis

In this section, we comprehensively evaluate and compare our algorithm with other multiview registration algorithms in terms of mapping accuracy and efficiency. To evaluate our algorithm, we selected the commonly used fine registration methods of lidar point clouds as the baseline methods, and also compared with state-of-the-art algorithms: ICP, point-to-plane ICP (ICP-Normal), GICP, LUM and BALM. They are all available open sourced algorithms in PCL or Github. However, the baseline methods (ICP, ICP-Normal, GICP) are not addressing the problem of associating all scans in the global map, so we use the same data association technique described in Sect. 3.1 to make a fair comparison while other state-of-the-art methods use their default parameters as available on their open source implementations.

4.1 Datasets Description

We use datasets from 7 different scenarios to evaluate the proposed method. Five scenes are from ETH dataset, which are Arch, Courtyard, Facade, Office and Trees. Heritage building and Bremen city are from WHU-TLS dataset and Robotic 3D Scan dataset respectively. The selected datasets include both indoor and outdoor scenes, as well as structured and unstructured scenes. These scenarios typically contain a large amount of point cloud data, measurement noise, and moving objects, which often pose significant challenges to the efficiency and accuracy of algorithms. The datasets offer the ground truth poses of each scan, which is aligned using a pre-placed artificial marker for quick evaluation on the effectiveness of algorithms.

The algorithms to be validated all require initial poses as input and continuously optimize the sensor pose through each iteration. In order to obtain the initial poses of these datasets, we chose to use HL-MRF [23] framework for rough registration. Similarly, this work has also been open-source on GitHub. All algorithms will use the results of the framework as the initial pose input for fair comparison. The detail of the datasets and the initial APE (Absolute Trajectory Error) are given in the supplementary materials.

4.2 Evaluation Criteria

The proposed algorithm is mainly aimed at the drift on trajectories in multi-view registration, expecting to construct a globally consistent point cloud map. Therefore, we use the absolute trajectory error (APE) to evaluate the accuracy of the algorithms. The APE can be calculated by the following equations:

$$\mathbf{F}_i = \mathbf{Q}_i^{-1}\mathbf{P}_i \tag{7}$$

where the $Q_i \in \mathrm{SE}(3)$ is the ground truth trajectory of i_{th} scan and $P_i \in \mathrm{SE}(3)$ is the trajectory estimated by the algorithms. So the RMSE and SSE are given by

$$\begin{aligned} RMSE(F_{i:n}) &= \left(\sum_{i=1}^{n}\left\|trans(F_i)\right\|^2\right)^{\frac{1}{2}} \cdot \\ SSE(F_{i:n}) &= \sum_{i=1}^{n}\left\|trans(F_i)\right\|^2 \end{aligned} \tag{8}$$

$trans()$ calculates the translation errors. Evo (See Footnote 1) package is used to evaluate the registration

4.3 Evaluation

(1) Mapping Consistent Analysis

We selected two scenarios with severe drift to illustrate the mapping quality refined by different methods. Figure 4 shows two scans in office with about 0.365 m drifting on the trajectory, and Fig. 5 show two scans in Bremen City with about 0.181 m drifting on the trajectory. The visual performance of different methods shows our method is capable of alleviating drift on the trajectory and outperforms the others.

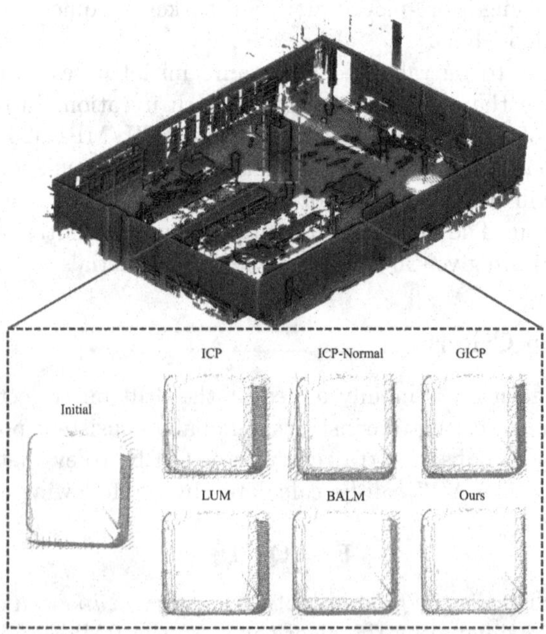

Fig. 4. The detail of office: the scan 1 (red) and 5 (green) are presented to show the drift caused by indirect registration (Initial). The refined point clouds of each method is presented in detail. (Color figure online)

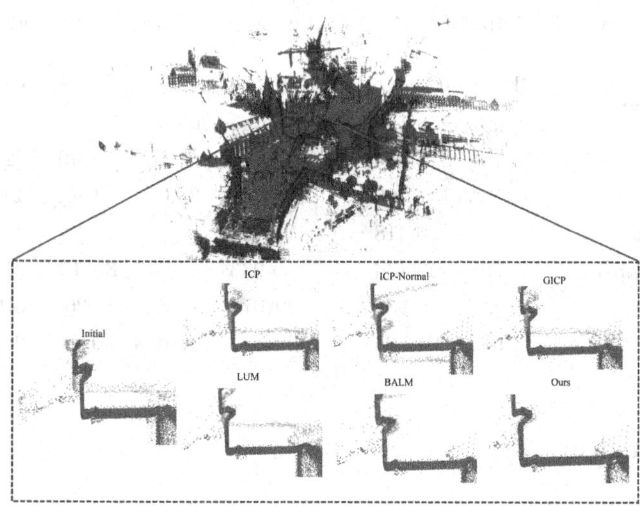

Fig. 5. The detail of Bremen city: the scan 1 (red) and scan 13 (green) are presented to show the drift caused by indirect registration (Initial). The refined point clouds of each method is presented in detail. (Color figure online)

(2) Quantitative Analysis

Table 1 shows the absolute trajectory error for different methods in 7 scenarios, the bold font represents the best result of all other methods and the font with underlined denotes the second-best result. It shows our method outperforms the other methods in 6 scenarios with one exception which will be discussed later. Table 2 shows the total optimization time cost of different methods, our method reaches the state-of-the-art level regarding time performance. The details on the error distribution is given in the supplementary materials using box plot.

Table 1. Absolute Trajectory Error (RMSE, meters) for different methods

Scene	ICP	ICP-Normal	GICP	LUM	BALM	Ours
1-Heritage building	0.062	0.061	0.050	0.046	<u>0.044</u>	**0.031**
2-Bremen City	0.101	0.062	0.137	0.092	<u>0.049</u>	**0.037**
3-Arch	0.011	0.008	0.008	0.011	<u>0.007</u>	**0.005**
4-Facade	0.017	**0.011**	<u>0.012</u>	0.020	0.018	0.014
5-Courtyard	0.038	0.034	<u>0.033</u>	0.035	0.035	**0.032**
6-Office	0.317	0.077	<u>0.070</u>	0.335	0.124	**0.019**
7-Trees	0.014	0.012	0.012	0.018	<u>0.011</u>	**0.009**

Table 2. Optimization time cost (second) for different methods

Scene	ICP	ICP-Normal	GICP	LUM	BALM	Ours
1-Heritage	216.996	258.186	295.689	209.333	**25.342**	28.407
2-Bremen City	276.101	312.614	353.871	267.505	**19.604**	22.648
3-Arch	38.585	181.768	199.443	39.011	12.408	**11.561**
4-Facade	16.061	20.418	25.354	17.703	13.454	**12.467**
5-courtyard	45.298	53.734	55.233	45.035	**8.512**	21.995
6-office	6.317	6.630	11.070	5.335	4.026	**3.697**
7-trees	45.989	64.703	116.012	40.018	**10.911**	24.009

5 Conclusion

This paper presented a multiview point cloud refinement framework to maintain global consistency in the point cloud map. Two novel features in the proposed framework are 1) the quaternion-based factor graph optimizaiton with point-to-plane factor and point-to-point factor. The two factors utilize both the strong constraint from structured plane surface and the complementary constraint from non-structured object. We derived the quaternion-based Jacobian matrix of the two factors and perform pose graph optimization. 2) the multi-scale voxel-grid optimization strategy to refine point clouds in a coarse-to-fine manner and gradually converge to a satisfying optimum. The number of plane patches were tracked to control the registration process. The developed framework is open sourced to benefit the community.

The proposed method was extensively verified in real-world datasets regarding the mapping quality, accuracy as well as efficiency. The experiments showed our method outperform other methods in accuracy and reached the state-of-the-art efficiency.

In the future, we would like to use the proposed framework in many other applications and platforms like UAV-based platform, lidar-based SLAM system, etc.

References

1. Agarwal, S., Mierle, K., Team TCS: Ceres Solver, March 2022
2. Bergevin, R., Soucy, M., Gagnon, H., Laurendeau, D.: Towards a general multi-view registration technique. IEEE Trans. Pattern Anal. Mach. Intell. **18**(5), 540–547 (1996)
3. Besl, P.J., McKay, N.D.: Method for registration of 3-D shapes. In: Sensor Fusion IV: Control Paradigms and Data Structures, vol. 1611, pp. 586–606. SPIE (1992)
4. Biber, P., Straßer, W.: The normal distributions transform: a new approach to laser scan matching. In: Proceedings 2003 IEEE/RSJ International Conference on Intelligent Robots and Systems, IROS 2003, (Cat. No. 03CH37453), vol. 3, pp. 2743–2748. IEEE (2003)

5. Borrmann, D., Elseberg, J., Lingemann, K., Nüchter, A., Hertzberg, J.: Globally consistent 3D mapping with scan matching. Robot. Auton. Syst. **56**(2), 130–142 (2008)
6. Bouaziz, S., Tagliasacchi, A., Pauly, M.: Sparse iterative closest point. In: Computer Graphics Forum, vol. 32, pp. 113–123. Wiley Online Library (2013)
7. Chen, Y., Medioni, G.: Object modelling by registration of multiple range images. Image Vis. Comput. **10**(3), 145–155 (1992)
8. Das, A., Waslander, S.L.: Scan registration using segmented region growing NDT. Int. J. Robot. Res. **33**(13), 1645–1663 (2014)
9. Ferrer, G.: Eigen-factors: plane estimation for multi-frame and time-continuous point cloud alignment. In: 2019 IEEE/RSJ International Conference on Intelligent Robots and Systems (IROS), pp. 1278–1284. IEEE (2019)
10. Gao, F., Wu, W., Gao, W., Shen, S.: Flying on point clouds: online trajectory generation and autonomous navigation for quadrotors in cluttered environments. J. Field Robot. **36**(4), 710–733 (2019)
11. Govindu, V.M., Pooja, A.: On averaging multiview relations for 3D scan registration. IEEE Trans. Image Process. **23**(3), 1289–1302 (2013)
12. Gressin, A., Mallet, C., Demantké, J., David, N.: Towards 3D LiDAR point cloud registration improvement using optimal neighborhood knowledge. ISPRS J. Photogramm. Remote. Sens. **79**, 240–251 (2013)
13. Hsiao, M., Westman, E., Zhang, G., Kaess, M.: Keyframe-based dense planar SLAM. In: 2017 IEEE International Conference on Robotics and Automation (ICRA), pp. 5110–5117. IEEE (2017)
14. Huang, X., Mei, G., Zhang, J., Abbas, R.: A comprehensive survey on point cloud registration. arXiv preprint arXiv:2103.02690 (2021)
15. Huber, D.F., Hebert, M.: Fully automatic registration of multiple 3D data sets. Image Vis. Comput. **21**(7), 637–650 (2003)
16. Kaess, M.: Simultaneous localization and mapping with infinite planes. In: 2015 IEEE International Conference on Robotics and Automation (ICRA), pp. 4605–4611. IEEE (2015)
17. Levinson, J., et al.: Towards fully autonomous driving: systems and algorithms. In: 2011 IEEE Intelligent Vehicles Symposium (IV), pp. 163–168. IEEE (2011)
18. Liu, Z., Liu, X., Zhang, F.: Efficient and consistent bundle adjustment on LiDAR point clouds. arXiv preprint arXiv:2209.08854 (2022)
19. Liu, Z., Zhang, F.: BALM: bundle adjustment for LiDAR mapping. IEEE Robot. Autom. Lett. **6**(2), 3184–3191 (2021)
20. Lu, F., Milios, E.: Globally consistent range scan alignment for environment mapping. Auton. Robot. **4**, 333–349 (1997)
21. Pulli, K.: Multiview registration for large data sets. In: Second International Conference on 3-D Digital Imaging and Modeling (Cat. No. pr00062), pp. 160–168. IEEE (1999)
22. Segal, A., Haehnel, D., Thrun, S.: Generalized-ICP. In: Robotics: Science and Systems, Seattle, WA, vol. 2, p. 435 (2009)
23. Wu, H., Yan, L., Xie, H., Wei, P., Dai, J.: A hierarchical multiview registration framework of TLS point clouds based on loop constraint. ISPRS J. Photogramm. Remote. Sens. **195**, 65–76 (2023)
24. Zhou, L., Koppel, D., Ju, H., Steinbruecker, F., Kaess, M.: An efficient planar bundle adjustment algorithm. In: 2020 IEEE International Symposium on Mixed and Augmented Reality (ISMAR), pp. 136–145. IEEE (2020)

SwinFusion: Channel Query-Response Based Feature Fusion for Monocular Depth Estimation

Pengfei Lai[1], Mengxiao Yin[1,2](✉), Yifan Yin[1], and Min Xie[1]

[1] School of Computer and Electronic Information, Guangxi University, Nanning, China
[2] Guangxi Key Laboratory of Multimedia Communications and Network Technology, Nanning 530004, China
ymx@gxu.edu.cn

Abstract. Given a single RGB image, Monocular Depth Estimation (MDE) predicts the depth of each pixel by utilizing both global and local information. Most MDE networks use simple skip connections, each skip connection only contain single-scale feature information. In this paper, we propose a multi-scale feature fusion module called Channel Query-Response Fusion (CQRF), which uses a channel query-response mechanism to filter out harmful or invalid channel information and enables soft fusion on the channels of multi-scale features. It can be inserted as an independent module into other MDE network architectures and enhance their performance. After passing through the CQRF module, the skip connections will contain valid channel information from all scales. Based on CQRF, we design a multi-scale feature fusion network SwinFusion, which achieves advanced results on public dataset. Additionally, we analyzed the effects of the number of utilized multi-scale features and different layer channels on the performance of MDE task.

Keywords: channel query-response · multi-scale features · feature fusion · skip connection · monocular depth estimation

1 Introduction

Estimating high-quality depth information from RGB images is crucial for tasks like 3D reconstruction and autonomous driving [1]. MDE requires utilizing both global and local information to predict the depth of each pixel, Some work has explored this [2–5]. Most MDE networks [6–8] adopt Encoder-Decoder structures with skip connections [9], Simple skip connections are partially helpful to supplement the details of the predicted depth map. However, single skip connection only provides single encoder layer features to the corresponding decoder layer, features from other layers of the encoder will not be utilized.

Our work was supported by the National Natural Science Foundation of China under Grant No. 61762007.

Q. Liu et al. (Eds.): PRCV 2023, LNCS 14426, pp. 246–258, 2024.
https://doi.org/10.1007/978-981-99-8432-9_20

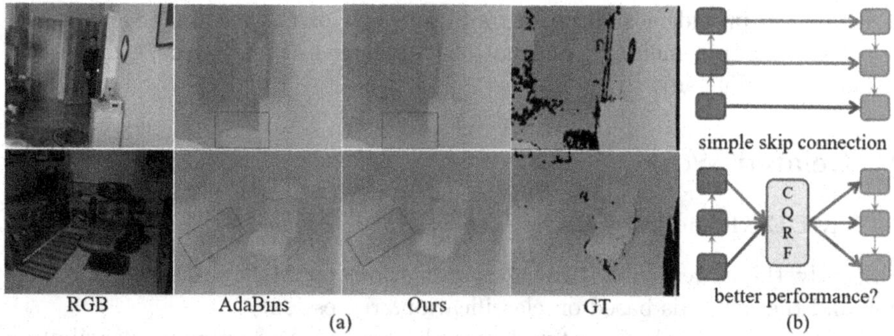

Fig. 1. We observe that AdaBins [10] can result in obvious incorrect depth predictions for regions with large color variations and significant differences in lighting intensity, like the carpet in the first row and the bed in the second row. A straightforward idea, as illustrated in Fig. 1(b), is to enhance the skip connections with multi-scale feature information, enabling each layer of the decoder to utilize features from multiple scales for depth prediction.

For decoder, although after up-sampling, the decoding layer input have the same resolution with skip connection input, there are significant differences in terms of semantic scale and receptive field size. Directly adding or concatenating (hard fusion) these two input feature maps may introduce redundant or harmful information, leading to poor feature fusion results.

In summary, single skip connection only utilizes one-scale features, and the hard fusion of encoder features and decoder features may lead to a decrease in fusion performance. In this paper, we design a multi-scale feature fusion module called CQRF based on a channel query-response mechanism, which enables the efficient utilization of multi-scale features from different encoder layers through each skip connection. By employing channel weighting, it achieves feature soft fusion, and enhancing the performance of the MDE network. Our main contributions are as follows:

1. We design a multi-scale feature fusion module CQRF based on channel query-response mechanism, which enhances single skip connection performance. CQRF module is an independent module that can be directly inserted between the encoder and decoder of other MDE networks. After adding the CQRF module to classic work AdaBins [10], all the metrics show a significant improvement.
2. With CQRF, we proposed a channel query-response mechanism-based multi-scale feature fusion network called SwinFusion for MDE tasks. SwinFusion achieved the best results on five metrics of the indoor dataset NYU-DepthV2 [11]. Compared to state-of-the-art models, the results of SwinFusion are improved by 3.5% and 6.3% in terms of rmse and abs-rel.
3. We design targeted experiments to investigate the impact of the number of multi-scale features and channels of features from different layers on MDE

tasks. We provide visualizations and analyses of the results. (See footnote Code and Supplementary Information will be released at https://github.com/lai-pf/SwinFusion)

2 Related Work

2.1 MDE Methods

Currently, there are mainly two types of methods for MDE tasks: direct regression and the methods based on classification®ression.

For regression methods, network directly predicts the continuous distribution of pixel depth. Eigen et al. [12] first uses Convolutional Neural Networks (CNN) to construct a dual-branch MDE network. Song et al. [13] construct Laplacian feature pyramids on multiple branches. Network progressively refines the coarse depth map in a top-down manner, layer by layer. The latest work NeW-CRFs [14] constructs a fully connected Conditional Random Field (CRF) network to improve the detail of predicted depth map.

Modeling the MDE task as a regression problem and directly predicting the fine-grained depths of pixels leads to limited depth prediction accuracy and convergence issues [1].

For classification®ression methods, DORN [15] first transform the MDE task from a regression problem to a K-classification problem by dividing the predicted depth range into K intervals in log space. Adabins [10] adaptively learns the width of depth interval bins through neural network and simultaneously predicts the classification probability of the pixel belonging to the depth interval. The final pixel depth prediction is a linear combination of the center of the bin and the pixel classification probability. However, BinsFormer [16] believes that the prediction of depth interval bins and interval classification probability based on the same layer of features, lacks effective interaction, leading to the confusion of global and fine-grained information.

Therefore, in SwinFusion, we try to decouple the prediction of depth interval bins and pixel classification probability maps as much as possible, using the global feature at the bottleneck to predict the depth interval bins and the high-resolution feature map of the last layer of the decoder to predict the classification probability.

2.2 Feature Fusion

UNet [17] architecture has been widely used for depth estimation task and image segmentation task. For UNet structure, skip connections are often used to alleviate information loss during down-sampling in the encoder. The decoding layer typically performs a direct concatenation or addition (hard fusion) of the input features and the skip connection input in the dimensionality. However, the hard fusion methods do not consider the mismatch of feature sets between the encoder and decoder, which leads to limited improvement or even degradation of the task performance.

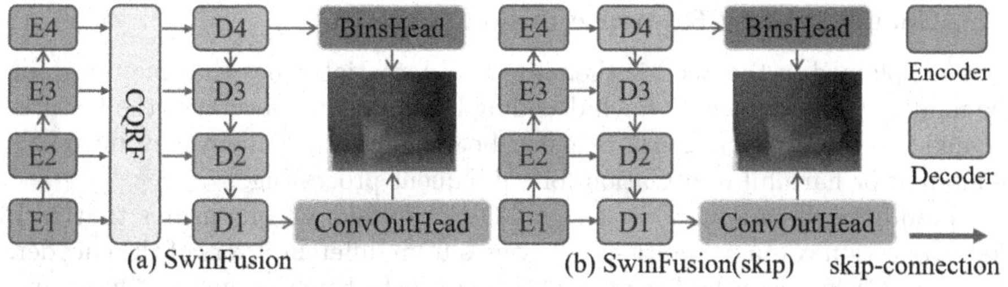

Fig. 2. Network architecture of SwinFusion and SwinFusion (skip)

Some works have attempted to design more effective feature fusion methods. AFF [18] uses the calculated weight matrix A and the matrix (1-A) to weight the original two scales features to achieve soft fusion of multi-scale features. The design of AFF is difficult to extend to the fusion of multiple layers and multiple scales features.

PPM [19] utilized convolutions with different kernel sizes to extract multi-scale features, while ASPP [20] used dilated convolutions with different kernel sizes to extract multi-scale features. Both methods directly concatenated the multi-scale features, leading to poor fusion results. Additionally, both methods constructed feature pyramids for a specific layer and did not effectively utilize multi-scale features from different layers.

3 Methodology

As shown in Fig. 2(a), our SwinFusion adopts an Encoder-CQRF-Decoder structure, consists of encoder, Channel Query-Response Fusion (CQRF) module, decoder, BinsHead and ConvOutHead.

CQRF filters out the harmful channel information and enables each skip connection between the encoder and the decoder to fuse multi-scale features. BinsHead and ConvOutHead predict depth intervals and pixel classification probabilities, respectively. Final depth prediction is the linear combination of the depth intervals and pixel classification probabilities.

3.1 Encoder and Decoder

For input image I, we use a pre-trained Swin Transformer [21] as encoder, output feature maps with resolutions of $\{\frac{1}{4}, \frac{1}{8}, \frac{1}{16}, \frac{1}{32}\}$ w.r.t. I and feature dimensions of {192,384,768,1536}.

SwinFusion uses simple convolution to up-sample for decoding. The decoder consists of convolution layers, batch normalization layers, and activation layers. This is an asymmetric design of a strong encoder and a simple decoder (Transformer-Encoder, Conv-Decoder), which displays the importance of building a good feature fusion bridge between different scale features.

3.2 Channel Query-Response Fusion (CQRF)

As we explained in the Sect. 1, there are two potential issues that may degrade the model's performance. 1. Each decoding layer can only leverages the features of single encoding layer. 2. Hard fusion before processing them may introduce redundant or harmful information for subsequent processing.

To address these issues, we design CQRF module, which ensures that each skip connection contains multi-scale features from different layers of the encoder. Besides, CQRF assigns higher weights to effective feature channels and filters out redundant feature channels with smaller weights. Each skip connection executes feature fusion processing before being fed into the decoder. This alleviates the learning difficulty of the decoder during the training process.

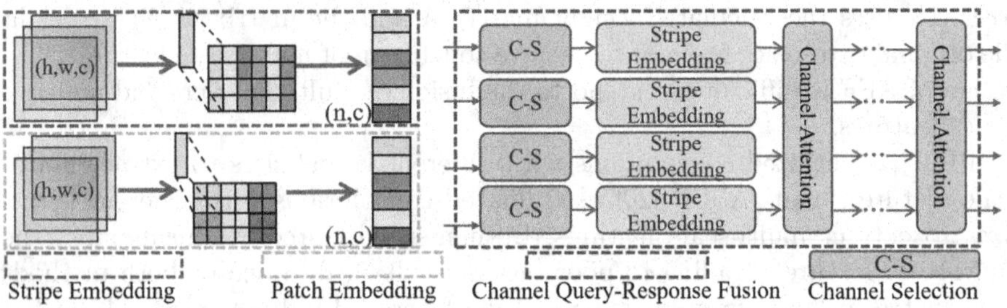

Fig. 3. Channel Query-Response Fusion (CQRF), h, w are the height and width of feature maps, c is the channel of features

As shown in Fig. 3, the CQRF module consists of Channel Selection layers, Stripe Embedding layers, and Channel Attention layers. The Channel Selection layers use convolution to decrease the number of channels for all input features by half as an initial selection of the effective channels.

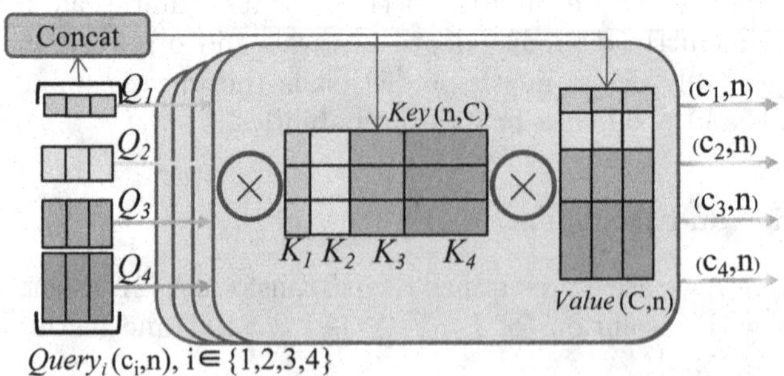

Fig. 4. Channel Attention

Stripe Embedding layers, as shown in Fig. 3, expand the receptive field of encoding information in the vertical direction, gathering richer depth-related visual cues from the vertical aspect of the image for subsequent depth prediction. Similar to patch embedding, stripe embedding can also reduce computational complexity.

Channel Attention [22] is shown in Fig. 4, which takes the stripe embedding features from *i-th* layer as Q_i with size $(c_i \times n)$, and concatenates them as Key and $Value$ in channel dimension with size $(n \times C)$. $C = \sum_{i=1}^{4} c_i$. K_i denotes the *i-th* layer's stripe embedding features, and the response of K_i channels to Q_i channels are calculated. The attention matrix represents the response of K_1, K_2, K_3, K_4 channels to Q_1, Q_2, Q_3, Q_4 channels. The performance of depth estimation mainly depends on the feature scales and channels with high response degrees. Channels with higher weights contribute more to the subsequent depth prediction process, while ineffective or harmful channel information is filtered or reduced in its impact on the subsequent prediction process in the form of low weights, thereby achieving soft fusion of multi-scale features.

3.3 BinsHead and ConvOutHead

Li et al. [16] suggested that if both depth interval and pixel classification probability predictions are based on the same layer features, global information and fine-grained information may be confused with each other, leading to difficulty in model learning. Taking inspiration from AdaBins [10] and Pixelformer [1], we use two lightweight output heads to decouple the prediction of depth interval division and pixel classification probability. By using feature maps from different layers, we aim to reduce the learning difficulty of the model.

BinsHead predicts the global depth intervals, consists of Global Average Pooling and multiple MLP layers, which takes the feature maps at the bottleneck with size $(C \times \frac{1}{32}H \times \frac{1}{32}W)$ as input. After passing through the Global Average Pooling layer, the size of output feature is $(C \times 1)$, then send it to MLP layers, the final output size is $(nbins \times 1)$, also denoted as $(nbins)$. C is the channels number, H,W denotes the height and width of input image. The depth interval numbers, $nbins$, is set to 256 in the paper.

ConvOutHead contains convolution layers and activation layers for predicting pixel classification probability. It takes the feature map from the last layer of the decoder as input, with a size of $(C \times \frac{1}{4}H \times \frac{1}{4}W)$. Passing through the convolution layers and activation layer, the output size is $(nbins \times \frac{1}{4}H \times \frac{1}{4}W)$.

The final predicted depth map is the linear combination of the depth intervals and pixel classification probabilities, with output size $(1 \times \frac{1}{4}H \times \frac{1}{4}W)$.

3.4 Training Loss

The depth intervals are predicting based on global-scale features. The supervised signal from ground truth (GT) can be directly fed to BinsHead. As shown in Eq. 1, we use only Scale-Invariant Loss (SILog) [12] for supervision, without the need for Chamfer loss to provide additional guidance.

$$SILogLoss = \alpha\sqrt{\frac{1}{n}\sum_{i=1}^{n} g_i^2 - \frac{\lambda}{n^2}(\sum_{i=1}^{n} g_i)^2} \quad (1)$$

$d_i^{'}$ represents the predicted depth value, d_i represents the corresponding ground truth value, n denotes the number of pixels with valid ground truth and $g_i = log_{d_i^{'}} - log_{d_i}$. The hyper-parameters α and λ are 10 and 0.85, respectively.

4 Experiments

4.1 Datasets and Evaluation Metrics

The NYU Depth V2 dataset contains 120K training samples and 654 testing samples with a resolution of 480 × 640 [11]. We used 50K samples from the 249 scenes officially split for training and 654 samples from 215 scenes for testing. In this paper, we evaluate our method using six metrics, including the Absolute Relative Error (abs-rel), Root Mean Squared Error ($rmse$), Average Log Error (log_{10}), Threshold Accuracy (σ_i) with thresholds {1.25, 1.25^2, 1.25^3 }, which are also used in the work [12].

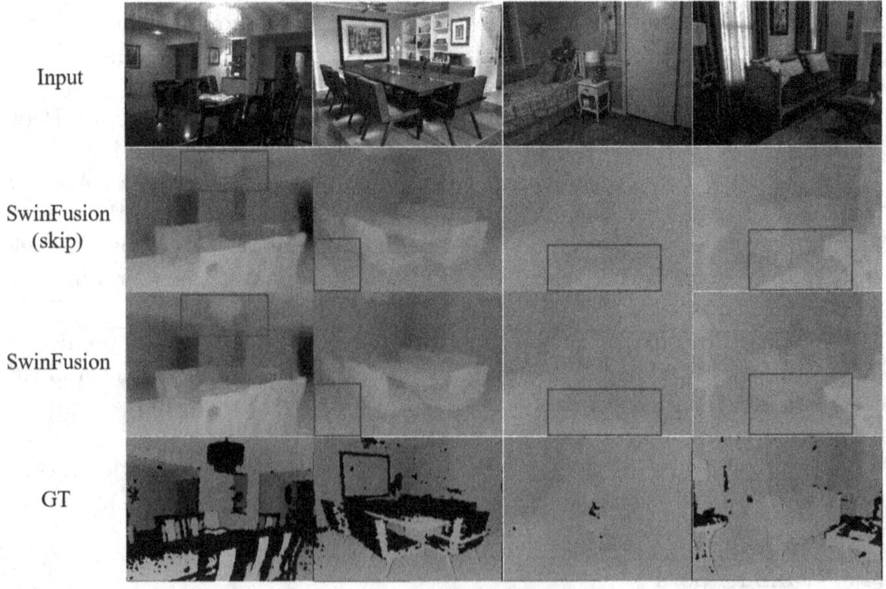

Fig. 5. Qualitative comparison of our proposed method SwinFusion on the dataset NYU Depth V2 against our baseline model SwinFusion (skip)

Table 1. The quantitative results compare with other methods using Swin Transformer backbone, pre-trained backbone used for the encoder is indicated in parentheses.

Methods	abs-rel ↓	rmse ↓	log_{10} ↓	σ_1 ↑	σ_2 ↑	σ_3 ↑
BinsFormer (Swin-Large) [16]	<u>0.094</u>	**0.330**	<u>0.040</u>	**0.925**	0.989	<u>0.997</u>
DepthFormer (Swin-Large) [23]	0.096	0.339	0.041	0.921	0.989	**0.998**
NewCRFs (Swin-Large) [14]	0.095	<u>0.334</u>	0.041	<u>0.922</u>	**0.992**	**0.998**
SwinFusion (skip)	**0.091**	**0.330**	**0.039**	**0.925**	<u>0.990</u>	<u>0.997</u>

Table 2. The quantitative results on NYU Depth V2.

Methods	abs-rel ↓	rmse ↓	log_{10} ↓	σ_1 ↑	σ_2 ↑	σ_3 ↑
Eigen (NIPS-2014) [12]	0.158	0.641	-	0.769	0.950	0.988
DORN (CVPR-2018) [15]	0.115	0.509	0.051	0.828	0.965	0.992
DAV (ECCV-2020) [24]	0.108	0.412	-	0.882	0.980	0.996
AdaBins (CVPR-2021) [10]	0.103	0.364	0.044	0.903	0.984	<u>0.997</u>
LocalBins (ECCV-2022) [25]	0.098	0.352	0.042	0.910	0.986	<u>0.997</u>
Naderi et al. (WACV-2022) [26]	0.097	0.444	0.042	0.897	0.982	0.996
NewCRFs (CVPR-2022) [14]	<u>0.095</u>	<u>0.334</u>	<u>0.041</u>	<u>0.922</u>	**0.992**	**0.998**
SwinFusion	**0.089**	**0.322**	**0.038**	**0.932**	<u>0.991</u>	**0.998**

4.2 Quantitative Results

Comparison Between SwinFusion (Skip) and Methods with Swin-Large Encoder. Our baseline model is SwinFusion (skip), which applied simple skip connection. Figure 2(b) shows SwinFusion (skip) structure. Tables 1 shows the quantitative results of SwinFusion (skip) on NYU Depth V2. Compare with state-of-the-art models that use the Swin-Large pre-trained model as the encoder, our baseline model achieves the best results in four metrics and the second-best results in two metrics. This shows that the performance of our baseline model is competitive, and qualitative results are shown in Fig. 5.

Comparison Between SwinFusion and State-of-Art Methods. SwinFusion doesn't apply a complex decoding structure to adapt to the encoded features. We separated the feature fusion from the decoder and replaced the skip connections in SwinFusion (skip) with our multi-scale feature fusion module CQRF. This allows each layer's features to interact with and query the encoded features at multiple scales, enabling the fusion of information across multiple scales. Tables 2 shows the comparison results on the NYU Depth V2 dataset. SwinFusion achieves the best results in five metrics. Compared with the best models LocalBins and NewCRFs, our SwinFusion model improves by 8.5% and 3.5% in rmse, and 9.1% and 6.3% in abs-rel.

Table 3. Quantitative results of SwinFusion with different Feature Fusion Methods.

Methods	abs-rel ↓	rmse ↓	log_{10} ↓	σ_1 ↑	σ_2 ↑	σ_3 ↑
SwinFusion (Skip)	<u>0.091</u>	0.330	<u>0.039</u>	0.925	<u>0.990</u>	<u>0.997</u>
SwinFusion (AFF)	<u>0.091</u>	0.332	<u>0.039</u>	0.925	<u>0.990</u>	<u>0.997</u>
SwinFusion (PPM)	<u>0.091</u>	<u>0.327</u>	<u>0.039</u>	<u>0.926</u>	**0.991**	**0.998**
SwinFusion	**0.089**	**0.322**	**0.038**	**0.932**	**0.991**	**0.998**

Table 4. Experimental results of ablating CQRF of AdaBins [10] and SwinFusion.

Methods	abs-rel ↓	rmse ↓	log_{10} ↓	σ_1 ↑	σ_2 ↑	σ_3 ↑
AdaBins-origin	0.103	0.364	0.044	0.903	0.984	**0.997**
AdaBins	0.106	0.369	0.044	0.898	0.983	0.996
Adabins (CQRF)	**0.099**	**0.360**	**0.042**	**0.903**	**0.985**	0.996
SwinFusion (Skip)	0.091	0.330	0.039	0.925	0.990	0.997
SwinFusion	**0.089**	**0.322**	**0.038**	**0.932**	**0.991**	**0.998**

4.3 Ablation Study

Results on Different Feature Fusion Methods. To demonstrate the performance of CQRF, we designed strong baseline SwinFusion (AFF) and SwinFusion (PPM) based on two feature fusion methods: AFF and PPM. More details of SwinFusion (AFF) and SwinFusion (PPM) are in supplementary materials.

Tables 3 shows the results. It can been seen that SwinFusion achieves the best results in all metrics. Compared to SwinFusion (Skip) and SwinFusion (PPM), SwinFusion improves the rmse by 2.4% and 1.5%. The experimental results indicate that the multi-scale feature fusion module CQRF brings better feature fusion capability than other feature fusion methods on MDE task.

Ablation Study of CQRF on SwinFusion and AdaBins. Adabins-origin is the results in original paper with batch_size = 16. Adabins is the result in our experimental environment, with batch_size = 12. We conducted ablation experiments on the CQRF module separately in AdaBins and SwinFusion. In AdaBins (CQRF), the skip connections were replaced with the CQRF module. As shown in Tables 4, AdaBins (CQRF) outperformed AdaBins by 2.4% in terms of rmse, while SwinFusion outperformed SwinFusion (Skip) by 2.4%. This indicates that the CQRF module effectively utilizes multi-scale features, enhancing each skip connection and it can be inserted as an independent module between the Encoder-Decoder of other network, improving the overall network performance.

Analysis of the Channel Attention Map in CQRF Module. The Channel Attention maps are visualized by matrix as shown in Fig. 6(a). We calculate

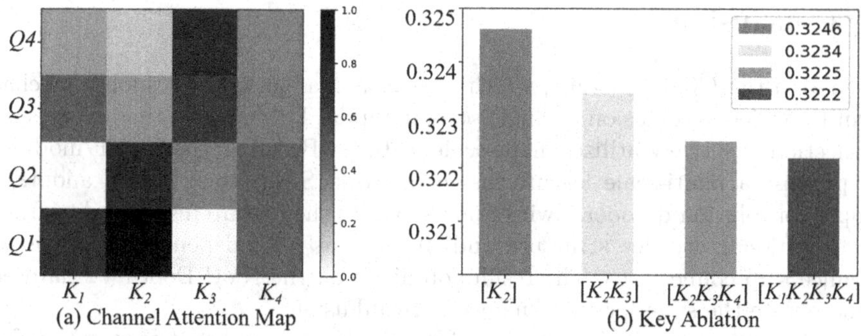

(a) Channel Attention Map (b) Key Ablation

Fig. 6. Channel Attention Map and Key Ablation

the correlation between $Query_i$ channels (Q_i) and Key channels (K_j), then we obtain the weight matrix for the Q_i channels and K_j channels. $i, j \epsilon$ {1,2,3,4}.

From the Channel Attention Map, we can observe that in the NYU Depth V2 dataset, K_2 channels have the highest response for the Q_1 channels, followed by K_1. The level of response reflects the network's dependence on different scale feature channels within the specified dataset. The weights of Q_2K_1, Q_4K_2, Q_1K_3 are close to 0, indicating that the channel information from the Key channels K_1, K_2, and K_3 are considered invalid or harmful information for the $Query$ channels Q_2, Q_4, and Q_1, the network tends to filter out these channels with low weights. These shows that the dependence of the MDE task on the channels from different layers of the encoder varies significantly.

Therefore, it is necessary to employ designs like CQRF to differentiate the feature channels, filter out ineffective or harmful channel information, and assign higher weights to the valid channel information.

Analysis of the Relationship Between Key Channel Number and MDE Task Performance. Our ablation experiments are performed on different numbers of multi-scale features concatenation, including $Key = [K_2]$, $Key = [K_2K_3]$, $Key = [K_2K_3K_4]$ and $Key = [K_1K_2K_3K_4]$. The evaluation metric is rmse.

As shown in Fig. 6(b), the experimental performances of using single-scale feature $Key = [K_2]$, using two-scale features $Key = [K_2K_3]$ and using three-scale features $Key = [K_2K_3K_4]$ are not as good as using all four scale features. As the number of Key channels, which interact with the $Query$ channels, increases, the network performance improves. This suggests that the interaction between feature channels of different scales is meaningful, and a higher number of channel interactions contribute to more accurate depth predictions.

5 Conclusion

We design the CQRF module, which utilizes a channel query-response mechanism to achieve soft fusion of multi-scale features. This ensures that every skip connection effectively utilizes multi-scale features. Based on the CQRF module, we propose a multi-scale feature fusion network, SwinFusion, which adopts a simple convolution decoder. SwinFusion separate the feature fusion ability from the traditional decoder structure and make it as an independent module to enhance skip connections. The results obtained on the NYU Depth V2 dataset demonstrate the superior performance of SwinFusion.

In the future, we plan to combine channel attention and spatial attention for more efficient multi-scale feature fusion. Additionally, it would be worthwhile to apply our multi-scale feature fusion module based on channel query-response mechanism to some pixel-level tasks.

References

1. Agarwal, A., Arora, C.: Attention attention everywhere: monocular depth prediction with skip attention. In: Proceedings of the IEEE/CVF Winter Conference on Applications of Computer Vision, pp. 5861–5870 (2023)
2. Lu, Y., Lu, G.: Self-supervised depth estimation from spectral consistency and novel view synthesis. In: 2022 International Joint Conference on Neural Networks, pp. 1–8 (2022)
3. Liu, D., Cui, Y., Yan, L., Mousas, C., Yang, B., Chen, Y.: DenserNet: weakly supervised visual localization using multi-scale feature aggregation. In: Proceedings of the AAAI Conference on Artificial Intelligence, pp. 6101–6109 (2021)
4. Yan, Z., Wang, K., Li, X., Zhang, Z., Li, J., Yang, J.: RigNet: repetitive image guided network for depth completion. In: Avidan, S., Brostow, G., Cissé, M., Farinella, G.M., Hassner, T. (eds.) Computer Vision, ECCV 2022. LNCS, vol. 13687, pp. 214–230. Springer, Cham (2022). https://doi.org/10.1007/978-3-031-19812-0_13
5. Yan, Z., Wang, K., Li, X., Zhang, Z., Li, J., Yang, J.: DesNet: decomposed scale-consistent network for unsupervised depth completion. In: Proceedings of the AAAI Conference on Artificial Intelligence, pp. 3109–3117 (2023)
6. Cheng, Z., Liang, J., Tao, G., Liu, D., Zhang, X.: Adversarial training of self-supervised monocular depth estimation against physical-world attacks. arXiv preprint arXiv:2301.13487 (2023)
7. Cheng, Z., et al.: Physical attack on monocular depth estimation with optimal adversarial patches. In: Avidan, S., Brostow, G., Cissé, M., Farinella, G.M., Hassner, T. (eds.) Computer Vision, ECCV 2022. LNCS, vol. 13698, pp. 514–532. Springer, Cham (2022). https://doi.org/10.1007/978-3-031-19839-7_30
8. Wang, K., et al.: Regularizing nighttime weirdness: efficient self-supervised monocular depth estimation in the dark. In: Proceedings of the IEEE/CVF International Conference on Computer Vision, pp. 16055–16064 (2021)
9. He, K., Zhang, X., Ren, S., Sun, J.: Deep residual learning for image recognition. In: Proceedings of the IEEE Conference on Computer Vision and Pattern Recognition, pp. 770–778 (2016)

10. Bhat, S.F., Alhashim, I., Wonka, P.: AdaBins: depth estimation using adaptive bins. In: Proceedings of the IEEE/CVF Conference on Computer Vision and Pattern Recognition, pp. 4009–4018 (2021)
11. Silberman, N., Hoiem, D., Kohli, P., Fergus, R.: Indoor segmentation and support inference from RGBD images. In: Fitzgibbon, A., Lazebnik, S., Perona, P., Sato, Y., Schmid, C. (eds.) ECCV 2012. LNCS, vol. 7576, pp. 746–760. Springer, Heidelberg (2012). https://doi.org/10.1007/978-3-642-33715-4_54
12. Eigen, D., Puhrsch, C., Fergus, R.: Depth map prediction from a single image using a multi-scale deep network. In: Advances in Neural Information Processing Systems, vol. 27 (2014)
13. Song, M., Lim, S., Kim, W.: Monocular depth estimation using Laplacian pyramid-based depth residuals. IEEE Trans. Circ. Syst. Video Technol. **31**(11), 4381–4393 (2021)
14. Yuan, W., Gu, X., Dai, Z., Zhu, S., Tan, P.: Neural window fully-connected CRFs for monocular depth estimation. In: Proceedings of the IEEE/CVF Conference on Computer Vision and Pattern Recognition, pp. 3916–3925 (2022)
15. Fu, H., Gong, M., Wang, C., Batmanghelich, K., Tao, D.: Deep ordinal regression network for monocular depth estimation. In: Proceedings of the IEEE Conference on Computer Vision and Pattern Recognition, pp. 2002–2011 (2018)
16. Li, Z., Wang, X., Liu, X., Jiang, J.: BinsFormer: revisiting adaptive bins for monocular depth estimation. arXiv preprint arXiv:2204.00987 (2022)
17. Ronneberger, O., Fischer, P., Brox, T.: U-Net: convolutional networks for biomedical image segmentation. In: Navab, N., Hornegger, J., Wells, W.M., Frangi, A.F. (eds.) MICCAI 2015. LNCS, vol. 9351, pp. 234–241. Springer, Cham (2015). https://doi.org/10.1007/978-3-319-24574-4_28
18. Dai, Y., Gieseke, F., Oehmcke, S., Wu, Y., Barnard, K.: Attentional feature fusion. In: Proceedings of the IEEE/CVF Winter Conference on Applications of Computer Vision, pp. 3560–3569 (2021)
19. Lin, T.-Y., Dollár, P., Girshick, R., He, K., Hariharan, B., Belongie, S.: Feature pyramid networks for object detection. In: Proceedings of the IEEE Conference on Computer Vision and Pattern Recognition, pp. 2117–2125 (2017)
20. Wang, Y., Liang, B., Ding, M., Li, J.: Dense semantic labeling with atrous spatial pyramid pooling and decoder for high-resolution remote sensing imagery. Remote Sens. **11**(1), 20 (2018)
21. Liu, Z., et al.: Swin transformer: hierarchical vision transformer using shifted windows. In: Proceedings of the IEEE/CVF International Conference on Computer Vision, pp. 10012–10022 (2021)
22. Wang, H., Cao, P., Wang, J., Zaiane, O.R.: UCTransNet: rethinking the skip connections in U-Net from a channel-wise perspective with transformer. In: Proceedings of the AAAI Conference on Artificial Intelligence, pp. 2441–2449 (2022)
23. Li, Z., Chen, Z., Liu, X., Jiang, J.: DepthFormer: exploiting long-range correlation and local information for accurate monocular depth estimation. arXiv preprint arXiv:2203.14211 (2022)
24. Huynh, L., Nguyen-Ha, P., Matas, J., Rahtu, E., Heikkilä, J.: Guiding monocular depth estimation using depth-attention volume. In: Vedaldi, A., Bischof, H., Brox, T., Frahm, J.-M. (eds.) ECCV 2020. LNCS, vol. 12371, pp. 581–597. Springer, Cham (2020). https://doi.org/10.1007/978-3-030-58574-7_35

25. Bhat, S.F., Alhashim, I., Wonka, P.: LocalBins: improving depth estimation by learning local distributions. In: Avidan, S., Brostow, G., Cissé, M., Farinella, G.M., Hassner, T. (eds.) Computer Vision, ECCV 2022. LNCS, vol. 13661, pp. 480–496. Springer, Cham (2022). https://doi.org/10.1007/978-3-031-19769-7_28
26. Naderi, T., Sadovnik, A., Hayward, J., Qi, H.: Monocular depth estimation with adaptive geometric attention. In: Proceedings of the IEEE/CVF Winter Conference on Applications of Computer Vision, pp. 944–954 (2022)

PCRT: Multi-branch Point Cloud Reconstruction from a Single Image with Transformers

Jiquan Bai[1], Zewei Yang[2], Yuchen Guan[1], and Qian Yu[2(✉)]

[1] School of Mechanical Engineering,
Jiangsu University of Technology, Changzhou, China
[2] School of Computer Engineering,
Jiangsu University of Technology, Changzhou, China
yuqian@jsut.edu.cn

Abstract. Reconstructing a 3D point cloud from a single image is challenging. Although previously developed methods have mainly relied on convolutional neural networks (CNNs) as backbones, recent advances in computer vision have demonstrated the high effectiveness of transformers. In this paper, we propose a new method, namely PCRT, which uses a transformer encoder to extract image features and another transformer decoder to obtain point cloud features. A set of linear layers then project the point cloud features into coordinates according to different branches. The point cloud reconstructed by PCRT has great visual quality, especially in non-smooth areas. Our experimental results show that PCRT achieved better performance than previous methods in single-view point cloud reconstruction tasks. Furthermore, we extends PCRT to achieve unsupervised semantic segmentation while reconstructing point clouds.

Keywords: Single-view Reconstruction · Point Cloud · Transformer · Unsupervised Semantic Segmentation

1 Introduction

Single-view 3D reconstruction is a powerful technique that utilizes a single image to generate a 3D model. It has recently gained significant attention in diverse fields and applications, such as in modeling and positioning 3D scenes for robots, rapidly constructing virtual scenes and character models, and reconstructing the tissue structure of patients.

Nonetheless, reconstructing a 3D model from a single image is challenging owing to the lack of geometric information. Traditional reconstruction methods [12,13,20] heavily depend on camera parameters and scene structure, requiring substantial image data and computational resources. In contrast, deep learning methods utilize neural networks to learn the mapping relationships directly between 2D images and 3D models, reducing the manual design effort required for handling complex scenes and irregular shapes. Previously proposed methods

Q. Liu et al. (Eds.): PRCV 2023, LNCS 14426, pp. 259–270, 2024.
https://doi.org/10.1007/978-981-99-8432-9_21

[6,23,25,27] use an auto-encoder to encode images into latent codes and decoders to reduce the latent codes into 3D shapes or use generative adversarial networks [14,16,22] to improve the reconstruction quality. However, these methods often rely on complex CNN pipelines and have limitations when dealing with the local details of objects.

Transformer [18] architecture has recently gained popularity in 3D computer vision and has achieved impressive results in several downstream tasks. In this paper, we propose a new 3D reconstruction model based on the transformer, namely PCRT, which represents the point cloud as a collection of multiple clusters in space and applies self-attention operations between them. We independently predict the coordinates of the points within each cluster using a set of linear projections, improving the reconstruction of non-smooth surfaces. Furthermore, we extend PCRT to segment different semantic parts of the point cloud during reconstruction. The contributions of our work are as follows:

1. We propose a new single-view 3D point cloud reconstruction method based on transformer, PCRT, which can focus on the attention information from different local clusters in the point cloud and use a set of linear projections to predict the coordinates of each point independently;
2. We compare PCRT with state-of-the-art methods, and the results show that PCRT achieved better performance across all evaluation metrics and had higher visual quality in the overall shape and fine details of reconstructed point clouds;
3. We extend the capability of the PCRT to segment different semantic parts of the reconstructed point cloud without additional supervision.

2 Related Works

In comparison with voxel and mesh representations, point clouds are simpler and more flexible, unconstrained by surface continuity, and suitable for 3D data of various shapes and sizes.

2.1 Single-View Reconstruction

The task of reconstructing a 3D model from a single image is challenging owing to the limited amount of structural information available. 3D-R2N2 [2] utilizes recurrent neural networks to learn the relationship between object images and their underlying 3D shapes from a vast number of 3D CAD models. PSGN [4] is the first point cloud generation network that predicts specific vertex coordinates from image features using an encoder-decoder structure. Furthermore, AtlasNet [6] represents a 3D shape as a collection of parametric surface elements. Methods [23,25,27] with an encoder-decoder structure similar to that of AtlasNet have also achieved impressive results. Pixel2Mesh [19] employs perceptual features extracted from the input image to progressively deform an ellipsoid, resulting in accurate geometry generation. DISN [21] utilizes predicted signed distance fields

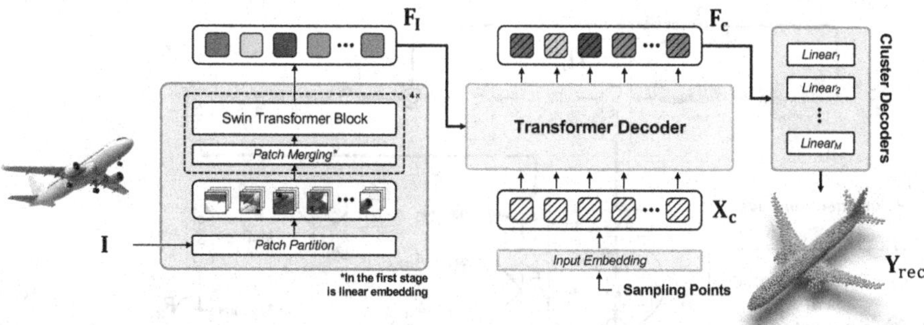

Fig. 1. The architecture of PCRT. Initially, the input image is encoded into features using the Swin-T. Following that, another transformer decoder is employed to extract features for each cluster by fusing the image features. The input to the transformer decoder is M identical embeddings, sampled from a sphere with a radius of 1. The features for each cluster are subsequently supplied to the corresponding linear decoder, producing a point set for each cluster. Finally, the point sets of the M clusters are concatenated together to obtain the final reconstructed point cloud.

to generate a detailed 3D mesh from a 2D image, resulting in high-quality output. OccNet [11] provides an implicit representation of the 3D surface by treating the decision boundary of a deep neural network classifier as a continuous surface.

Transformer has been recently applied to single-view reconstruction tasks. 3D-VRVT [9] uses a vision transformer (ViT) [3] to extract regional features, which are then used by a specially designed voxel decoder to generate 3D voxels from encoded image features. 3D-RETR [15] builds on 3D-VRVT by adding another transformer decoder to decode image features and obtain voxel features. These voxel features are fed to a CNN decoder to produce the final results.

2.2 Transformer

The transformer was originally proposed for natural language processing tasks. It utilizes self-attention, which enables the model to focus on different parts of the input sequence when computing the output. This attention mechanism allows the Transformer to capture long-range dependencies and achieve impressive results in tasks including machine translation and language understanding.

In 2D image-understanding tasks related to single-view reconstruction, ViT divides the image into fixed-size patches and obtains patch embeddings through linear transformations prior to feeding them into the transformer for feature extraction and classification. DeiT [17] is an image transformer that learns through knowledge distillation, requiring less data and computational resources. Further, the Swin Transformer [10] restricts self-attention computation to nonoverlapping local windows, enabling cross-window connections for greater efficiency.

For point cloud tasks, Point Transformer [26] enables the exchange of information between local feature vectors using subtractive relationships and position

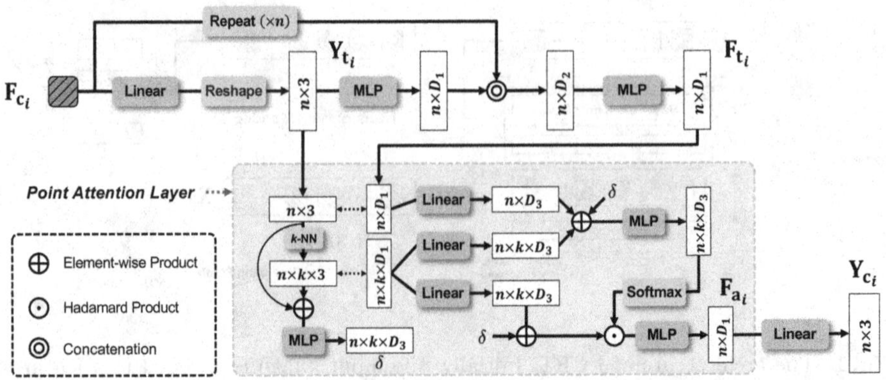

Fig. 2. Attention Cluster Decoder (ACD) for semantic part reconstruction.

encoding and generates new feature vectors for all data points as output. PCT [7] enhances input embedding by incorporating the furthest point sampling and nearest-neighbor search to better capture the local context in the point cloud.

3 Methods

In this section, we delve into the specifics of the network architecture of PCRT and present the details of the joint loss function.

3.1 Network Architecture

The network architecture of PCRT is shown in Fig. 1. It starts by processing an image $\mathbf{I} \in \mathbb{R}^{W \times H \times 3}$, using a Swin-T (the tiny version of Swin Transformer) to obtain its features $\mathbf{F_I} \in \mathbb{R}^{(\frac{W}{32} \times \frac{W}{32}) \times D}$. The Swin-T processes image features hierarchically manner by merging image patches multiple times and using shifted windows based self-attention. The dimension of all sub-layers and embedding layers in the transformer component of PCRT is D.

We define the M input embeddings of the transformer decoder as the M clusters of the target point cloud $\mathbf{Y}_{\text{rec}} \in \mathbb{R}^{N \times 3}$, where N represents the total number of points in the target point cloud. To overcome the strong 2-manifold constraint in the traditional CNN-based methods, PCRT predicts each point independently in the local cluster, which can accurately reconstruct more fine-grained object details. We sample $n = \frac{N}{M}$ points from a sphere with a radius of 1 and utilize a linear layer $(n \times 3, D)$ to project these points into an embedding of dimension D. The M clusters share the same embedding and are employed as the input $\mathbf{X_c} \in \mathbb{R}^{M \times D}$ to the transformer decoder. For the sampling method, we use the Fibonacci sphere algorithm [5], which maps the planar Fibonacci lattice $\mathcal{E} = \left\{(a_i, b_i) \mid a_i = \frac{i}{\phi}\%1, b_i = \frac{i}{n}, 0 < i \leq n\right\}$ onto the sphere via two equal-area

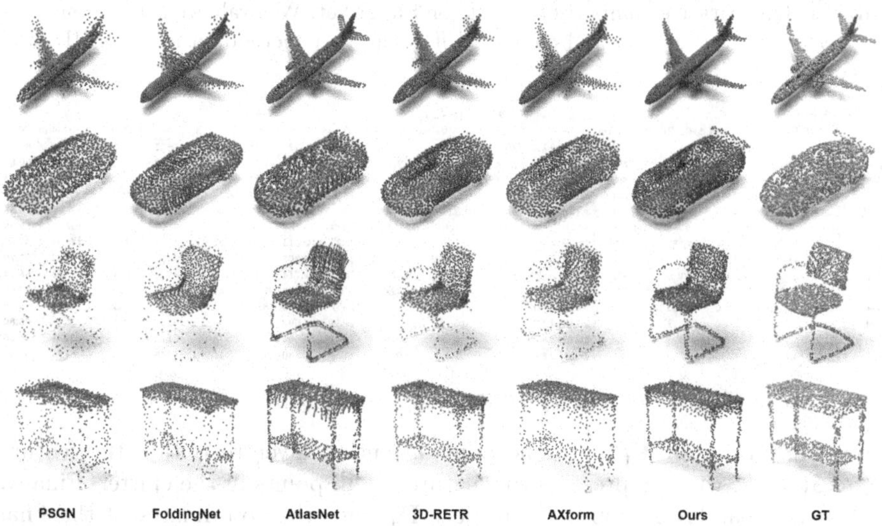

Fig. 3. Visualization results on ShapeNet. The reconstruction quality of our method is significantly better than that of other methods, especially for non-smoothed local geometric details.

projections. Following the projections, the coordinates of the sampling points $\mathcal{P} = \{(x_i, y_i, z_i) \mid 0 < i \leq n\}$ are as follows:

$$z_i = \frac{2i-1}{n-1}, x_i = \sqrt{1-z_i^2} \cdot \cos(2\pi i \phi), y_i = \sqrt{1-z_i^2} \cdot \sin(2\pi i \phi), \quad (1)$$

where $\phi = \frac{\sqrt{5}-1}{2}$, is the golden ratio.

We initialize a learnable positional encoding and use it as an input to the transformer decoder along with the embeddings discussed. Additionally, during the cross-attention stage of the transformer decoder, we also add a learnable positional encoding to the image features output by the encoder, which is necessary to improve the reconstruction quality. The output of transformer decoder is $\mathbf{F_c} \in \mathbb{R}^{M \times D}$, which represents the features of the internal points of each cluster. We define a linear layer $(D, n \times 3)$ for each cluster and project the features directly onto the coordinate $\mathbf{Y}_{\mathbf{c}_i} \in \mathbb{R}^{n \times 3}$ of the points in that cluster. Finally, we concatenate the sets of points from M clusters to obtain the reconstructed point cloud $\mathbf{Y}_{\text{rec}} \in \mathbb{R}^{N \times 3}$.

To achieve unsupervised semantic segmentation during reconstruction, we designed the attention cluster decoder (ACD) with inspiration from the Point Transformer [26] and used it to extend the linear layers above. The structure of the ACD is shown in Fig. 2. Given a feature $\mathbf{F}_{\mathbf{c}_i} \in \mathbb{R}^D$ of a cluster, we use a linear layer $(D, n \times 3)$ to decode and reshape it into the points $\mathbf{Y}_{\mathbf{t}_i} \in \mathbb{R}^{n \times 3}$, which we refer to as transition points. We then use an MLP $(3, 256, 256)$ to project transition points to the feature and concatenate it with $\mathbf{F}_{\mathbf{c}_i}$. Another MLP $(256 + D, 256, 256)$ processes the concatenating features into $\mathbf{F}_{\mathbf{t}_i} \in \mathbb{R}^{n \times 256}$. We

Table 1. Quantitative comparison results on ShapeNet. We evaluated the reconstruction results of our method and compared it with other methods using CD, HD, and F-Score.

Category	Chamfer Distance ($\times 10^3, \downarrow$)						Hausdorff Distance ($\times 10, \downarrow$)						F-Score ($\%, \uparrow$)					
	PSGN	FoldingNet	AtlasNet	3D-RETR	AXform	Ours	PSGN	FoldingNet	AtlasNet	3D-RETR	AXform	Ours	PSGN	FoldingNet	AtlasNet	3D-RETR	AXform	Ours
plane	3.276	2.762	2.341	2.353	2.328	**2.066**	2.363	2.002	1.621	1.656	1.851	**1.531**	68.04	71.22	78.55	77.29	75.60	**80.84**
bench	6.364	4.009	3.912	3.128	3.593	**2.955**	2.943	2.174	1.943	1.925	2.184	**1.764**	47.60	59.30	63.53	64.23	61.29	**69.80**
cabinet	5.831	4.598	4.351	4.393	4.396	**3.917**	2.872	2.156	1.738	1.848	1.984	**1.619**	29.04	41.47	47.68	43.23	40.33	**52.99**
car	4.437	3.719	3.651	3.316	3.331	**3.123**	2.062	1.674	1.505	1.505	1.566	**1.456**	46.69	56.16	58.51	58.58	57.95	**63.39**
chair	6.360	5.037	4.688	4.525	4.691	**4.156**	2.651	2.460	2.099	2.066	2.239	**1.956**	35.45	45.06	51.14	48.45	45.72	**54.26**
monitor	8.565	6.241	6.018	5.988	5.380	**5.370**	3.043	2.084	1.946	2.008	2.113	**1.784**	31.59	41.93	47.25	43.64	42.29	**52.28**
lamp	12.73	10.75	11.50	10.83	**9.873**	9.962	3.569	2.979	2.808	2.804	3.033	**2.514**	34.15	37.72	47.60	43.12	41.84	**47.89**
speaker	11.23	8.045	9.553	**7.610**	8.617	7.820	3.361	2.841	2.370	2.520	2.679	**2.267**	18.46	28.69	34.31	31.34	25.42	**37.70**
firearm	2.915	2.056	2.453	1.910	1.872	**1.731**	3.929	2.035	1.785	1.873	2.126	**1.714**	75.29	80.90	83.23	82.61	82.08	**85.24**
couch	6.937	5.008	4.869	4.486	4.878	**4.387**	2.925	2.333	2.047	1.892	2.333	**1.834**	27.72	40.61	44.82	43.23	38.15	**49.59**
table	5.508	4.461	4.310	4.153	4.003	**3.826**	2.876	2.707	2.290	2.227	2.431	**2.159**	48.56	56.92	60.09	59.27	58.91	**63.19**
cellphone	5.851	3.683	3.330	3.092	3.638	**2.716**	2.176	1.442	1.243	1.181	1.436	**1.092**	42.29	56.46	61.74	59.92	55.55	**66.89**
watercraft	5.752	4.058	3.718	3.323	3.594	**3.315**	3.832	2.256	1.984	2.058	2.204	**1.894**	46.57	55.78	61.96	61.18	57.99	**64.65**
mean	6.099	4.725	4.655	4.324	4.365	**4.030**	2.876	2.334	2.014	2.004	2.206	**1.880**	44.47	53.17	58.15	56.52	54.44	**61.56**

input both $\mathbf{F}_{\mathbf{t}_i}$ and $\mathbf{Y}_{\mathbf{t}_i}$ into the point attention layer to obtain the feature $\mathbf{F}_{\mathbf{a}_i} \in \mathbb{R}^{n \times 256}$, which represents the features of all points in the cluster. Finally, we use a linear layer $(256, 3)$ to project $\mathbf{F}_{\mathbf{a}_i}$ into the coordinates of the final points. We repeat this process for M clusters and concatenate the resulting points to obtain the reconstructed point cloud $\mathbf{Y}_{\text{rec}} \in \mathbb{R}^{N \times 3}$. The point attention layer in ACD is a self-attention module that utilizes a subtraction relation and the self-attention output of point $\mathbf{x}_i$ is calculated as follows:

$$\text{Attention}(\mathbf{x}_i) = \sum_{\mathbf{x}_j \in \mathcal{X}(i)} \text{softmax}\left(\gamma\left(\varphi(\mathbf{x}_i) - \psi(\mathbf{x}_j) + \delta\right)\right) \odot \left(\omega(\mathbf{x}_j) + \delta\right), \quad (2)$$

where the set $\mathcal{X}(i)$ comprises the k-nearest neighbors of $\mathbf{x}_i$ $(k = 16)$. The self-attention module is only applied within the local neighborhood of each point. The MLPs φ, ψ, and ω are used to project "query," "key," and "value," respectively. γ is an MLP that maps the self-attention scores. δ is the positional encoding.

Table 2. The impact of the number of clusters on the evaluation results. We choose five clusters to build the network architecture.

Number of Clusters	1	5	15	25	125
CD ($\times 10^3, \downarrow$)	4.048	**4.030**	4.090	4.092	4.061
HD ($\times 10, \downarrow$)	1.901	**1.880**	1.900	1.892	1.896
F-Score ($\%, \uparrow$)	**61.77**	61.56	61.12	61.73	61.57

The loss function of PCRT consists of two components, namely Chamfer Distance (CD) and repulsion loss [24]. We use the CD metric to calculate the similarity between the ground-truth point cloud $\mathcal{S}_{\text{gt}}$ and the reconstructed point cloud $\mathcal{S}_{\text{rec}}$. The CD is computed by finding point $\mathbf{q}$ in $\mathcal{S}_{\text{rec}}$ with the smallest distance to each point $\mathbf{p}$ in $\mathcal{S}_{\text{gt}}$ and averaging all distances. The similarity from

$\mathcal{S}_{\text{rec}}$ to $\mathcal{S}_{\text{gt}}$ can also be calculated using this method. The CD can be expressed using the following equation:

$$\mathcal{L}_{\text{cd}}\left(\mathcal{S}_{\text{gt}}, \mathcal{S}_{\text{rec}}\right) = \frac{1}{\left|\mathcal{S}_{\text{gt}}\right|} \sum_{\mathbf{p}\in\mathcal{S}_{\text{gt}}} \min_{\mathbf{q}\in\mathcal{S}_{\text{rec}}} \|\mathbf{p}-\mathbf{q}\|_2^2 + \frac{1}{\left|\mathcal{S}_{\text{rec}}\right|} \sum_{\mathbf{q}\in\mathcal{S}_{\text{rec}}} \min_{\mathbf{p}\in\mathcal{S}_{\text{gt}}} \|\mathbf{p}-\mathbf{q}\|_2^2, \quad (3)$$

Our experiments showed that points in specific clusters tended to gather in local regions. We utilize repulsion loss to eliminate these gathered blocks of points and reconstruct a more uniform point cloud. The repulsion loss can be expressed as follows:

$$\mathcal{L}_{\text{rep}} = -\sum_{i=0}^{\widehat{N}} \sum_{i'\in\mathcal{K}(i)} \|\mathbf{x}_{i'} - \mathbf{x}_i\| \, e^{-\frac{\|\mathbf{x}_{i'}-\mathbf{x}_i\|^2}{h^2}}, \quad (4)$$

where the points in the set of the k-nearest neighbors of each point $\mathbf{x}_i$ are subjected to a distance penalty to prevent the local gathering of points ($k = 16$). $e^{(\cdot)}$ is a fast-decaying weight function, and h is a constant with a value of 0.03.

Table 3. Our model using Swin-T achieves the best results and remains in the top-three in most metrics after replacement with ResNet. The model used for the segmentation task outperforms almost all previous methods.

Method	Backbone	CD ($\times 10^3, \downarrow$)	HD ($\times 10, \downarrow$)	F-Score (%, $\uparrow$)
PSGN [4]	Custom Conv	6.099	2.876	44.47
FoldingNet [23]	ResNet-18	4.725	2.334	53.17
AtlasNet [6]	ResNet-18	4.655	2.014	58.15
3D-PCN [27]	ResNet-18	4.673	2.217	53.26
3D-VRVT [9]	ViT-B/16	4.250	2.003	58.67
3D-RETR [15]	DeiT-Ti	4.324	2.004	56.52
AXform [25]	ResNet-18	4.365	2.206	54.44
Ours	Swin-T	**4.030**	**1.880**	**61.56**
Ours	ResNet-18	4.250	2.023	60.04
Ours-ACD	Swin-T	4.196	2.003	58.89

We train the network end-to-end by minimizing the joint loss function below.

$$\mathcal{L} = \mathcal{L}_{\text{cd}} + \alpha\mathcal{L}_{\text{rep}}, \quad (5)$$

where α is the weight parameter for balancing CD and repulsion loss and is set to 0.02.

4 Experiments

In this section, we introduce the dataset and the parameter settings during training. Furthermore, we evaluate the performance of PCRT through a series of experiments.

Fig. 4. Visualization results of point cloud segmentation and reconstruction. PCRT-ACD provides a better segmentation and reconstruction of different parts of the point cloud without semantic labeling supervision. Meanwhile, the reconstruction results of PCRT-ACD have better space consistency, and the color distribution of different point clouds from the same category is similar.

4.1 Dataset and Implementation Details

We used 13 categories from ShapeNet as the dataset to train PCRT and obtain rendered RGB images from 3D-R2N2 [2]. We randomly sampled 2,500 points on each full 3D model as the ground truth. Simultaneously, we randomly selected one RGB image from the corresponding rendered images and resized it to 224×224 pixels as the input image.

All experiments were conducted using a batch size of 32 and were trained for 200 epochs. The dimension D was set to 768. We utilized the AdamW optimizer with a weight decay of 10^{-4}. The learning rate for the Swin-T was set to a constant of 10^{-5}. The initial learning rates for the transformer decoder and the M linear layers were set to 10^{-4} and 10^{-3}, respectively, and were decreased to 0.5 times the previous stage at the 100th, 130th, and 160th epochs. The parameter α, which balances the CD and repulsion loss, was set to 0.02. The single-view reconstruction task and the unsupervised semantic segmentation task use the same training settings.

Table 4. Results of the evaluation of space consistency on ShapeNet. The evaluation metric was CD multiplied by 10^2.

Methods	plane	bench	cabinet	car	chair	monitor	lamp	speaker	firearm	couch	table	cellphone	watercraft	mean
Ours-ACD	**1.170**	**3.336**	8.383	0.698	**5.307**	**4.505**	**15.16**	10.51	0.894	**2.875**	6.373	**1.722**	**4.428**	**5.028**
Axform [25]	1.307	3.340	**7.440**	**0.555**	5.478	5.506	18.32	**9.840**	**0.740**	2.940	**5.292**	2.834	9.226	5.602
K-means++ [1]	35.81	64.94	62.99	57.61	38.16	52.48	44.96	64.21	41.47	60.57	51.69	57.46	43.70	35.81

4.2 Single-View Reconstruction

We evaluated the performance of PCRT in the single-view reconstruction task and compared the results with the previous methods PSGN [4], FoldingNet [23], AtlasNet [6], 3D-RETR [15], and AXform [25]. For the methods above and PCRT with the cluster number $M = 5$, we evaluated the CD, Hausdorff distance (HD), and F-Score metrics between the reconstructed point clouds and the ground truth. Table 1 presents the quantitative comparison results, with the bolded values indicating the best results in a given category or in the mean. The PCRT outperforms other methods in the mean and nearly all categories for the three evaluation criteria. The point clouds reconstructed by the different methods are shown in Fig. 3. The comparison clearly shows that the point clouds reconstructed by methods such as AtlasNet with the strong 2-manifold constraint are bloated in the non-smooth local region, resulting in many invalid topological correlations between individual points. In contrast, PCRT predicts individual points within a single cluster, allowing for a more accurate reconstruction of local details.

Next, we conducted several experiments and tested different settings for the number M of clusters. As shown in Table 2, all settings produced good results. However, we ultimately choose $M = 5$. The results suggested that when M is too large, the self-attention calculation between clusters generates a significant amount of invalid noise information, which can affect the reconstruction of internal points within the clusters. In fact, most regular objects in the 13 categories of ShapeNet do not require many clusters, and most real objects are semantically segmented into clusters between 2 and 6.

4.3 Unsupervised Semantic Part Reconstruction

By extending the existing model without additional semantic labels, we can achieve high-quality 3D point cloud reconstruction and enable the reconstructed point cloud to express richer semantic information.

The ACD brings self-clustering and space consistency features to PCRT, allowing PCRT to segment the different semantic parts. We rendered the clusters in the reconstructed point cloud using different colors for a qualitative comparison. As shown in Fig. 4 despite the ground truth being disordered, PCRT-ACD reconstructs 3D point clouds with intuitive semantics for multiple object categories without requiring additional supervised information. Table 3 shows that the evaluation results for the PCRT-ACD continue to be excellent and outperform those of the previous method.

Our evaluation of the space consistency properties of point clouds reconstructed using the PCRT-ACD is presented in Table 4. We randomly selected 100 samples for each category, each with five semantic parts, and we calculated the CD between the same semantic part across different samples. The resulting average was used to assess the space consistency of each part. To make comparisons, we evaluated our model alongside AXform [25], which is a multi-branch architecture similar to ours. Simultaneously, we used the K-Means++ [1] to process point clouds decoded by a single linear layer and as the benchmark.

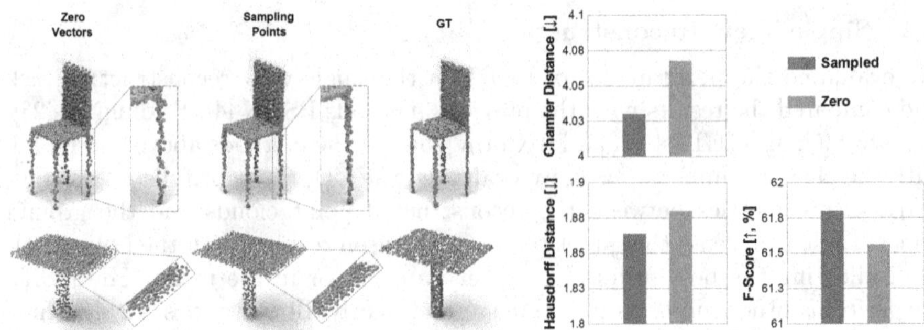

Fig. 5. (left)The reconstruction using sphere point embedding results in smoother object edges and significantly better reconstruction quality than zero-vector embedding. (right)Initializing embeddings using spherical points outperforms using zero vectors for multiple evaluation metrics.

4.4 Ablation Study

Image Feature Extractor. We conducted experiments using the ResNet-18 [8] to replace Swin-T as the image feature extractor for the ablation comparison with other methods. As shown in Table 3, the model still achieves the top-three results for most evaluation criteria, although its performance is slightly degraded. These experimental results show that the image feature extractor had no decisive effect on PCRT. The custom use of the transformer decoder and the cluster decoder we designed was thus proved to be effective.

Embedding Initialization. To demonstrate the validity of sampling points on the sphere to initialize the input to the transformer decoder, we attempted to initialize all clusters using the zero vectors. As shown in Fig. 5(left), the point cloud reconstructed using the Fibonacci sphere algorithm initialization has a distribution of internal points closer to the ground truth and smoother at the edges. In contrast, the point cloud reconstructed from the model trained with zero vector initialization is rough and loose, with many discrete points at the edges of the point cloud. Fig. 5(right) shows the quantitative comparison results of the two initialization methods. The sampling point initialization also outperforms the zero-vector initialization for each evaluation metric.

5 Conclusion

In this paper, we proposed PCRT, a single-view 3D point cloud reconstruction model based on transformer architecture. Unlike traditional CNN-based networks, which tend to focus only on features in local regions, PCRT can accurately capture the attentional information between different clusters of points in a point cloud and independently predict the coordinates of the points to reconstruct a higher-quality 3D point cloud. Furthermore, it can segment different semantic regions to a certain extent during the reconstruction process. Our experiments demonstrated the validity of our method, but PCRT still has some limitations. For example, it currently has a fixed number of clusters and cannot accommodate the number of semantic regions for all categories of objects. We aim to make it flexible and adaptable to the input image, which will be the focus of our future research.

Acknowledgments. This work was supported by the National Natural Science Foundation of China (Grant Nos. 61902159), Dual Creative Doctors of Jiangsu Province; and Postgraduate Research & Practice Innovation Program of Jiangsu Province (20810112212).

References

1. Arthur, D., Vassilvitskii, S.: k-means++: the advantages of careful seeding. Technical report, Stanford (2006)
2. Choy, C.B., Xu, D., Gwak, J.Y., Chen, K., Savarese, S.: 3D-R2N2: a unified approach for single and multi-view 3D object reconstruction. In: Leibe, B., Matas, J., Sebe, N., Welling, M. (eds.) ECCV 2016. LNCS, vol. 9912, pp. 628–644. Springer, Cham (2016). https://doi.org/10.1007/978-3-319-46484-8_38
3. Dosovitskiy, A., et al.: An image is worth 16 × 16 words: transformers for image recognition at scale. arXiv preprint arXiv:2010.11929 (2020)
4. Fan, H., Su, H., Guibas, L.J.: A point set generation network for 3D object reconstruction from a single image. In: Proceedings of the IEEE Conference on Computer Vision and Pattern Recognition, pp. 605–613 (2017)
5. González, Á.: Measurement of areas on a sphere using Fibonacci and latitude-longitude lattices. Math. Geosci. **42**, 49–64 (2010)
6. Groueix, T., Fisher, M., Kim, V.G., Russell, B.C., Aubry, M.: A papier-mâché approach to learning 3D surface generation. In: Proceedings of the IEEE Conference on Computer Vision and Pattern Recognition, pp. 216–224 (2018)
7. Guo, M.H., Cai, J.X., Liu, Z.N., Mu, T.J., Martin, R.R., Hu, S.M.: PCT: point cloud transformer. Comput. Vis. Media **7**, 187–199 (2021). https://doi.org/10.1007/s41095-021-0229-5
8. He, K., Zhang, X., Ren, S., Sun, J.: Deep residual learning for image recognition. In: Proceedings of the IEEE Conference on Computer Vision and Pattern Recognition, pp. 770–778 (2016)
9. Li, X., Kuang, P.: 3D-VRVT: 3D voxel reconstruction from a single image with vision transformer. In: 2021 International Conference on Culture-oriented Science & Technology (ICCST), pp. 343–348. IEEE (2021)
10. Liu, Z., et al.: Swin transformer: hierarchical vision transformer using shifted windows. In: Proceedings of the IEEE/CVF International Conference on Computer Vision, pp. 10012–10022 (2021)
11. Mescheder, L., Oechsle, M., Niemeyer, M., Nowozin, S., Geiger, A.: Occupancy networks: learning 3D reconstruction in function space. In: Proceedings of the IEEE/CVF Conference on Computer Vision and Pattern Recognition, pp. 4460–4470 (2019)
12. Newcombe, R.A., Fox, D., Seitz, S.M.: DynamicFusion: reconstruction and tracking of non-rigid scenes in real-time. In: Proceedings of the IEEE Conference on Computer Vision and Pattern Recognition, pp. 343–352 (2015)

13. Newcombe, R.A., et al.: KinectFusion: real-time dense surface mapping and tracking. In: 2011 10th IEEE International Symposium on Mixed and Augmented Reality, pp. 127–136. IEEE (2011)
14. Pan, X., Dai, B., Liu, Z., Loy, C.C., Luo, P.: Do 2D GANs know 3D shape? Unsupervised 3D shape reconstruction from 2D image GANs. arXiv preprint arXiv:2011.00844 (2020)
15. Shi, Z., Meng, Z., Xing, Y., Ma, Y., Wattenhofer, R.: 3D-RETR: end-to-end single and multi-view 3D reconstruction with transformers. arXiv preprint arXiv:2110.08861 (2021)
16. Shu, D.W., Park, S.W., Kwon, J.: 3D point cloud generative adversarial network based on tree structured graph convolutions. In: Proceedings of the IEEE/CVF International Conference on Computer Vision, pp. 3859–3868 (2019)
17. Touvron, H., Cord, M., Douze, M., Massa, F., Sablayrolles, A., Jégou, H.: Training data-efficient image transformers & distillation through attention. In: International Conference on Machine Learning, pp. 10347–10357. PMLR (2021)
18. Vaswani, A., et al.: Attention is all you need. In: Advances in Neural Information Processing Systems, vol. 30 (2017)
19. Wang, N., Zhang, Y., Li, Z., Fu, Y., Liu, W., Jiang, Y.-G.: Pixel2Mesh: generating 3D mesh models from single RGB images. In: Ferrari, V., Hebert, M., Sminchisescu, C., Weiss, Y. (eds.) ECCV 2018. LNCS, vol. 11215, pp. 55–71. Springer, Cham (2018). https://doi.org/10.1007/978-3-030-01252-6_4
20. Wei, X., Zhang, Y., Li, Z., Fu, Y., Xue, X.: DeepSFM: structure from motion via deep bundle adjustment. In: Vedaldi, A., Bischof, H., Brox, T., Frahm, J.-M. (eds.) ECCV 2020. LNCS, vol. 12346, pp. 230–247. Springer, Cham (2020). https://doi.org/10.1007/978-3-030-58452-8_14
21. Xu, Q., Wang, W., Ceylan, D., Mech, R., Neumann, U.: DISN: deep implicit surface network for high-quality single-view 3D reconstruction. In: Advances in Neural Information Processing Systems, vol. 32 (2019)
22. Yang, B., Wen, H., Wang, S., Clark, R., Markham, A., Trigoni, N.: 3D object reconstruction from a single depth view with adversarial learning. In: Proceedings of the IEEE International Conference on Computer Vision Workshops, pp. 679–688 (2017)
23. Yang, Y., Feng, C., Shen, Y., Tian, D.: FoldingNet: point cloud auto-encoder via deep grid deformation. In: Proceedings of the IEEE Conference on Computer Vision and Pattern Recognition, pp. 206–215 (2018)
24. Yu, L., Li, X., Fu, C.W., Cohen-Or, D., Heng, P.A.: PU-Net: point cloud upsampling network. In: Proceedings of the IEEE Conference on Computer Vision and Pattern Recognition, pp. 2790–2799 (2018)
25. Zhang, K., Yang, X., Wu, Y., Jin, C.: Attention-based transformation from latent features to point clouds. In: Proceedings of the AAAI Conference on Artificial Intelligence, vol. 36, pp. 3291–3299 (2022)
26. Zhao, H., Jiang, L., Jia, J., Torr, P.H., Koltun, V.: Point transformer. In: Proceedings of the IEEE/CVF International Conference on Computer Vision, pp. 16259–16268 (2021)
27. Zhao, Y., Birdal, T., Deng, H., Tombari, F.: 3D point capsule networks. In: Proceedings of the IEEE/CVF Conference on Computer Vision and Pattern Recognition, pp. 1009–1018 (2019)

Progressive Point Cloud Generating by Shape Decomposing and Upsampling

Deli Shi and Kun Sun(✉)

Hubei Key Laboratory of Intelligent Geo-Information Processing, School of Computer Sciences, China University of Geosciences, Wuhan 430078, China
{dannystone,sunkun}@cug.edu.cn

Abstract. How to generate realistic and precise 3D point clouds of different shapes has become a hot topic for many researchers. However, due to the complexity and discreteness of point cloud data, generating high-resolution complete point clouds from implicit vectors is still very challenging. It is difficult to obtain high-resolution points in one step. Besides, it is also hard to recover local details of a shape. To tackle these two challenges, we propose a new point cloud generation network. It learns to decompose the partial structures and extract the local information from them. We use this local information to guide the generation of local shapes. To control the quality of point clouds during the procedure, we progressively generate dense point clouds through iterative upsampling, in which the next step is performed based on the priority of the previous step. To be specific, We first generate a coarse point cloud and then decompose it into multiple local parts. Each of them is then mapped to a latent local feature. For each local area, we gradually improve the resolution of the shape by multiple upsampling layers. Each upsampling layer combines local features, global features, and point-wise features to predict two offsets for each point to increase the number of points. Our network can progressively generate uniformly multi-resolutions point clouds, with realistic and detailed local structures. Experiment results demonstrate that our model's generation ability is better than other GAN-based generation methods.

Keywords: Point cloud generation · Generative adversarial network · Upsampling network

1 Introduction

Among a variety of 3D related researches [1–4], 3D shape analysis plays a very important role in many disciplines and industries. It helps people better understand the construction and design of 3D objects. The most commonly used data structures for representing three-dimensional shapes are point clouds, voxels, and 3D meshes. 3D point cloud data is a set of three-dimensional coordinate information in Euclidean space. Compared to voxels and meshes, point clouds are

Q. Liu et al. (Eds.): PRCV 2023, LNCS 14426, pp. 271–282, 2024.
https://doi.org/10.1007/978-981-99-8432-9_22

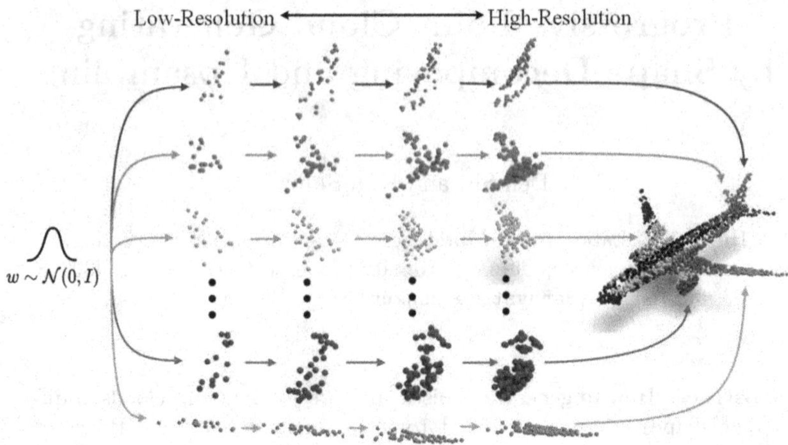

Fig. 1. The generating process of different local parts. Our method first generates a sparse point cloud with multiple local parts. And then gradually expands the resolution of different local point clouds that contain more dense local point clouds.

more convenient for geometric and shape transformation, while also having the advantages of lower computational cost and easier storage. However, acquiring point clouds with instruments such as laser scanners is often expensive and complicated. Therefore, learning modern neural networks to generate point clouds of certain shapes has drawn the attention of more and more researchers.

A point cloud generative network learns a potential map to transform a random noise distribution into the target shape distribution. Generative Adversarial Network (GAN) [5] is widely used for this task. Compared with other methods such as Variational Auto-Encoder (VAE) [6], flow-base model [7], and diffusion model [8], GAN has lower training cost and faster generating speed. Unfortunately, converting a random vector to a point cloud is still challenging due to its unstructured and unordered properties. Hui proposed a deconvolution point cloud generation framework PDGN [9]. It first generates a coarse point cloud containing a small number of points, then uses K-Nearest Neighbor bilateral interpolation in coordinate and feature space to gradually output higher-resolution point clouds. This method effectively overcomes the complexity of the whole point cloud. However, PDGN only combines the information of several nearest neighbors for each point, without fully considering the potential local semantic structure information, which may cause to generate some fuzzy local structures. Some studies introduce the idea of local block features, such as WarpingGAN [10]. It twists multiple 3D priors into different local regions of the 3D shape under the guidance of local structure perception semantics. EditVAE [11] learns the segmentation of point cloud shape by converting latent local features into point representations and shape primitives in an unsupervised manner. By considering latent local structure information, these methods can generate more clear local shapes. However, they all directly generate shapes with the guidance of local codes. Due to the discreteness and complexity of point cloud data, it is often difficult to directly generate high-resolution point clouds, which may lead to some shape defects and errors.

To address the above issues, we propose a novel point cloud generation network that can produce high-resolution results in a coarse-to-fine structure. A point cloud shape can be seen as the combination of multiple continuous local parts, each containing certain structural and semantic information. We view the generating process of point clouds as the progressive upsampling procedure of multiple local structures, with the guidance of both global and local features. The flowchart of our method is shown in Fig. 1, in which different colors represent different local parts. Our model takes a random Gaussian noise as input. We transform it into a rough point cloud and divide it into multiple parts to extract local features. Then we utilize the information of local structures to enhance the generation of local point cloud shapes. As for an airplane shape, each of the parts such as the left wing, right wing, turbine, and tail wing, are encoded by an independent feature vector. Then the resolution of local point clouds is gradually increased through several upsampling layers. Here we fused latent local codes into point cloud upsampling layers to guide the generation of local structures in different resolutions. We concate these local point clouds together as the output shape of our model. With the guidance of local perception information, our model can generate high-resolution point clouds with fine details.

In summary, the contributions of our work are as follows: (1) We propose a novel point cloud generation network, which progressively produces high-resolution shapes in a coarse-to-fine manner via several upsampling operations. (2) We design a new upsampling layer that integrates point-wise features, local part features, and global shape features for generating denser points. (3) Extensive experiments show that our model achieves competitive results when compared with other popular GAN-based approaches.

2 Related Work

2.1 Deep Learning on Point Cloud

Deep learning techniques are often applied to analysis and process point cloud. PointNet [12] extracts the point cloud features through Shared MLPs and uses max-pooling to obtain the global shape feature. To capture more local structure information, PointNet++ [13] iteratively combines each point with its neighbor points to enhance the network's local perception ability in multiple dimensions. To learn the spatial relationships of each point in the local structure, DGCNN [14] extracts the edge features between each point and its neighbor points. Then these features are aggregated through a pooling layer as the local area feature around each point. PCT [15] and PointTransformer [16] introduce Transformer [17] into point cloud learning tasks, they calculate correlation weights between different points to mining important regions in point cloud shapes. SnowflakeNet [18] proposes a new point cloud deconvolution method that multiplies the number of sets by predicting multiple different offsets for each point. It captures and inherits the shape features of parent points through point-wise splitting, and then uses a skip-transformer to achieve communication between different upsampling layers.

2.2 GAN-Based Point Cloud Generative Models

In the two-dimensional image generation field, GAN [5] has achieved significant results, so some point cloud generation studies also use GAN for training. L-GAN [19] is the first GAN-based point cloud generation network. It is trained by two-stage, first learning an auto-encoder to compress point cloud shapes into high-dimensional global shape features. Then it trains a latent vector generator that generates global features from random noise. To learn local shape information, TreeGAN [20] introduces a tree structure graph convolutional network that combines the information of ancestor points to create new points. PDGN [9] adopts a coarse-to-fine method, which first generates sparse point clouds and then gradually expands the point cloud density through a deconvolute network. SP-GAN [21] utilizes global spherical priors and local latent codes to achieve global shape modeling and structure-aware shape manipulation. WarpingGAN [10] combines multiple local structure-aware semantics to deform multiple 3D point set priors into shapes.

2.3 Energy-Based Point Cloud Generative Models

Some studies introduce energy functions to simulate the motion of point sets from random noise sampling to specific shapes. PointFlow [7] proposes a point cloud generation network based on two standardized flows. One learns to create shape features and the other uses these features to generate point cloud shapes from Gaussian distribution. To better generate local structures in the flow model, ChartPointFlow [22] uses label-based mapping to distinguish different local structures, so each complex shape would be unsupervisly divided into multiple parts. ShapeGF [23] generates point clouds by sampling points from Gaussian distribution and approaching them onto a specific object surface. In recent years, diffusion models have achieved significant results in two-dimensional image generation tasks [24]. Some researchers use it to generate 3D point clouds. By continuously adding noise to the point cloud, points will gradually transition from a shape distribution to a random Gaussian distribution. DPM [8] constructs this process as a Markov chain. It uses a flow model to generate latent shape features and restore shapes from noise sets. PVD [25] and Gecco [26] respectively use PVCNN [27] and SetTransformer [28] to learn the denoising steps from Gaussian noise to 3D shape. On the basis of PVD, LION [29] proposed a point cloud diffusion model in latent space. It first trains a VAE to encode point cloud shapes into features and latent points, and then uses two different diffusion models to learn the denoising process of features and latent points, respectively.

3 Method

3.1 OverView

Our 3D point cloud generator is shown in Fig. 2 (a). The input is a random vector sampled from the Gaussian distribution. The output is a point cloud shape containing 2048 coordinates in Euclidean space. Our network consists of two modules: a coarse point cloud generator and three point cloud upsampling layers. First, the random vector is converted into a global vector and a low-resolution sparse point cloud P_1 through a coarse point cloud generator. Here we divide P_1 into different local parts and map them into local features. In the upsampling layer, two offsets are predicted for each point to divide each point into two new points. We combine global vector and local features with the coordinates of points, to gradually improve the resolution of point clouds.

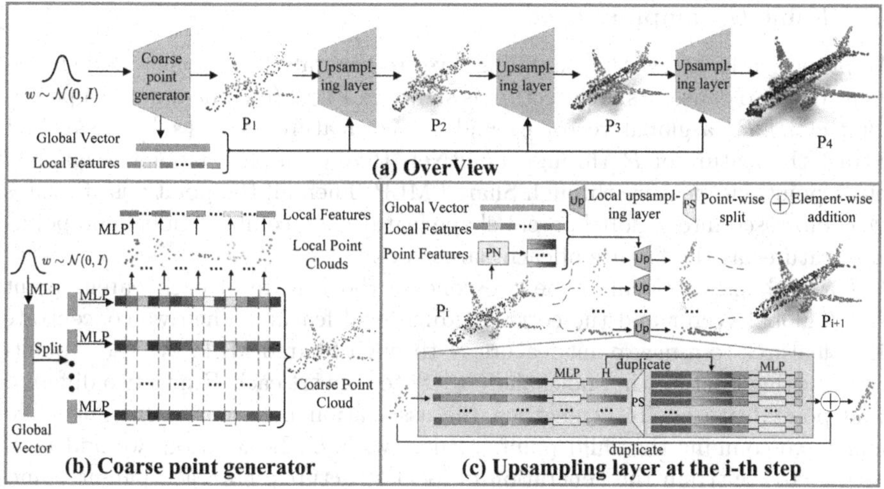

Fig. 2. The overall of our network. P_1, P_2, P_3, P_4 contains 256, 512,1024, 2048 points, respectively.

3.2 Coarse Point Generator

The architecture of our coarse point generator is shown in Fig. 2 (b). It aims to convert a random vector w into a global vector f_g, n local features, and a coarse point cloud P_1 with n local parts. Because the input vector w is randomly sampled from the Gaussian distribution, which is different from the distribution of shape features. So we use MLP to transform w into a global vector f_g.

Our point cloud decompose strategy aims to generate n local parts, each containing m points. We first split f_g into m sub-vectors with the same number of channels. Here we use Shared MLPs to transform m vectors into m subsets.

In Fig. 2 (*b*), they are marked by m rows of colored horizontal bars. Each block represents a point. They are evenly distributed on the surface of the shape. Coarse point cloud P_1 can be defined as $P_1 = \{q_i\}_{i=1}^m$. Each subset q_i contains n points, which can be defined as $q_i = \{p_j^i\}_{j=1}^n$. In Fig. 2 (*b*), we use the same color to mark the points located in the same sequence. They are generated by shared MLPs with the same weights. So they have similar semantic information and close distance. We concatenate points in the same sequence together as n local parts. Therefore P_1 can also be defined as $P_1 = \{l_j\}_{j=1}^n$. Each part contains m points, which can be defined as $l_j = \{p_i^j\}_{i=1}^m$. Unlike other methods that use Farthest Point Sampling and K-Nearest Neighbor to split each point cloud into local parts. Our splitting operation does not require any calculation or comparison of Euclidean distances between each point. In addition, we map n local parts to n local features $S = \{s_j\}_{j=1}^n$ through simple PointNet. Different local structures are represented by different local features.

3.3 Point Upsampling Layer

We propose a local-aware upsampling layer to generate local point cloud shapes from sparse to dense. Its architecture is shown in Fig. 2 (*c*). The input is a parent point cloud P_i, a global vector g_f and n local features $S = \{s_j\}_{j=1}^n$. We first extract the feature of P_i through PointNet. Here each parent point is mapped into a point-wise feature through Shared MLP. Then all the point-wise features are compressed into a global parent shape feature. We combine global and point-wise features as the feature of each parent point.

For each parent local shape, we concate the feature of each parent point with a global vector and the corresponding local feature. Then we concentrate these features to a new point feature H through Shared MLP. We use a point-wise splitting operator in SnowflakeNet [18] to convert each H into two different child point features. To strengthen the association between parent and child points, we combine the child point features with H. In addition, we add local features to constrain the generation of local structures. For each parent point, we convert these concated features into two offsets through Shared MLP and add the residuals to the input parent point. Each of them corresponds to two offsets. By adding these two offsets to the coordinates, we can split each parent point into two child points, thereby doubling the number of parent point clouds.

3.4 Loss Function

We use the generative adversarial loss to train our point cloud generation network. To ensure the realness of the point cloud shape during upsampling, our generator outputs multi-resolution point clouds. We use multiple discriminators to distinguish point clouds in different resolutions. The loss of generator $\mathcal{L}_{gen}$ and the loss of discriminator $\mathcal{L}_{disc}$ are defined as follows:

$$\mathcal{L}_{gen} = -\sum_{l=1}^{M} \mathbb{E}_{X_l \sim P_{X_l}} \left[D_l \left(X_l \right) \right] \tag{1}$$

$$\mathcal{L}_{disc} = \sum_{l=1}^{M} \mathbb{E}_{X_l \sim P_{X_l}} [D_l (X_l)] - \sum_{l=1}^{M} \mathbb{E}_{Y_l \sim P_{Y_l}} [D_l(Y_l)] + \mathcal{L}_{sp} \tag{2}$$

where M is the number of resolutions. X_l means the point cloud from l-th generated point cloud set P_{X_l}, and Y_l means the point cloud from l-th reference point cloud sets P_{Y_l}. The l-th discriminator is defined as D_l. To ensure the stability of point cloud shapes at different resolutions, we use the shape preservation loss L_{sp} in PDGN [9] to constrain the shape differences between point clouds at different resolutions. The definition of L_{sp} is as follow:

$$\mathcal{L}_{sp} = \sum_{l=1}^{M-1} \sum_{m=l+1}^{M} d(X_l, X_m) \tag{3}$$

where X_l and X_m are two generated point clouds with different resolutions. d is the Chamfer distance of the means and covariances between two point clouds.

4 Experiments

4.1 Experiment Settings

Dataset. Same as the previous point cloud generation methods, we use the ShapeNet [30] dataset, which contains 35708 shapes for training and 5158 shapes for testing. We sampled 256, 512, 1024, and 2048 points from each shape surface as reference point clouds in different resolutions. To normalize different shapes, we center each point cloud to the origin and scale all points to a $[-1, 1]$ cube.

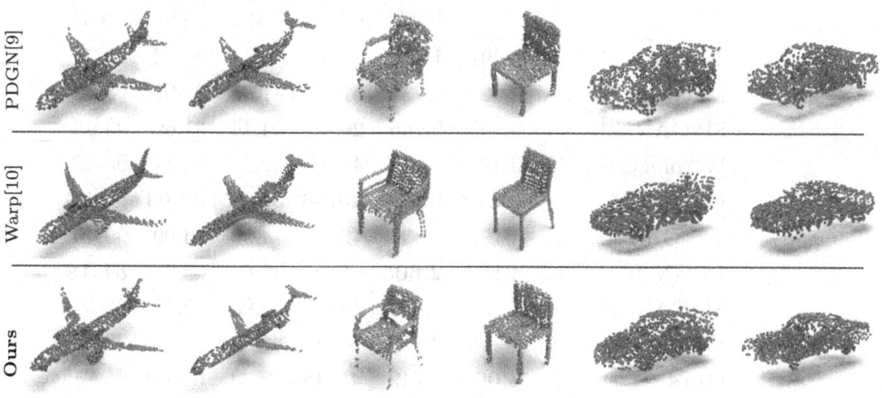

Fig. 3. Examples of generated point clouds from different generation methods. We compared with other two GAN-based methods: PDGN and WarpingGAN. Shapes on different rows are from different methods, each column shows similar point cloud shapes.

Evaluation Metrics. We measure the distance between two point clouds using Chamfer Distance(CD) and Earth Move Distance(EMD) respectively. Like other point cloud generation methods, we use Minimum matching distance(MMD), Coverage(COV), and 1-nearest neighbor accuracy(1-NNA) as our evaluation metrics. MMD calculates the average distance between each point cloud in the generated set and the nearest neighbor in the reference set. COV measures the scores that match the reference point cloud set and the generated point cloud set. 1-NNA classifies each point cloud into a reference set or a generated set based on its nearest sample label. Ideally, if two point sets are identical, 1-NNA will converge to 50%.

Implementation Details. We trained our model on an A5000 GPU, using Adam to optimize training. We set the learning rates of the generator and discriminator to $1{\times}10^4$. The input of our framework is a 128-D random vector sampled from the Gaussian distribution. Each global vector contains 512 channels, we use an MLP to transform the global vector and evenly split it into 16 local vectors. The value of 16 is selected empirically through experiments. We gradually increase the resolution of the point cloud from 256 to 512, 1024, and 2048 points. In the upsampling layer, the number of channels for global, local, and point-wise features are 512, 64, and 128 respectively. We use Mitsuba2 [31] to visualize point clouds.

4.2 Evaluation of Point Cloud Generation

Table 1. Generation metrics on Airplane, Chair, Car. MMD-CD is multiplied with 1×10^3, MMD-EMD is multiplied with 1×10^2. Both the training and testing shapes are individually normalized.

Category	Model	MMD (↓)		COV (%, ↑)		1-NNA (%, ↓)	
		CD	EMD	CD	EMD	CD	EMD
Airplane	TreeGAN [20]	0.391	1.679	40.99	8.64	94.32	99.75
	PDGN [9]	0.422	0.739	37.78	31.85	94.07	90.86
	SP-GAN [21]	0.403	0.766	26.42	24.44	94.69	93.95
	WarpingGAN [10]	0.407	1.122	41.23	22.47	95.31	98.52
	Ours	**0.373**	**0.718**	**43.46**	**38.27**	**88.64**	**90.00**
Chair	TreeGAN [20]	4.176	4.537	42.3	14.05	74.09	99.40
	PDGN [9]	4.170	**2.605**	44.41	39.88	69.26	**81.19**
	SP-GAN [21]	4.208	2.620	40.03	32.93	72.58	83.69
	WarpingGAN [10]	4.295	2.757	40.79	**43.20**	77.49	89.20
	Ours	**4.008**	2.742	**47.43**	42.30	**69.03**	85.73
Car	TreeGAN [20]	1.147	1.461	32.95	19.60	90.77	99.86
	PDGN [9]	1.176	1.002	32.95	36.08	90.06	89.91
	SP-GAN [21]	1.168	1.021	34.94	31.82	87.36	85.94
	WarpingGAN [10]	**1.165**	1.015	35.80	36.65	90.48	91.9
	Ours	1.166	**0.923**	**37.22**	**43.18**	**87.22**	**85.09**

Quantitative Comparison. Since our method belongs to the GAN-based framework, we compare it with other GAN-based methods for fairness, including Tree-GAN [20], PDGN [9], WarpingGAN [10] and SP-GAN [21]. We train and test Tree-GAN, PDGN, and WarpingGAN on the three most commonly used categories: Airplane, Chair, and Car, with the same data processing method. The result of SP-GAN is from LION [29], which trained and tested on ShapeNet too. In the experimental results shown in Table 1, our method achieves the best generation performance in most metrics of the Airplane and Car categories. And in the Chair category, our method outperforms others in CD-based metrics. From the visualization result in Fig. 3, it can be seen that our method can generate more refined local structures while ensuring the diversity of shapes.

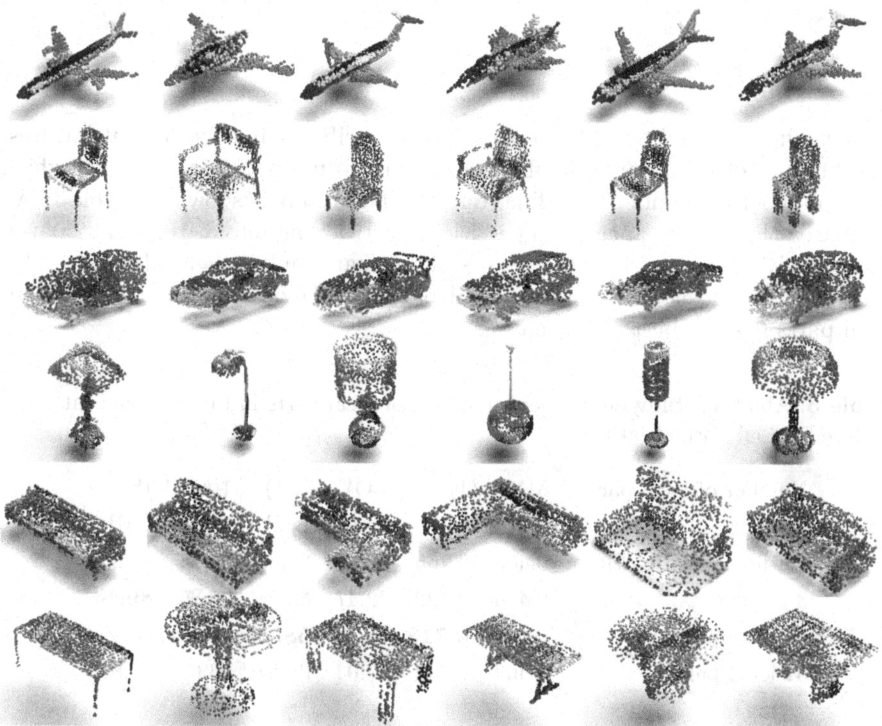

Fig. 4. Visualization examples of 6 categories, shown from top to bottom: "Airplane", "Chair", "Car", "Lamp", "Sofa", and "Table". Different colors represent different local structures.

Visual Results. In Fig. 4, we provide more visual results for 6 different categories generated by our model. The different colors represent the different local parts divided by our network. It can be seen that within the same category, areas marked by the same color have similar semantic information. It indicates that our model can effectively learn local information of the point cloud on multiple categories. At the same time, it ensures the diversity and realness of shapes.

Efficiency Comparison. We have compared the inference time with other GAN-based methods TreeGAN [20], PDGN [9], WarpingGAN [10] and diffusion-based methods DPM [8], PVD [25], LION [29]. It's worth noting that we have tested other diffusion-based methods. The evaluation shows that they receive much better results. However, due to the need for multiple time steps denoise process, the generation speed of diffusion models is slower than ours (Table 2).

Table 2. Comparison of the inference time of different generation methods.

Model	TreeGAN	PDGN	WarpingGAN	DPM	PVD	LION	Ours
Inference (s)	0.017	0.14	**0.0056**	22.8	29.9	27.1	0.018

4.3 Ablation Study

We set different numbers of local partitions to verify the impact of local features. The number of local parts should satisfy that its multiples equals 2048, which is the number of points of our final output. The result is shown in Table 3. We compared the performance of no partition, partitioning into 8, 16, and 32 parts, respectively. The experimental results show that combining local features for generation is better than without local partition. Dividing each shape into 16 local parts has the best performance.

Table 3. Ablation study on different numbers of local parts in the Airplane category, with individual normalization.

Number of local parts	MMD (↓)		COV (%, ↑)		1-NNA (%, ↓)	
	CD	EMD	CD	EMD	CD	EMD
Without local partition	0.403	0.765	40.99	30.62	90.74	92.84
8 local parts	0.406	0.722	42.47	33.58	90.86	**89.14**
16 local parts	**0.373**	**0.718**	**43.46**	**38.27**	**88.64**	90.00
32 local parts	0.404	0.742	37.04	34.57	93.09	91.85

5 Conclusion

In this work, we introduce a point generation method to generate high-resolution point clouds with fine details. It first generates a coarse shape and divides it into multiple local parts to extra local features, and then generates a dense local shape from each of them. In this way, the local details are preserved. We adopt a local-aware upsampling layer, which gradually increases the resolution of local point clouds with the guidance of global and local information. The point cloud generation experiments on the ShapeNet dataset have demonstrated that the performance of our model is superior to existing GAN-based methods.

Acknowledgment. We thank the reviewers for their efforts. This work is supported by the National Natural Science Foundation of China No.62176242, also in part by No.61906177.

References

1. Sun, K., Tao, W., Qian, Y.: Guide to match: multi-layer feature matching with a hybrid gaussian mixture model. IEEE Trans. Multimedia **22**(9), 2246–2261 (2020)
2. Chen, Z., Sun, K., Yang, F., Guo, L., Tao, W.: Sc^2-pcr++: rethinking the generation and selection for efficient and robust point cloud registration. IEEE Trans. Pattern Anal. Mach. Intell. **45**, 12358–12376 (2023)
3. Sun, K., Tao, W.: A center-driven image set partition algorithm for efficient structure from motion. Inf. Sci. **479**, 101–115 (2019)
4. Sun, K., Jinhong, Yu., Tao, W., Li, X., Tang, C., Qian, Y.: A unified feature-spatial cycle consistency fusion framework for robust image matching. Inf. Fusion **97**, 101810 (2023)
5. Goodfellow, I., et al.: Generative adversarial networks. Commun. ACM **63**(11), 139–144 (2020)
6. Kim, J., Yoo, J., Lee, J., Hong, S.: Setvae: learning hierarchical composition for generative modeling of set-structured data. In: Proceedings of the IEEE/CVF Conference on Computer Vision and Pattern Recognition, pp. 15059–15068 (2021)
7. Yang, G., Huang, X., Hao, Z., Liu, M.Y., Belongie, S., Hariharan, B.: Pointflow: 3D point cloud generation with continuous normalizing flows. In: Proceedings of the IEEE/CVF International Conference on Computer Vision, pp. 4541–4550 (2019)
8. Luo, S., Hu, W.: Diffusion probabilistic models for 3D point cloud generation. In: Proceedings of the IEEE/CVF Conference on Computer Vision and Pattern Recognition, pp. 2837–2845 (2021)
9. Hui, L., Xu, R., Xie, J., Qian, J., Yang, J.: Progressive point cloud deconvolution generation network. In: Vedaldi, A., Bischof, H., Brox, T., Frahm, J.-M. (eds.) ECCV 2020. LNCS, vol. 12360, pp. 397–413. Springer, Cham (2020). https://doi.org/10.1007/978-3-030-58555-6_24
10. Tang, Y., Qian, Y., Zhang, Q., Zeng, Y., Hou, J., Zhe, X.: Warpinggan: warping multiple uniform priors for adversarial 3D point cloud generation. In: Proceedings of the IEEE/CVF Conference on Computer Vision and Pattern Recognition, pp. 6397–6405 (2022)
11. Li, S., Liu, M., Walder, C.: Editvae: unsupervised parts-aware controllable 3D point cloud shape generation. In: Proceedings of the AAAI Conference on Artificial Intelligence, vol. 36, pp. 1386–1394 (2022)
12. Qi, C.R., Su, H., Mo, K., Guibas, L.J.: Pointnet: deep learning on point sets for 3D classification and segmentation. In: Proceedings of the IEEE Conference on Computer Vision and Pattern Recognition, pp. 652–660 (2017)
13. Qi, C.R., Yi, L., Su, H., Guibas, L.J.: Pointnet++: deep hierarchical feature learning on point sets in a metric space. Adv. Neural Inf. Process. Syst. **30** (2017)
14. Wang, Y., Sun, Y., Liu, Z., Sarma, S.E., Bronstein, M.M., Solomon, J.M.: Dynamic graph cnn for learning on point clouds. ACM Trans. Graph. (tog) **38**(5), 1–12 (2019)
15. Guo, M.H., Cai, J.X., Liu, Z.N., Mu, T.J., Martin, R.R., Hu, S.M.: PCT: point cloud transformer. Comput. Visual Media **7**, 187–199 (2021)

16. Zhao, H., Jiang, L., Jia, J., Torr, P.H.S., Koltun, V.: Point transformer. In: Proceedings of the IEEE/CVF International Conference on Computer Vision, pp. 16259–16268 (2021)
17. Vaswani, A., et al.: Attention is all you need. Adv. Neural Inf. Process. Syst. **30** (2017)
18. Xiang, P., et al.: Snowflakenet: point cloud completion by snowflake point deconvolution with skip-transformer. In: Proceedings of the IEEE/CVF International Conference on Computer Vision, pp. 5499–5509 (2021)
19. Achlioptas, P., Diamanti, O., Mitliagkas, I., Guibas, L.: Learning representations and generative models for 3D point clouds. In: International Conference on Machine Learning, pp. 40–49. PMLR (2018)
20. Shu, D.W., Park, S.W., Kwon, J.: 3D point cloud generative adversarial network based on tree structured graph convolutions. In: Proceedings of the IEEE/CVF International Conference on Computer Vision, pp. 3859–3868 (2019)
21. Li, R., Li, X., Hui, K.-H., Chi-Wing, F.: SP-GAN: sphere-guided 3D shape generation and manipulation. ACM Trans. Graph. (TOG) **40**(4), 1–12 (2021)
22. Kimura, T., Matsubara, T., Uehara, K.: Chartpointflow for topology-aware 3D point cloud generation. In: Proceedings of the 29th ACM International Conference on Multimedia, pp. 1396–1404 (2021)
23. Cai, R., et al.: Learning gradient fields for shape generation. In: Vedaldi, A., Bischof, H., Brox, T., Frahm, J.-M. (eds.) ECCV 2020. LNCS, vol. 12348, pp. 364–381. Springer, Cham (2020). https://doi.org/10.1007/978-3-030-58580-8_22
24. Ho, J., Jain, A., Abbeel, P.: Denoising diffusion probabilistic models. Adv. Neural. Inf. Process. Syst. **33**, 6840–6851 (2020)
25. Zhou, L., Du, Y., Wu, J.: 3D shape generation and completion through point-voxel diffusion. In: Proceedings of the IEEE/CVF International Conference on Computer Vision, pp. 5826–5835 (2021)
26. Tyszkiewicz, M.J., Fua, P., Trulls, E.: Gecco: geometrically-conditioned point diffusion models. arXiv preprint arXiv:2303.05916 (2023)
27. Liu, Z., Tang, H., Lin, Y., Han, S.: Point-voxel CNN for efficient 3D deep learning. Adv. Neural Inf. Process. Syst. **32** (2019)
28. Lee, J., Lee, Y., Kim, J., Kosiorek, A., Choi, S., Teh, Y.W.: Set transformer: a framework for attention-based permutation-invariant neural networks. In: International Conference on Machine Learning, pp. 3744–3753. PMLR (2019)
29. Zeng, X., et al.: Lion: latent point diffusion models for 3D shape generation. Adv. Neural Inf. Process. Syst. **35**, 10021–10039 (2022)
30. Chang, A.X., et al.: Shapenet: an information-rich 3D model repository. arXiv preprint arXiv:1512.03012 (2015)
31. Nimier-David, M., Vicini, D., Zeltner, T., Jakob, W.: Mitsuba 2: a retargetable forward and inverse renderer. ACM Trans. Graph. (TOG) **38**(6), 1–17 (2019)

Three-Dimensional Plant Reconstruction with Enhanced Cascade-MVSNet

He Ren[1], Jianzhong Zhu[1], Liufeng Chen[2], Xue Jiang[1], Kai Xie[1], and Ruifang Zhai[1](✉)

[1] Huazhong Agricultural University, Wuhan 430070, China
rfzhai@mail.hzau.edu.cn

[2] Qianxin Security Technology (Wuhan) Co., Ltd., Wuhan 430000, China

Abstract. Three-dimensional reconstruction is an important method for recovering the morphological structure of plants, and a complete and accurate 3D point cloud of plants can better reflect the phenotypic parameters of plants, such as plant height, volume of the plant and leaf area. In order to obtain more complete and accurate plant point clouds, this paper proposes a series of enhancements based on the Cascade-MVSNet network. To improve the reconstruction of weakly textured areas of plants, a lightweight attention mechanism is incorporated into the feature extraction stage. To enhance the valid points of the plant point clouds and suppress invalid points, focal loss is employed as the loss function, treating the depth estimation problem as a classification task to obtain more accurate depth information. Additionally, to improve the efficiency of the model, we substitute standard convolutions with depth-wise separable convolutions, reducing the number of parameters and computational complexity while maintaining performance. The above work is applied to the plant dataset and the reconstructions showed a significant reduction in invalid points and a clearer reconstructed point cloud.In addition to conducting research on the plant dataset, we also evaluate our approach on the publicly available DTU dataset. Experimental results demonstrate a noticeable increase in reconstruction completeness on the DTU dataset, achieving competitive performance overall.

Keywords: Mulit-View Stereo · 3D reconstruction · Noise constraint · Plant Phenomics

1 Introduction

Obtaining the three-dimensional(3D) morphological structure of plants is of great significance for extracting phenotypic parameters. A complete and accurate 3D point cloud of plants can better reflect their phenotypic parameters, such as plant height, volume of the plant and leaf area. In order to generate

This work was supported by the National Key R&D Program of China (2022YFD2002304).

Q. Liu et al. (Eds.): PRCV 2023, LNCS 14426, pp. 283–294, 2024.
https://doi.org/10.1007/978-981-99-8432-9_23

high-quality plant point clouds in a cost-effective manner, Multi-View Stereo (MVS) is an effective method.

Multi-View Stereo is a technique in the field of computer vision and photo grammetry that aims to reconstruct a 3D scene from multiple views of images. Its principle is based on the concepts of disparity and triangulation, and the depth information of the scene is recovered by analyzing the disparity relationships among the multi-view images.

The reconstruction principle of MVS can be divided into five stages. Firstly, features are extracted from multiple-view images using techniques such as approaches Scale-Invariant Feature Transform (SIFT) [13,15] or Speeded-Up Robust Features (SURF) [1,13], focusing on keypoints or corners. These features are used to establish correspondences between different views through feature point matching, determining the correspondence relationships. Next, depth map estimation is performed by triangulating the correspondences between views to estimate the depth information for each pixel in the images [4]. This can be achieved through dense stereo matching or by utilizing geometric constraints derived from the camera parameters. After obtaining the initial depth map, refinement techniques are applied to enhance the accuracy of depth estimation. Optimization methods encompass both global and local optimization algorithms. In the fourth stage, the depth map is converted into 3D point coordinates using triangulation methods, resulting in a 3D point cloud representation of the scene. Finally, additional processing steps, such as illumination estimation [6,12] and texture mapping [5,8], can be applied to the generated 3D point cloud to further enhance its quality and appearance.

In recent years, the wide application of deep learning in the field of computer vision has also penetrated into the field of MVS. Deep learning has been widely applied in the field of computer vision and has also made its way into the MVS domain. Researchers have proposed end-to-end MVS reconstruction methods using Convolutional Neural Networks (CNN), such as MVSNet [20], MVSCRF [18], etc. These methods leverage deep learning techniques to extract image features and predict depth maps, which have certain advantages over traditional reconstruction approaches in terms of performance and reconstruction quality.

In the field of crop phenotyping, the use of deep learning methods for 3D reconstruction of crops can lead to better plant models and establish a foundation for extracting more accurate phenotypic parameters. Among the existing MVS networks, Cascade-MVSNet(CasMVSNet) [3] stands out for its balance between efficiency and reconstruction quality. However, it suffers from issues such as too many stray points at the reconstruction edges and incomplete reconstruction of regions with weak textures. In our research on a self-built plant dataset, we observed that plants have more complex edges compared to commonly seen regular geometric objects in public datasets. Moreover, plant leaves exhibit similar textures and less distinctive features. These characteristics exacerbates the aforementioned issues during plant reconstruction using CasMVSNet. During the process of generating the reconstructed point cloud, we aim to obtain more valid points while suppressing invalid points. Invalid points include redundant points, noisy points, and incorrectly generated background points at the edges. Therefore, in this paper, we propose improvements to the CasMVSNet network, referred to as E-CasMVSNet, focusing on three main aspects: feature extraction, loss function constraints, and efficiency enhancement. The primary contributions of this paper are as follows:

- We introduce a lightweight attention mechanism in the feature extraction module to enhance the network's feature extraction capability and improve the reconstruction performance of regions with weak textures.
- We adopt focal loss as the network's loss function, treating the depth estimation problem as a classification task. This approach enables better depth estimation and yields more accurate depth information.
- To enhance the efficiency of the model, we replace standard convolutions with depth-wise separable convolutions. This substitution reduces the number of parameters and computational complexity, thereby improving the overall efficiency of the model.

2 Related Work

In recent years, with the rapid development of deep learning technology, it has been introduced into the field of 3D reconstruction, demonstrating advantages in terms of efficiency and reconstruction accuracy compared to traditional methods.

In 2018, Yao et al. proposed an end-to-end deep learning architecture named MVSNet, designed to estimate depth maps from multiple-view images and subsequently reconstruct the 3D structure of the scene. MVSNet first performs feature extraction on the multi-view images using a feature extraction network. It then constructs a cost volume to capture the matching relationships between each pixel and other view images. By applying 3D convolutions to the cost volume, initial depth maps and probability maps are generated. These initial depth maps are further optimized to improve depth estimation accuracy and enhance reconstruction precision. Finally, the depth maps are fused to obtain a reconstructed 3D point cloud. MVSNet significantly improves the completeness of the reconstructed point cloud. However, it consumes large amounts of GPU memory, making it challenging to reconstruct high-resolution scenes.

After the introduction of MVSNet, researchers have continued to explore further along this line. R-MVSNet [21] utilizes recurrent neural networks to process the 3D cost volume, which reduces GPU memory consumption but increases computational time. P-MVSNet [14] introduces pixel-level illumination estimation to enhance the quality and robustness of multi-view stereo reconstruction, but the illumination estimation module requires additional training and computation, making it dependent on the dataset. CasMVSNet employs a coarse-to-fine strategy to infer depth maps, enabling the reconstruction of high-resolution scenes while significantly reducing memory consumption and computational time. However, it falls short in reconstructing scenes with indistinct features and weak textures.

Therefore, in this paper, based on CasMVSNet, we incorporate lightweight attention modules and focal loss to abtain better reconstruction results. Additionally, we introduce depth-wise separable convolutions to enhance the efficiency of the network.

3 Methodology

This section describes the incorporation of a lightweight attention module in the feature extraction stage of CasMVSNet (E-CasMVSNet), the utilization of focal loss as a constraint in the loss function, and the enhancement of model computational efficiency through the use of depth-wise separable convolutions. The network architecture of this paper is illustrated in Fig. 1.

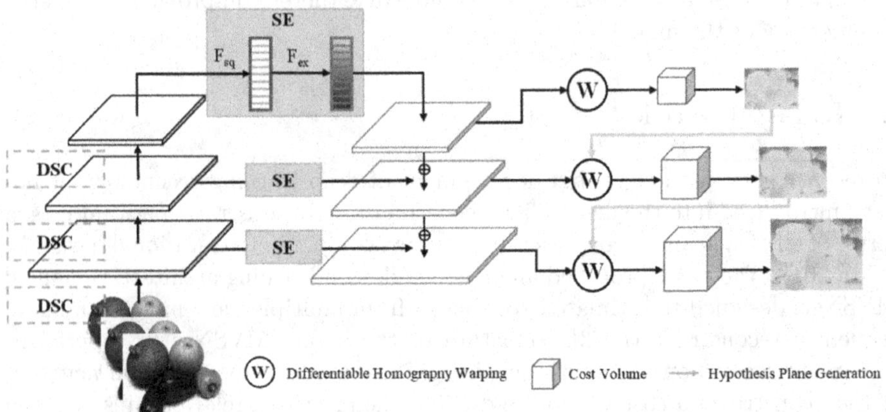

Fig. 1. Network architecture of E-CasMVSNet.

3.1 Network Overview

E-CasMVSNet employs a Feature Pyramid Network (FPN) [10] in the feature extraction stage to extract image features at different scales. The convolution used for feature extraction is depth-wise separable convolution. To capture global information, a lightweight Squeeze Excitation (SE) attention module [7] is incorporated to enhance the communication between multi-scale feature channels. Once the feature extraction is completed, a cost volume is constructed by performing homography matrix transformation on the feature maps. The cost volume is then regularized to obtain the initial depth map. To improve the accuracy of depth estimation, focal loss [11] is employed during the depth map refinement process as a constraint. Finally, the depth maps are fused to generate the reconstructed point cloud.

3.2 Squeeze Excitation Module

CasMVSNet first applies FPN to extract multi-scale feature maps at different spatial resolutions, but it overlooks the interdependencies of global information, which fails to capture crucial information for subsequent depth inference tasks. This limitation becomes particularly evident when dealing with images containing indistinct features, resulting in suboptimal reconstruction performance.

In the feature extraction module, E-CasMVSNet incorporates a lightweight SE attention module (as shown in Fig. 2) between multi-scale feature channels, which aims to improve the quality of the feature representation by explicitly modelling the interdependencies among channels throughout the entire feature hierarchy. This is achieved without introducing additional computational overhead. The SE module learns global information and selectively emphasizes informative features while suppressing less important ones. The SE module consists of two main operations: Squeeze and Excitation. The Squeeze operation compresses each feature channel's spatial resolution to a single value z through global average pooling. The formula describing this operation is as follows:

$$z_c = F_{sq} = \frac{1}{H \times W} \sum_{i=1}^{H} \sum_{j=1}^{W} u_c(i, j) \tag{1}$$

where u denotes the feature map obtained through convolution, and c denotes the number of channels. After the Squeeze operation gets the statistics of each channel, all the channel statistics are combined to get a $1 \times 1 \times c$ feature vector. Then, the Excitation operation reduces the number of channels to c/r using fully connected layers. After applying the ReLU activation function for feature mapping, another fully connected layer restores the number of channels to c. Finally, the sigmoid activation function is applied to obtain the importance weights for each feature channel, which are then reweighted to each channel. The calculation formula is shown as follows:

$$s = F_{\text{ex}} = \sigma\left(W_2 \delta\left(W_1 z\right)\right) \tag{2}$$

where s denotes the weight value of each channel, δ refers to the ReLU function, W_1 and W_2 denotes the two fully connected layers and σ denotes the sigmoid activation function.

3.3 Depth-Wise Separable Convolution

In the feature extraction stage, E-CasMVSNet differs from CasMVSNet by using depth-wise separable convolution instead of standard convolution, as shown in Fig. 3. Depthwise separable convolution decomposes the standard convolution into two steps: depthwise convolution and pointwise convolution. The depthwise convolution performs convolution operations on each input channel separately, followed by linear combinations of channels using the pointwise convolution. This approach allows operations to be performed within and between channels,

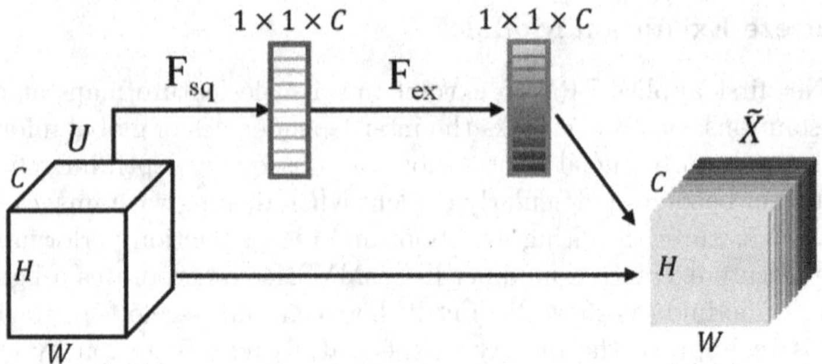

Fig. 2. The structure of Squeeze Excitation Module.

reducing parameters and computational complexity while maintaining approximate performance, thus making the network more lightweight.

On the other hand, depthwise separable convolution exhibits strong feature learning capabilities. By focusing on intra-channel and inter-channel feature representations separately, depthwise separable convolution has a larger receptive field, enabling better learning of spatial and channel-related features from the input data. This contributes to enhancing the representational power of the features.

3.4 Loss Function

The SE attention module enhances the network's feature extraction capability by extracting more features, particularly in the presence of weak texture features. However, it also leads to the generation of excessive points, including redundant and repetitive points, as well as background points erroneously generated at the reconstructed edge regions. These points, referred to as invalid points, impose a burden on depth estimation and subsequent processing. To reduce unnecessary computations and the number of points generated while achieving more accurate edge depth estimation, this paper formulates the depth estimation problem as a classification task.

The edge regions of objects often exhibit depth uncertainty and ambiguity, with a scarcity of positive samples and a predominance of negative samples from the background class. Focal Loss has a key property of mitigating class imbalance issues. Therefore, in this work, focal loss is employed as the loss function to provide stronger constraints on the ambiguous regions. The focal loss for each depth estimation stage is defined as follows:

$$Loss = \sum_{\mathbf{p}\in\{\mathbf{p}_v\}} -\left(1 - P^{(\tilde{d})}(\mathbf{p})\right)^{\gamma} \log\left(P^{(\tilde{d})}(\mathbf{p})\right) \tag{3}$$

where $P^d(\mathbf{p})$ denotes the predicted probability of the depth hypothesis d at pixel $\mathbf{p}$, and $\tilde{d}$ denotes the depth value closest to the ground truth. $\mathbf{p}_v$ denotes

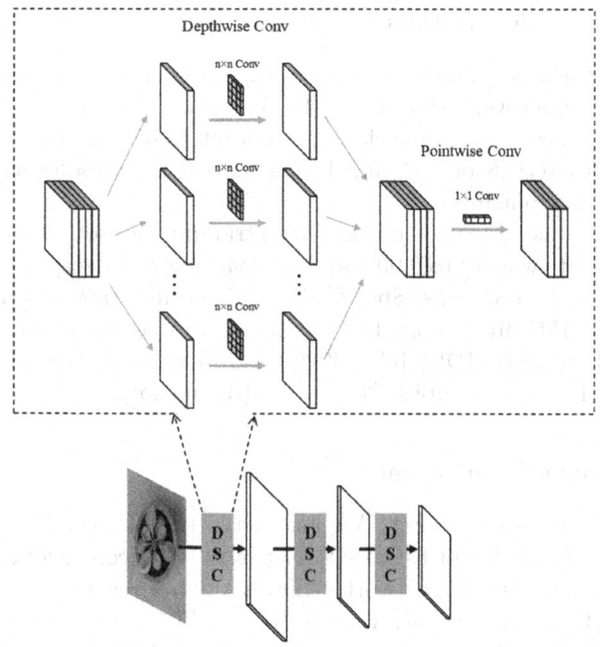

Fig. 3. The structure of Depth-wise Separable Convolution.

the subset of pixels with valid ground truth. When γ is equal to 0, the focal loss is equivalent to cross-entropy loss.

4 Experiments

4.1 Datasets

This paper used the publicly available dataset DTU [9] for training and evaluated the performance on both the self-made plant dataset and DTU dataset. The DTU dataset was divided into 79 training scans, 18 validation scans, and 22 evaluation scans, following the setup of MVSNet.

Cabbage Dataset. The self-made plant dataset focused on 14 cabbage plants and captured RGB images of the plants in natural lighting conditions at a frequency of one week per cycle.

DTU Dataset. The DTU dataset, was captured under controlled laboratory conditions with a fixed camera trajectory, consisting of 128 scans and 49 views across seven different lighting conditions.

4.2 Implementation Details

In this paper, the input image resolution, the number of input views, and the depth sampling count were set to 640×512, 5, and 192, respectively. The number of depth hypotheses planes in each stage was set to 48, 32, and 8, with corresponding depth intervals of 4, 2, and 1. The feature map resolutions were 1/16, 1/4, and 1 of the input image.

The training was performed on the entire training dataset for 16 epochs. The initial learning rate was set to 0.001, and the learning rate decayed exponentially with the number of iterations. Specifically, the learning rate was halved at the 10th, 12th, and 16th iterations. The batch size was set to 4, and the training was conducted on an NVIDIA RTX 3090 GPU. The entire training phase took approximately 14 h and occupied 20 GB of GPU memory.

4.3 Experimental Performance

Testing on Cabbage Dataset. We conducted tests on the Cabbage dataset using both CasMVSNet and E-CasMVSNet, and the reconstruction results are shown in Fig. 4. Visually, our network achieves a better reconstruction compared to the original CasMVSNet, with fewer artifacts. The reconstructed point cloud of our model exhibits clearer details along the edges of plant leaves compared to the original CasMVSNet and CasMVSNet with SE module, reducing the false generation of background points. Overall, our model effectively addresses the issue of boundary artifacts in plant point clouds reconstruction.

Evaluation on DTU Dataset. To further validate the effectiveness of our model, we also conducted experiments on the publicly available DTU dataset. The visualization results are shown in Fig. 5. From the figure, it can be observed that our method effectively reduces the generation of background points at the edges of non-plant objects. It also performs well in reconstructing weak texture areas, with fewer surface artifacts and clearer results.

We evaluated our method on the DTU dataset using official evaluation metrics, and the results are shown in Table 1. The metrics include Accuracy and Completeness, which are official metrics provided by the dataset. Accuracy measures the average absolute point-to-point distance from the MVS reconstruction to the ground truth, while Completeness measures the opposite direction. Overall represents the average of Accuracy and Completeness. E-CasMVSNet shows a significant improvement in reconstruction completeness and achieves competitive performance overall.

4.4 Ablation Study

To verify the effectiveness of each module in our method, we conducted ablation experiments on the DTU dataset. In these experiments, all hyperparameters were set to be the same as in the main method.

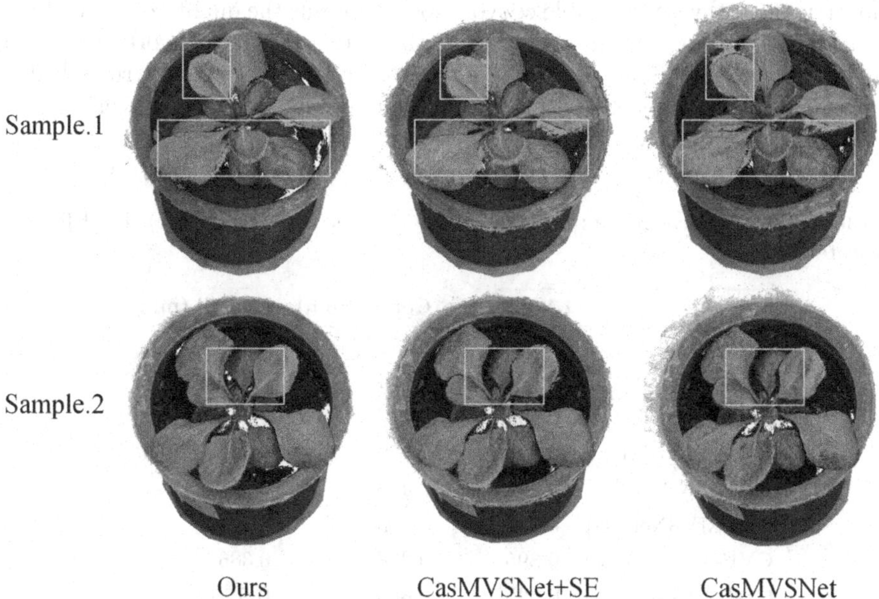

Fig. 4. Comparison of reconstructed results with CasMVSNet, on two cabbage samples.

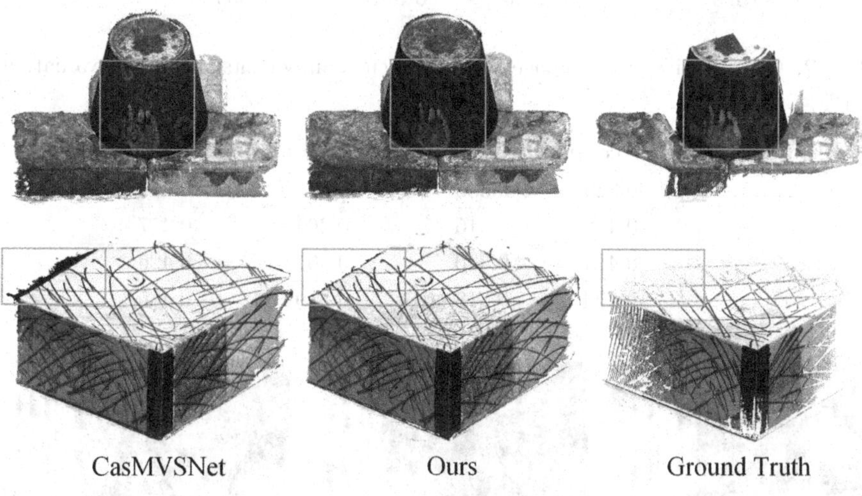

Fig. 5. Comparison of reconstructed results with CasMVSNet, on DTU evaluation set.

The experimental results for each module are presented in Table 2, and the visualizations are shown in Fig. 6. The network model with the inclusion of the SE module generates denser and more complete point clouds. Additionally, the lightweight nature of the SE module reduces the training time of the model. The adoption of depthwise separable convolution maintains the quality of the trained model while improving the efficiency of the network. The network, with the inclusion of the SE module, has a denser point cloud while the reconstructed edge part also incorrectly generates some background points, and focal loss effectively mitigates this problem.

Table 1. Quantitative results on DTU evaluation set (lower is better). Bold figures indicate the best.

Method	Acc. (mm)	Comp. (mm)	Overall (mm)
Gipuma [2]	**0.283**	0.873	0.578
COLMAP [16]	0.400	0.664	0.532
MVSNet [20]	0.396	0.527	0.462
R-MVSNet [21]	0.385	0.459	0.422
P-MVSNet [14]	0.406	0.434	0.420
Fast-MVSNet [22]	0.336	0.403	0.370
CVP-MVSNet [19]	0.395	0.378	0.386
CasMVSNet [3]	0.325	0.385	**0.355**
AA-RMVSNet [17]	0.376	0.339	0.357
Vis-MVSNet [23]	0.369	0.361	0.365
Ours	0.400	**0.318**	0.359

Table 2. Quantitative performance with different components on DTU evaluation dataset.

SE	DSConv	F.L	Acc. (mm)	Comp. (mm)	Overall (mm)	Run-time (s)
			0.325	0.385	0.355	0.492
✓			0.412	0.316	0.364	0.457
✓	✓		0.424	0.310	0.367	0.426
✓	✓	✓	0.400	0.318	0.359	0.435

Fig. 6. Comparison of reconstructed results with different modules.

5 Conclusion

In this paper, we have proposed a series of improvements based on the CasMVSNet network, referred to as E-CasMVSNet. E-CasMVSNet enhances the feature extraction capability of the network through the SE module. It utilizes depth-wise separable convolution to maintain the quality of the generated point cloud while reducing the computational complexity and improving the efficiency of the model. Additionally, it treats the depth estimation problem as a classification task and employs focal loss to reduce the generation of redundant points and erroneous background points along the edges. Experimental results on the DTU dataset and a self-made plant dataset demonstrate the effectiveness and improvement achieved by our approach.

References

1. Bay, H., Ess, A., Tuytelaars, T., Van Gool, L.: Speeded-up robust features (surf). Comput. Vision Image Underst. **110**(3), 346–359 (2008). https://doi.org/10.1016/j.cviu.2007.09.014. https://www.sciencedirect.com/science/article/pii/S1077314207001555
2. Galliani, S., Lasinger, K., Schindler, K.: Massively parallel multiview stereopsis by surface normal diffusion. In: 2015 IEEE International Conference on Computer Vision (ICCV), pp. 873–881 (2015). https://doi.org/10.1109/ICCV.2015.106
3. Gu, X., Fan, Z., Dai, Z., Zhu, S., Tan, F., Tan, P.: Cascade cost volume for high-resolution multi-view stereo and stereo matching (2020)
4. Hartley, R.I., Sturm, P.: Triangulation. Comput. Vis. Image Underst. **68**(2), 146–157 (1997). https://doi.org/10.1006/cviu.1997.0547
5. Heckbert, P.S.: Survey of texture mapping. IEEE Comput. Graph. Appl. **6**(11), 56–67 (1986). https://doi.org/10.1109/MCG.1986.276672
6. Hold-Geoffroy, Y., Sunkavalli, K., Hadap, S., Gambaretto, E., Lalonde, J.F.: Deep outdoor illumination estimation (2018)
7. Hu, J., Shen, L., Albanie, S., Sun, G., Wu, E.: Squeeze-and-excitation networks (2019)
8. Igehy, H., Eldridge, M., Proudfoot, K.: Prefetching in a texture cache architecture. In: Proceedings of the ACM SIGGRAPH/EUROGRAPHICS Workshop on Graphics Hardware, HWWS 1998, p. 133-ff. Association for Computing Machinery, New York (1998). https://doi.org/10.1145/285305.285321
9. Jensen, R., Dahl, A., Vogiatzis, G., Tola, E., Aanæs, H.: Large scale multi-view stereopsis evaluation. In: 2014 IEEE Conference on Computer Vision and Pattern Recognition, pp. 406–413 (2014). https://doi.org/10.1109/CVPR.2014.59
10. Lin, T.Y., Dollár, P., Girshick, R., He, K., Hariharan, B., Belongie, S.: Feature pyramid networks for object detection (2017)
11. Lin, T.Y., Goyal, P., Girshick, R., He, K., Dollár, P.: Focal loss for dense object detection (2018)
12. Liu, Y., Shen, S.: Self-adaptive single and multi-illuminant estimation framework based on deep learning (2019)
13. Lowe, D.: Object recognition from local scale-invariant features. In: Proceedings of the Seventh IEEE International Conference on Computer Vision, vol. 2, pp. 1150–1157 (1999). https://doi.org/10.1109/ICCV.1999.790410

14. Luo, K., Guan, T., Ju, L., Huang, H., Luo, Y.: P-MVSNet: learning patch-wise matching confidence aggregation for multi-view stereo. In: 2019 IEEE/CVF International Conference on Computer Vision (ICCV), pp. 10451–10460 (2019). https://doi.org/10.1109/ICCV.2019.01055
15. Rublee, E., Rabaud, V., Konolige, K., Bradski, G.: ORB: an efficient alternative to sift or surf. In: 2011 International Conference on Computer Vision, pp. 2564–2571 (2011). https://doi.org/10.1109/ICCV.2011.6126544
16. Schönberger, J.L., Zheng, E., Frahm, J.-M., Pollefeys, M.: Pixelwise view selection for unstructured multi-view stereo. In: Leibe, B., Matas, J., Sebe, N., Welling, M. (eds.) ECCV 2016. LNCS, vol. 9907, pp. 501–518. Springer, Cham (2016). https://doi.org/10.1007/978-3-319-46487-9_31
17. Wei, Z., Zhu, Q., Min, C., Chen, Y., Wang, G.: AA-RMVSNet: adaptive aggregation recurrent multi-view stereo network (2021)
18. Xue, Y., et al.: MVSCRF: learning multi-view stereo with conditional random fields. In: 2019 IEEE/CVF International Conference on Computer Vision (ICCV), pp. 4311–4320 (2019). https://doi.org/10.1109/ICCV.2019.00441
19. Yang, J., Mao, W., Alvarez, J.M., Liu, M.: Cost volume pyramid based depth inference for multi-view stereo (2020)
20. Yao, Y., Luo, Z., Li, S., Fang, T., Quan, L.: MVSNet: depth inference for unstructured multi-view stereo (2018)
21. Yao, Y., Luo, Z., Li, S., Shen, T., Fang, T., Quan, L.: Recurrent MVSNet for high-resolution multi-view stereo depth inference. In: 2019 IEEE/CVF Conference on Computer Vision and Pattern Recognition (CVPR), pp. 5520–5529 (2019). https://doi.org/10.1109/CVPR.2019.00567
22. Yu, Z., Gao, S.: Fast-MVSNet: sparse-to-dense multi-view stereo with learned propagation and gauss-newton refinement (2020)
23. Zhang, J., Yao, Y., Li, S., Luo, Z., Fang, T.: Visibility-aware multi-view stereo network (2020)

Learning Key Features Transformer Network for Point Cloud Processing

Guobang You, Yikun Hu, Yimei Liu, Haoyan Liu, and Hao Fan(✉)

College of Computer Science and Technology, Ocean University of China, Qingdao, China

fanhao@ouc.edu.cn

Abstract. Due to the unordered and irregular nature of point cloud data, it is challenging for neural networks to learn from it. Attention mechanisms have shown promising results in point cloud processing. It is also inherently permutation-invariant when dealing with a set of points, which makes it ideal for point cloud learning. In this paper, an attention-based encoder-decoder architecture called KFT-Net (Key Features Transformer) is designed for point cloud classification and segmentation tasks. KFT-Net has improved upon previous methods by capturing long-range contextual information, preserving important attention scores, and utilizing convolutional neural networks to extract local features. Additionally, to enhance the computational efficiency, we introduce the Top-k operation into the attention mechanism and utilize the average pooling operation to improve attention score calculation and feature extraction efficiency. Extensive experiments validate the effectiveness of the KFT-Net, demonstrating impressive performance in point cloud classification and segmentation tasks.

Keywords: Point cloud · Self-attention · Feature extraction · Classification · Segmentation

1 Introduction

Point cloud data is represented in coordinate form, where each point contains positional information and other attributes, such as normals and colors. Point cloud classification and segmentation tasks are crucial in various applications. However, point cloud data is often sparse, irregular, and noisy, which makes it challenging to extract meaningful information using traditional methods.

This paper proposes an encoder-decoder architecture based on self-attention [20] called KFT-Net, which aims to improve point cloud classification and segmentation performance. KFT-Net can enhance the ability of long-range dependency capturing, global feature modeling, and local feature extraction. To improve the vanilla attention mechanism, the Top-k operation is incorporated in the attention matrix, and the average pooling operation is applied on $K(key)$ and $V(values)$ to improve attention score calculation and feature extraction efficiency. These include:

Q. Liu et al. (Eds.): PRCV 2023, LNCS 14426, pp. 295–306, 2024.
https://doi.org/10.1007/978-981-99-8432-9_24

Top-k Operation. We use the Top-k operation, where at each position the top k elements with the highest attention weight are selected. By ignoring lower weights, the model can focus on the most important parts for processing. To normalize and preserve the effectiveness of the chosen attention weights, we normalize the highest weight values such that their sum is equal to or close to 1. This normalization ensures that the attention probabilities have the same properties. Using the filtered attention weights, we perform a weighted sum over the corresponding value vector for each location. This operation directs the model's attention to values related to key information. Thus, it helps to create a more accurate representation.

Average Pooling Operation. We also perform an average pooling operation before calculating the attention matrix, we perform the average pooling operation on K and V to alleviate computational overhead and focus on higher attention weights. This operation involves a weighted average between attention weights and value vectors, thereby allowing the model to emphasize key features and disregard irrelevant or noisy influences. Moreover, by summarizing multiple value vectors while preserving important information, the average pooling operation helps capture global information and generate more comprehensive and representative representations.

Experimental results show that KFT-Net achieves good performance in point cloud classification and segmentation tasks. The main contributions of this paper are summarized as follows:

1) We propose a novel transformer-based point cloud learning framework called KFT-Net, which is designed to better handle the irregularity and disorder of point cloud data.
2) We perform a Top-k operation on the attention matrix, which allows us to extract the most crucial point cloud features and improve learning efficiency compared to the original self-attention module.
3) We apply the average pooling operation on K and V, which can reduce the number of points considered during attention computation. By aggregating multiple points into a representative point, it helps to capture global information and summarize the overall context of the entire point cloud.

2 Related Work

2.1 Point-Based Deep Learning

Deep learning has shown remarkable performance in point cloud classification and segmentation. PointNet [13] is introduced as a novel framework for directly processing unordered point clouds without pre-processing or transformations. In the years following PointNet, many techniques such as Pointnet++ [14], PointCNN [10], and PointConv [23] are developed to improve the accuracy of point cloud classification and segmentation. PointWeb [28] addresses the challenges of arbitrary point order and complex feature extraction through innovative techniques. PointGrid [9] converts point cloud data into a structured grid representation to achieve efficient feature extraction for various tasks. PointNeXt [16] improves the PointNet++ framework by introducing Inverted Residual Bottleneck and separable MLPs, achieving superior results in point cloud classification and segmentation tasks.

2.2 Transformer for Point Cloud

Point cloud processing faces challenges in preserving spatial relationship and capturing long-range dependencies. To address these challenges, attention mechanisms are proposed. Self-attention [20] is designed to capture global and local features simultaneously. Transformers have shown impressive performance in both the field of NLP (Natural Language Processing) [11,20] and CV (Computer Vision) [2,7,21]. PointASNL [27] use an adaptive sampling strategy to exploit higher-order local structures in graphs. PCT [4] hierarchically extracts local features using point-to-cluster operations. PT [15] aggregates contextual information using prism-based pooling methods. Pointformer [29] encodes point relationships using multi-head attention mechanisms after translating point clouds to continuous tensors.

3 Point Cloud Processing with KFT-Net

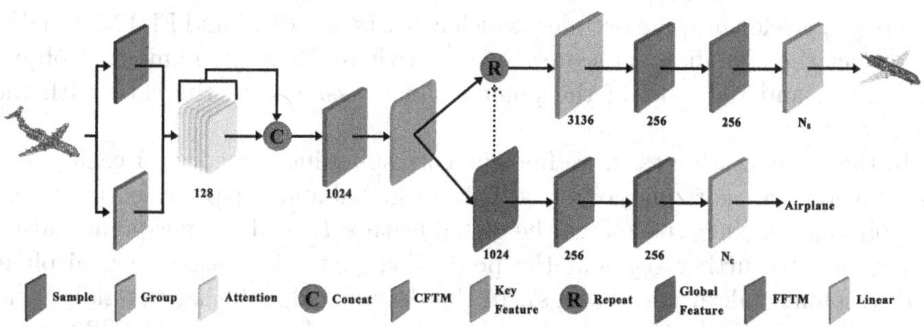

Fig. 1. KFT-Net architecture. The encoder consists of a Group module, a Sample module, and Contextual-Aware Attention Module (CAAM). The decoder is mainly composed of a Coarse Feature Transfer Module (CFTM) and Fine Feature Transfer Module (FFTM), which outputs segmentation and classification results. And the number of output channels is marked below each module. CFTM combines Linear, BatchNorm and ReLU layers. FFTM combines Linear, BatchNorm, ReLU and dropout layers.

3.1 Overall Architecture

The overall architecture of KFT-Net is shown in Fig. 1. KFT-Net is an encode-decoder architecture. The encoder consists of a Group module, a Sample module, and Contextual-Aware Attention Module (CAAM). The decoder is mainly composed of a Coarse Feature Transfer Module (CFTM) and Fine Feature Transfer Module (FFTM), which produces the final feature representation for segmentation and classification results.

Given the input point cloud $P \in \mathbb{R}^{N \times d}$, where N points have d-dimensional features, the encoder learns d_c-dimensional reduction embedding feature $F_c \in \mathbb{R}^{N \times d_c}$ through the Group and Sample modules [4]. The per-point feature representation $F_o \in \mathbb{R}^{N \times d_o}$ is obtained by concatenating the CAAM outputs.

$$\begin{aligned} F_1 &= CAAM_1(F_c), \\ F_i &= CAAM_i(F_{i-1}), i = 2, 3, 4, 5, 6, \\ F_o &= concat(F_1, F_2, F_3, F_4, F_5, F_6) \cdot W_o, \end{aligned} \tag{1}$$

where $CAAM_i$ represents the i-th CAAM layer, each layer has the same output dimension and W_o is the learnable parameters. A detailed explanation of the CAAM layer will be elaborated in Sect. 3.2. Various implementations of input embedding and attention mechanisms will be described in detail below. We concatenate CAAM outputs to learn per-point feature representation. Thus, global feature representation F_g is extracted based on the learned per-point features.

We classify the point cloud P into N_c specific categories (e.g. airplane, car, chair) using the extracted global features for classification tasks. The global features F_g are fed into the decoder, which consists of CFTM and FFTM. Finally, we predict the classification score $S \in \mathbb{R}^{N_c}$ (where N_c is the number of object categories), and the label of the point cloud is assigned to the class with the maximal score.

In the segmentation task, we find the corresponding category for each point. As it involves categorizing points, we extract global and per-point features from the point cloud. Then, we merge the global feature F_g and the per-point feature F_o together to further segment the point cloud into N_c classes (e.g. airplane head, airplane tail, airplane wings). In the decoder, after fusing the global features and per-point features obtained from the encoder, we use a CFTM and a FFTM. Finally, we apply a linear layer for score prediction. Specifically, after the decoding process, we can obtain the predicted scores $S \in \mathbb{R}^{N \times Nc}$ for each point, and we determine the label of the point based on the highest score.

3.2 Contextual-Aware Attention Module

When dealing with point clouds, the extracted point is regarded as a word, and the entire point cloud is treated as a sentence. In this case, the point cloud P is embedded into a d_c-dimensional space represented by $F_c \in \mathbb{R}^{N \times d_c}$, where N is the number of points and d_c is set to 128 for computational efficiency. In our KFT-Net, we propose Contextual-Aware Attention Module (CAAM) to

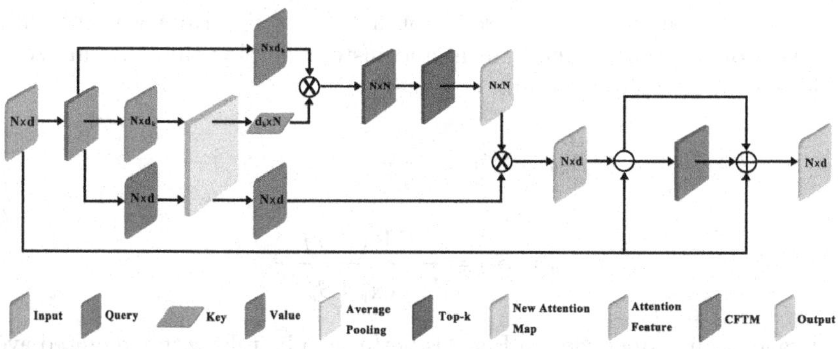

Fig. 2. Architecture of Contextual-Aware Attention Module.

enhance point cloud processing, where CAAM is decomposed of self-attention mechanism, Top-K operation and average pooling operation. The structure of CAAM is illustrated in Fig. 2.

Self-Attention Mechanism. The self-attention mechanism involves transforming each element in the input sequence into three vectors: *query*, *key*, and *value*. The similarity between the *query* vector and *key* vector is then calculated to obtain attention weights. Finally, the attention weights are multiplied with the *value* vector and summed up. Referring to literature [20], let Q, K, V be the query matrix, key matrix, and value matrix generated by linear transformation of the input features $F_{in} \in \mathbb{R}^{N \times d_c}$ as follows:

$$\begin{aligned} Attn(Q, K, V) = F_{in} \cdot (W_q, W_k, W_v), \\ Q, K \in \mathbb{R}^{N \times d_k}, V \in \mathbb{R}^{N \times d_c}, \\ W_q, W_k \in \mathbb{R}^{d_c \times d_k}, W_v \in \mathbb{R}^{d_c \times d_c}, \end{aligned} \tag{2}$$

where $Attn$ is the vanilla self-attention mechanism, W_q , W_k and W_v are shared learnable linear transformation operations, d_k is the dimension of the *query* vector and *key* vector, and d_c is the dimension of the *value* vector.

Average Pooling Operation. To obtain features F_{sa}, we perform an average pooling [7] operation on K' and V'. The specific operation is as follows:

$$F_{sa}(Q', K', V') = Attn\left(F_{in} \cdot W_q, P_t\left(F_{in} \cdot W_k\right), P_t\left(F_{in} \cdot W_v\right)\right), \tag{3}$$

where $Attn$ is the vanilla self-attention mechanism. P_t is an average pooling operation with stride t for transforming the spatial dimension to reduce computational cost. Then, we can use the query and key matrices to calculate the attention weights through matrix dot-product:

$$\bar{A}_{sa} = \bar{S}_{i,j} = Q' \cdot K'^T. \tag{4}$$

These weights are then normalized to obtain $A_{sa} = S_{i,j}$. Here, we depart from the conventional attention mechanism and instead employ the l_1-norm used in the [4] network for regularization:

$$\begin{aligned} \tilde{A}_{sa} &= \tilde{S}_{i,j} = \frac{\bar{S}_{i,j}}{\sqrt{d'_k}}, \\ A_{sa} &= S_{ij} = \frac{\exp\left(\tilde{S}_{i,j}\right)}{\sum_k \exp\left(\tilde{S}_{i,k}\right)}. \end{aligned} \tag{5}$$

Average pooling operation in Fig. 2 is performed by taking the weighted average of the value vectors using the corresponding attention weights to obtain the transformed output. After the operation of weighted average pooling operation, we obtain the pooled K' and V'. The K' summarizes the feature representations of keys with higher attention weights, while the V' summarizes the feature representations of values with higher attention weights. In this way, during the calculation process of the attention matrix A and V', each value vector V' is multiplied by its corresponding attention weight, and then all the weighted value vectors are summed up. The final result is a weighted average value vector that represents the aggregated information of the query vector Q'.

Top-K Operation. After implementing the average pooling operation on self-attention mechanism, an attention weight matrix A_{sa} is obtained. Different from the random dropout strategy that randomly drops scores, an adaptive selection of the Top-k operation contributing scores on matrix A_{sa} is implemented here, aiming to retain the most important components and discard irrelevant ones. Here, k is a tunable parameter that dynamically controls the sparsity level, which is formally obtained through the weighted average of appropriate scores, we set $k = N/2$, where N is the number of elements in the input sequences. Therefore, for softmax computation, only the top k values within the range $[\delta 1, \delta 2]$ are normalized for each row of matrix A_{sa}. For other elements below the top k scores, their probabilities are replaced with 0 using a scatter function at the given indices. This dynamic selection shifts attention from dense to sparse, which is written by the following equation:

$$A_{topk} = \left[T_k(S)\right]_{i,j} = \begin{cases} S_{i,j}\ ,\ S_{i,j} \in Top - k\,(row, j)\,, \\ 0\ ,\ otherwise. \end{cases} \tag{6}$$

where T_k represents the Top-k operation.

After adding the Top-k operation in Fig. 2, we can select specific elements with the highest attention weights. By dropping lower attention weights, noise or irrelevant information could be restrained, and it is of benefit to improve the model's robustness and accuracy. For large-scale point cloud data, we can reduce computational costs and storage requirements by retaining more important information. The output feature F_r is the weighted sum of the value vectors using the refined attention weights:

$$F_r = A_{topk} \cdot V'. \tag{7}$$

3.3 Output

In the final output, we refer to the methods of [24] and [4]. As shown in Fig. 2, during the decoding process, we perform element-wise subtraction between encoded features F_r and the input features F_{in}. The subtracted result is then passed into the CFTM and FFTM.

We shift offsets and add them to query features, which boosts sensitivity to other points and helps in calculating similarities. This operation is performed on all query features and yields attention distributions for each one. Offset vectors enhance positional information and make the attention mechanism more flexible and accurate, ultimately improving performance in point cloud classification and segmentation tasks.

4 Experiments

We conduct extensive experiments on three public benchmarks to evaluate the effectiveness of the KFT-Net. We use the same softmax cross-entropy loss function as [4]. KFT-Net is trained with the Adam (Adaptive Moment Estimation) optimizer using a learning rate of 10^{-4} and weight decay of 10^{-3}.

Table 1. ModelNet40 shape classification. We compare our model with other networks. 'P' denotes the number of input points, and 'N' denotes the corresponding normals.

Method	Input	Accuracy
PointNet [13]	P, N	89.4
PointMLP [12]	P	91.1
Kd-Net [5]	P	91.5
PointNet++ [14]	P, N	91.9
PointGrid [9]	P	92.0
PointWeb [28]	P	92.2
PointCNN [10]	P	92.4
PointConv [23]	P, N	92.5
KPConv [18]	P	92.7
DGCNN [24]	P	92.9
PointASNL [27]	P	92.9
PCT [4]	P	92.9
Ours	**P**	**93.4**

4.1 Shape Classification on ModelNet40

We use ModelNet40 [25] to evaluate the performance on point cloud classification tasks. The dataset is divided into a training set and a test set, following a standard evaluation protocol for a fair comparison. Each object is uniformly sampled to 1,024 points using the same sampling strategy as [13]. Our model is trained for 250 epochs with an initial learning rate of 10^{-4}, employing cosine decay per epoch and a batch size of 16. The experimental results are shown in Table 1. Compared to [13] and [4], KFT-Net makes a 4.0% and 0.5% improvement respectively. KFT-Net achieves the best result of 93.4% overall accuracy.

Table 2. Comparison on the ShaperNet part segmentation dataset. 'pIoU' means part-average Intersection-over-Union.

Method	pIoU	airplane	bag	cap	car	chair	earphone	guitar	knife	lamp	laptop	motorbike	mug	pistol	rocket	skateboard	table
Kd-Net [5]	82.3	80.1	74.6	74.3	70.3	88.6	73.5	90.2	87.2	81.0	94.9	57.4	86.7	78.1	51.8	69.9	80.3
PointNet [13]	83.7	83.4	78.7	82.5	74.9	89.6	73.0	91.5	85.9	80.8	95.3	65.2	93.0	81.2	57.9	72.8	80.6
PointNet++ [14]	85.1	82.4	79.0	87.7	77.3	90.8	71.8	91.0	85.9	83.7	95.3	71.6	94.1	81.3	58.7	76.4	82.6
DGCNN [24]	85.2	84.0	83.4	86.7	77.8	90.6	74.7	91.2	87.5	82.8	95.7	66.3	94.9	81.1	63.5	74.5	82.6
SFCNN [17]	85.4	83.0	83.4	87.0	80.2	90.1	75.9	91.1	86.2	84.2	**96.7**	69.5	94.8	82.5	59.9	75.1	82.9
PointConv [23]	85.5	–	–	–	–	–	–	–	–	–	–	–	–	–	–	–	–
SGPN [22]	85.5	80.4	78.6	78.8	71.5	88.6	78.0	90.9	83.0	78.8	95.8	**77.8**	93.8	87.4	60.1	**92.3**	**89.4**
PT [15]	85.7	–	–	–	–	–	–	–	–	–	–	–	–	–	–	–	–
PCT [4]	85.8	84.8	81.3	88.5	**81.2**	90.9	71.5	91.2	87.9	**86.3**	95.6	63.3	95.8	83.2	61.3	77.6	83.7
PointCNN [10]	85.8	84.1	**86.5**	86.0	80.8	90.6	**79.7**	**92.3**	88.4	85.3	96.1	77.2	95.2	**84.2**	64.2	80.0	83.0
PointASNL [27]	86.0	84.1	84.7	87.9	79.7	92.2	73.7	91.0	87.2	84.2	95.8	74.4	95.2	81.0	63.0	76.3	83.2
Ours	**86.2**	**85.0**	82.4	**88.4**	77.8	**92.5**	72.0	91.4	**88.6**	84.3	95.6	71.9	**95.9**	82.0	**64.8**	80.7	82.9

4.2 Object Part Segmentation on ShapeNet

We conduct point cloud segmentation experiments using the ShapeNet [26] dataset. According to [13], all models are downsampled to 2048 points, preserving point-to-part annotations. Table 2 shows the class-wise segmentation results. The evaluation metric used is partial mean pIoU, which provides overall and per-object category performance metrics. The results indicate that our KFT-Net outperforms [13] and [4] by 2.5% and 0.4%, respectively. KFT-Net achieves the best performance with pIoU of 86.2%. Figure 3 illustrates segmentation examples using KFT-Net.

Table 3. Comparison on the S3DIS semantic segmentation dataset tested on Area 5. 'OA' means overall point accuracy. 'mIoU' means Intersection-over-Union.

Method	OA	mIoU	ceiling	floor	wall	beam	column	window	door	chair	table	bookcase	sofa	board	clutter
PointNet [13]	78.5	49.6	88.8	97.3	69.8	**0.1**	3.9	46.3	10.8	59.0	52.6	5.9	40.3	26.4	33.2
PointNet++ [14]	–	53.2	90.8	96.5	74.1	0.0	5.8	43.6	25.4	69.2	76.9	21.5	55.6	49.3	41.9
DGCNN [24]	84.1	56.1	–	–	–	–	–	–	–	–	–	–	–	–	–
SPG [8]	85.5	62.1	89.4	96.9	78.1	0.0	**42.8**	48.9	61.6	84.7	75.4	**69.8**	52.6	2.1	52.2
PointCNN [10]	88.1	65.4	92.3	98.2	79.4	0.0	17.6	22.8	**62.1**	74.4	80.6	31.7	66.7	62.1	**56.7**
PointWeb [28]	87.3	66.7	92.0	98.5	79.4	0.0	21.1	59.7	34.8	76.3	**88.3**	46.9	**69.3**	64.9	52.5
PCT [4]	87.6	67.1	92.5	98.4	**80.6**	0.0	19.4	61.6	48.0	**76.6**	85.2	46.2	67.7	**67.9**	52.3
Ours	**88.3**	**67.9**	**92.8**	**98.5**	80.0	0.0	20.6	**63.5**	51.2	75.8	84.1	47.1	68.3	66.4	51.6

4.3 Semantic Segmentation on S3DIS

The S3DIS [1] dataset is a dataset used for 3D segmentation and understanding of indoor scenes. We used the S3DIS dataset for the task of semantic segmentation of indoor scenes. We use the same method as [13] for semantic segmentation of the S3DIS dataset, but we made improvements in the network model. Our KFT-Net outperforms [13] in terms of mIoU, and OA by 18.3% and 9.8% respectively. At the same time, it is faster in terms of speed. The speed improvement is due to the use of Top-k operation in the SA block of KFT-Net, which reduces computation. As the model expands, the performance of KFT-Net can gradually improve. As shown in Table 3, our KFT-Net achieves the best results with mIoU and OA reaching 67.9% and 88.3%.

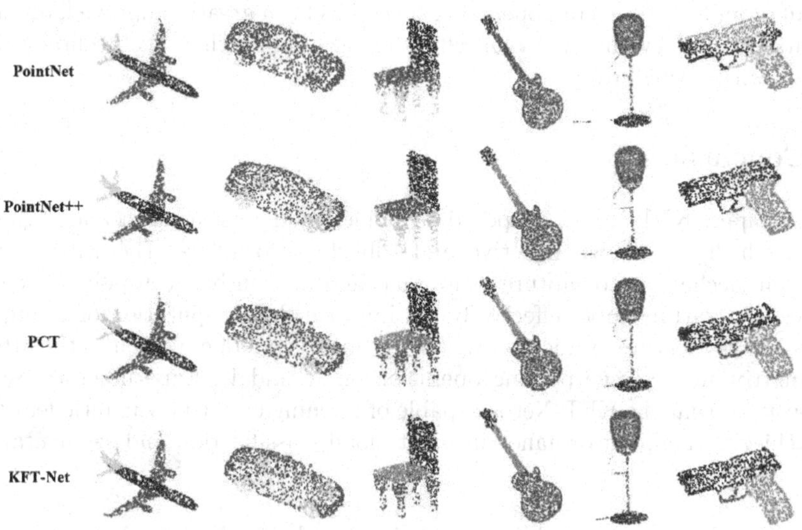

Fig. 3. Qualitative results of PointNet, PointNet++, PCT, KFT-Net.

4.4 Ablation Studies

Effectiveness of Top-k Operation. With the Top-k operation, we focus the attention mechanism on the most relevant elements in the input sequence while discarding irrelevant elements. This reduces complexity and improves accuracy. As shown in Table 4, the Top-k operation achieves accuracy improvements on all three tasks, which verifies the advantages of the proposed Top-k operation. For example, 0.7% Acc has been improved on ModelNet40, and 0.6% pIoU has been increased on ShapeNet.

Table 4. Ablation study on ModelNet40, ShapeNet and S3DIS benchmarks with different modules.

Datasets	Baseline	+Top-k	+Average pooling	KFT-Net
ModelNet40 Acc(%)	92.9	93.6	92.2	93.4
ShapeNet pIoU(%)	85.8	86.4	85.4	86.2
S3DIS mIoU(%)	67.1	68.2	66.5	67.9

Effectiveness of Average Pooling Operation. We leverage the average pooling operation to narrow the focus to only the most relevant key-value pairs for the computation. This increases speed and maintains high-quality attention. As shown in Table 4, after adding average pooling operation, the accuracy rate has a small drop, but the overall operation speed has been greatly improved, and the benefits brought by the operation efficiency are higher than the accuracy rate benefits with a small drop.

5 Conclusion

In this paper, KFT-Net is proposed to point cloud classification and segmentation, which has shown effective and robust performance. By utilizing the attention mechanism to capture contextual information between points, we can extract key features more effectively. To improve the computation of attention scores and key feature extraction, we incorporate Top-k operation into the attention matrix and average pooling operation on K and V. Experimental results demonstrate that the KFT-Net is capable of learning excellent semantic features and achieving high performance in point cloud classification and segmentation tasks.

Acknowledgment. This work is supported in part by the National Natural Science Foundation of China (Grant No. 42106193, 41927805).

References

1. Armeni, I., et al.: 3D semantic parsing of large-scale indoor spaces. In: Proceedings of the IEEE Conference on Computer Vision and Pattern Recognition (2016)
2. Chen, X., et al.: Learning a sparse transformer network for effective image deraining. In: Proceedings of the IEEE/CVF Conference on Computer Vision and Pattern Recognition (2023)
3. Dai, Z., et al.: Transformer-xl: attentive language models beyond a fixed-length context. arXiv preprint arXiv:1901.02860 (2019)
4. Guo, M.-H., et al.: Pct: point cloud transformer. Comput. Visual Media **7**, 187–199 (2021)
5. Klokov, R., Lempitsky, V.: Escape from cells: deep kd-networks for the recognition of 3d point cloud models. In: Proceedings of the IEEE International Conference on Computer Vision (2017)

6. Lee, J., et al.: BioBERT: a pre-trained biomedical language representation model for biomedical text mining. Bioinformatics **36**(4), 1234–1240 (2020)
7. Li, J., et al.: Next-vit: next generation vision transformer for efficient deployment in realistic industrial scenarios. arXiv preprint arXiv:2207.05501 (2022)
8. Landrieu, L., Simonovsky, M.: Large-scale point cloud semantic segmentation with superpoint graphs. In: Proceedings of the IEEE Conference on Computer Vision and Pattern Recognition (2018)
9. Le, T., Duan, Y.: Pointgrid: a deep network for 3D shape understanding. In: Proceedings of the IEEE Conference on Computer Vision and Pattern Recognition (2018)
10. Li, Y., et al.: Pointcnn: convolution on x-transformed points. Adv. Neural Inf. Process. Syst. **31** (2018)
11. Lin, Z., et al.: A structured self-attentive sentence embedding. arXiv preprint arXiv:1703.03130 (2017)
12. Ma, X, et al.: Rethinking network design and local geometry in point cloud: a simple residual MLP framework. arXiv preprint arXiv:2202.07123 (2022)
13. Qi, C.R., et al.: Pointnet: deep learning on point sets for 3D classification and segmentation. In: Proceedings of the IEEE Conference on Computer Vision and Pattern Recognition (2017)
14. Qi, C.R., et al.: Pointnet++: deep hierarchical feature learning on point sets in a metric space. Adv. Neural Inf. Process. Syst. **30** (2017)
15. Pan, X., et al.: 3D object detection with pointformer. In: Proceedings of the IEEE/CVF Conference on Computer Vision and Pattern Recognition (2021)
16. Qian, G., et al.: Pointnext: revisiting pointnet++ with improved training and scaling strategies. Adv. Neural Inf. Process. Syst. **35**, 23192–23204 (2022)
17. Rao, Y., Lu, J., Zhou, J.: Spherical fractal convolutional neural networks for point cloud recognition. In: Proceedings of the IEEE/CVF Conference on Computer Vision and Pattern Recognition (2019)
18. Thomas, H., et al.: Kpconv: flexible and deformable convolution for point clouds. In: Proceedings of the IEEE/CVF International Conference on Computer Vision (2019)
19. Tchapmi, L., et al.: Segcloud: semantic segmentation of 3D point clouds. In: 2017 International Conference on 3D vision (3DV). IEEE (2017)
20. Vaswani, A., et al.: Attention is all you need. Adv. Neural Inf. Process. Syst. **30** (2017)
21. Wu, B., et al.: Visual transformers: token-based image representation and processing for computer vision. arXiv preprint arXiv:2006.03677 (2020)
22. Wang, W,, et al.: SGPN: similarity group proposal network for 3d point cloud instance segmentation. In: Proceedings of the IEEE Conference on Computer Vision and Pattern Recognition (2018)
23. Wu, W., Qi, Z., Li, F.: Pointconv: deep convolutional networks on 3d point clouds. In: Proceedings of the IEEE/CVF Conference on Computer Vision and Pattern Recognition (2019)
24. Wang, Y., et al.: Dynamic graph cnn for learning on point clouds. ACM Trans. Graph. (tog) **38**(5), 1–12 (2019)
25. Wu, Z., et al.: 3D shapenets: a deep representation for volumetric shapes. In: Proceedings of the IEEE Conference on Computer Vision and Pattern Recognition (2015)
26. Yi, L., et al.: A scalable active framework for region annotation in 3d shape collections. ACM Trans. Graph. (ToG) **35**(6), 1–12 (2016)

27. Yan, X., et al.: Pointasnl: robust point clouds processing using nonlocal neural networks with adaptive sampling. In: Proceedings of the IEEE/CVF Conference on Computer Vision and Pattern Recognition (2020)
28. Zhao, H., et al.: Pointweb: enhancing local neighborhood features for point cloud processing. In: Proceedings of the IEEE/CVF Conference on Computer Vision and Pattern Recognition (2019)
29. Zhao, H., et al.: Point transformer. In: Proceedings of the IEEE/CVF International Conference on Computer Vision (2021)

Unsupervised Domain Adaptation for 3D Object Detection via Self-Training

Di Luo(✉)

College of Computer Science, Tianjin Key Laboratory of Network and Data Security Technology, NanKai University, Tianjin, China
Rody1911641@gmail.com

Abstract. 3D object detection based on point clouds plays a crucial role in autonomous driving. High quality detection results can provide reliable basis for subsequent stages such as trajectory prediction and path planning. Although many advanced 3D object detectors currently exist, when employing them to another domain, there is often a huge performance drop. In addition, existing domain adaptation methods for 3D object detection only focus on one or two observable factors such as scale mismatch and density variation that lead to domain shift and they do not take some invisible factors (weather, road condition and sensor type, etc.) into account. Therefore, we attempt to propose a self-training pipeline for unsupervised domain adaptation on 3D object detection. Firstly, we pretrain the detectors with a specific data processing paradigm which includes object random scaling, random beam re-sampling and etc. Then, we employ mean-teacher framework which includes cross-domain student model and target-only teacher model. We employ adversarial learning in student model, enforcing the student model to learn domain-invariant features. This process could further eliminate the invisible factors that lead to domain shift. Furthermore, in order to further obtain high-quality pseudo labels, we apply different data augmentation strategy and mutual learning between student model and teacher model. In addition, we adopt domain statistics normalization to ensure a stable training behavior. Extensive experiments under three different adaptation tasks demonstrate the effectiveness of our method.

Keywords: Autonomous driving · Point cloud · 3D object detection · Unsupervised domain adaptation

1 Introduction

With the rapid development of LiDARs for autonomous vehicles, a great deal of datasets have been collected [1,4,15,23]. Based on these large-scale datasets and advances in deep learning, 3D object detection [5,7,19–21,28,30–32] on point clouds has achieved remarkable success. However, many open source datasets were captured with fixed sensors in a particular city on sunny day (*e.g.*KITTI, Germany) and these detectors are often trained on one dataset. When test

Q. Liu et al. (Eds.): PRCV 2023, LNCS 14426, pp. 307–318, 2024.
https://doi.org/10.1007/978-981-99-8432-9_25

in other scenarios, it will suffer a large performance drop. In order to solve this problem, for one thing, people devote to collect large real-world human-annotated datasets, but it is expensive and time-consuming. For another, developing algorithms that can transfer the knowledge learned from one labeled dataset (source domain) to another unlabeled dataset (target domain) is wildly accepted recently.

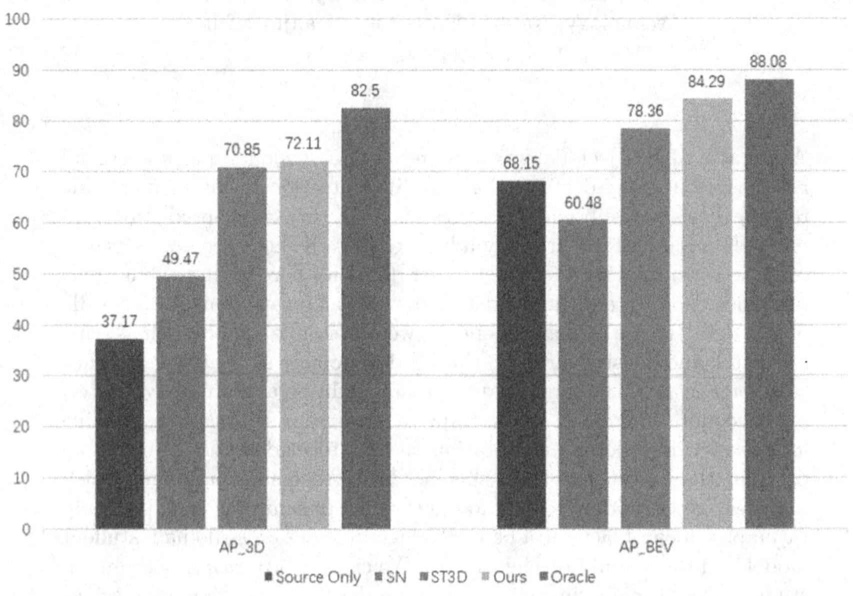

Fig. 1. Performance of our method on nuScenes → KITTI task using PV-RCNN [19], compared to other UDA (Source Only, ST3D [29]) and weakly-supervised (SN [25]) approachs and Oracle.

In contrast to intensive studies of 2D scenarios [2,8,9,18,33,34], few works have been introduced to unsupervised domain adaptation (UDA) for 3D detection. In 2D domain adaptation, images are regular arrangement of pixels. Domain gap originates from the lighting, color and texture variations. On the contrary, point clouds are dispersed and discrete. Points contain geometric information, domain shift in 3D comes from the bounding box scale mismatch [25], density variation [26], missing point [27] and some hidden factors. Therefore, popular 2D unsupervised domain adaptation methods are not readily transferable to 3D detection. But some ideas are universal. Pseudo label based self-training is the most straightforward way to close the domain gap in domain adaptation. To explore the potential of self-training on the target domain for improved detection performance in 3D UDA, ST3D [29] and Hu [6] all use pseudo label based self-training. Despite the accuracy gains, the pseudo label generated from pre-trained model in source domain is usually full of false positives because of the large domain shift, beacuse they only takes scale mismatch or density variation into account. As a result, it still has many false positives and leads to sub-optimal adaptation performance.

There are many reasons for the different distribution between source domain and target domain. SN [25] first proposed that car size is an important cause of inter-domain differences. Subsequently, Wei [26] proposed that point cloud density is also a key factor. They adopt various methods to deal with these directly. However, it is undeniable that there are many other factors for differences between domains, such as weather, road conditions, sensor types and other hidden factors. We can hardly design a solution for every cause. Therefore, the importance of adversarial feature learning to learn domain-invariant features is obvious.

Specifically, inspired by [9], we propose a self-training framework for 3D unsupervised domain adaptation. Our method leverages mean-teacher framework and adversarial learning to mitigate the domain shift and gain high-quality pseudo labels. First, we pretrain detector with ROS [29] and RSBS [6] to overcome the scale mismatch and density variation between source and target domain. Then, we employ mean-teacher framework. Our model consists of target-only teacher model and cross-domain student model. For student model, we apply adversarial learning introducing a discriminator with gradient reverse layer [3] to learn domain-invariant representation. In this way, we further eliminate the invisible factors that lead to domain shift. The student model is supervised by ground truth bounding boxes (source domain) and pseudo labels generated from the teacher model (target domain). In addition, a threshold adaptive pseudo label filtering method is proposed, gradually increasing the positive and negative threshold during the training process to further purify pseudo labels. We also apply neural statistics normalization to achieve a more stable training behavior, following mlcnet [11]. With all the techniques, we outperform all the existing works. We summarize the contributions of this paper as follows:

1) We design a data processing paradigm specific to cross domain object detection tasks, which includes object random scaling, object random rotation, weak-strong augmentation, etc. It can greatly solve the problem of object scale mismatch and density variation between domains, help reduce inter-domain differences, and enhance the generalization of the model.
2) We combine mean-teacher and adversarial learning in self-training pipeline. By using this strategy, the student model learns domain-invariant features between the source and target domains, reducing the inter-domain differences caused by hidden factors, and thus obtaining high-quality pseudo labels.
3) We validate our method on several datasets. Our method is capable of dealing with the domain shift and outperform the existing SOTA by a large margin as shown in Fig. 1.

2 Related Work

Lidar-based 3D detection methods mainly come from three categories: point-based methods, grid-based methods and point-voxel based methods. Thanks to PointNet [16] and PointNet++ [17], many previous works [13,14,21] directly extracting features from point cloud. Grid-based 3D object detectors first transform the irregular point cloud into a regular cylinder [7] or bird's eye view (BEV) [28] consisting of voxels. And then, traditional 2D convolution can be directly used. Methods based on point-voxel represent a new trend in 3D object detection. The method that fused points and voxels has great advantages in both accuracy and speed. Currently, point-voxel based methods have also been developed for one-stage [5,10] and two-stage object detection [12,19].

Unsupervised domain adaptation for 3D object detection aims at transfer the detector trained on source domain to unlabeled target domain. Wang *et al* [25] proposed that the challenge of 3D UDA lies in the differnet size of objects in different scenarios. They figure that there is large gap in the distribution of length, width and height of objects between Waymo, nuScenes and KITTI. Therefore, they normalize the object sizes of the source and target domain by statistical normalization (SN). Although experiments proved that SN is simple and effective, it will use the statistical information of the target domain, which is not suitable for UDA. To solve this issue, ST3D [29] proposed a pseudo-label based self-training method. They employ quality-aware triplet memory bank to improve the quality of pseudo-labels for 3D object detection. In addition, due to the difference of sensors, the density of points is different. It is specific and import among many domain-variant factors. Wei *et al* [26] proposed a density downsampling method based knowledge distillation, which gradually downsampled the high-density data to align the density of source and target domains. And DTS [6] introduce random beam re-sampling, which can not only downsample but also upsample point cloud.

3 Method

3.1 Problem Definition

In this work, we consider the 3D object detection under the unsupervised domain adaptation. We have the access to one labeled data (source domain) $\mathcal{D}_s = \{x_s^i, y_s^i\}_{i=1}^{N_i}$ and one unlabeled data (target domain) $\mathcal{D}_t = \{x_t^i\}_{i=1}^{N_t}$, where N_s and N_t are the number of samples of source and target domains. x_i^s and y_i^s denote the i-th point cloud scene and the corresponding label in the source domain. y_i^s contains the bounding box locations B_i^s and categories C_i^s of all objects in the i-th point cloud. The bounding box B_i^s is uniquely determined by its size (l, w, h), center coordinates (x, y, z) and heading angle θ. Similarly, x_i^s is the i-th unlabeled sample of the target domain. Our goal is to train a model based on $\mathcal{D}_s$ and $\mathcal{D}_t$ and maximize the performance on $\mathcal{D}_t$. The main architecture of our method is illustrated in Fig. 2.

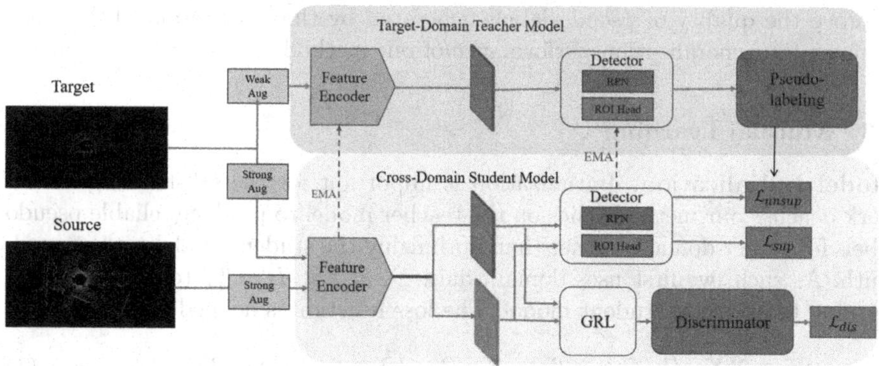

Fig. 2. Overview of our proposed self-training pipeline. We first pretrain 3D object detector with ROS and RSBS to overcome the scale mismatch and density variation. And then, we use the pretrained model to initialize the teacher model and student model. Our mean-teacher framework is composed of target-only teacher model and cross-domain student model. The input to the student model is point cloud that has been processed with strong augmentation while point cloud with weak augmentation as inputs of the teacher model. The teacher model is learned by the exponential moving average (EMA) of weights from student model. The parameters of student model are updated by the pseudo labels generated by the teacher model (target domain) and the groundtruth (source domain). We employ adversarial learning in student model, enforcing the student model to learn domain-invariant features. In addition, we employ neural statistics normalization to achieve a more stable training behavior.

3.2 Data Augmentation

Following ST3D [29] and DTS [6], we pretrained the model in source domain with random object scaling and random beam re-sampling. And in mutual learning, we employ weak-strong augmentation strategy.

Our method is composed of two modules: target-only teacher model and cross-domain student model. The teacher model is learned by the exponential moving average (EMA) of the weights from the student model. The parameters of the student model are updated by the pseudo labels generated by the teacher model (target domain) and the groundtruth (source domain). In order to enhance the effectiveness of the mean-teacher framework and obtain accurate pseudo labels for target domain, we employ weak-strong data augmentation. The input to the student model is point cloud that has been processed with strong augmentation while point cloud with weak augmentation as inputs of the teacher model. Point cloud data is composed of a seriers of point coordinates, which contains plenty of geometric information. As such, strong augmentation includes random object scaling and random object rotation, which will significantly change the coordinates of objetcs. Weak augmentation includes random beam re-sampling and random world scaling. These data process methods barely change the coordinates of the object. Weak-strong augmentation can not only improve the quality of pseudo-labels generated by the teacher model, but also enhance the generalization performance of our method on the target domain.

3.3 Multual Learning

Model Initialization. Initialization is important for the self-training framework beacuse our method relies on the teacher model to produce reliable pseudo labels for target domain, rather than optimizing the student model with groundtruth. As such, we first uses domain data $D_s = \{x_s^i, B_s^i, C_s^i\}$ to do supervised learning to optimize student model. The loss function is defined as follow:

$$\mathcal{L}(X_s, B_s, C_s) = \mathcal{L}_{rpn}(X_s, B_s, C_s) + \mathcal{L}_{rcnn}(X_s, B_s, C_s), \quad (1)$$

where $\mathcal{L}_{rpn}$ is the loss to generate candidate boxes in the first stage. $\mathcal{L}_{rcnn}$ is the loss of the second stage fine-tuning in PV-RCNN. Detail refers to [19].

Optimize Student Model with Target Pseudo-labels. Since the target domain groundtruth is not available, our mean-teacher framework employs the pseudo-labels egnerated by the teacher model on target domain to optimize the network. The loss as follow:

$$\mathcal{L}(X_t, \tilde{B}_t, \tilde{C}_t) = \mathcal{L}_{rpn}(X_t, \tilde{B}_t, \tilde{C}_t) + \mathcal{L}_{rcnn}(X_t, \tilde{B}_t, \tilde{C}_t), \quad (2)$$

where $\tilde{B}_t$ and $\tilde{C}_t$ are box labels and class labels poduced by the teacher model.

Periodically Update Teacher from Student. In order to obtain high-quality pseudo-labels, in addition to threshold adptation, Exponential Moving Average (EMA) is also applied to update the teacher model by copying the parameters of the student model. The update formula is as follow:

$$\theta_t \leftarrow \alpha\theta_t + (1 - \alpha)\theta_s, \quad (3)$$

where θ_t and θ_s denote the network parameters of the teacher and student respectively.

3.4 Threshold Adaptation Based Predicted IoU

To evaluate the localization quality of pseudo labels, we employ 3D object detector with a lightweight IoU regression prediction head. The features after the RoI Pooling pass through two fully connected layers to directly predict the intersection over union (IoU) between the bounding box and groundtruth. A sigmoid function is added after the fully connected layer to normalize the output to $[0, 1]$. During model training, the IoU head is optimized with the following cross-entropy loss:

$$\mathcal{L}_{IoU} = -\hat{u}logu - (1 - \hat{u})log(1 - u). \quad (4)$$

Table 1. Dataset overview.

Dataset	Size	LiDAR Type	Beam Angles	Object Type	Rain/Night	Points Per Scene
KITTI	14999	64-beam	[−24°, 4°]	8	No/No	118624
nuScenes	34149	32-beam	[−16°, 11°]	23	Yes/Yes	24966
Waymo	192484	64-beam	[−18°, 2°]	4	Yes/Yes	160139

As describe in [29], IoU regression can supervised the performance of detection, and the IoU regression has better localization quality than the classification regression.

In self-supervised training, the score of IoU prediction is used as an evaluation metric for the quality of pseudo labels. Those with IoU scores higher than threshold δ are chosen as pseudo labels. In the experiment, it is observed that more and more pseudo labels are generated as the training proceeds. Therefore, the idea of progressive training is adopted: after a certain number of epochs, the threshold δ is increased, which can be multiplied by α or added to a fixed value β, etc.

In addition, threshold based filtering of pseudo labels leads to some extreme cases there is no label in some scenes. To address this issue, we follow the idea of posotive and negative labels in ST3D [29]. But different from it, we do not adopt the tedious pseudo-label update method. Specifically, if the IoU prediction score T_i of the bounding box B_i is bigger than the positive threshold T_{pos}, it is used as the pseudo label to optimize the model. If the IoU predict score T_i is less than T_{pos} but bigger than T_{neg}, it is used as the negative label to supervise the model trainging, but the gradient is not backpropagated.

4 Experiments

4.1 Experimental Setup

Datasets. We conduct experiments on three widely used autonmous driving datasets: KITTI [4], nuScenes [1] and Waymo [23]. Our experiments are based two aspects. 1) Large datasets migrate to small datasets (Waymo → KITTI). 2) Adapting cross domains with different LiDAR-beam densities (Waymo → nuScenes, nuscenes → KITTI). The details of datasets are given in Table 1.

Comparison Methods. We compare our method with four methods: (1) **Source only**: the detector is only pretrained on the source domain and then transferred to the target domain. (2) **SN** [25]. (3) **ST3D** [29] is the SOTA domain adaptation method on 3D object detection. (4) **Oracle** indicates the fully supervised model trained on the target domain.

Experimental Seetings. Followed ST3D, we adopt KITTI evaluation metric for evaluating our methods on the commonly used car category (vehicle in

Waymo Open Dataset). We report the average precision (AP) over 40 recall positions, and the IoU thresholds are 0.7 for both the bird's eye view (BEV) IoUs and 3D IoUs. We also adopt Closed Gap [29] to demonstrate the effectiveness on domain adaptation, which is defined as $ClosedGap = \frac{AP_{model} - AP_{sourceOnly}}{AP_{oracle} - AP_{SourceOnly}} \times 100\%$.

Implementation Details. We validate our method on two detection backbones SECOND-IoU [28] and PV-RCNN [19]. SECOND-IoU indicates the SECOND detector with an extra IoU head to estimate the IoU between the bounding boxes and their GTs. We adopt the training settings of the popular point cloud detection codebase OpenPcdet [24] to pretrain our detectors on the source domain with ROS and RSBS. For fair comparison, we adopt the same parameters as ST3D [29] and DTS [6]. The EMA smoothing coefficient hyperparameter α is set to 0.99. T_{pos} is set to 0.65 in PV-RCNN and 0.70 in SECOND-IoU. T_{neg} is set to 0.35 in two detectors.

Table 2. Performance comparison of different methods on different domain adaptation tasks. We report AP_{BEV} and AP_{3D} over 40 recall positions of the car category at IoU=0.7. We indicate the best adaptation results by **bold**. Closed Gap [29]= $\frac{AP_{model} - AP_{sourceOnly}}{AP_{oracle} - AP_{SourceOnly}} \times 100\%$.

Task	Method	SECOND-IOU		PV-RCNN	
		AP_{BEV}/AP_{3D}	Closed GAP	AP_{BEV}/AP_{3D}	Closed GAP
Waymo → KITTI	Source Only	67.64/27.48	–	61.18/22.01	–
	SN	78.96/59.20	+72.33%/+69.00%	79.78/63.60	+66.91%/+68.76%
	ST3D	82.19/61.83	+92.97%/+74.72%	**84.10**/64.78	+82.45%/+70.71%
	Ours	**82.79/62.70**	+96.81%/+76.62%	82.01/**65.41**	+74.93%/+71.75%
	Oracle	83.29/73.45	–	88.98/82.50	–
Waymo → nuScenes	Source Only	32.91/17.24	–	34.50/21.47	–
	SN	33.23/18.57	+1.69%/+7.54%	34.22/22.29	–1.50%/+4.80%
	ST3D	35.92/20.19	+15.87%/+16.73%	36.42/22.99	+10.32%/+8.89%
	Ours	**36.85/21.05**	+20.77% /+21.61%	**36.84/23.36**	+12.57%/+11.06%
	Oracle	51.88/34.87	–	53.11/38.56	–
nuScenes → KITTI	Source Only	51.84/17.92	–	68.15/37.17	–
	SN	40.03/21.23	-37.55%/+5.96%	60.48/49.47	–36.82%/+27.13%
	ST3D	75.94/54.13	+76.63%/+59.50%	78.36/70.85	+49.02%/+74.30%
	Ours	**79.52/58.53**	+88.01%/+73.13%	**84.29/72.11**	+77.72%/+77.08%
	Oracle	83.29/73.45	–	88.98/82.50	–

4.2 Main Results and Comparisons

We compare our method with Source Only, SN [25] and ST3D [29]. We compare with other approaches in terms of three settings, the Weakly-supervised DA, the Unsupervised DA (UDA) and the Oracle. We indicate the best adaptation results by **bold**.

As shown in Table 2, our method outperforms all compared methods by large margins on all settings in AP_{3D}. Especially when the target domain is KITTI, our method makes great progress. Compared with the current SOTA ST3D [29], we improve on Waymo → KITTI and nuScenes → KITTI tasks by a margin of around 3% ∼5% in AP_{3D} and 5% ∼ 7% in AP_{BEV}. In addition, we also achieves about 5% performance improvement compared with ST3D in Waymo → nuScenes, which is closer to fully supervised learning. The overall experimental results verify the effectiveness of our method in different domain adaptation tasks and its strong generalization across different 3D detectors.

Further, although we make some progress in Waymo → nuScenes, there is a significant performance gap compared with oracle. We figure that nuScenes is a large and complex dataset. It collects many scenarios of severe weather conditions and road conditions. Besides, the performance of backbone detector on large datasets is also one of the limiting factors.

4.3 Ablation Studies

In this section, we conduct extensive ablation experiments to investigate the componements of our method. And the influence of threshold selection is further analyzed. All experiments are conducted with the 3D detector PV-RCNN on the task of nuScenes → KITTI.

Component Ablation Studies. In Table 3, ST represents naive self-training, DA represents data augmentation paradigm, including ROS and RBRS as mentioned in Sect. 3.2. M-T means the mean-teacher framework and Ad means the adversarial learning. TA represents threshold adaptation as mentioned in Sect. 3.4.

Table 3. Component ablation studies.

Method	AP_{BEV}/AP_{3D}
Source Only (baseline)	68.15/37.17
ST	66.24/54.93
ST+DA	70.29/60.36
ST+DA+M-T	79.64/67.75
ST+DA+M-T+Ad	80.01/70.29
ST+DA+M-T+Ad+TA	**84.29/72.11**
Oracle	88.98/82.50

The results reported in Table 3 validate the effectiveness of our individual components. The AP_{3D} outperforms Source Only by 34.94% and naive self-training by 17.18%. The data augmentation paradigm and mean-teacher framework provide 5.43% and 7.39% performance gains, respectively. Adversarial feature learning and threshold adaptation separately further yield around 2.5% and 1.8% improvements, respectively.

Ablation Studies on Threshold. The selection of threshold in self-supervised learning often directly determines the quality of detection results. Therefore, we further carry out ablation studies on positive and negative threshold to explore the influence of threshold on detection results. The experimental results are shown in Table 4.

The choice of the threshold has little effect on detection results, around 1.0%. In other words, our method is robust to the threshold. Thanks to the strategy of adaptive adjustment of the threshold in Sect. 3.4: as the number of iterations increases, the teacher model generates more and more bounding boxes, and we automatically adjust the threshold to filter out the true positive for student model. This leads to better detection results with different positive and negative thresholds.

Table 4. Ablation studies on threshold.

T_{neg}	T_{pos}	AP_{BEV}/AP_{3D}
0.3	0.7	83.45/71.58
0.3	0.6	82.49/70.78
0.3	0.65	**84.29/72.11**
0.35	0.65	84.19/71.13
0.4	0.65	83.25/71.34

5 Conclusion

We propose a novel self-training pipeline that combines the mean-teacher framework with adversarial feature learning for unsupervised domain adaptive 3D object detection from point clouds. Our method involves specific data augmentation, multual learning between teacher model and student model, adversarial learning and threshold adapatation based predicted IoU to address fundamental challenges stemming from UDA on point-based 3D object detection. Experiments demonstrate that our method advance the current state of art.

References

1. Caesar, H., et al.: nuscenes: a multimodal dataset for autonomous driving. In: Proceedings of the IEEE/CVF Conference on Computer Vision and Pattern Recognition, pp. 11621–11631 (2020)
2. Ganin, Y., Lempitsky, V.: Unsupervised domain adaptation by backpropagation. In: International Conference on Machine Learning, pp. 1180–1189. PMLR (2015)

3. Ganin, Y., et al.: Domain-adversarial training of neural networks. J. Mach. Learn. Res. **17**(1), 2096–2030 (2016)
4. Geiger, A., Lenz, P., Urtasun, R.: Are we ready for autonomous driving? The kitti vision benchmark suite. In: 2012 IEEE Conference on Computer Vision and Pattern Recognition, pp. 3354–3361. IEEE (2012)
5. He, C., Zeng, H., Huang, J., Hua, X.S., Zhang, L.: Structure aware single-stage 3D object detection from point cloud. In: Proceedings of the IEEE/CVF Conference on Computer Vision and Pattern Recognition, pp. 11873–11882 (2020)
6. Hu, Q., Liu, D., Hu, W.: Density-insensitive unsupervised domain adaption on 3D object detection. arXiv preprint arXiv:2304.09446 (2023)
7. Lang, A.H., Vora, S., Caesar, H., Zhou, L., Yang, J., Beijbom, O.: Pointpillars: fast encoders for object detection from point clouds. In: Proceedings of the IEEE/CVF Conference on Computer Vision and Pattern Recognition, pp. 12697–12705 (2019)
8. Li, G., Li, X., Wang, Y., Wu, Y., Liang, D., Zhang, S.: PseCo: pseudo labeling and consistency training for semi-supervised object detection. In: Avidan, S., Brostow, G., Cisse, M., Farinella, G.M., Hassner, T. (eds.) ECCV 2022. LNCS, vol. 13669, pp. 457–472. Springer, Heidelberg (2022). https://doi.org/10.1007/978-3-031-20077-9_27
9. Li, Y.J., et al.: Cross-domain adaptive teacher for object detection. In: Proceedings of the IEEE/CVF Conference on Computer Vision and Pattern Recognition, pp. 7581–7590 (2022)
10. Liu, Z., Tang, H., Lin, Y., Han, S.: Point-voxel cnn for efficient 3D deep learning. Adv. Neural Inf. Process. Syst. **32** (2019)
11. Luo, Z., et al.: Unsupervised domain adaptive 3d detection with multi-level consistency. In: Proceedings of the IEEE/CVF International Conference on Computer Vision, pp. 8866–8875 (2021)
12. Mao, J., Niu, M., Bai, H., Liang, X., Xu, H., Xu, C.: Pyramid r-cnn: towards better performance and adaptability for 3D object detection. In: Proceedings of the IEEE/CVF International Conference on Computer Vision, pp. 2723–2732 (2021)
13. Miao, Z., et al.: Pvgnet: a bottom-up one-stage 3D object detector with integrated multi-level features. In: Proceedings of the IEEE/CVF Conference on Computer Vision and Pattern Recognition, pp. 3279–3288 (2021)
14. Pan, X., Xia, Z., Song, S., Li, L.E., Huang, G.: 3D object detection with pointformer. In: Proceedings of the IEEE/CVF Conference on Computer Vision and Pattern Recognition, pp. 7463–7472 (2021)
15. Patil, A., Malla, S., Gang, H., Chen, Y.T.: The h3d dataset for full-surround 3d multi-object detection and tracking in crowded urban scenes. In: 2019 International Conference on Robotics and Automation (ICRA), pp. 9552–9557. IEEE (2019)
16. Qi, C.R., Su, H., Mo, K., Guibas, L.J.: Pointnet: deep learning on point sets for 3d classification and segmentation. In: Proceedings of the IEEE Conference on Computer Vision and Pattern Recognition, pp. 652–660 (2017)
17. Qi, C.R., Yi, L., Su, H., Guibas, L.J.: Pointnet++: deep hierarchical feature learning on point sets in a metric space. Adv. Neural Inf. Process. Syst. **30** (2017)
18. Saito, K., Ushiku, Y., Harada, T., Saenko, K.: Strong-weak distribution alignment for adaptive object detection. In: Proceedings of the IEEE/CVF Conference on Computer Vision and Pattern Recognition, pp. 6956–6965 (2019)
19. Shi, S., et al.: Pv-rcnn: point-voxel feature set abstraction for 3D object detection. In: Proceedings of the IEEE/CVF Conference on Computer Vision and Pattern Recognition, pp. 10529–10538 (2020)
20. Shi, S., et al.: PV-RCNN++: point-voxel feature set abstraction with local vector representation for 3D object detection. Int. J. Comput. Vision **131**, 531–551 (2022)

21. Shi, S., Wang, X., Li, H.: Pointrcnn: 3D object proposal generation and detection from point cloud. In: Proceedings of the IEEE/CVF Conference on Computer Vision and Pattern Recognition, pp. 770–779 (2019)
22. Shi, W., Rajkumar, R.: Point-GNN: graph neural network for 3D object detection in a point cloud. In: Proceedings of the IEEE/CVF Conference on Computer Vision and Pattern Recognition, pp. 1711–1719 (2020)
23. Sun, P., et al.: Scalability in perception for autonomous driving: waymo open dataset. In: Proceedings of the IEEE/CVF Conference on Computer Vision and Pattern Recognition, pp. 2446–2454 (2020)
24. Team, O.D.: Openpcdet: an open-source toolbox for 3d object detection from point clouds (2020). https://github.com/open-mmlab/OpenPCDet
25. Wang, Y., et al.: Train in germany, test in the USA: making 3D object detectors generalize. In: Proceedings of the IEEE/CVF Conference on Computer Vision and Pattern Recognition, pp. 11713–11723 (2020)
26. Wei, Y., Wei, Z., Rao, Y., Li, J., Zhou, J., Lu, J.: Lidar distillation: bridging the beam-induced domain gap for 3d object detection. In: Computer Vision-ECCV 2022: 17th European Conference, Tel Aviv, Israel, 23–27 October 2022, Proceedings, Part XXXIX, pp. 179–195. Springer, Heidelberg (2022). DOI: https://doi.org/10.1007/978-3-031-19842-7_11
27. Xu, Q., Zhou, Y., Wang, W., Qi, C.R., Anguelov, D.: SPG: unsupervised domain adaptation for 3D object detection via semantic point generation. In: Proceedings of the IEEE/CVF International Conference on Computer Vision, pp. 15446–15456 (2021)
28. Yan, Y., Mao, Y., Li, B.: Second: sparsely embedded convolutional detection. Sensors **18**(10), 3337 (2018)
29. Yang, J., Shi, S., Wang, Z., Li, H., Qi, X.: St3d: self-training for unsupervised domain adaptation on 3D object detection. In: Proceedings of the IEEE/CVF Conference on Computer Vision and Pattern Recognition, pp. 10368–10378 (2021)
30. Yang, Z., Sun, Y., Liu, S., Jia, J.: 3dssd: Point-based 3D single stage object detector. In: Proceedings of the IEEE/CVF Conference on Computer Vision and Pattern Recognition. pp. 11040–11048 (2020)
31. Yang, Z., Sun, Y., Liu, S., Shen, X., Jia, J.: Ipod: Intensive point-based object detector for point cloud. arXiv preprint arXiv:1812.05276 (2018)
32. Yin, T., Zhou, X., Krahenbuhl, P.: Center-based 3D object detection and tracking. In: Proceedings of the IEEE/CVF Conference on Computer Vision and Pattern Recognition (CVPR), pp. 11784–11793 (2021)
33. Zhu, J.Y., Park, T., Isola, P., Efros, A.A.: Unpaired image-to-image translation using cycle-consistent adversarial networks. In: Proceedings of the IEEE International Conference on Computer Vision, pp. 2223–2232 (2017)
34. Zhu, X., Pang, J., Yang, C., Shi, J., Lin, D.: Adapting object detectors via selective cross-domain alignment. In: Proceedings of the IEEE/CVF Conference on Computer Vision and Pattern Recognition, pp. 687–696 (2019)

Generalizable Neural Radiance Field with Hierarchical Geometry Constraint

Qinyi Zhang[1,2], Jin Xie[1,2(✉)], and Hang Yang[1,2]

[1] PCA Lab, Key Lab of Intelligent Perception and Systems for High-Dimensional Information of Ministry of Education, Nanjing University of Science and Technology, Nanjing, China
{zqy-katrina,csjxie,hangyang}@njust.edu.cn

[2] Jiangsu Key Lab of Image and Video Understanding for Social Security, School of Computer Science and Engineering, Nanjing University of Science and Technology, Nanjing, China

Abstract. The emergence of the generalized neural radiance field models has greatly expanded the applicability of novel view synthesis task. However, existing methods primarily rely on 2D features obtained from neighboring views to assist the rendering, which often leads to degradation in unseen viewpoints. Some approaches introduce cost volumes in the model to provide geometric priors, but such coarse geometric information fails to constrain the model effectively, resulting in rendered images with abundant artifacts. In this work, we propose to provide the model with hierarchical geometry constraints to achieve better rendering results. We introduce cascaded MVSNet to provide hierarchical scene structural features for inferring the underlying scene geometry. It constrains the model's rendering under unseen viewpoints. Also, the hierarchical features provide fine representations that helps the reconstruction of details. Additionally, we utilize the hierarchical depth maps generated by cascade MVSNet to constrain the sampling process, ensuring that the sampled points are concentrated near the scene surface. Such sampling strategy filtering out numerous useless sampling points and improving both the sampling efficiency and rendering quality. Unlike previous generalized methods, we employ a new weight function proposed by Neus to eliminate inherent biases in the density field, disentangling the prediction of color and density with a signed distance field. Extensive experiments have demonstrated that our proposed method significantly improves the rendering quality and outperforms previous methods.

Keywords: Novel view synthesis · Neural radiance field · Generalization

1 Introduction

Given a set of posed images, novel view synthesis enables the recovery of images from arbitrary poses. This task can be applied to many downstream tasks, such

Q. Liu et al. (Eds.): PRCV 2023, LNCS 14426, pp. 319–330, 2024.
https://doi.org/10.1007/978-981-99-8432-9_26

as robot navigation, pose estimation and visual localization. Recently, this long-standing problem has made significant process due to the emergence of neural radiance field (NeRF) [14] and its subsequent work.

However, NeRF is a per-scene optimization model which requires retraining when synthesizing images for new scene. This limitation greatly restricts the widespread application of NeRF. Existing representative generalization models, such as PixelNeRF [22], IBRNet [19] and NeuRay [13], primarily rely on image-based rendering techniques to achieve generalization effects. These models utilize 2D features obtained from neighboring views to assist in generating novel views. However, they do not take into account the 3D structural information of the scene, which can result in degraded renderings in novel viewpoints, especially in the case of limited view inputs. MVSNeRF [1] innovatively introduces multi-view stereo method into the generalized model. It constructs a cost volume and provides geometry prior to help the model synthesize novel views across scenes. Nevertheless, the geometric features used by MVSNeRF are relatively coarse that fail to provide effective guidance and constraints for rendering, resulting in rendered images with limited details and abundant artifacts.

In an ideal scenario, if we know the exact geometry information such as depth of the pixel to be rendered, we can obtain the surface point in target viewpoint and its exact features on adjacent views, thus achieve better rendering results with correct and unobstructed features. However, due to the generalization setting, it is not feasible to obtain the exact geometric information of each new scene. So, we propose to optimize the model's geometric perception to help the synthesis of images from unseen viewpoints.

In this work, we optimize the model's geometric perception by providing hierarchical geometry constraints. Specifically, we introduce CasMVSNet [7], which constructs multi-scale cost volumes in a coarse-to-fine manner to provide hierarchical and accurate geometry prior. We fuse the multi-scale volume features generated from the cost volumes, obtaining a more comprehensive perception of the scene geometry. This effectively imposes guidance and geometric constraints on unseen viewpoints, enabling the reconstruction of fine details with less artifacts. Such hierarchical features derived from variance-based cost volume also encode the appearance variations resulting from changes in viewpoints and scene geometry, which are discriminative. When predicting the volume density, we combine 3D coordinates with the corresponding hierarchical volume features to enhance the geometry at the sampled points, thereby predicting more accurate SDF values.

In addition, the model acquires the required information through sampling, making the accuracy of sampling crucial for capturing scene geometry. Previous works require extensive sampling points along the ray to ensure good rendering quality due to the lack of geometric priors. Empirically, only points near the scene surface can contribute to the final rendering result. Such approach generates a large number of useless points. We propose a novel hierarchical sampling method constrained by cascaded MVSNet, which can concentrate the sampling points near the scene surface and reduce unnecessary sampling points. This strategy effectively enhances the rendering quality while reducing more than half of the sampling points.

Furthermore, all those generalized models overlook the fundamental issue present in NeRF: the weight function used by NeRF cannot guarantee the precise concentration of volume density near the scene surface due to the lack of sufficient level set constraints, which will introduce biases in the perception of scene geometry. To alleviate this issue, we replace the original weight function with the optimized one in Neus [18] and disentangle the predictions for color and density.

In summary, our main contributions are as follows:

- We introduce the CasMVSNet to obtain fine and comprehensive geometry prior to constrain and guide the generalized model in rendering.
- We introduce a novel sampling approach that constrains more useful sampling points near the scene surface, which can improve rendering performance while reducing more than half of the sampling points.
- Extensive experiments have demonstrated that our proposed method significantly improves the rendering quality and outperforms previous methods.

2 Related Work

2.1 Novel View Synthesis

Novel view synthesis is a long standing problem in computer vision, which involves predicting new views of a scene from a given collection of images. Early works [5,12] achieved photo-realistic results by constructing a 4D light filed but required densely sampled views. When given sparse input views, recent works turn to explicit 3D representations like point clouds [20], mesh [17], multiplane images [15] to seek solutions. But such approaches have limited performance because of their representation.

More recently, NeRF [14] has brought a new implicit scene representation, modeling a volumetric radiance field by learning an implicit function and using standard volumetric rendering [10] and alpha compositing techniques [16] to generate photo-realistic views from novel viewpoints. Many following works [4, 22,23] have attempted to improve NeRF from various aspects.

2.2 Neural Scene Representation

With the success of NeRF, neural representation becomes a popular way to represent scenes for its compactness and flexibility. Though NeRF is capable of rendering photo-realistic images, volume rendering used in NeRF is lack of constraints, resulting in poor scene geometry learned by the model. Many works have used extra geometric information to supervise the optimization process of NeRF. [21] combines volumetric radiance fields with point clouds to approximate scene geometry, [2] uses sparse 3D points generated by structure-from-motion as a free depth supervision.

Neural Implicit Surface Representation. Compared to rendering tasks, surface reconstruction tasks require high-fidelity scene geometry. Implicit surface reconstruction methods [3,18], represented by Neus [18], combine the advantages of surface rendering and volume rendering, constraining the scene space as a signed distance function and training as volume rendering for its robustness. NeuS defines a conversion function from SDF to volume density, to construct a better weight function of volume rendering to obtain accurate scene surface without introducing additional geometric supervision.

Generalization Volume Rendering. NeRF is a per-scene optimization method, and training on each new scene usually takes a long time. To address this issue, recent works [1,9,13,19,22] have introduced generalized radiance field models. The common motivation behind such methods is to blend features extracted from a set of nearby input views. MVSNeRF [1] first proposes to add volume feature to improve the perception of scene geometry and to increase the generalization ability of the model. GeoNeRF [9] and NeuRay [13] consider the occlusion of nearby input views and filter the invisible features by constructing a cost volume modeling the visibility of the sampling points on the input views.

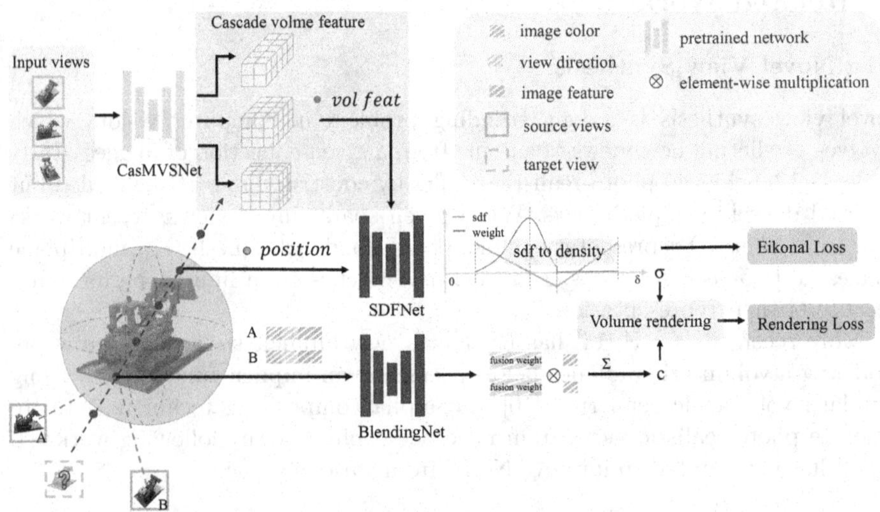

Fig. 1. Overview of the model.Our model consists of two main components: feature generation and rendering. Firstly, a pretrained cascaded MVSNet is used to generate multi-scale geometry features to impose geometry constraints. Then, our novel hierarchical sampling method is used to sample along the ray and aims to constrains the sampling points near the scene surface. Finally, these sampling points, along with their corresponding features, are processed through two networks for volume rendering. We disentangle the prediction of color and density with a signed distance field to eliminate inherent biases in the density field.

3 Method

Given a set of input views $\{I_k\}_{k=1}^N$ with known camera poses $\{R_k\}_{k=1}^N$, our goal is to synthesize a novel view corresponding to the query pose R_q. Figure 1 shows the architecture of our model.

As is analyzed previously, we will present our approach for improving the geometric perception by providing hierarchical geometry constraints from three perspectives. Firstly, we introduce a novel representation of opaque density in Sect. 3.1. Then, we present how to obtain fine multi-scale geometric features by introducing cascaded MVSNet in Sect. 3.2. Finally, we propose a novel and efficient sampling strategy guided by cascaded MVS in Sect. 3.4.

3.1 Preliminaries

Neural Radiance Fields. A Neural Radiance Field (NeRF) [14] represents a scene as a 5D function mapping by training a multi-layer perception (MLP) to predict the volume density and view-dependent color of 3D points in space. More specially, given a 3D point $\mathbf{x} \in \mathbb{R}^3$ and viewing direction $\mathbf{d} \in \mathbb{R}^2$, NeRF fits a mapping function from 3D coordinates and viewing directions to volume density σ and emitted radiance $\mathbf{c}$ like this: $f_\theta : (\mathbf{x}, \mathbf{d}) = (\sigma, \mathbf{c})$, where θ denotes MLP parameters.

Volume Rendering. To render an expected pixel, a ray $\mathbf{r}(t) = \mathbf{o} + t\mathbf{d}$ is cast from the camera origin $\mathbf{o}$ along the direction $\mathbf{d}$ to pass through the pixel, where $t \in [t_n, t_f]$ represents the distance to the origin. The color can be rendered using volume rendering formulation as:

$$\hat{C}(r) = \int_{t_n}^{t_f} w(\mathbf{r}(t))c(\mathbf{r}(t), \mathbf{d})dt, \tag{1}$$

where w is a weight function depends on the scene geometry. Ideally, it will reach its peak at the scene surface. So, we choose to adopt the weight function schema proposed by NeuS [18] for its good properties of unbiased and occlusion perception:

$$w(t) = T(t)\rho(t), \tag{2}$$

where $T(t)$ is the accumulated transmittance checking for scene occlusion from t_n to t_f:

$$T(t) = \exp(-\int_{\mathrm{t_n}}^{\mathrm{t_f}} \rho(\tau)\mathrm{d}\tau) \tag{3}$$

and $\rho(t)$ is the opaque density of each point on a ray to represent scene geometry. $f_s : \mathbb{R}^3 \to \mathbb{R}$ is a sign distance function and $f_s(\mathbf{r}(t))$ represents the SDF value of the point at distance t. Φ_s is the Sigmoid function. $\rho(t)$ defines a transformation from SDF to opaque density:

$$\rho(t) = max(\frac{-\frac{\mathrm{d}\Phi_s}{\mathrm{dt}}(f_s(\mathbf{r}(t)))}{\Phi_s(f_s(\mathbf{r}(t)))}, 0) \tag{4}$$

3.2 Multi-scale Geometric Feature

We input N views with size H × W, and set one of them as reference and the others as sources. We construct a cascaded cost volume at reference view following CasMVSNet [7]. First, all input images go through a **F**eature **P**yramid **N**etwork to generate three scales of image features f_k^l. The top-level features contain high-level semantic information, while the low-level features contain finer representation. The features concatenate with original image color to obtain image feature map F_k of each input view, which are utilized in blending color.

$$f_k^l = \text{FPN}(I_k) \in \mathbb{R}^{\frac{H}{2^l} \times \frac{W}{2^l} \times 2^l C}, l \in \{0, 1, 2\} \tag{5}$$

Following the extraction of image features, we leverage the corresponding extrinsic and intrinsic to construct a homography matrix to warp source feature maps into hypothesis planes of the reference view to form feature volumes. We use variance-based cost metric to aggregate multiple feature volumes to one cost volume. We construct three scales of cost volumes v^l where $l \in \{2, 1, 0\}$. In the first stage ($l = 2$), the hypothesis range R^l covers the entire depth range of the scene with a maximum number of hypothesis planes P^l, corresponding to the lowest spatial resolution of the cost volume. In the subsequent stages, we narrow the hypothesis range based on the predicted depth map D^l from the previous stage, reducing the number of planes while increasing the spatial resolution of the cost volume. Finally, a 3D hourglass network $\mathcal{M}_{3D}^l$ is utilized to regularize the cost volumes and generate multi-scale geometric features and depth maps. Concatenating the three scales of 3D features yields the multi-scale volume features V.

$$V^l \in \mathbb{R}^{\frac{H}{2^l} \times \frac{W}{2^l} \times P^l \times C}, D^l \in \mathbb{R}^{\frac{H}{2^l} \times \frac{W}{2^l} \times 1} = \mathcal{M}_{3D}^l(v^l), l \in \{2, 1, 0\} \tag{6}$$

3.3 Volume Rendering

In our model, opaque density is transformed from a SDF value, while colors are generated through a blending network. Specifically, the 3D coordinates of sampled points x_i and their corresponding multi-scale geometric features V_i are inputted into an Signed Distance Field Network $\mathcal{M}_{sdf}$, which predicts the SDF value and normal vector of each point. The opaque density is then obtained by converting the SDF using Eq. (4). The combination of coordinates and associated multi-scale volume features provides the model with stronger generalization ability.

$$s = \mathcal{M}_{sdf}(x_i, V_i), s \rightarrow \sigma \tag{7}$$

The input to the blending network $\mathcal{M}_{blender}$ consists of the RGB values c_i and image features F_i of the source images corresponding to the sampled points, as well as the viewpoint differences Δd between the source and target views. We use a blending network similar to IBRNet [19] to obtain the final color values c for each sampled point (without ray transformer).

$$c = \mathcal{M}_{blender}(c_i, F_i, \Delta d) \tag{8}$$

Finally, we use volume rendering in Eq. (1) to generate the colors $\hat{C}$ for each pixel.

Scene Normalization. To facilitate the learning process of the signed distance function (SDF) network, we normalize all scenes in the dataset by constraining the points coordinate range to $(-1, 1)$. Specifically, we select the two farthest points from the eight vertices that form the viewing frustum in the world coordinate system across all input and target views. The line connecting these two points serves as the diameter of a unit circle, which is used to normalize the scene within this unit circle. Experimental results demonstrate that scene normalization effectively enhances the performance of the model.

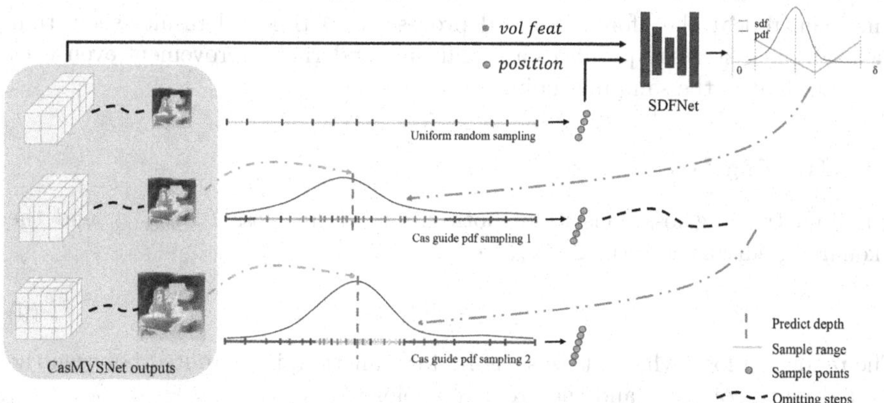

Fig. 2. Hierarchical sampling guided by Cascaded MVSNet: By utilizing the depth maps generated from cascaded MVSNet, we progressively narrow down the sampling range along the ray. New sampling points are allocated based on the probability density distribution derived from previous sampling. Such approach effectively constrains the sampling points near the scene surface, reducing plenty of empty samples and enhancing both the sampling efficiency and rendering quality.

3.4 Hierarchical Sampling Guided by Cascade-MVS

We utilize the concept of hypothesis range reduction from CasMVSNet to narrow down the sampling range along each ray towards the scene surface, leveraging depth maps generated by cascaded MVSNet. This sampling strategy ensures that even a limited number of sampling points are effective and meaningful in contributing to the final rendering result. Figure 2 illustrates our sampling strategy.

Specifically, we first perform a basic sampling on all rays and sample N_b points uniformly with a sample range from the near plane to far plane. Then we apply twice hierarchical sampling, each with N_h sample points. In the first hierarchical sampling, we define the sampling range to include N_{h_1} sample points ($N_h < N_{h_1}$). Based on the depth map $D^{l=1}$ generated by cascaded MVSnet, we can obtain the predicted depths p_i for all rays. Then we take the sample point closest to the predicted depth as the center and expand $1/2N_{h_1}$ points on both sides, which is the sampling range for the first hierarchical sampling. Although the predicted depth we obtained is the scene depth under reference view, it still has great reference value due to the small difference between the reference view and target view. Considering such differences, we still generate required N_h sampling points based on the probability density distribution of opaque density on each ray. This strategy narrows the sampling range while enabling generation of sampling points based on the target view's scene geometry. In the second hierarchical sampling, we further restrict the sampling range by including N_{h_2} sampling points ($N_h < N_{h_1} < N_{h_2}$) based on the predicted depth map $D^{l=0}$. And then repeat the aforementioned process. Experimental results show that such sampling strategy produces a significant rendering improvement even with less than half of the sampling points.

3.5 Training Loss

Our final training loss consists of both the rendering loss $\mathcal{L}_{rendering}$ and the eikonal regularization term $\mathcal{L}_{eikonal}$.

$$\mathcal{L}_{rendering} = \|C_{gt} - \hat{C}\|_2^2 \tag{9}$$

The rendering loss, which utilizes the L2 loss function, is computed between the ground truth color C_{gt} and the predicted color $\hat{C}$.

$$\mathcal{L}_{eikonal} = \frac{1}{\|\mathbb{X}\|} \sum_{x \in \mathbb{X}} (\|\nabla \mathcal{M}_{sdf}(x)\|_2 - 1)^2 \tag{10}$$

Eikonal term [6] is applied on sampled points to regularize the SDF values derived from $\mathcal{M}_{sdf}$, where x is the sample point, $\mathbb{X}$ is the set of all sample points, $\nabla \mathcal{M}_{sdf}(x)$ is the gradient of $\mathcal{M}_{sdf}$ corresponding to sample point x.

4 Experiments

Dataset. We only train our model on the DTU [8] dataset. We divide the dataset into 88 training scenes and 16 test scenes as the same as PixelNeRF [22]. To validate the generalization of our model, we also test on the Realistic Synthetic NeRF dataset [14] whose view distributions are completely different

Table 1. Quantitative results of novel view synthesis.

Method	Settings	Synthetic Data (NeRF [14])			Real Data (DTU [8])		
		PSNR↑	SSIM↑	LPIPS↓	PSNR↑	SSIM↑	LPIPS↓
PixelNeRF [22]	No per-scene optimization	7.39	0.658	0.411	19.31	0.789	0.382
IBRNet [19]		22.44	0.874	0.195	26.04	0.917	0.190
MVSNeRF [1]		23.62	0.897	0.176	26.63	**0.931**	0.168
GeoNeRF [9]		25.86	0.916	0.111	**28.95**	0.923	**0.116**
Ours		**26.14**	**0.936**	**0.098**	27.31	0.918	0.293
NeRF [14]	Per-scene optimization	**30.63**	**0.962**	<u>0.093</u>	27.01	0.902	0.263
IBRNet$_{ft}$ [19]		25.62	0.939	0.110	**31.35**	**0.956**	<u>0.131</u>
MVSNeRF$_{ft}$ [1]		27.07	0.931	0.168	28.50	0.933	0.179
Ours$_{ft}$		<u>27.14</u>	<u>0.941</u>	**0.088**	<u>29.33</u>	<u>0.936</u>	**0.098**

from our training set. For each testing scene, we select 4 as testing views and take the remains for per-scene fine-tuning. We render images with a resolution of 512×640 for DTU and 800×800 for NeRF Synthetic.

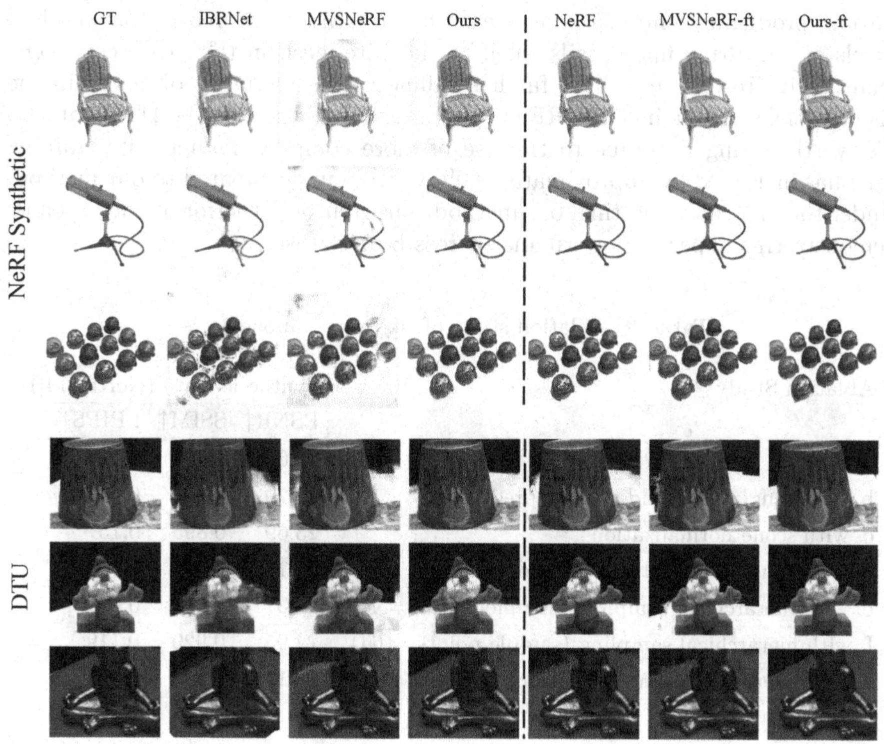

Fig. 3. Visual rendering results comparison. On the left, we present the generalization results of each model, while on the right, we show the fine-tuning results.

Implementation Details. We use three inputs views as the setup of sparse input. We employ a pretrained cascaded MVSNet to generate multi-scale feature maps and depth maps. For 2D features f^l, we use C = 32, while for 3D features v^l, we use C = 8. The cascaded MVSNet is trained together with the model. In terms of hierarchical sampling, we collect $N_b = 48$ sampling points during the basic sampling, followed by two rounds hierarchical sampling with $N_h = 16$ sampling points each, resulting in a total of 80 sampling points. The first sampling range is define as $N_{h_1} = 32$, and the second sampling range as $N_{h_2} = 24$. Our model is trained on a single TITAN RTX GPU. We randomly sample 1024 pixels per training iteration and use Adam [11] optimizer with a learning rate of 5×10^{-4}.

4.1 Experimental Results

We compared our method with PixelNeRF [22], IBRNet [19], MVSNeRF [1], GeoNeRF [9] and vanilla NeRF [14] on DTU dataset and NeRF Synthetic dataset. Our visualization results are presented in Fig. 3. While Table 1 shows the quantitative comparison based on PSNR, SSIM and LPIPS. All experiments were conducted with the same setting as described in this paper. From Fig. 3, it is evident that our method outperforms other approaches in terms of generalization, producing rendered images with fewer artifacts. Moreover, our method excels at capturing fine details, such as the wire-mesh in the MIC scene. The quantitative results in Table 1 further validate the superiority observed in the visualizations. Though GeoNeRF outperforms our method on the DTU dataset, it's worth noting that due to the use of more complex modules, its training time has increased by approximately 60% GPU hours compared to our method. Under the fine-tuning setting, our method consistently outperforms the baseline, demonstrating superior performance across both datasets.

Table 2. Ablation study of the key components.

Ablation Study	Synthetic Data (NeRF [14])		
	PSNR↑	SSIM↑	LPIPS↓
a. baseline [1]	23.62	0.897	0.176
b. disentangle color and density prediction	22.70	0.887	0.147
c. with scene normalization	23.69	0.895	0.157
d. with multi-scale volume feature	25.11	0.920	0.115
e. with hierarchical sampling (sample points = 192)	26.37	0.937	0.092
f. with hierarchical sampling (sample points = 64)	24.95	0.920	0.119
g. with hierarchical sampling (sample points = 80)	26.14	0.936	0.098

4.2 Ablation Study

We conducted ablation experiments on the NeRF Synthetic dataset to validate the effectiveness of key points proposed in this paper, as presented in Table 2. The results demonstrate that our approach benefits significantly from the utilization of multi-scale feature maps and hierarchical sampling, both of which contribute to improving rendering quality. Notably, when color and density predictions are solely disentangled using the SDF network, there is a noticeable decrease in performance, highlighting the importance of scene normalization. Additionally, we performed ablation experiments on the number of sampling points in hierarchical sampling. Based on the experimental results, we determined that sampling 80 points per ray strikes a balance between the number of sampling points and rendering quality.

5 Conclusion

In this paper, we argue that a generalized rendering model requires strong scene geometry perception to achieve more realistic rendering results. To this end, we provide the model with hierarchical geometry constraints. Specifically, we introduce cascaded MVSNet to provide more comprehensive scene geometry priors, aiding the generalized model in focusing on accurate features and reconstructing fine details. Additionally, we present a novel hierarchical sampling approach that restricts sampling points near the scene surface, thereby filtering out numerous useless sampling points and improving both the sampling efficiency and rendering quality. Finally, we optimize the weight function in volume rendering to ensure that the weight of volume density peaks out at the scene surface. Our experimental results and ablation studies demonstrate the effectiveness of our proposed method.

References

1. Chen, A., et al.: MVSNeRF: fast generalizable radiance field reconstruction from multi-view stereo. In: Proceedings of the IEEE/CVF International Conference on Computer Vision, pp. 14124–14133 (2021)
2. Deng, K., Liu, A., Zhu, J.Y., Ramanan, D.: Depth-supervised NERF: fewer views and faster training for free. In: Proceedings of the IEEE/CVF Conference on Computer Vision and Pattern Recognition, pp. 12882–12891 (2022)
3. Fu, Q., Xu, Q., Ong, Y.S., Tao, W.: Geo-Neus: geometry-consistent neural implicit surfaces learning for multi-view reconstruction. Adv. Neural. Inf. Process. Syst. **35**, 3403–3416 (2022)
4. Garbin, S.J., Kowalski, M., Johnson, M., Shotton, J., Valentin, J.: FastNeRF: high-fidelity neural rendering at 200fps. In: Proceedings of the IEEE/CVF International Conference on Computer Vision, pp. 14346–14355 (2021)
5. Gortler, S.J., Grzeszczuk, R., Szeliski, R., Cohen, M.F.: The lumigraph. In: Proceedings of the 23rd Annual Conference on Computer Graphics and Interactive Techniques, pp. 43–54 (1996)

6. Gropp, A., Yariv, L., Haim, N., Atzmon, M., Lipman, Y.: Implicit geometric regularization for learning shapes. arXiv preprint arXiv:2002.10099 (2020)
7. Gu, X., Fan, Z., Zhu, S., Dai, Z., Tan, F., Tan, P.: Cascade cost volume for high-resolution multi-view stereo and stereo matching. In: Proceedings of the IEEE/CVF Conference on Computer Vision and Pattern Recognition, pp. 2495–2504 (2020)
8. Jensen, R., Dahl, A., Vogiatzis, G., Tola, E., Aanæs, H.: Large scale multi-view stereopsis evaluation. In: Proceedings of the IEEE Conference on Computer Vision and Pattern Recognition, pp. 406–413 (2014)
9. Johari, M.M., Lepoittevin, Y., Fleuret, F.: GeoNeRF: generalizing NeRF with geometry priors. In: Proceedings of the IEEE/CVF Conference on Computer Vision and Pattern Recognition, pp. 18365–18375 (2022)
10. Kajiya, J.T., Von Herzen, B.P.: Ray tracing volume densities. ACM SIGGRAPH Comput. Graph. **18**(3), 165–174 (1984)
11. Kingma, D.P., Ba, J.: Adam: a method for stochastic optimization. arXiv preprint arXiv:1412.6980 (2014)
12. Levoy, M., Hanrahan, P.: Light field rendering. In: Proceedings of the 23rd Annual Conference on Computer Graphics and Interactive Techniques, pp. 31–42 (1996)
13. Liu, Y., et al.: Neural rays for occlusion-aware image-based rendering. In: Proceedings of the IEEE/CVF Conference on Computer Vision and Pattern Recognition, pp. 7824–7833 (2022)
14. Mildenhall, B., Srinivasan, P.P., Tancik, M., Barron, J.T., Ramamoorthi, R., Ng, R.: NeRF: representing scenes as neural radiance fields for view synthesis. Commun. ACM **65**(1), 99–106 (2021)
15. Penner, E., Zhang, L.: Soft 3d reconstruction for view synthesis. ACM Trans. Graph. (TOG) **36**(6), 1–11 (2017)
16. Porter, T., Duff, T.: Compositing digital images. In: Proceedings of the 11th Annual Conference on Computer Graphics and Interactive Techniques, pp. 253–259 (1984)
17. Shih, M.L., Su, S.Y., Kopf, J., Huang, J.B.: 3D photography using context-aware layered depth inpainting. In: Proceedings of the IEEE/CVF Conference on Computer Vision and Pattern Recognition, pp. 8028–8038 (2020)
18. Wang, P., Liu, L., Liu, Y., Theobalt, C., Komura, T., Wang, W.: NeuS: learning neural implicit surfaces by volume rendering for multi-view reconstruction. arXiv preprint arXiv:2106.10689 (2021)
19. Wang, Q., et al.: IBRNet: learning multi-view image-based rendering. In: Proceedings of the IEEE/CVF Conference on Computer Vision and Pattern Recognition, pp. 4690–4699 (2021)
20. Wiles, O., Gkioxari, G., Szeliski, R., Johnson, J.: SynSin: end-to-end view synthesis from a single image. In: Proceedings of the IEEE/CVF Conference on Computer Vision and Pattern Recognition, pp. 7467–7477 (2020)
21. Xu, Q., et al.: Point-NeRF: point-based neural radiance fields. In: Proceedings of the IEEE/CVF Conference on Computer Vision and Pattern Recognition, pp. 5438–5448 (2022)
22. Yu, A., Ye, V., Tancik, M., Kanazawa, A.: pixelNeRF: neural radiance fields from one or few images. In: Proceedings of the IEEE/CVF Conference on Computer Vision and Pattern Recognition, pp. 4578–4587 (2021)
23. Zhang, K., Riegler, G., Snavely, N., Koltun, V.: NeRF++: analyzing and improving neural radiance fields. arXiv preprint arXiv:2010.07492 (2020)

ACFNeRF: Accelerating and Cache-Free Neural Rendering via Point Cloud-Based Distance Fields

Xinjie Yang, Xiaotian Sun, and Cheng Wang(✉)

Fujian Key Laboratory of Sensing and Computing for Smart Cities, School of Informatics, Xiamen University, Xiamen, China
{xjyang,sunxt2020}@stu.xmu.edu.cn, cwang@xmu.edu.cn

Abstract. Neural radiance fields offer a remarkable avenue for realistic scene rendering and novel view synthesis. Nevertheless, challenges such as sluggish training times, protracted inference durations, and limitations in handling large-scale scenes persist. To address the bottleneck of slow inference in NeRF, our propose ACFNeRF leveraging point cloud to train a distance field, improving NeRF's sampling strategy, and substantially bolstering its inference speed. Our approach achieves an impressive inference rate of 150 frames per second, enabling real-time rendering within room-scale scenes. Comprehensive experimentation validates our method's superiority, demonstrating a notable 10–20x acceleration over existing NeRF acceleration techniques under cache-free conditions.

Keywords: NeRF · Distance Fields · Point Cloud

1 Introduction

The quest for achieving high-fidelity, real-time rendering of scenes is a fundamental pursuit within the domains of computer graphics and computer vision, with wide-ranging applications in virtual reality (VR), augmented reality (AR), metaverse, and beyond. Image-based rendering [1,7,11] leverages data from a collection of 2D images to represent a 3D model or scene, enabling the synthesis of novel views. Within this realm, neural rendering [7,11] has emerged as a disruptive breakthrough, harnessing the capabilities of neural networks to generate rendered images directly.

The most compelling recent work in neural rendering is NeRF [11]. It is capable of learning an implicit volumetric representation of a scene given the images, allowing for photo-realistic rendering of the scene from novel views. NeRF represents scenes in a fully-connected deep network, whose input is a single continuous 5D coordinate (3D location and 2D viewing direction), and whose output is the volume density and view-dependent color at that location. Then it uses classic volume rendering techniques to project the output densities and colors into an image. Despite its impressive ability to generate realistic renderings, NeRF is burdened by several limitations, including slow training and inference, and failure in large scenes.

Q. Liu et al. (Eds.): PRCV 2023, LNCS 14426, pp. 331–342, 2024.
https://doi.org/10.1007/978-981-99-8432-9_27

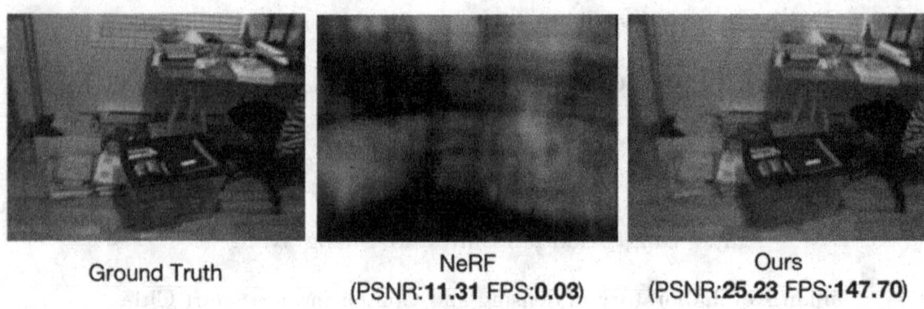

Fig. 1. Comparing on the Scannet dataset [2], our method not only improves the rendering quality compared to NeRF, but also the inference speed is nearly 5000 times faster than NeRF.

Our work primarily focuses on addressing the slow inference speed of NeRF. In NeRF, the volumetric representation relies on MLP, and for each pixel in a rendered image, a large number of sampled points along the ray through that pixel are used as input. Consequently, rendering a single pixel necessitates numerous neural network evaluations to predict the volume density and color of each sampled point. For example, rendering a 640×480 image with 256 points sampled on each ray would require approximately 0.7864 billion neural network calls. This means that rendering the image on a high-end GPU would take several seconds, falling short of the real-time rendering requirement.

Several recent approaches have made efforts to accelerate NeRF rendering [3,4,8,9,12,14,15,20,23], achieving faster speeds compared to the original NeRF model. However, real-time rendering remains a challenge. Many of these approaches trade off memory overhead to speed up rendering by pre-caching information for each query point in space [3], which is not feasible for rendering room-scale scenes due to the significant cost.

To achieve real-time rendering, our proposed ACFNeRF uses point cloud as a geometric prior. Unlike previous approaches [21], we do not directly employ geometric to guide NeRF training and inference. Instead, we train a distance field based on the point cloud. The distance field serves as an implicit representation of the 3D scenes, predicting the distance from each point in space to the nearest surface. Compared to explicit geometric prior, distance fields offer advantages such as continuity, fast inference, and low storage requirements. With the distance field, we concentrate the points sampled on each ray near the surface of the scene, thus reducing the computational effort considerably. Additionally, we optimize the distance field further using the volume density of the NeRF output. Experimental results demonstrate that our method achieves a remarkable inference speed of 150 frames per second, enabling real-time rendering of room-scale scenes. Without caching, our proposed ACFNeRF is nearly 5000 times faster than the original NeRF and outperforms existing NeRF acceleration techniques by 10–20 times, achieving optimal inference speed (see Fig. 1).

In summary, our main contributions can be summarised as follows:

- A novel approach is presented for training distance fields to represent scenes using point cloud.
- We propose a method for training and inference of NeRF using distance fields. Our proposed ACFNeRF enables optimal inference speed in cache-free conditions.

2 Related Work

2.1 Neural Scene Representations

The use of neural networks for scene representation has gained significant attention in recent research. While explicit representations such as point cloud [18] and depth maps [5] are intuitive, they can be storage-intensive and limited in resolution for rendering [16]. Many works have explored implicit representations of scenes [6,9,10,13], the classical one being signed distance fields, which use MLPs to map the coordinates of points in continuous space to signed distance values that represent their distance to the nearest surface.

The advancement of differentiable rendering has led to the development of various methods utilizing implicit representations to learn object geometry and appearance. For example, the early SRNs [17] directly mapped spatial coordinates to feature representations and used a learnable ray marcher to accumulate rays for obtaining the final RGB values.

The emergence of NeRF [11] achieved impressive rendering results by directly optimising the parameters of the 5D scene representation for the synthesis of novel views. However, it required numerous MLP calls for each sampled point on every ray, making real-time rendering unattainable.

2.2 Acceleration of Neural Radiance Fields

To address the slow rendering issue of the NeRF [11], several research efforts have attempted to solve this problem. These solutions can be categorized into two main types.

One approach is to improve the rendering speed at the expense of a large amount of memory [3,4,20,23], they speed up the rendering by caching the information of each query point in the space. However, caching NeRF would have a complexity of $O(n^5)$. This means that rendering even a small object would require hundreds of terabytes of cache, which is impractical. Therefore, these approaches attempt to optimize the NeRF model and reduce the cache size. However, they still require hundreds of gigabytes of cache to represent a room-scale scene. In contrast, our proposed ACFNeRF only requires 7M, offering a significant reduction in storage requirements.

Another way is to use external geometric data [9,12,15,21]. The original NeRF samples a large number of points along the ray spread over meaningless blank regions in space, and with a geometric prior one can improve efficiency by skipping the blank regions. For example, NSVF [9]combines MLP with voxel meshes, PointNeRF [21] uses point cloud as geometric priors, and Dense Depth Priors for NeRF [15] and DoNeRF [12] incorporate depth information. These approaches concentrate the sampled points near the scene surface, thereby reducing the number of sampled points and improving efficiency.

3 Method

In this section, we present our proposed ACFNeRF, which focuses on enhancing the inference speed of novel view synthesis in room-scale scenes. Our method comprises two key components: optimizing distance fields using point cloud and jointly optimizing the distance field and radiance field using images.

3.1 Distance Fields Based on Point Cloud

In order to obtain the ground truth of points in space to the nearest scene surface, we incorporate point cloud. Which can be acquired by 3D sensors such as LiDAR or reconstructed from dense multi-views using methods such as MVS [22]. The point cloud is represented as a set of 3D points $\mathbf{P} = \{P_i | i = 1, ..., N\}$, where each point P_i is a vector of its coordinates (x, y, z).

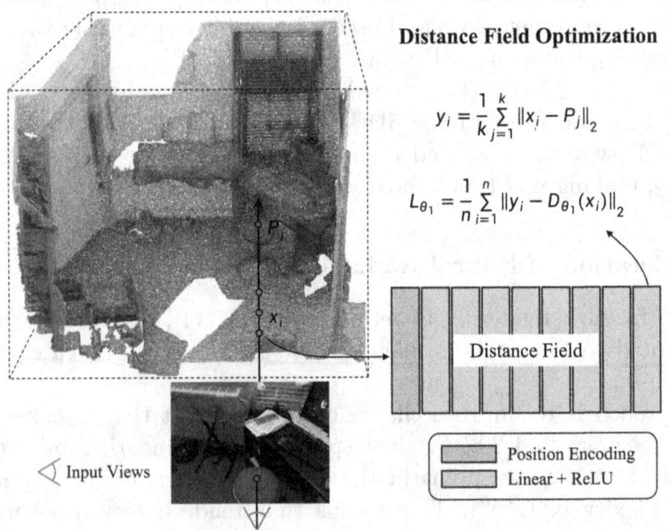

Fig. 2. Overview of the training distance field scene representation. We feed position coordinates to the MLP by sampling position coordinates along the camera ray to predict distance values to the nearest surface. We can optimize our scene representation by minimizing the residuals between the distance values predicted by the network and the ground truth distance values.

We train the distance fields through the point cloud as an implicit representation of the scene, expressing the continuous scene as a 3D vector-valued function, where the input is the coordinate (x, y, z) of a point in space, and the output is the distance d_{pre} from that point to the nearest surface. We input rays to the network and sample uniformly distributed points along these rays. The spatial coordinates of these sampled points are denoted as $\mathbf{x} = \{\mathbf{x_i} | i = 1, ..., n\}$. Using an MLP network, we approximate this continuous 3D scene function D_{θ_1} and optimize its weights θ_1 to map the coordinates of points in continuous space to unsigned distance values representing their distances to the nearest surface. Figure 2 visualises the pipeline of the training distance field.

We optimize our scene representation by minimizing the residuals between the distance values predicted by the network and the ground truth distance values. This is achieved through the following steps:

$$\mathcal{L}_{\theta_1}(\mathbf{x}, \mathbf{P}) = \frac{1}{n} \sum_{i=1}^{n} |y_i - D_{\theta_1}(\mathbf{x_i})| \tag{1}$$

$$y_i = \frac{1}{k} \sum_{j=1}^{k} \sqrt{\mathbf{x_i} - P_j} \tag{2}$$

where y_i represents the ground truth distance from the sampled point $\mathbf{x_i}$ to the nearest surface of the scene. To obtain the ground truth distance, we utilize the KD-Tree KNN algorithm to calculate the k nearest neighbors to the sampled point $\mathbf{x_i}$ within the point cloud. We then compute the average Euclidean distance from that point to its nearest neighbors, which serves as the ground truth distance from the point to the nearest surface of the scene.

3.2 Radiance Fields with Distance Fields

Given a set of room-scale RGB images $\{I_i\}_{i=1}^{N}$, our objective is to render high-fidelity images in real-time from unseen perspectives. Similar to NeRF [11], we represent the scene as a continuous 5D function F_{θ_2} encoded by an MLP network. By sampling a set of 3D points along rays, we use the positional coordinates $\mathbf{x} = (x, y, z)$ and their corresponding 2D viewing directions $\mathbf{d} = (\theta, \phi)$ as inputs to the neural network. The network outputs sets of color $\mathbf{c} = [r, g, b]$ and density σ for each point. These colors and densities are then accumulated using traditional volume rendering techniques to generate the final image.

$$[\mathbf{c}, \sigma] = F_{\theta_2}(\gamma(\mathbf{x}), \mathbf{d}) \tag{3}$$

To achieve efficient rendering of room-scale neural representations, we utilize distance fields to guide the training and inference of NeRF. In particular, for each pixel, we define a ray $\mathbf{r}(t) = \mathbf{o} + t\mathbf{d}$ passing through that pixel, where $\mathbf{o}$ is the camera position and $t_k \in [t_n, t_f]$ represents the sampled positions along the ray between the near and far planes. The distance field D_{θ_1} provides the distance value from each point on the ray to the nearest surface of the scene. Importantly, querying the distance field is a fast operation (Fig. 3).

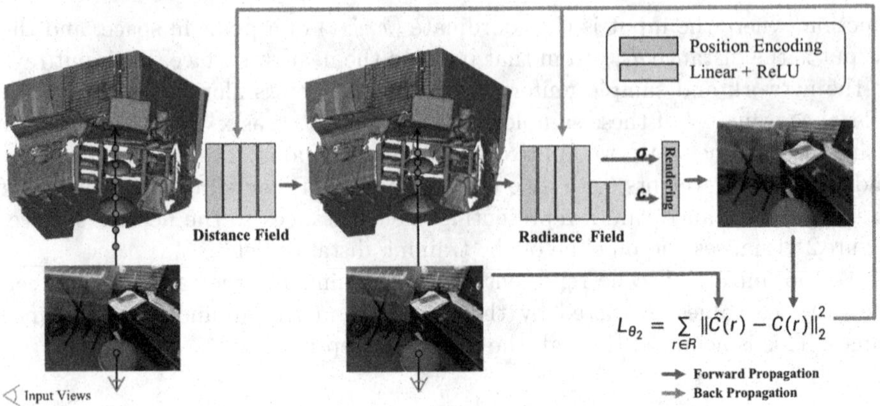

Fig. 3. An overview of distance fields guiding NeRF training and inferencing.

We further filtered out the S points with the smallest distance values, which are our ideal sample locations. These points are concentrated near the scene surface. The positions and orientations of these S points are then fed to the network.

$$\begin{aligned} \hat{\mathbf{C}}(\mathbf{r}) &= \sum_{s=1}^{S} w_s \mathbf{c}_s \\ \text{where} \quad w_s &= T_s \left(1 - \exp\left(-\sigma_s \delta_s\right)\right), \\ T_s &= \exp\left(-\sum_{s'=1}^{s} \sigma_{s'} \delta_{s'}\right), \\ \delta_s &= t_{s+1} - t_s. \end{aligned} \tag{4}$$

To optimize the network weights θ_2, we compute the residuals between the rendered image and the real image. These residuals are then used to update the network weights, enabling the network to better approximate the scene's geometry and appearance.

$$\mathcal{L}_{\theta_2} = \sum_{\mathbf{r} \in R} \|\hat{\mathbf{C}}(\mathbf{r}) - \mathbf{C}(\mathbf{r})\|_2^2 \tag{5}$$

In addition to predicting the color of each pixel, we also calculate the depth of each ray to further optimize the ability of the distance field to represent the scene. This is particularly useful when dealing with raw point cloud that may contain voids or missing information.

By incorporating depth information, we can enhance the accuracy and completeness of the distance field representation. Depth values provide additional cues about the scene's geometry, allowing us to better estimate the distances to the nearest surface.

Integrating depth information into the optimization process helps refine the distance field and improves its ability to characterize the scene accurately. This, in turn, enhances the rendering quality and fidelity of the generated images.

$$\hat{d}(\mathbf{r}) = \sum_{s=1}^{S} w_s t_s \tag{6}$$

$$\hat{d}(\mathbf{s}) = \left| \sum_{s'=1}^{s} w_{s'} t_{s'} - \hat{d}(\mathbf{r}) \right| \tag{7}$$

$$\mathcal{L}_{\theta_1} = \sum_{\mathbf{s} \in S} \| \hat{d}(\mathbf{s}) - D_{\theta_1}(\mathbf{s}) \|_2^2 \tag{8}$$

4 Experiments

4.1 Implementation Details

Training Distance Fields: The distance field is modeled as an MLP network consisting of 8 fully connected layers with 256 channels each and activated using RELU. For the input position coordinates, the position encoding $\gamma(\mathbf{x})$ is applied, and we use a position encoding function referenced to NeRF [11], which maps the input to a higher dimensional space and allows a better fit to data containing high frequency variations. Among them, we set $L = 4$:

$$\gamma(p) = \left(\sin\left(2^0 \pi p\right), \cos\left(2^0 \pi p\right), \cdots, \sin\left(2^{L-1} \pi p\right), \cos\left(2^{L-1} \pi p\right)\right) \tag{9}$$

To determine the ground truth distance of each point to the scene surface, a KD-Tree KNN (K-Nearest Neighbors) algorithm is employed. When performing the KNN search, we set the value of k to 24. This means that for each point, the algorithm identifies the 24 nearest neighbors in the point cloud, and the average Euclidean distance from the point to these neighbors is considered as the ground truth distance from that point to the nearest surface of the scene.

Training Neural Radiance Fields: For the input position coordinates and observation directions, position encoding is also applied as Eq. 9. The location-related network is modeled using an 8-layer MLP with 256 hidden channels per layer and activated using RELU, while jump connections are applied at layer 5. While the view-related network is modeled using a 2-layer MLP with 128 hidden channels per layer. During the rendering process, eight points are sampled near the surface on each ray.

Our proposed ACFNeRF achieves real-time rendering of room-scale scenes with resolution 640 × 480 on NVIDIA RTX 3090, achieving an impressive inference speed of about 150 frames per second.

4.2 Results

In our experiments, we evaluated the performance of ACFNeRF on two datasets: NeRF synthetic 360° [11] and Scannet [2]. These datasets were chosen to evaluate the quality of rendering and inference speed of ACFNeRF in comparison to other methods. It is important to note that all the methods we compared in our experiments were without the use of caching.

Table 1. Tests on the NeRF synthetic 360° dataset compare image quality, inference speed, and memory requirements.

Method	Image Generation Quality			Speed [FPS] ↑	Memory [MB] ↓
	PSNR ↑	SSIM ↑	LPIPS ↓		
NeRF [11]	30.23	0.946	0.050	0.04	**5**
FastNeRF (no-cache) [3]	29.16	0.936	0.053	0.03	28
NeRF-SH (no-cache) [23]	31.57	0.953	0.047	14	12
JAXNeRF+ Deferred [4]	30.55	0.952	0.049	0.01	18
KiloNeRF [14]	31.00	0.920	0.060	50	160
AutoInt (8 sections) [8]	25.55	0.911	0.170	0.6	**5**
DIVeR32 (RT) [20]	**32.12**	**0.958**	**0.033**	47	68
ours	30.17	0.938	0.051	**142**	7

To evaluate the rendering quality, we compare the outputs to ground truth using Peak Signal to Noise Ratio (PSNR), Structural Similarity (SSIM) [19] and perceptual LPIPS [24]. To evaluate the inference speed, we calculate the number of frames per second (FPS) generated.

The NeRF synthetic 360° dataset consists of 360-degree views of complex objects. Although it does not represent room-scale scenes, it is commonly used as a benchmark to compare methods that aim to improve the inference speed of NeRF. The images were rendered at a resolution of 800 × 800, following previous work. The experimental results, presented in Table 1, demonstrate that

Table 2. Quantitative results on Scannet, rendering 640 × 480 resolution images.

Method	scene-710		scene-758		scene-781	
	PSNRs↑	Speed [FPS]↑	PSNR↑	Speed [FPS]↑	PSNR↑	Speed [FPS]↑
NeRF [11]	11.42	0.03	11.31	0.03	11.51	0.03
Point-NeRF [21]	**26.36**	0.51	**26.73**	0.37	**26.20**	0.46
Depth Priors [15]	21.03	0.19	21.36	0.11	21.47	0.11
DONeRF [12]	15.54	19.49	15.97	18.96	16.60	18.57
ours	24.64	**148.85**	25.23	**147.70**	25.52	**147.56**

ACFNeRF achieves a significant speedup in inference while maintaining high-quality rendering. Additionally, ACFNeRF's network weights require only 7M of memory.

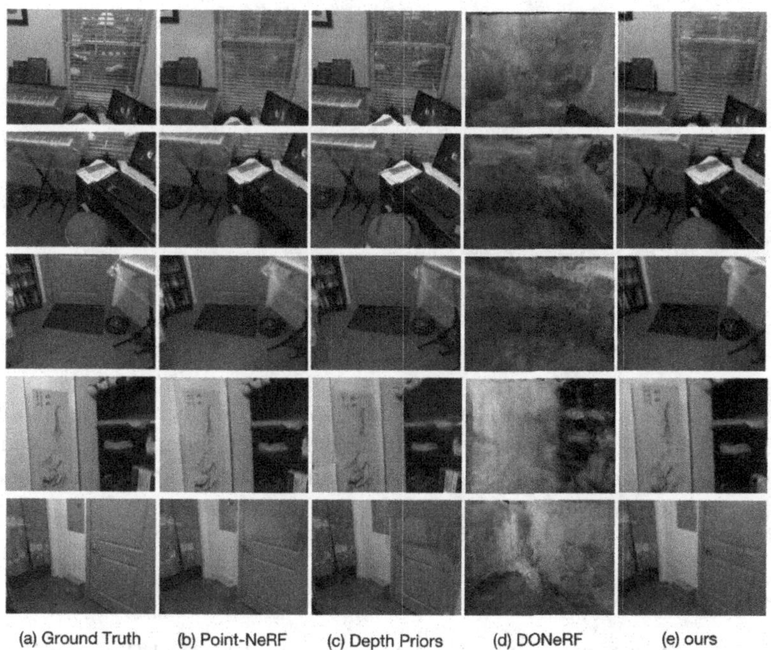

Fig. 4. Qualitative results on Scannet scene-710.

Scannet is a room-scale scene dataset. Specifically, we selected three test scenes (710, 758, and 781) from Scannet, each containing 150–200 images. We chose these three scenarios to follow the setup of previous work so that we can compare them fairly. The quantitative results of our comparison with other methods, namely PointNeRF [21], Dense Depth Priors for NeRF [15], and DONeRF [12], are presented in Table 2. These methods also aim to improve efficiency by incorporating external geometric data without caching. The experimental results demonstrate that ACFNeRF achieves competitive image quality while significantly enhancing the inference speed. For a qualitative comparison, please refer to Fig. 4.

Overall, the experimental results validate the effectiveness of ACFNeRF in terms of both rendering quality and inference speed, showcasing its competitiveness and advancements in the field of neural scene representation and rendering.

4.3 Ablation Studies

We provide an ablation study to explore the relationship between the number of sampled points, the running speed and reconstruction quality. Our ablation experiments were performed on the Scannet scene-710 (See Table 3).

Table 3. Ablation of the relationship between the number of sampled points, the running speed and reconstruction quality.

Samples per ray	PSNR ↑	Speed [FPS] ↑
2	20.97	182.35
4	22.50	168.42
8	24.64	148.85
16	25.02	86.10

From Table 3, we can see that too few points will degrade the quality of the rendered image, while too many points will affect the inference speed. We consider the balance between image quality and inference speed, and choose 8 points per ray for inference.

5 Conclusion

In this paper, ACFNeRF is introduced as a method for real-time rendering of room-scale scenes without using caching, rendering realistic images at nearly 150 Hz on consumer hardware. By incorporating point cloud to train distance fields, ACFNeRF takes full advantage of the implicit representation of the distance fields: fast and memory-efficient. Although our approach adds training volume, it represents a significant step towards enabling real-time neural rendering.

References

1. Chang, Y., Guo-Ping, W.: A review on image-based rendering. Virtual Real. Intell. Hardw. **1**(1), 39–54 (2019)
2. Dai, A., Chang, A.X., Savva, M., Halber, M., Funkhouser, T., Nießner, M.: ScanNet: richly-annotated 3d reconstructions of indoor scenes. In: Proceedings of the IEEE Conference on Computer Vision and Pattern Recognition, pp. 5828–5839 (2017)
3. Garbin, S.J., Kowalski, M., Johnson, M., Shotton, J., Valentin, J.: FastNeRF: high-fidelity neural rendering at 200fps. In: Proceedings of the IEEE/CVF International Conference on Computer Vision, pp. 14346–14355 (2021)

4. Hedman, P., Srinivasan, P.P., Mildenhall, B., Barron, J.T., Debevec, P.: Baking neural radiance fields for real-time view synthesis. In: Proceedings of the IEEE/CVF International Conference on Computer Vision, pp. 5875–5884 (2021)
5. Huang, P.H., Matzen, K., Kopf, J., Ahuja, N., Huang, J.B.: DeepMVS: learning multi-view stereopsis. In: Proceedings of the IEEE Conference on Computer Vision and Pattern Recognition, pp. 2821–2830 (2018)
6. Jiang, Y., Ji, D., Han, Z., Zwicker, M.: SDFDiff: differentiable rendering of signed distance fields for 3d shape optimization. In: Proceedings of the IEEE/CVF Conference on Computer Vision and Pattern Recognition, pp. 1251–1261 (2020)
7. Karras, T., Laine, S., Aila, T.: A style-based generator architecture for generative adversarial networks. In: Proceedings of the IEEE/CVF Conference on Computer Vision and Pattern Recognition, pp. 4401–4410 (2019)
8. Lindell, D.B., Martel, J.N., Wetzstein, G.: AutoInt: automatic integration for fast neural volume rendering. In: Proceedings of the IEEE/CVF Conference on Computer Vision and Pattern Recognition, pp. 14556–14565 (2021)
9. Liu, L., Gu, J., Zaw Lin, K., Chua, T.S., Theobalt, C.: Neural sparse voxel fields. Adv. Neural. Inf. Process. Syst. **33**, 15651–15663 (2020)
10. Mescheder, L., Oechsle, M., Niemeyer, M., Nowozin, S., Geiger, A.: Occupancy networks: learning 3d reconstruction in function space. In: Proceedings of the IEEE/CVF Conference on Computer Vision and Pattern Recognition, pp. 4460–4470 (2019)
11. Mildenhall, B., Srinivasan, P.P., Tancik, M., Barron, J.T., Ramamoorthi, R., Ng, R.: NeRF: representing scenes as neural radiance fields for view synthesis. Commun. ACM **65**(1), 99–106 (2021)
12. Neff, T., et al.: DONeRF: towards real-time rendering of compact neural radiance fields using depth oracle networks. In: Computer Graphics Forum, vol. 40, pp. 45–59. Wiley Online Library (2021)
13. Park, J.J., Florence, P., Straub, J., Newcombe, R., Lovegrove, S.: DeepSDF: learning continuous signed distance functions for shape representation. In: Proceedings of the IEEE/CVF Conference on Computer Vision and Pattern Recognition, pp. 165–174 (2019)
14. Reiser, C., Peng, S., Liao, Y., Geiger, A.: KiloNeRF: speeding up neural radiance fields with thousands of tiny MLPs. In: Proceedings of the IEEE/CVF International Conference on Computer Vision, pp. 14335–14345 (2021)
15. Roessle, B., Barron, J.T., Mildenhall, B., Srinivasan, P.P., Nießner, M.: Dense depth priors for neural radiance fields from sparse input views. In: Proceedings of the IEEE/CVF Conference on Computer Vision and Pattern Recognition, pp. 12892–12901 (2022)
16. Sitzmann, V., Thies, J., Heide, F., Nießner, M., Wetzstein, G., Zollhofer, M.: DeepVoxels: learning persistent 3d feature embeddings. In: Proceedings of the IEEE/CVF Conference on Computer Vision and Pattern Recognition, pp. 2437–2446 (2019)
17. Sitzmann, V., Zollhöfer, M., Wetzstein, G.: Scene representation networks: continuous 3d-structure-aware neural scene representations. Adv. Neural Inf. Process. Syst. **32** (2019)
18. Wang, J., Sun, B., Lu, Y.: MVPNet: multi-view point regression networks for 3d object reconstruction from a single image. In: Proceedings of the AAAI Conference on Artificial Intelligence, vol. 33, pp. 8949–8956 (2019)
19. Wang, Z., Bovik, A.C., Sheikh, H.R., Simoncelli, E.P.: Image quality assessment: from error visibility to structural similarity. IEEE Trans. Image Process. **13**(4), 600–612 (2004)

20. Wu, L., Lee, J.Y., Bhattad, A., Wang, Y.X., Forsyth, D.: DIVeR: real-time and accurate neural radiance fields with deterministic integration for volume rendering. In: Proceedings of the IEEE/CVF Conference on Computer Vision and Pattern Recognition, pp. 16200–16209 (2022)
21. Xu, Q., et al.: Point-NeRF: point-based neural radiance fields. In: Proceedings of the IEEE/CVF Conference on Computer Vision and Pattern Recognition, pp. 5438–5448 (2022)
22. Yao, Y., Luo, Z., Li, S., Fang, T., Quan, L.: MVSNet: depth inference for unstructured multi-view stereo. In: Ferrari, V., Hebert, M., Sminchisescu, C., Weiss, Y. (eds.) ECCV 2018. LNCS, vol. 11212, pp. 785–801. Springer, Cham (2018). https://doi.org/10.1007/978-3-030-01237-3_47
23. Yu, A., Li, R., Tancik, M., Li, H., Ng, R., Kanazawa, A.: PlenOctrees for real-time rendering of neural radiance fields. In: Proceedings of the IEEE/CVF International Conference on Computer Vision, pp. 5752–5761 (2021)
24. Zhang, R., Isola, P., Efros, A.A., Shechtman, E., Wang, O.: The unreasonable effectiveness of deep features as a perceptual metric. In: Proceedings of the IEEE Conference on Computer Vision and Pattern Recognition, pp. 586–595 (2018)

OctPCGC-Net: Learning Octree-Structured Context Entropy Model for Point Cloud Geometry Compression

Xinjie Wang[1], Hanyun Wang[2](✉), Ke Xu[1], Jianwei Wan[1], and Yulan Guo[1]

[1] College of Electronic Science and Technology, National University of Defense Technology, Changsha, China
{wangxinjie,xuke,wanjianwei,yulan.guo}@nudt.edu.cn

[2] School of Surveying and Mapping, Information Engineering University, Zhengzhou, China
why.scholar@gmail.com

Abstract. In Point Cloud Geometry Compression (PCGC), an accurate context entropy model is necessary to reduce spatial redundancy. The octree-based auto-regressive context entropy model has great potential to explore large-scale context dependency. However, over-concentrated attention maps and instability of training process usually occur in large-scale context entropy models. To address these problems, we propose a novel OctPCGC-Net for PCGC based on deep learning framework. Specifically, we introduce a scaled cosine attention method in a large-scale context entropy model to alleviate the problem of over-concentrated attention maps caused by self-attention mechanism, thereby improving the model's prediction accuracy. In order to improve the stability of model training, we further introduce a residual post normalization strategy to alleviate the phenomenon of accumulating activation scores as the network deepens, which makes the activation scores of different layers smoother and more stable. Experimental results show that compared with the state-of-the-art large-scale auto-regressive entropy models, our method saves 6.3%, 8.7%, and 6.3% bitrates in terms of Bjøntegaard Delta Bit Rate (BDBR) on benchmark datasets SemanticKITTI, 8iVFB, and Owlii, respectively. Additionally, our method also achieves higher reconstruction quality (D1 PSNR) and smaller Chamfer distance (CD) under similar bits per point (BPP) on SemanticKITTI dataset.

Keywords: Point cloud geometry compression · Octree context · Entropy model · Scaled cosine attention

1 Introduction

Point cloud is a crucial data structure for representing 3D scenes and is widely utilized in 3D vision systems like autonomous driving and immersive applications. The amount of points obtained per second by laser scanners usually contain millions of points, and such a massive amount of data makes it difficult to store and transmission. Therefore, developing efficient point cloud compression techniques can reduce memory footprints and transmission bandwidth and is also necessary for practical point cloud applications.

Q. Liu et al. (Eds.): PRCV 2023, LNCS 14426, pp. 343–355, 2024.
https://doi.org/10.1007/978-981-99-8432-9_28

Recently, deep learning methods have promoted the development of point cloud compression [1–9]. The typical approach involves training an octree-based entropy model to estimate the occupancy symbol distributions for each octree node. The point cloud is first organized as an octree, and then the occupancy symbols are encoded losslessly into a bitstream using an entropy coder such as the arithmetic coder [10]. An accurate entropy model is crucial to reduce the cross entropy between the estimated and ground truth distribution corresponding to actual bitrates. Many researchers have tried to improve accuracy by designing context structures [1–3, 6] that can increase capacity and introduce references from high-resolution octree representations. Notably, the recently proposed OctAttention method [6] incorporates context from hundreds of previously decoded sibling nodes (at the same octree level) and their ancestor nodes. In designing large-scale context entropy models based on octree, adding sibling nodes to expand the context can provide more information for reasoning. However, some nodes in the context are redundant. A practical approach uses a self-attention mechanism to find similarities and strong dependencies between nodes, filter out irrelevant nodes, and emphasize valid nodes [6]. The OctAttention employs the dot product to compute the similarity between two nodes, which is implemented by taking the dot product of the features of the two nodes and dividing it by the square root of the feature dimension. However, the attention mechanism based on dot product results in the attention map frequently concentrating on only a few nodes.

To address this issue, inspired by the work [26], we propose to replace the dot product attention with the scaled cosine attention, which computes the cosine similarity between two nodes and then divides it by a learned scalar. Scaled cosine attention makes the attention scores independent of the amplitude of the input features and less likely to be in extreme cases. Using cosine similarity as a metric ensures that the similarity between two nodes falls within the range of 0 to 1, and the attention map does not suffer from over-concentration issues since the similarity values are distributed on a uniform scale. In subsequent ablation studies, it will be shown that scaled cosine attention can further enhance the model's prediction accuracy.

On the other hand, directly adding the output of residual units back to the main branch significantly increases the inconsistency of cross-layer activation amplitude [26]. As a result, activation values accumulated layer by layer significantly increase in deeper layers compared to early layers. To alleviate this issue, we introduce a new normalization method called residual post normalization in the octree contextual entropy model. Precisely, we move the layer normalization (LN) layer from the beginning of each residual unit to the end, providing gentler activation values for the entire network layer. Through post-normalization, the output of the attention or MLP is added to the shortcut after being layer normalized. The residual post normalization layer minimizes the accumulation of activation scores as the network gets deeper, and the activation scores of different layers become smoother and more stable.

We evaluate the proposed model on three different point cloud compression datasets: SemanticKITTI [11], Owlii [12], and 8iVFB [13]. The comparison results with state-of-the-art methods demonstrate the superiority of the proposed method. In summary, our main contributions are summarized as follows:

- We introduce the scaled cosine attention mechanism and residual post normalization layer into the octree context entropy model, which improves the context model's training stability and the occupied nodes' prediction accuracy.
- The proposed approach achieves lower BPP costs with equal reconstruction quality and superior lossy/lossless compression performance while maintaining a practical coding speed suitable for real-world applications.

2 Related Work

2.1 Traditional Point Cloud Compression

In recent years, tree structure has been adopted as an essential representation for point cloud compression [14, 20, 21]. The pioneering work is the G-PCC (geometry-based point cloud compression) method developed by the MPEG group [14], which employs an octree-structure for static point cloud compression. In Google's open-source compression software Draco [27], a KD-tree structure [28] is used to compress point clouds. However, these methods are all based on hand-crafted techniques and cannot be optimized end-to-end.

2.2 Deep Learning-Based Point Cloud Compression

Several data structures and models have been developed to improve the rate-distortion performance for deep learning-based point cloud compression. Pioneering works such as [7, 15, 16] create analysis and synthesis transformations for converting point clouds into latent representations. These transformations are learned directly from point clouds using the PointNet backbone [17, 18] or on voxel structures utilizing 3D convolution. However, preserving high-frequency information in latent vectors is challenging since distortions are inevitable during these transformations. Therefore, these structures are more suitable for lossy compression at low bitrates.

Learning an entropy model is another effective pipeline for reducing bitrates without introducing distortion. Building an informative context is crucial as the inference of the entropy model relies on the context. A typical entropy model predicts the binary occupancy status of the voxels based on the context of several adjacent voxels [4]. Similarly, an entropy model can be predicted on the octree by estimating occupancy symbol distributions. Early tree-structured entropy models leverage contexts from ancestor nodes [1–3]. Later works introduce adjacent voxels with the exact resolution as the currently coded octree node to provide high-resolution contexts [19]. Recently, a novel approach replaced neighboring voxels with sibling and ancestor nodes to extend the contexts [6], yielding state-of-the-art performance. However, its compression performance is limited by the expensive dot product self-attention mechanism.

2.3 Transformer in Point Cloud Processing Tasks

In recent years, Transformer has achieved great success in Natural Language Processing (NLP), Computer Vision (CV), and other fields. Different from the traditional approach of Convolutional Neural Networks (CNNs) and Recurrent Neural Networks (RNNs),

Transformer is a neural network architecture based on attention mechanisms. In the field of NLP, its core idea is to capture critical information in sentences through Multi-head Self-Attention (MSA) operation and residual connections, which can better capture long-distance dependencies and reduce the training difficulty of RNNs. As a result, it can achieve high-quality sentence-level representation learning.

Transformer models have attracted widespread attention in 3D point cloud processing and achieved impressive results in various 3D tasks. Point Transformer [22] is an advanced point cloud semantic segmentation framework based on Transformer and utilizes vector attention instead of self-attention to reduce computation complexity. Point Cloud Transformer [23] extracts neighborhood relationships within the point cloud using a neighborhood feature extraction module. It introduces an offset attention mechanism to calculate the offset between the input and output of the self-attention module.

3 The Proposed Approach

3.1 A Brief Review of OctAttention

OctAttention efficiently compresses sparse and dense point cloud geometry information by utilizing large-scale context based on octree structure [6]. The overall framework of OctAttention is shown in Fig. 1. Specifically, the input point cloud is first encoded into an octree structure, with each octree node characterized by distinct features such as occupancy code, level, and octant. In order to capture dependencies between neighboring nodes, OctAttention constructs a context window (shown in red) consisting of 3 layers of ancestors (shown in green). During the encoding process, a node (shown in purple) is processed by feeding the context window into a masked context attention module. The module learns to model the occupancy code distribution using a multi-layer perceptron (MLP). Finally, an arithmetic encoder uses the estimated distribution to encode the serialized occupancy code into a final compressed bit stream.

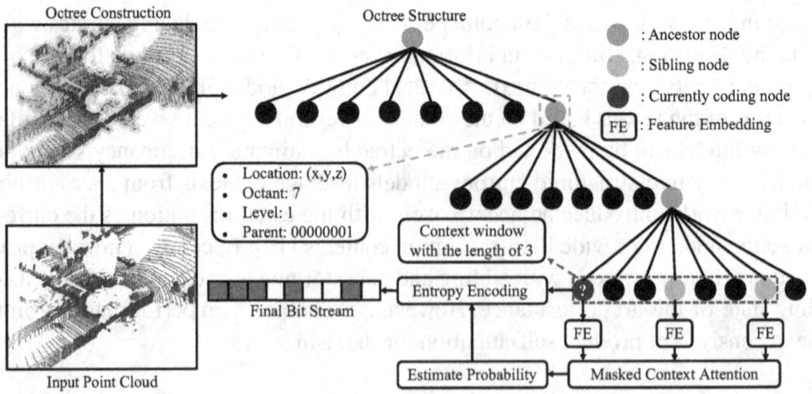

Fig. 1. The overview of the OctAttention. (Color figure online)

The quality of compression performance mainly depends on the accuracy of the node occupancy prediction by the Transformer to extract the features of parent and child nodes. Under the same reconstruction quality, that is, the same octree depth, if the octree node occupancy prediction is more accurate, the 'Final Bit Stream' in the network will be shorter, the number of bits spent will be less, and the compression performance will be better.

3.2 OctPCGC-Net

Similar to the approach in [6], we utilize the features from sibling nodes and their ancestors to model a large-scale context. To explore the dependencies between different nodes in a large-scale context, we adopt the attention mechanism [25] to model the dependencies among the nodes. The similarity between two nodes is generally computed using a scaled dot product in the original self-attention mechanism. However, when the number of parameters is increased in the context entropy model, the learned attention maps are often dominated by a few nodes, leading to over-concentrated attention maps and an unstable training process [26].

On the other hand, most vision Transformers adopt a layer normalization (LN) layer at the beginning of each block. [26] points out that when enlarging the model capacity, a significant increase in activation values is observed in deeper layers. In the pre-normalization configuration, the output activation values of each residual block are directly merged back to the main branch, and the amplitude of the main branch becomes increasingly large in deeper layers. The significant differences in amplitude between different layers lead to an unstable training process.

In order to improve the stability of the training of the large-scale context entropy model, inspired by [26], we propose to utilize the scaled cosine attention and residual post normalization to replace the dot-product attention and the pre-normalization used in the original transformer. The differences in attention mechanism between the proposed method and the original OctAttention are shown in Fig. 2.

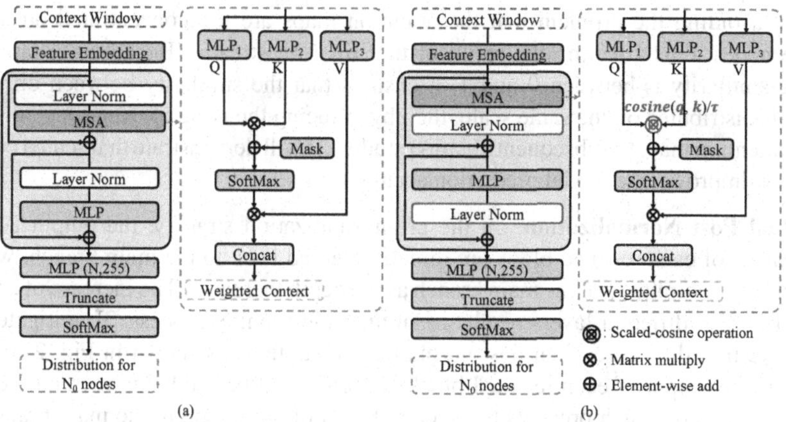

Fig. 2. Comparisons of attention mechanism between the proposed method and the original OctAttention. (a) Dot-product MSA in [6]. (b) Scaled-cosine MSA in our approach.

Similar to [6], we first employ an embedding layer to increase the feature dimension. Then we introduce a Scaled Cosine Attention and Residual Post Normalization operation to extract features for every node in Transformer layer. The OctPCGC-Net block can be formally defined as:

$$X_1 = LN(MSA(FE)) + FE \tag{1}$$

$$X_2 = LN(MLP(X_1)) + X_1 \tag{2}$$

where FE is the output of the Feature Embedding layer. X_1 denotes the input before the first MSA layer. X_2 denotes the outputs of the Transformer module.

Finally, the context is fed into three layers of MSA and MLP and then outputs a 255-dimensional probability prediction vector for the current encoding node n_i:

$$\tilde{p}_i(\cdot|\boldsymbol{f}_i; \mathbf{w}) = SoftMax(MLP(MutiHead^{(3)}(\boldsymbol{f}_i, i))) \tag{3}$$

where high-dimensional vector $\boldsymbol{f}_i$ denotes the concatenation of embedding features of sibling nodes and features of its ancestors. $\mathbf{w}$ is the weight of the context entropy model.

Scaled Cosine Attention. The original self-attention uses a scaled dot product to calculate the similarity between two nodes, defined as the dot product of the two nodes' features divided by the feature dimension's square root. However, as described before, this kind of attention causes the attention maps to be dominated by a few octree nodes and makes it difficult to capture more helpful information. To address this issue, we leverage scaled cosine attention, which calculates the cosine similarity between two nodes and then divides it by a learnable scalar τ:

$$sim(q_i, k_j) = \cos(q_i, k_j)/\tau \tag{4}$$

where τ is a learnable scalar, non-shared across heads and layers, usually set larger than 0.01. The cosine function is naturally normalized and thus can have milder attention values, avoiding the problem that the attention maps are too concentrated on a few octree nodes, thus affecting the result of the 'final bit stream'. Because the range of cosine similarity is between 0 and 1, it ensures that the similarity between different nodes is distributed on the same scale, thereby reducing the issue of over-concentration in the attention map. In subsequent ablation studies, it will demonstrate that scaled cosine attention improves the model prediction accuracy.

Residual Post Normalization. In the pre-normalization strategy, the output activation values of each residual block are directly merged back to the main branch, which increases the amplitude of the main branch as the depth increases. The increase of amplitudes between different layers results in an unstable training process. To mitigate this issue, we introduce a residual post normalization strategy, as shown in Fig. 2. In this strategy, the output of each residual block is normalized before being merged back to the main branch, which prevents the accumulation of amplitude in the main branch as the depth increases. Therefore, in the training stage of the proposed model, an additional layer normalization (LN) operation is introduced for the main branch of each Transformer block to stabilize the training process further.

3.3 Learning

The optimization objective of the octree-based entropy model is to minimize the total bitrates that are required to communicate the octree nodes. This bitrate is determined by calculating the cross-entropy between the estimated distribution $\tilde{p}(x)$ and the ground truth distribution $p(x)$ at each node:

$$\ell = -\sum_{i} \log \tilde{p}_i\left(x_i | \boldsymbol{f}_i; \mathbf{w}\right) \tag{5}$$

where $\tilde{p}_i\left(x_i | \boldsymbol{f}_i; \mathbf{w}\right)$ is the estimated distribution by the proposed context entropy model.

4 Experiments

4.1 Experimental Settings

Sparse LiDAR Point Cloud Dataset. SemanticKITTI [11] is a high-volume LiDAR Point cloud dataset widely employed in autonomous driving applications. Collected using a Velodyne HDL-64E sensor, the dataset includes 43552 scans comprising a staggering 4549 million points. Following the principle proposed in OctAttention [6], a normalization step is performed by setting the raw data to $[-1, 1]^3$ as a reference point cloud. Sequences 00 to 10 are utilized to train the model, and some sequences are randomly selected from sequences 11 to 21 to evaluate the model.

Dense Object Point Cloud Dataset. Owlii [12] is a dynamic human mesh dataset that contains four dynamic Full Bodies subject sequences. The dataset offers both 11-bit precisions. Microsoft Voxelized Upper Bodies (MVUB) [24] is a dynamic voxelized point cloud dataset containing five half-body subject sequences with 9-bit and 10-bit precision. Similarly, the MPEG 8i Voxelized Full Bodies (8iVFB) [13] dataset provides complete human shape point clouds with 10-bit and 12-bit precision. The training phase, consistent with the OctAttention [6] methodology, focuses on the Andrew10, David10, and Sarah10 sequences from MVUB and the Soldier10 and Longdress10 sequences from 8iVFB. We select various point cloud sequences with different resolutions for the testing stage, including Basketballplayer11, Dancer11, Exercise11, and Model11 from the Owlii dataset, as well as Boxer9/10, Thaidancer9/10, Loot10, and Redandblack10 from the 8iVFB dataset.

Baselines. To demonstrate the efficacy of the proposed entropy model, we conduct comparison experiments with OctAttention [6]. We reproduce the results of the OctAttention using its publicly accessible source code. Additionally, to provide a comprehensive evaluation, we compare our approach with the well-known hand-crafted compression method MPEG G-PCC (TMC13 v14.0) [14].

Training and Testing Strategy. For LiDAR dataset compression, we train a single model with a max octree depth of 12, as this allows our entropy model to learn the distribution of all levels in one model. We truncate the octree over $8-12$ levels during testing to evaluate our entropy model at different bitrates. We train a single model for object dataset compression using the octree sequence data converted from point clouds at depths 9 and 10. All experiments are conducted on an Intel I5-11600 CPU and one NVIDIA 2080TI GPU. To ensure a fair comparison and demonstrate the superiority of our proposed method, we follow the training settings of OctAttention, using a batch size of 32, 8 epochs, three layers, and four heads in MSA, and the Adam optimizer with a learning rate of 1e−3. It takes one day to train the model on Object Dataset and two days on LiDAR Dataset.

Evaluation Metrics. For LiDAR dataset compression, following MPEG standards [14], we utilize the point-to-point PSNR (D1 PSNR) and Chamfer distance (CD) to measure lossy geometry compression distortion. We follow the official default configuration in TMC13 for object dataset compression to measure the lossless geometry compression performance with bits per point (BPP). To compare different compression methods in sparse and dense point clouds, we adopt Bjøntegaard Delta Bit Rate (BDBR) [29] as the Evaluation Metric, Representing the Average Bitrate Reductions When These Methods' Reconstructed Quality (e.g., D1 PSNR) is the Same. Unless Otherwise Specified, All Bitrates Are Obtained by Averaging Over All Sequences. Note that BDBR is a Relative Metric for Lossy/lossless Compression. A Negative BDBR Indicates a Better Test Method Performance Than the Baseline Method.

4.2 Experiment Results

Results for Sparse LiDAR Point Cloud Compression. The Rate-Distortion (RD) curves of LiDAR compression are shown in Fig. 3. Because the deeper octree (D = 12) obtains finer leaf nodes and finds accurate geometric information, it means that the quality of the reconstructed point cloud depends on the maximum depth of the octree, so the meaning of different points refers to the bit overhead at different octree depths. We present a visual comparison of compression distortions among different methods at similar bitrates in Fig. 4. The results demonstrate that our method obtains superior reconstruction performance compared with other baselines. Our method obtains higher D1 PSNR and lower CD under similar BPP. For example, OctPCGC-Net outperforms G-PCC by roughly 3 dB D1 PSNR improvements. Furthermore, Table 1 shows the BDBR gains achieved at the same octree depth, and we can observe that our method outperforms other baselines at all bitrates. On average, our approach achieves −26.50% and −6.34% BDBR gains on SemanticKITTI compared with G-PCC and OctAttention, respectively. In addition, from Table 1 we can observe that our method achieves the highest BDBR gain when truncating the octree at low resolutions (e.g., D = 8). This is because our method avoids the over-concentration of attention maps dominated by a few nodes. As a result, it can capture more contextual node information.

Results for Dense Object Point Cloud Compression. Table 2 presents the BPP and BDBR of the lossless compression on object point clouds. Our proposed method outperforms OctAttention (saving 7.48% bitrates in terms of BDBR) and yields an average BDBR gain of −37.20% over G-PCC. It is easy to observe from Table 2 that our method achieves the highest BDBR gain for the same human point clouds with lower geometric accuracy (e.g., Boxer9 and Thaidancer9). This is due to the sparser distribution of points in lower geometric accuracy point clouds compared to high geometric accuracy point

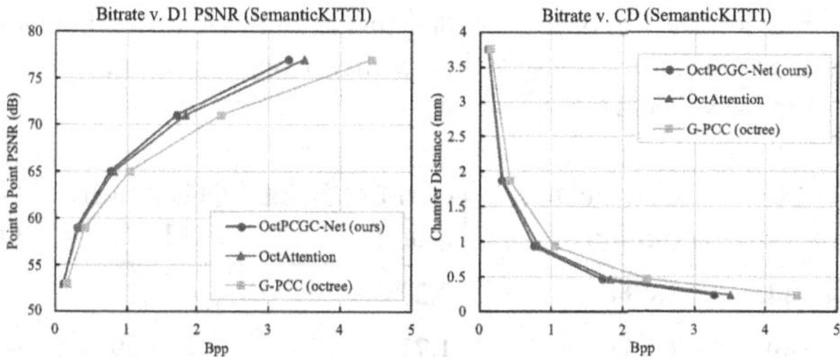

Fig. 3. Rate-distortion (RD) performance of different methods on SemanticKITTI.

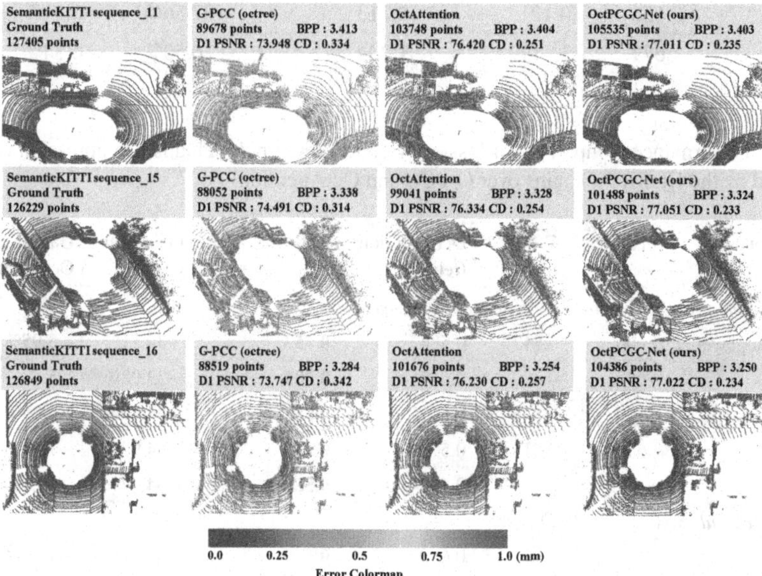

Fig. 4. Visualized compression results of G-PCC (octree), OctAttention, and our method on the SemanticKITTI dataset.

clouds. Our method avoids the issue of over-concentrated attention maps and allows it to capture more contextual node information.

4.3 Ablation Study and Analysis

We conduct the ablation study to demonstrate the effectiveness of the proposed residual post normalization and scaled cosine attention approaches based on the Transformer backbone. From Table 3, it can be seen that under the same Transformer, the compression performance (BPP) is gradually improved by adding two methods step by step, which proves that both the residual normalization and cosine attention methods can improve the output results of the network. The simultaneous addition can maximize the improvement on sparse LiDAR point clouds and dense human body point clouds. The results further

Table 1. Performance comparison of lossy coded sparse point clouds on SemanticKITTI. Each octree depth represents the same reconstruction quality. Compression ratio is measured by the BDBR (%) gains over G-PCC and OctAttention.

Max Octree Depth D	G-PCC (octree)	OctAttention (retrain)	OctPCGC-Net (ours)	Gain over G-PCC	Gain over OctAttention
12	4.444	3.494	**3.277**	−26.26%	−6.21%
11	2.339	1.833	**1.717**	−26.59%	−6.33%
10	1.042	0.813	**0.761**	−26.97%	−6.40%
9	0.414	0.324	**0.304**	−26.57%	−6.17%
8	0.156	0.121	**0.113**	−27.56%	−6.61%
Average	1.679	1.317	**1.234**	−26.50%	−6.34%

Table 2. Performance comparison of losslessly coded dense point clouds. Compression ratio is measured by the BDBR (%) gains over G-PCC and OctAttention.

Point Clouds	Frames	G-PCC (octree)	OctAttention (retrain)	OctPCGC-Net (ours)	Gain over G-PCC	Gain over OctAttention
Owlii Dynamic Human Textured Mesh Sequence Dataset						
Basketballplayer11	600	0.847	0.588	**0.553**	−34.71%	−5.95%
Dancer11	600	0.880	0.617	**0.581**	−33.98%	−5.83%
Exercise11	600	0.846	0.592	**0.555**	−34.40%	−6.25%
Model11	600	0.815	0.572	**0.530**	−34.97%	−7.34%
Average	–	0.847	0.592	**0.555**	−34.47%	−6.29%
8i Voxelized Full Bodies (8iVFB) Dataset						
Boxer9	1	0.957	0.610	**0.549**	−42.63%	−10.00%
Boxer10	1	0.943	0.605	**0.546**	−42.10%	−9.75%
Loot10	300	0.947	0.624	**0.573**	−39.49%	−8.17%
Redandblack10	300	1.086	0.754	**0.701**	−35.45%	−7.03%
Thaidancer9	1	0.995	0.661	**0.599**	−39.80%	−9.38%
Thaidancer10	1	0.986	0.661	**0.610**	−38.13%	−7.72%
Average	–	0.992	0.652	**0.596**	−39.92%	−8.66%

illustrate the effectiveness of the residual post normalization strategy and scaled cosine attention mechanism. Specifically, for LiDAR point clouds, compared with the original Transformer, the BDBR gains for each level from $D = 8$ to $D = 12$ are −5.0%, −6.2%, −6.4%, −6.3%, and −6.2%, respectively. For two types of human body point clouds at different geometry precision, compared with the original Transformer, the BDBR gains are −8.6% and −6.2%, respectively.

During the training, we found that the whole curve's fluctuation is smaller than that of the OctAttention network, and the training process is more stable and smooth. To emphasize the difference between the training process with to without residual post normalization and the effect of the two refinements on reconstruction quality, we show a visual comparison of the training process as shown in Fig. 5.

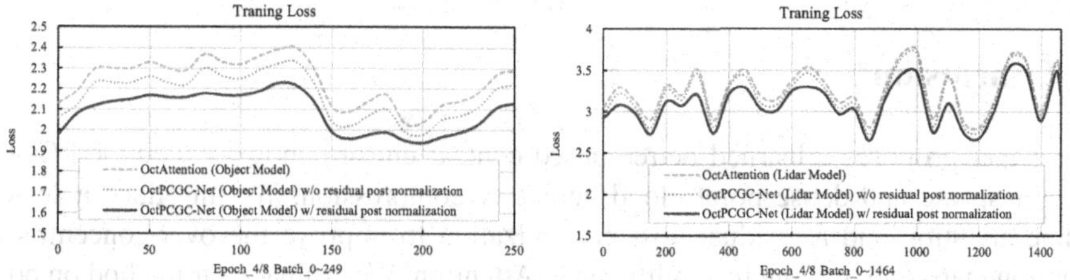

Fig. 5. OctPCGC-Net versus OctAttention in training process on two types of datasets.

Table 3. Lossy/ Lossless compression performance ablation study.

Backbone	residual post normalization	scaled cosine attention	Bpp on SemanticKITTI					Bpp on 8iVFB/Owlii
			D = 8	D = 9	D = 10	D = 11	D = 12	
Transformer			0.122	0.330	0.815	1.824	3.481	0.652/0.592
Transformer		√	0.121	0.323	0.804	1.805	3.414	0.642/0.582
Transformer	√		0.119	0.318	0.783	1.776	3.361	0.612/0.573
Transformer	√	√	**0.113**	**0.304**	**0.761**	**1.717**	**3.277**	**0.596/0.555**

Table 4. Average runtime of different methods on object point clouds.

Methods	G-PCC (octree)	G-PCC (trisoup)	OctAttention (retrain)	OctPCGC-Net (ours)
Average encoding time (s)				
Owlii_vox11	7.26	21.15	1.47	2.71
8iVFB_vox10	3.05	5.59	0.74	1.37
8iVFB_vox9	0.81	1.37	0.44	0.63
Average decoding time (s)				
Owlii_vox11	1.89	14.61	3241	3413
8iVFB_vox10	0.78	3.41	1193	1201
8iVFB_vox9	0.21	0.83	302	315

4.4 Runtime

To evaluate the efficiency of the OctPCGC-Net, we have listed the encoding and decoding times at various object point clouds in Table 4. Although the encoding time of our method is equivalent to G-PCC, the decoding time is longer than the G-PCC. This is because both the proposed method and OctAttention are based on global attention mechanisms for large-scale context modeling of long-range dependencies. As described in [6], its complexity is quadratic to the length of the context.

5 Conclusion

This paper proposes a learned octree-based context entropy model called OctPCGC-Net for sparse and dense point cloud geometry compression. We introduce a scaled cosine attention and a residual post normalization to improve the over-concentrated attention maps and training instability in OctAttention. We evaluate our method on both the LiDAR and object point cloud datasets. The results show that our method achieves better compression performance than the state-of-the-art methods on both datasets.

References

1. Biswas, S., Liu, J., Wong, K., Wang, S. L., Urtasun, R.: MuSCLE: multi sweep compression of LiDAR using deep entropy models. In: Advances in Neural Information Processing Systems, pp. 22170–22181 (2020)
2. Huang, L.L., Wang, S.L., Wong, K., Liu, J., Urtasun, R.: OctSqueeze: octree-structured entropy model for LiDAR compression. In: IEEE/CVF Conference on Computer Vision and Pattern Recognition, pp. 1313–1323 (2020)
3. Que, Z.Z., Lu, G., Xu, D.: VoxelContext-Net: an octree based framework for point cloud compression. In: IEEE/CVF Conference on Computer Vision and Pattern Recognition, pp. 6042–6051 (2021)
4. Nguyen, D.T., Quach, M., Valenzise, G., Duhamel, P.: Lossless coding of point cloud geometry using a deep generative model. IEEE Trans. Circuits Syst. Video Technol. **31**(12), 4617–4629 (2021)
5. Wang, J.Q., Zhu, H., Liu, H.J., Ma, Z.: Lossy point cloud geometry compression via end-to-end learning. IEEE Trans. Circuits Syst. Video Technol. **31**(12), 4909–4923 (2021)
6. Fu, C.Y., Li, G., Song, R., Gao, W., Liu, S.: OctAttention: Octree-based large-scale contexts model for point cloud compression. In: AAAI Conference on Artificial Intelligence, pp. 625–633 (2022)
7. He, Y., Ren, X.L., Tang, D.H., Zhang, Y.D., Xue, X.Y., Fu, Y.W.: Density-preserving deep point cloud compression. In: IEEE/CVF Conference on Computer Vision and Pattern Recognition, pp. 2333–2342 (2022)
8. Fang, G.C., Hu, Q.Y., Wang, H.Y., Xu, Y.L., Guo, Y.L.: 3DAC: learning attribute compression for point clouds. In: IEEE/CVF Conference on Computer Vision and Pattern Recognition, pp. 14799–14808 (2022)
9. Wang, J.Q., Ding, D.D., Li, Z., Feng, X.X., Cao, C.T., Ma, Z.: Sparse tensor-based multiscale representation for point cloud geometry compression. IEEE Trans. Pattern Anal. Mach. Intell. **45**(7), 9055–9071 (2023)
10. Witten, I.H., Neal, R.M., Cleary, J.G.: Arithmetic coding for data compression. Commun. ACM **30**(6), 520–540 (1987)
11. Behley, J., et al.: SemanticKITTI: a dataset for semantic scene understanding of LiDAR sequences. In: IEEE/CVF International Conference on Computer Vision, pp. 9296–9306 (2019)

12. Xu, Y., Lu, Y., Wen, Z.Y.: Owlii Dynamic human mesh sequence dataset. ISO/IEC JTC1/SC29/WG11 m41658, 120th MPEG Meeting, Macau, October (2017)
13. Eugene, D., Bob, H., Taos, M., Philip, A.C.: 8i Voxelized Full Bodies - A Voxelized Point Cloud Dataset. ISO/IEC JTC1/SC29 Joint WG11/WG1 (MPEG/JPEG) input document WG11M40059/WG1M74006 (2017)
14. Schwarz, S., et al.: Emerging MPEG standards for point cloud compression. IEEE J. Emerg. Sel. Topics Circuits Syst. **9**(1), 133–148 (2019)
15. Huang, T.X., Liu, Y.: 3D point cloud geometry compression on deep learning. In: ACM International Conference on Multimedia, pp. 890–898 (2019)
16. Quach, M., Valenzise, G., Dufaux, F.: Learning convolutional transforms for lossy point cloud geometry compression. In: IEEE International Conference on Image Processing, pp. 4320–4324 (2019)
17. Qi, C.R., Su, H., Mo, K., Guibas, L.J.: PointNet: deep learning on point sets for 3D classification and segmentation. In: IEEE/CVF Conference on Computer Vision and Pattern Recognition, pp. 652–660 (2017)
18. Qi, C.R., Yi, L., Su, H., Guibas, L.J.: PointNet++: deep hierarchical feature learning on point sets in a metric space. In: Advances in Neural Information Processing Systems, pp. 5099–5108 (2017)
19. Kaya, E.C., Tabus, I.: Neural network modeling of probabilities for coding the octree representation of point clouds. In: IEEE International Workshop on Multimedia Signal Processing, pp. 1–6 (2021)
20. Queiroz, R.L.D., Chou, P.A.: Compression of 3D point clouds using a region-adaptive hierarchical transform. IEEE Trans. Image Process. **25**(8), 3947–3956 (2016)
21. Queiroz, R.L.D., Chou, P.A.: Transform coding for point clouds using a gaussian process model. IEEE Trans. Image Process. **26**(7), 3507–3517 (2017)
22. Zhao, H.S., Jiang, L., Jia, J.Y., Torr, P.H.S., Koltun, V.: Point transformer. In: IEEE/CVF International Conference on Computer Vision, pp. 16239–16248 (2021)
23. Guo, M.H., Cai, J.X., Liu, Z.N., Mu, T.J., Martin, R.R., Hu, S.M.: PCT: point cloud transformer. Comput. Vis. Media **7**(2), 187–199 (2021)
24. Charles, L., Qin, C., Sergio, O.E., Philip, A.C.: Microsoft voxelized upper bodies - a voxelized point cloud dataset. ISO/IEC JTC1/SC29 Joint WG11/WG1 (MPEG/JPEG) input document m38673/M7201 (2016)
25. Vaswani, A., et al.: Attention is all you need. In: Advances in Neural Information Processing Systems, pp. 6000–6010 (2017)
26. Liu, Z., et al.: Swin Transformer V2: scaling up capacity and resolution. In: IEEE/CVF Conference on Computer Vision and Pattern Recognition, pp. 11999–12009 (2022)
27. Google: Draco 3d data compression (2017). https://github.com/google/draco
28. Devillers, O., Gandoin, P.M.: Geometric compression for interactive transmission. In: IEEE Visualization, pp. 319–326 (2000)
29. Bjontegaard, G.: Calculation of average PSNR differences between rd-curves. In: ITU-T SG 16/Q6, 13th VCEG Meeting. document VCEG-M33 (2001)

Multi-modal Feature Guided Detailed 3D Face Reconstruction from a Single Image

Jingting Wang, Cuican Yu, and Huibin Li(✉)

School of Mathematics and Statistics, Xi'an Jiaotong University, Xi'an, China
huibinli@xjtu.edu.cn

Abstract. Reconstructing a 3D face model with high-quality geometry and texture from a single face image is ill-conditioned and challenging. On the one hand, many methods heavily rely on a large amount of training data, which is not easy to obtain. On the other hand, position local features of a face surface can not reflect the global information of an entire face. Due to these challenges, existing methods can hardly reconstruct detailed geometry and realistic textures. To address these issues, we propose a multi-modal feature guided 3D face reconstruction method, named MMFG, which does not require any training data and can generate detailed geometry from a single image. Specifically, we represent the reconstructed 3D face as a signed distance field, and propose to combine the position local feature and multi-modal global features to reconstruct a detailed 3D face. To obtain region-aware information, a Swin Transformer is used as our global feature extractor to extract multi-modal global feature from the rendered multi-view RGB images and depth images. Furthermore, considering the different effects of RGB and depth information on albedo and shading, we use the global features from different modal to guide the recovery of BRDF component respectively during differentiable rendering. Experimental results demonstrate that the proposed method can generate more detailed 3D faces, achieving state-of-the-art results on texture reconstruction and competitive results on shape reconstruction on the NoW dataset.

Keywords: 3D face reconstruction · multi-modal feature guided · differentiable rendering

1 Introduction

In recent years, metaverse and digital humans have garnered public attentions, offering novel experiences for individuals in daily lives and entertainment. One crucial technique of digital human technology is 3D face reconstruction, which has received widespread attentions. Because it is inherently ill-conditioned, creating a realistic 3D face from a single image is a difficult problem in the field

This work was supported in part by the National Natural Science Foundation of China (NSFC) under Grant (No. 61976173), Shaanxi Fundamental Science Research Project for Mathematics and Physics (Grant No. 22JSY011) and the MoE-CMCC Artificial Intelligence Project (No. MCM20190701).

Q. Liu et al. (Eds.): PRCV 2023, LNCS 14426, pp. 356–368, 2024.
https://doi.org/10.1007/978-981-99-8432-9_29

of computer vision and graphics. The reconstruction accuracy is greatly influenced by the given prior knowledge and developed physical models. This issue is resolved by the development of deformable 3D face models [1], which use statistical parametric models to represent 3D faces. These models go on to become the mainstream methods for template-based 3D face reconstruction. This breakthrough has led to numerous applications for 3D face reconstruction, including 3D digital humans, face generation, face editing, and face recognition [2].

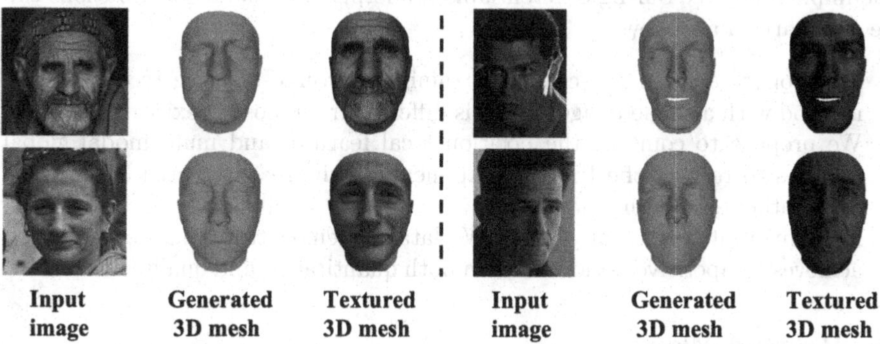

Fig. 1. Reconstruct results of the proposed method. Given a single face image, the proposed method can reconstruct a 3D face model with personalized geometric details and high-quality textures.

While the widely used deformable 3D face model, i.e., 3DMM (3D Morphable Model), has limitations such as obvious linearization, and weak personalization, unclear identity information, poor detail recovery and realism. To overcome these challenges, various methods [3,4] have been developed to reconstruct detailed facial geometry and have achieved enhanced performance. The state-of-the-art methods [5,6] employ neural networks to learn displacement maps that capture specific feature details. However, these methods require a large amount of training data. To reconstruct detailed 3D faces with a single image, the recent method PBIDR [7] represents reconstructed 3D face as an implicit surface, and optimizes it by sampling points on the implicit surface with local features to recover their BRDF components. They have achieved state-of-the-art 3D face reconstruction results from a single image and without any other training data. Considering that multi-view RGB images and depth images of a 3D face contain more information of geometry and texture, which may contribute to reconstructed 3D face, such as personalized details of the geometry or the lighting and shading of the texture, combine local features from the implicit surface and multi-modal features from RGB images and depth images could be explored.

In this paper, we propose a novel method for 3D face reconstruction from a single image, guided by multi-modal global features and position local features. The reconstruction results are shown in Fig. 1. First of all, a coarse 3D face is reconstructed from a single image based on 3DMM, and multi-view RGB images and depth images can be obtained from it. To achieve multi-modal features, we introduce a Swin Transformer [8] to extract region-aware global features from multi-view RGB images and depth images. Then, we propose a detailed 3D face reconstruction framework, in which the reconstructed 3D face is represented by a signed distance field (SDF) [9] and position local features of the implicit surface are extracted by a neural network. Finally, the position local features and multi-modal global features are combined to recover the BRDF (Bidirectional Reflectance Distribution Function) components to recover texture and optimize the implicit surface during differentiable rendering. Our main contributions can be summarized as follows:

- We propose a novel low-cost and training data-free 3D face reconstruction method with a single image, which is different from most of existing methods.
- We propose to combine the position local features and multi-modal global features to recover the BRDF components of the reconstructed 3D face via differentiable rendering.
- Experimental results on the NoW dataset demonstrate that our method achieves competitive performance in both quantitative and qualitative results.

2 Related Work

The task of reconstructing 3D faces from images can be traced back to the work of Park [10], which is a pioneering method for reconstructing 3D faces from multi-view images. 3D face reconstruction with a single image is more challenging and has made substantial progress by statistical deformable 3D face models, such as 3DMM [1], as well as deep convolutional neural networks. Existing methods can be divided into two categories, i.e., implicit space coding and explicit space regression. The implicit space coding based methods focus on optimizing the basis coefficients of the basic statistical deformable 3D face models. These methods can usually offer robustness in face topology changes but lack facial details. Explicit space regression based methods describe 3D faces as multiple data in explicit space and directly regress them using end-to-end deep networks [11–13]. These methods typically can capture more personalized 3D facial details and are robust to illumination and occlusion.

In recent years, neural implicit representations [9,14,15] has become a more effective and suitable 3D geometric representation, demonstrating excellent performance in surface and volume modeling. For example, predict signed, unsigned distances and calculate implicit surfaces [9,15] by projecting sampled queries onto the surface. Utilize structured local features to partition shape elements, grids, or octrees is another neural implicit representation [16–18]. However, existing techniques face challenges in achieving high fidelity and high diversity. Although i3DMM [19] is the first implicit 3D morphable model for human heads, it suffers from significant quality issues in representing facial regions. IMFace [20] introduces an INR-based 3DMM that encodes complex variations of facial shape through two explicitly disentangled deformation fields, enabling learning fine-grained and semantically meaningful representations. Ren et al. [7] represent reconstructed 3D face as an implicit surface by the signed distance field, and reconstruct the detailed 3D face by a introduced render named physical-based implicit differentiable render.

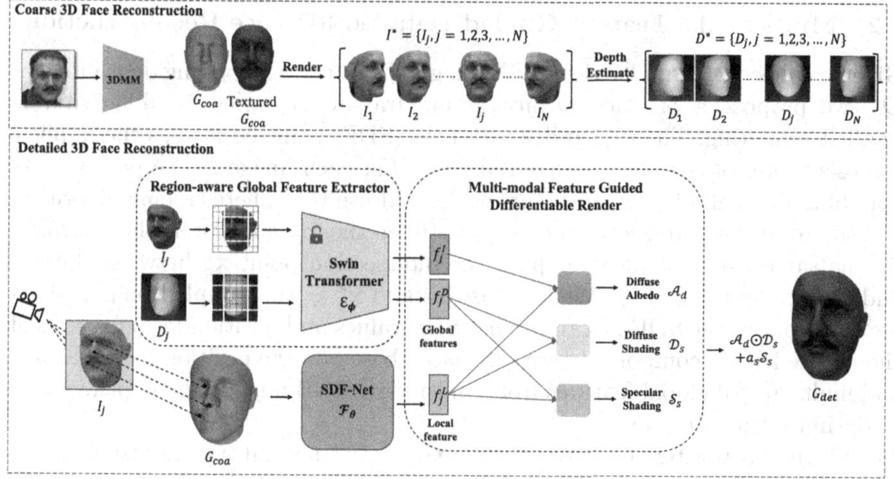

Fig. 2. The framework of the proposed method.

3 Method

As the framework shown in Fig. 2, the proposed method consists of two stages, namely coarse 3D face reconstruction and detailed 3D face reconstruction. At the first stage, given a face image, a coarse textured 3D face model is reconstructed by using the 3DMM fitting method. This coarse 3D face model is rendered into multi-view RGB images, and their corresponding depth images are estimated. At the second stage, we use a region-aware global feature extractor [8] to extract multi-modal global features from the multi-view RGB images and depth images. Then, we regard the coarse 3D face model G_{coa} as the initial implicit surface. The SDF-Net is used to predict the SDF values and position local features of the sampled points on the implicit surface. Finally, to recover the texture and optimize SDF-Net, local and global features are combined to calculate the BRDF components of the sampled points. Thus, detailed 3D face geometry can be achieved from the optimized SDF-Net.

3.1 Coarse 3D Face Reconstruction

The goal of coarse 3D face reconstruction is to generate an initial, roughly geometric 3D face model from a given image and obtain its high-resolution facial texture map. We first use the 3DMM algorithm [21] to generate a coarse geometric 3D face G_{coa} from the given image. Then, G_{coa} and the given image are used to project its high-resolution face texture UV-map [22]. We then render the textured coarse 3D face model into multi-view 2D face images

$I^* = \{I_j, j = 1, 2, ..., N\}$ with N camera poses and estimate their corresponding depth images $D^* = \{D_j, j = 1, 2, ..., N\}$ using an off-the-shelf depth estimation algorithm [23].

3.2 Multi-modal Feature Guided Detailed 3D Face Reconstruction

Given a collection of multi-view RGB images I^* and corresponding depth images D^*, we propose a 3D face reconstruction framework based on differentiable implicit rendering. The signed distance field (SDF) is introduced as the implicit representation of the reconstructed 3D face. For each image I_j, we perform ray matching for each pixel p on the image I_j, and use the sphere tracing algorithm [24] to obtain the intersection point $\mathbf{x}_p$ between the ray and the implicit surface. At the same time, we also compute the intersection point $\mathbf{x}_p^*$ between the ray and the coarse mesh G_{coa}. We then use $\mathbf{x}_p$ and $\mathbf{x}_p^*$ as the sampled points which are the input of the SDF-Net to predict SDF values and position local features of them. The BRDF components are calculated based on the position local features and multi-modal global features from multi-view RGB images and depth images to optimize the SDF-Net.

Multi-modal Global Feature Extracting. As a special type of visual entity, a face exhibits strong connections between its different and disjoint parts. The relationship between different or disjoint parts can also reflect a person's individual information more globally. For the same face, RGB images generated from different views can differ significantly. Given these challenges, we introduce a region-aware global feature extractor that can learn cross-region information to capture the structure and texture relationship between non-adjacent pixels. To this end, we adopt a pre-trained Swin Transformer [8] as our global feature extractor $\mathcal{E}_\phi$. Specifically, Swin Transformer uses the hierarchical transformer to construct pyramidal features at different scales, and in each layer uses shift windows to restrict the self-attention computation to non-overlapping local windows, while increasing the information interaction between different windows through cross-window connections. In our method, the Swin Transformer $\mathcal{E}_\phi$ keep frozen.

Considering that RGB images have limitations in representing geometric information accurately. Instead, multi-modal data can provide more prior information. For example, RGB images can reflect the light reflection and environmental factors, and the depth images can reflect the geometric information of the face, which is helpful for the recovery of face shading information. Therefore, in our method, both the RGB images and depth images are used to provide the global features. Specifically, each RGB image I_j and its corresponding depth image D_j are fed into the global feature extractor $\mathcal{E}_\phi$ respectively to obtain the global features, i.e., $f_j^I = \mathcal{E}_\phi(I_j)$, $f_j^D = \mathcal{E}_\phi(D_j)$, which are then used as the guidence to estimate the BRDF components, and further to optimize the signed distance field.

Position Local Feature Extracting. Inspired by the implicit neural rendering [25], we use a signed distance field (SDF) as the representation of reconstructed face surface. Specifically, a deep neural network $\mathcal{F}_\theta$ with learnable parameters θ, referred to SDF-Net is adopted to approximate a signed distance function $\mathcal{F} : \mathbf{x} \in \mathbb{R}^3 \mapsto s \in \mathbb{R}$. Therefore, we obtain an approximate implicit surface

$\mathcal{G} = \{\mathbf{x} \mid \mathcal{F}(\mathbf{x};\theta) = 0\}$, where $\mathbf{x}$ represents the coordinates of the query point in 3D space. With points $\mathbf{x}_p$ sampled from implicit surface $\mathcal{G}$ as the input, SDF-Net outputs the signed distance s on the surface. Here, the coarse 3D face model G_{coa} is the initial implicit surface and the sample points $\mathbf{x}_p^*$ on G_{coa} are also input to SDF-Net to predict SDF values, which are used to optimize the SDF-Net. In this work, we adopt the SDF-Net also output the point-wise feature $f^L \in \mathbb{R}^{m\times c}$ of m sampled points, each point has a c-dimensional feature. f^L is extracted from the position information of the points, so it is called position local feature.

Multi-modal Feature Guided Differentiable Rendering. Previous publicly available works mostly used the neural volume rendering to describe the object's light reflection and texture. However, human face has complex skin reflection thus it is difficult to obtain realistic rendering results. Therefore, we employ the Physically-Based Rendering (PBR) to accurately predict the complex skin reflection of human face. Basically, the outgoing radiance L_o at surface position $\mathbf{x}$ along with direction ω_o can be expressed as:

$$L_o(\mathbf{x},\omega_o) = \underbrace{\frac{\mathbf{a}(\mathbf{x})}{\pi}\int_{S^2} L_i(\omega_i)(\omega_i\cdot\mathbf{n})^+ \,\mathrm{d}\omega_i}_{\text{diffuse term}} + \underbrace{\alpha\int_{S^2} L_i(\omega_i) f_s(\mathbf{x},\omega_i,\omega_o)(\omega_i\cdot\mathbf{n})^+ \,\mathrm{d}\omega_i}_{\text{specular term}} \tag{1}$$

where $\mathbf{a}(\mathbf{x})$ means the albedo of the surface point $\mathbf{x}$, $L_i(\omega_i)$ denotes the incident radiance from direction ω_i, f_s describes the skin BRDF and $(\omega_i\cdot\mathbf{n})^+$ means the half-cosine function between surface normal $\mathbf{n}$ and ω_i over sphere S^2. Naturally, we consider using the neural BRDF to learn complex appearance components (i.e., diffuse and specular). However, calculating BRDF components based on ray tracing sampling points only contains local information hardly effectively maintain good light consistency throughout the entire face. Therefore, we propose to use the multi-modal global features to guide the calculation of neural BRDFs. Specifically, we consider L_o in Equation (1) consisting of four components: $\mathcal{A}_d$, $\mathcal{D}_s$, a_s, and $\mathcal{S}_s$, representing diffuse albedo, diffuse shading, specular albedo, and specular shading, respectively. Therefore, Equation (1) at point $\mathbf{x}_p$ can be rewritten as:

$$L_{\mathbf{x}_p}(\mathbf{v}) = \mathcal{A}_d(\mathbf{x}_p, f_j^L, f_j^I)\mathcal{D}_s(\mathbf{n}_p^*, f_j^D) + a_s\mathcal{S}_s(\mathbf{n}_p, \mathbf{v}, f_j^D), \tag{2}$$

where we use the gradient at the SDF function of the intersection point $\mathbf{x}_p$ between the ray and the implicit surface as the specular reflection normal to optimize details: $\mathbf{n}_p = \nabla_{\mathbf{x}}\mathcal{F}_\theta(\mathbf{x}_p)$, the smooth diffuse normal $\mathbf{n}_p^*$ comes from the gradient of the SDF function of the intersection point between the ray and G_{coa} obtained through barycentric sampling. During the learning process, we adhere to a prior that the directions of $\mathbf{n}_p$ and $\mathbf{n}_p^*$ should be close to each other. $\mathbf{v}$ represents the view direction. We will approximate $L_{\mathbf{x}_p}(\mathbf{v})$ as the color at the point of $\mathbf{x}_p$. Based on above settings, we define the BRDF components as follows:

Diffuse Albedo: $\mathcal{A}_d(\mathbf{x}_p, f_j^L, f_j^I) \in \mathbb{R}^3$ means diffuse albedo which is related to surface point $\mathbf{x}_p$ and the local feature f_j^L. We propose to use the global RGB feature f_j^I as an additional condition to generate the diffuse albedo. This is due to that f_j^I contains lighting-related information, which can be used as a guide for albedo information to better recover the same lighting as the input image. **Diffuse Shading:** $\mathcal{D}_s(\mathbf{n}_p^*, f_j^L, f_j^D) \in \mathbb{R}$ means diffuse shading which related to the normal of the intersection of the ray and coarse mesh G_{coa}, local feature f_j^L and global depth feature f_j^D. We propose to utilize the global depth feature f_j^D as an additional condition to generate the diffuse shading due to that it can provide information on the geometric structure and thus could better guide the 3D face reconstruction with the same shading conditions as the input image. **Specular Albedo:** $a_s \in \mathbb{R}$ denotes a specular albedo. It can be assumed to be constant, and we hypothesize that it will increase at milestone epochs. **Specular Shading:** $\mathcal{S}_s(\mathbf{n}_p, \mathbf{v}, f_j^L, f_j^D) \in \mathbb{R}$ denotes a specular shading. It is learned from the normal intersection of the ray and the implicit surface, view direction $\mathbf{v}$, local feature f_j^L and global depth feature f_j^D. Similarly, we propose to use the global depth feature f_j^D to offer more geometric information.

Based on above design, we can see that only the normal of SDF is related to high-frequency facial details, guiding the SDF to represent more accurate 3D facial details. After optimization, we can represent a detailed mesh G_{det} as follows

$$G_{det} = G_{coa} + \sum_{\mathbf{x} \in \mathcal{V}} \mathbf{n}_{\mathcal{V}}(\mathbf{x}) \mathcal{F}_\theta(\mathbf{x}), \tag{3}$$

where $\mathcal{V}$ represents the set of facial vertices, $\mathbf{n}_{\mathcal{V}}(\mathbf{x})$ denotes the normal, and $\mathcal{F}_\theta(\mathbf{x})$ represents the SDF value at the point $\mathbf{x}$. Thus, the second term can be seen as a learned displacement map, which can be added to the initial coarse mesh G_{coa} to obtain the final detailed mesh G_{det}.

3.3 Training Loss

Considering a set of pixels $\mathcal{P}$ sampled from an image I_j through a direction $\mathbf{v}$, define $\mathcal{P}^*$ as a subset of $\mathcal{P}$ which contains all the pixels $\mathbf{x}_p$ and $\mathbf{x}_p^*$ that project onto the same intersection. Our framework is trained with the following loss functions to achieve accurate texture and geometry, which is the same as the loss mentioned in [7], where α_1, α_2, α_3, α_4 are hyper-parameters:

$$\mathcal{L}_{total} = \mathcal{L}_{photo} + \alpha_1 \mathcal{L}_{mask} + \alpha_2 \mathcal{L}_{sdf} + \alpha_3 \mathcal{L}_{reg} + \alpha_4 \mathcal{L}_{normal}. \tag{4}$$

Photometric Loss. We predict a final color $\mathbf{c}_j^p$ of $p \in \mathcal{P}$ by a photometric loss which is defined as the error of the ground-truth color I_j^p at pixel p and the predicted color $\mathbf{c}_j^p$ at pixel p as: $\mathcal{L}_{photo} = 1/|\mathcal{P}| \cdot \sum_{p \in \mathcal{P}^*} \left|\mathbf{M}_p^j \left(I_j^p - \mathbf{c}_j^p\right)\right|$, where $\mathbf{M}_p^j$ is the mask of the pixel p on image I_j, $\mathbf{c}_j^p$ can be approximated by $L_{\mathbf{x}_p}(\mathbf{v})$. **Mask Loss.** The silhouette error is constrained by a mask loss as $\mathcal{L}_{mask} = 1/(\alpha|\mathcal{P}|) \cdot \sum_{p \in \mathcal{P} \setminus \mathcal{P}^*} \left(1 - \mathbf{M}_p^j\right) \mathcal{L}_{CE}\left(\mathbf{M}_p^j, \mathbf{s}_{i,\alpha}(p)\right)$, where α is a hyper-parameter, $\mathcal{L}_{CE}$ is the binary cross-entropy loss, and the predicted silhouette $\mathbf{s}_{j,\alpha}(p) = \mathrm{sigmoid}(-\alpha \min \mathcal{F}_\theta(\mathbf{x}_p))$. **Eikonal Loss.** To regularize the spatial gradients of the fields in the network have a unit norm, Eikonal loss is employed as: $\mathcal{L}_{sdf} = \mathbb{E}_{\mathbf{x}} \left(\|\nabla_{\mathbf{x}} \mathcal{F}_\theta(\mathbf{x})\| - 1\right)^2$. **Registration Loss.** In addition, we present a registration loss $\mathcal{L}_{reg}$ to regularize the details of the implicit surface with respect to the coarse mesh $\mathcal{G}_{coa}$. Suppose a ray $\mathbf{r}$ intersects implicit surface $\mathcal{G}$ at $\mathbf{x}_p$ and intersects G_{coa} at $\mathbf{x}_p^*$. By providing $\mathbf{x}_p^*$ as input to $\mathcal{F}_\theta$ and utilizing its SDF value and specular normal, we can determine the nearest neighbor $\mathbf{x}_p^{*\dagger}$ as: $\mathbf{x}_p^{*\dagger} = \mathbf{x}_p^* - \left|\mathcal{F}_\theta\left(\mathbf{x}_p^*\right)\right| \cdot \frac{\nabla \mathcal{F}_\theta(\mathbf{x}_p^*)}{\left\|\nabla \mathcal{F}_\theta(\mathbf{x}_p^*)\right\|_2}$. Then registration loss is defined as: $\mathcal{L}_{reg} = \frac{1}{\mathcal{P}} \sum_{p \in \mathcal{P}^*} \left\|\mathbf{x}_p - \mathbf{x}_p^{*\dagger}\right\|_2$.

Normal Loss. In order to adhere to a prior that the direction $\mathbf{n}_p^{*\dagger}$ not differs too much from $\mathbf{n}_p$, a normal loss is introduced as $\mathcal{L}_{normal} = \frac{1}{\mathcal{P}} \sum_{p \in \mathcal{P}^*} (1 - \frac{\mathbf{n}_p \cdot \mathbf{n}_p^{*\dagger}}{\max\left(\|\mathbf{n_p}\|_2 \cdot \left\|\mathbf{n}_p^{*\dagger}\right\|_2, 1e^{-8}\right)})$.

4 Experiments

4.1 Experimental Setup

Dataset. We conducted our experiment on CelebAMask-HQ [26], FFHQ [27] and NoW datasets [28]. CelebAMask-HQ is a large-scale high-resolution face dataset with fine-grained mask labels, which consists of over 30,000 face images with a resolution of 512×512. FFHQ is a high-quality face dataset containing 70,000 high-definition face images at 1024×1024 resolution, with obvious variations in age, race, image background, face attributes, as well as face peripheral accessories. The validation set of NoW dataset consists of 20 subjects with ground-truth 3D face scans and corresponding landmarks.

Network Architecture. The global feature extractor is implemented as a pre-trained Swin Transformer with parameters and structure are frozen. The SDF-Net is implemented as a MLP with 8 layers, including a skip-connection from the input to the fourth layer. Softplus is used as the activation function.

For the networks of multi-modal feature guided differentiable rendering, we use three MLPs with 5-layer, 2-layer, and 2-layer respectively to predict $\mathcal{A}_d$, $\mathcal{D}_s$ and $\mathcal{S}_s$ respectively. To get a multi-modal feature, we concate f^L with f^I or f^D, and use a linear layer to change the dimensions of the combined features.

Implementation Details. All experiments were conducted on a GeForce RTX 3090 GPU, taking 2 h for each input image. The Adam optimizer was implemented, with 400 epochs utilized for training. We set $\alpha_1 = 100$, $\alpha_2 = 0.1$, $\alpha_3 = 5$, $\alpha_4 = 1$ and $\alpha = 50$, and share the same hyper-parameter settings for all subjects. For every subject, we use 17 render views to produce the set of RGB images at the coarse stage.

4.2 Main Results

To demonstrate the quality of the reconstruction results, we firstly conducted experiments on a subset of data randomly selected from CeleAMask-HQ and FFHQ to compare the texture quality. The metrics of PSNR is used to measure the mean squared error, and SSIM is used to measure the luminance, contrast, and structure of two images. Then, we compare the shape quality on the NoW dataset with the metric of absolute distance error (ADE) of all vertices.

Comparing Reconstructed Texture Quality. As shown in Table 1, our method achieves the state-of-the-art performance in both PSNR and SSIM, outperforming PBIDR and several other methods for 3D face reconstruction from a single image. Our results retain realistic facial textures and highly conform to the lighting conditions when the photos were taken. The texture improvement achieved by our method is attributed to the multi-modal global features used to guide BRDF component optimization. RGB features provide the same lighting conditions, and depth features provide the same shadow conditions as the input image. Multi-modal global features provide additional texture and geometric information for better texture reconstruction.

Table 1. Comparisons of the reconstructed texture quality.

	PSNR ↑		SSIM ↑	
	mean	std	mean	std
PBIDR [7]	22.183	1.969	0.9986	0.0034
Deep3D [4]	20.413	2.057	0.9980	0.0006
OSTeC [22]	21.675	1.687	0.9984	0.0006
MMFG (ours)	**24.089**	**1.357**	**0.9995**	**0.0002**

Comparing Reconstructed Shape Quality. As the reconstructed shape on NoW dataset [28] listed in Table 2, our method outperforms all compared baseline methods on ADE metric. Qualitative results are also illustrated and compared in Fig. 3. We can find that our method can accurately reconstruct facial shapes, expressions, and geometrical details.

It should be emphasized that our method and PBIDR do not need any training data. In Fig. 4, personalized facial details of reconstructed 3D face mesh are compared especially, such as wrinkles, and contours, including eye and mouth corners. We can find that more geometrical details of facial surface can be reconstructed by our method, because we use a region-aware extractor to extract the relationship of the non-adjacent region of human face to get more personalized global feature. These global feature not only can guide a realistic texture

Table 2. Comparisons of the reconstructed facial shape on NoW dataset.

	ADE ↓		
	median	mean	std
Deep3D [4]	1.22	1.51	1.21
3DDFA-V2 [29]	1.23	1.57	1.39
PBIDR [7]	1.23	1.53	1.31
MMFG (ours)	**1.20**	**1.47**	1.27

reconstruct, but also can optimize the SDF-Net to learn more detailed and personalized geometry.

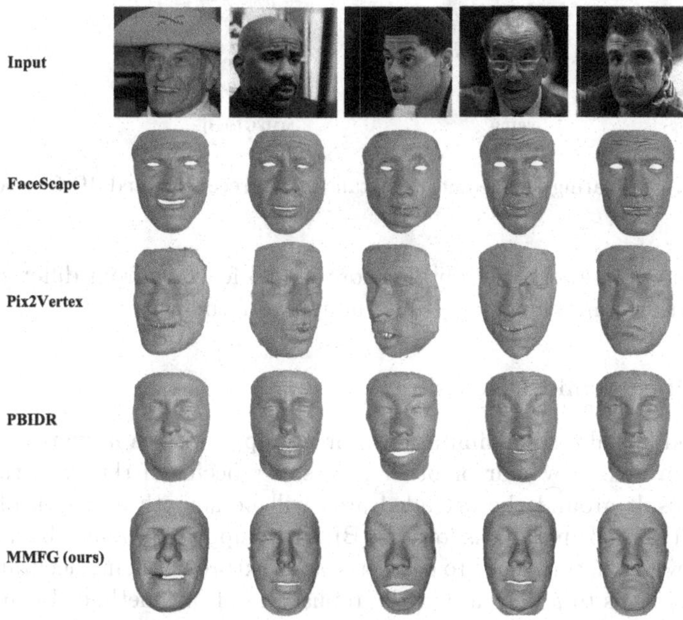

Fig. 3. Comparisons to single-image detailed face reconstruction methods. From top to bottom: Inputs, FaceScape [30], Pix2Vertex [3], PBIDR [7] and MMFG (ours).

4.3 Ablation Study

In this subsection, we will conduct more experiments to show the effectiveness of the global features in our proposed method. In particular, we removed the global depth feature and the global RGB feature respectively, and compare the reconstructed texture in Table 3. The decreased results on PSNR and SSIM demonstrate that both the global RGB feature and depth feature in our framework are important for 3D face reconstruction. This could be attributed to the

Table 3. The effectiveness of the global RGB and depth features.

	PSNR ↑		SSIM ↑	
	mean	std	mean	std
Full model	24.089	1.357	0.9995	0.0002
W/o f_j^I	23.373	1.872	0.9997	0.0018
w/o f_j^D	22.575	1.677	0.9996	0.0009

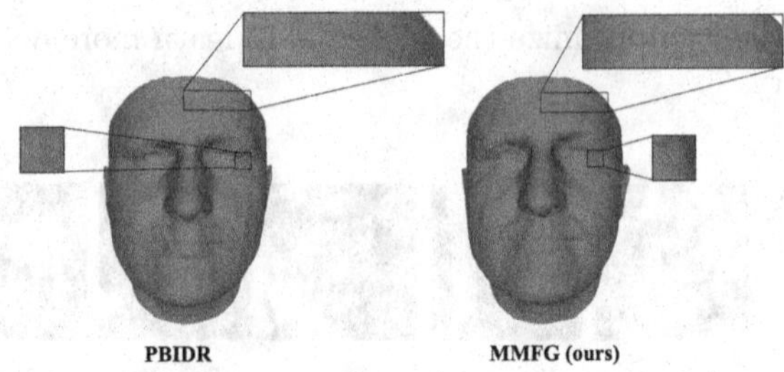

Fig. 4. Comparing the geometric details of the reconstructed 3D face scan.

fact that our framework can obtain more global features from different views, thereby further improving the reconstruction performance.

4.4 Limitation and Discussion

Our method has also some limitations, for example, when a large part od facial occlusions are exist by hair or other accessories occluded the face, the reconstruction result around the occluded area will be smooth and lack of details. And when the ambient light is low, the BRDF components cannot be accurately predicted, which in turn leads to geometry and texture not being accurately optimized. This leads to a decrease in the robustness of our method. In the future, on the one hand, we will consider using more accurate illumination models to predict the texture and geometry and thus better guide the optimization of the reconstruction results. On the other hand, we will also consider improving the generalization and robustness of our method for the data collected in the wild.

5 Conclusion

In this paper, we proposed a novel method for 3D face reconstruction from a single image. First, a coarse 3D face model is obtained by 3DMM, and it is rendered into multi-view RGB images. These RGB images with their corresponding depth images are then used to extract the multi-modal global features. The position local feature extract by SDF-Net and multi-modal global feature are used to guide the BRDF recovery and optimize the implicit surface predict by SDF-Net, starting from the coarse 3D face. Our method does not need any training data and can achieve competitive reconstructed texture and geometry on the NoW dataset.

References

1. Vetter, T., Blanz, V.: Estimating coloured 3d face models from single images: an example based approach. ECCV **1407**, 499–513 (1998)
2. Hu, L., Saito, S., Wei, L., Nagano, K., Seo, J., Fursund, J., Sadeghi, I., Sun, C., Chen, Y.C., Li, H.: Avatar digitization from a single image for real-time rendering. ToG **36**(6), 1–14 (2017)
3. Sela, M., Richardson, E., Kimmel, R.: Unrestricted facial geometry reconstruction using image-to-image translation. In: ICCV, pp. 1576–1585 (2017)
4. Deng, Y., Yang, J., Xu, S., Chen, D., Jia, Y., Tong, X.: Accurate 3d face reconstruction with weakly-supervised learning: From single image to image set. In: CVPR Workshops (2019)
5. Feng, Y., Feng, H., Black, M.J., Bolkart, T.: Learning an animatable detailed 3d face model from in-the-wild images. ToG **40**(4), 1–13 (2021)
6. Zielonka, W., Bolkart, T., Thies, J.: Towards metrical reconstruction of human faces. In: Avidan, S., Brostow, G., Cissé, M., Farinella, G.M., Hassner, T. (eds.) Computer Vision, ECCV 2022, LNCS, vol. 13673, pp. 250–269. Springer, Cham (2022). https://doi.org/10.1007/978-3-031-19778-9_15
7. Ren, X., Lattas, A., Gecer, B., Deng, J., Ma, C., Yang, X.: Facial geometric detail recovery via implicit representation. In: FG, pp. 1–8. IEEE (2023)
8. Liu, Z., et al.: Swin transformer: hierarchical vision transformer using shifted windows. In: ICCV, pp. 10012–10022 (2021)
9. Park, J.J., Florence, P., Straub, J., Newcombe, R.: DeepSDF: learning continuous signed distance functions for shape representation. In: CVPR, pp. 165–174 (2019)
10. Parke, F.I.: Measuring three-dimensional surfaces with a two-dimensional data tablet. Comput. Graph. **1**(1), 5–7 (1975)
11. Jackson, A.S., Bulat, A., Argyriou, V., Tzimiropoulos, G.: Large pose 3d face reconstruction from a single image via direct volumetric CNN regression. In: ICCV, pp. 1031–1039 (2017)
12. Feng, Y., Wu, F., Shao, X., Wang, Y.: Joint 3d face reconstruction and dense alignment with position map regression network. In: ECCV, pp. 534–551 (2018)
13. Zeng, X., Peng, X., Qiao, Y.: Df2net: a dense-fine-finer network for detailed 3d face reconstruction. In: ICCV, pp. 2315–2324 (2019)
14. Gropp, A., Yariv, L., Haim, N., Atzmon, M., Lipman, Y.: Implicit geometric regularization for learning shapes. arXiv preprint arXiv:2002.10099 (2020)
15. Mescheder, L., Oechsle, M., Niemeyer, M., Nowozin, S.: Occupancy networks: learning 3d reconstruction in function space. In: CVPR, pp. 4460–4470 (2019)
16. Genova, K., Cole, F., Sud, A., Sarna, A., Funkhouser, T.: Local deep implicit functions for 3d shape. In: CVPR, pp. 4857–4866 (2020)
17. Ibing, M., Lim, I., Kobbelt, L.: 3d shape generation with grid-based implicit functions. In: CVPR, pp. 13559–13568 (2021)
18. Takikawa, T., et al.: Neural geometric level of detail: Real-time rendering with implicit 3d shapes. In: CVPR, pp. 11358–11367 (2021)
19. Yenamandra, T., et al.: i3dmm: Deep implicit 3d morphable model of human heads. In: CVPR, pp. 12803–12813 (2021)
20. Zheng, M., Yang, H., Huang, D., Chen, L.: ImFace: a nonlinear 3d morphable face model with implicit neural representations. In: CVPR, pp. 20343–20352 (2022)
21. Gecer, B., Ploumpis, S., Kotsia, I., Zafeiriou, S.: GANFIT: generative adversarial network fitting for high fidelity 3d face reconstruction. In: CVPR, pp. 1155–1164 (2019)

22. Gecer, B., Deng, J., Zafeiriou, S.: OSTeC: one-shot texture completion. In: CVPR, pp. 7628–7638 (2021)
23. Ranftl, R., Lasinger, K., Hafner, D., Schindler, K., Koltun, V.: Towards robust monocular depth estimation: mixing datasets for zero-shot cross-dataset transfer. TPAMI **44**(3), 1623–1637 (2020)
24. Jiang, Y., Ji, D., Han, Z., Zwicker, M.: SDFDiff: differentiable rendering of signed distance fields for 3d shape optimization. In: CVPR, pp. 1251–1261 (2020)
25. Peng, S., et al.: Animatable neural radiance fields for modeling dynamic human bodies. In: ICCV, pp. 14314–14323 (2021)
26. Lee, C.H., Liu, Z., Wu, L., Luo, P.: MaskGAN: towards diverse and interactive facial image manipulation. In: CVPR, pp. 5549–5558 (2020)
27. Karras, T., Laine, S., Aila, T.: A style-based generator architecture for generative adversarial networks. In: CVPR, pp. 4401–4410 (2019)
28. Sanyal, S., Bolkart, T., Feng, H., Black, M.: Learning to regress 3D face shape and expression from an image without 3D supervision. In: CVPR, pp. 7763–7772 (2019)
29. Guo, J., Zhu, X., Yang, Y., Yang, F., Lei, Z., Li, S.Z.: Towards fast, accurate and stable 3d dense face alignment. In: Vedaldi, A., Bischof, H., Brox, T., Frahm, J.-M. (eds.) ECCV 2020. LNCS, vol. 12364, pp. 152–168. Springer, Cham (2020). https://doi.org/10.1007/978-3-030-58529-7_10
30. Yang, H., et al.: FacEscape: a large-scale high quality 3d face dataset and detailed riggable 3d face prediction. In: CVPR, pp. 601–610 (2020)

Character Recognition

Advanced License Plate Detector in Low-Quality Images with Smooth Regression Constraint

Jiefu Yu[1], Dekang Liu[1], Tianlei Wang[1], Jiangmin Tian[1], Fangyong Xu[2], and Jiuwen Cao[1(✉)]

[1] Machine Learning and I-health International Cooperation Base of Zhejiang Province, Hangzhou Dianzi University, Hangzhou, China
{jmtian,jwcao}@hdu.edu.cn

[2] Zhejiang Jiaguang Information Technology Company Limited, Hangzhou, China

Abstract. Improving the detection performance of license plate detectors in low-quality images is one of the core goals of license plate recognition community. While low-quality images encompasses various forms, this paper primarily focuses on two issues of *blurring and distortion* which have been empirically identified as the main challenges encountered by systems deployed in practical scenarios. And based on the designed **S**mooth **R**egression **C**onstraint from the perspective of detection heads, we investigate and propose a novel **L**icense **P**late **D**etector, namely SrcLPD. Specifically, we observe that corner-based detection head is robust to low-quality image with geometric distortion, while center-based regression methods are insensitive to blurred images. Therefore, we propose a novel soft-coupling loss to integrate the two types of bounding box regression strategies into a unified framework to instantiate smooth regression constraint. Extensive experiments verify that the proposed SrcLPD explicitly combines the advantages of the dual heads. SrcLPD outperforms 4 subsets of the CCPD benchmark, especially for blurry low-quality images, surpassing the previous method by 0.7% and achieving 97.6% detection precision.

Keywords: Intelligent transportation · License plate detection · Smooth regression constraint · Deep learning · Convolutional neural network

1 Introduction

Automatic license plate recognition (ALPR) is of great significance for intelligent transportation systems [1,21,27]. Detection and recognition are two core tasks of ALPR, and the detection is a prerequisite for accurate subsequent recognition [5,20]. The current popular license plate detection (LPD) algorithms achieve near-saturated performance on benchmarks [5,27,28]. However, the robustness of license plate detector is challenging in low-quality images, such as distorted deformation, uneven illumination, or occlusion. Formally, the hard cases above mainly compromise the integrity of the license plate detection bounding box

Q. Liu et al. (Eds.): PRCV 2023, LNCS 14426, pp. 371–382, 2024.
https://doi.org/10.1007/978-981-99-8432-9_30

(BBox). Meanwhile, we suspect that the noisy labels are incorrectly aligned with incomplete detection and induce inflated numerical metrics under the evaluation protocol of IoU > 0.7 [27]. In this paper, we consider the topic of more precise bounding boxes from the perspective of more reasonable coordinate regression. The proposed method effectively pushes the detection performance of license plate detector in low-quality images.

Corner or center point regression approaches are mainstream in current literatures. The former is based on object detection algorithms, such as YOLO [16], CenterNet [30], etc., supplemented by rotation angle prediction of the license plate [26] or region rectangular descriptor prediction [15]. The latter predicts the center point of license plate region and then predicts vectors pointing to the four corners [1,7]. Generally, the above-mentioned methods lie in category of single pixel type (*corner or center*) regression-based license plate detection. Because of the separate adoption of corner or center points, these methods show their own emphasis and are not suitable for various forms of hard cases.

The hard cases in LPD are broadly categorized into blurred and distorted. As illustrated in Fig. 1, compared with the corner-based competitors, the center-based methods are sensitive to geometric interference and the detection boxes are inaccurate. The blurred license plates are affected by the surrounding environment resulting in obscure visual features, the center-based regression is robust to these hard cases. However, the detection performance of corner-based approaches suffer from severe attenuation owing to the presence of blurred corner features. The above observations inspire us to integrate the corner- and center-based methods into one framework to achieve complementary advantages. Therefore, to further enhance the performance limit of license plate detector in low-quality images, we investigate dual pixel type regression for LPD which implicitly integrates the advantages of both regression types. Furthermore, an additional soft-coupling loss is proposed to improve training stability to adjust the association of the two regressions.

Overall, the key contributions of this paper can be summarized as follows:

- We propose a novel smooth regression constraint (SRC) consisting of a dual-branch prediction head for corner or center point-based regression and a soft-coupling loss (SCL) to piece-wisely supersize the dual-branch predictions.
- The soft-coupling loss (SCL) pushes the performance limit and robustness of license plate detector in low-quality images and stabilize the training process.
- An advanced license plate detector named SrcLPD is developed based on the smooth regression constraint which exhibits superior detection performance on both blurred and distorted interference. Removable center-based branch at inference drives the flexibility of the SrcLPD.

2 Related Work

Early license plate detection methods [19,20,26] are always motivated by anchor-based object detection approaches, such as Faster RCNN [18], YOLO [16,17]. Instead of rectangle BBox as license plate boundary descriptor, the geometric

Fig. 1. The qualitative results of the single pixel type regression based LPD. CDM and CETCO indicate the acronyms for **c**orner **d**etection **m**odule and **ce**nter-**t**o-**co**rner regression respectively. The detection results corresponding to two type regression strategies as well as GT are drawn by green, yellow, and red. (a)–(d) are distortion cases, and (e)–(f) are samples where the license plate area is not obvious due to non-geometric interference such as lighting, blurring, and occlusion. (Color figure online)

transformation matrix from rectangle anchor to quadrangle BBox generally is modeled as learnable parameters [19,20]. Parameters of affine transformation and perspective transformation matrix are learned to handle plane and spatial distortion license plate object. Currently, the flexible paradigm of converting LPD to keypoint detection is becoming more and more popular [1,7,15,25,29].

Specifically, Zhang et al. [29] establish a lightweight architecture based on corners rather than rectangular boxes and propose a multi-constrained Gaussian distance loss to improve corner localization accuracy. Qin et al. [15] develop a unified convolutional neural network (CNN) where the corner points based quadrangle BBox and rectangle BBox are detected simultaneously. Meanwhile, another CNN branch is designed to achieve license plate character recognition. Jiang el al. [7] develop LPDNet to detect license plate rectangle BBox and four corners. Moreover, centrality loss is proposed to drive performance improvement in complex environments. Overall, the above approaches follow a coarse-to-fine paradigm. Concretely, the corners are first coarsely located and the heatmaps are generated. Then the corner coordinates are refined by regression based on the activated pixels.

Although corner-based methods are robust to geometric distortion perturbations, their performance decay significantly in other kinds of low-quality images such as blurry and uneven lighting where the visual features of license plate corners are weak. To enhance the representation quality of deep features, Wang

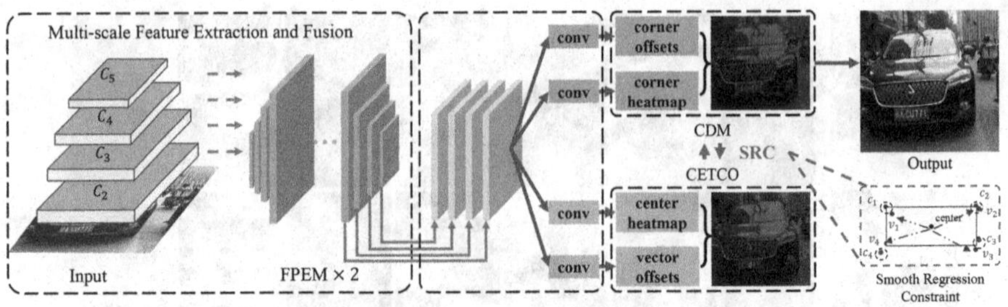

Fig. 2. The overall pipeline of the proposed SrcLPD. For the output example, the red and green BBox represent GT and final prediction. (Color figure online)

et al. [25] propose a p-norm squeeze-and-excitation attention mechanism that exhibits superior detection performance on CCPD benchmark [27].

Compared with corner points, the visual feature salience assumption of the license plate center point is usually satisfied by almost all images. Therefore, the pipeline of center point heatmap prediction and vector offsets regression is attractive. Fan et al. [1] modify the object detection approach CenterNet [30] to LPD with superior performance on solving blurred images.

Our SrcLPD integrates corner- and center-based license plate detection mechanisms. The predictions from structurally decoupled dual branch task-specific heads are further coupled by the designed smooth regression constraint. The advantages of these two types of corner coordinate regression strategies are implicitly combined. SrcLPD achieves significant detection performance and superior robustness in low-quality images.

3 Proposed SrcLPD

As illustrated in Fig. 2, SrcLPD consists of three components: (1) multi-scale features extraction and fusion, (2) corner detection module (CDM) and (3) center-to-corner branch (CETCO). The CETCO will be removed at inference to lightweight the model. The CDM and CETCO are integrated and optimized by the smooth regression constraint.

3.1 Multi-scale Features Extraction and Fusion

Following the previous works [1,14,15], ResNet-50 [4] is adopted as the backbone. Given colored input image $\mathbf{I} \in \mathbb{R}^{h \times w \times 3}$, where h, w are height and width predefined for training. Multi-scale features $\mathbf{P}_i, i \in \{2, 3, 4, 5\}$ are then obtained. Different from [15,25] that directly applying FPN [10] into the LPD model, we further consider the small object attribute of license plate region. Inspired by [22–24], a variant of FPN, named the feature pyramid enhancement module (FPEM) is first introduced into the LPD pipeline for better feature fusion. Specifically, FPEM processes features with two phases, i.e., up-scale enhancement and down-scale enhancement. The enhanced features $\mathbf{F}_i$ are element-wise addition results of the input pyramid features $\mathbf{P}_i$ and the processed features by the up-scale and down-scale stages.

3.2 Decoupled Dual Branch Prediction Heads

Enhanced pyramid features $\mathbf{F}_i, i \in \{2,3,4,5\}$ cover low-level and high-level semantic information, we combine these features by upsampling and concatenating. To maintain efficient training and inference, a convolution layer with 3×3 kernel size adjust and fuse the concatenated features. The final input to the head is $\mathbf{F} \in \mathbb{R}^{128\times128\times64}$.

Furthermore, inspired by advanced object detection algorithms [2,11], we attach additional convolution layers (3×3 kernel size) followed by batch normalization [6] layer and a rectified linear units (ReLU) [3] to instantiate information decoupling and achieve task-specific prediction heads, i.e., corner heatmaps, corner offsets, center heatmap, and vector offsets. The details of all prediction heads are described in Sect. 3.3.

3.3 Smooth Regression Constraint

The core component of SrcLPD is smooth regression constraint which consists of the corner detection module (CDM) for direct prediction of four corner coordinates, the center-to-corner regression (CETCO) for indirect prediction, and the soft-coupling loss (SCL) for supervision between the dual branches. Each sub-component is described in detail below.

CDM: Corner Detection Module. The CDM contains two sub-branches, i.e., corner heatmaps $\mathbf{F}_{co}^h \in \mathbb{R}^{\frac{h}{4}\times\frac{w}{4}} \times 4$ for coarse locating and corner offsets $\mathbf{F}_{co}^o \in \mathbb{R}^{\frac{h}{4}\times\frac{w}{4}\times 8}$ for coordinates refinement. The channel number of $\mathbf{F}_{co}^h$ is adjusted to 4 by a 1×1 convolution, then the sigmoid activation function is applied to classify the foreground corner points and background pixels. The 8 channel $\mathbf{F}_{co}^o$ refers to the refinement correction of 4 corner coordinates regressed from the $\mathbf{F}_{co}^h$ with respect to the x- and y-axis directions. $\mathbf{F}_{co}^o$ compensates for the discretization error caused by the stride and makes the prediction corner points accurate. Following [9,30], for each corner point $p_i \in \mathbb{R}^2, i \in \{1,2,3,4\}$, the corresponding approximate coordinates on low-resolution map with size $\frac{h}{4} \times \frac{w}{4}$ are computed via $\tilde{p}_i = \lfloor \frac{p_i}{4} \rfloor$. Then 4 ground truth corner points are splatted on a heatmap $\mathbf{Y} \in [0,1]^{\frac{h}{4}\times\frac{w}{4}\times 4}$ using a Gaussian kernel, i.e., $\mathbf{Y} = \exp\left(-\frac{(x-\tilde{p}_i^x)^2+(y-\tilde{p}_i^y)^2}{2\sigma_p^2}\right)$, here σ_p denotes an object size-adaptive standard deviation [9,30]. The supervision of $\mathbf{F}_{co}^o$ comes from $\frac{p_i}{4} - \tilde{p}_i$.

CETCO: Center-to-corner Regression. In constrained scenarios, CDM serves as an advanced license plate detector. However, in low-quality images with blur, uneven illumination, or occlusion, the performance of CDM decays significantly. Therefore, CETCO is integrated into SrcLPD. Similar to CDM, two sub-branches (heatmap and offset) are included in CETCO. Center point heatmap $\mathbf{F}^h_{ce} \in \mathbb{R}^{\frac{h}{4}\times\frac{w}{4}\times 1}$ and center-to-corner offsets $\mathbf{F}^o_{ce} \in \mathbb{R}^{\frac{h}{4}\times\frac{w}{4}\times 8}$ are predicted. The ground truth generalization is similar to CDM. However, the $\mathbf{F}^o_{ce}$ is obtained via $\frac{p_i}{4} - \frac{c}{4}$ and indicates the offsets from each corner point p_i to center point c. The Gaussian kernel also participates in the ground truth generalization of $\mathbf{F}^h_{ce}$, but the offsets refinement of the center point are removed for efficient computation.

SCL: Soft-Coupling Loss. The key idea behind the smooth regression constraint is to minimize the difference between the two branch predictions. Intuitively, consistency objective function such as mean squared error (MSE) loss is trivial solution. However, as above mentioned, different regression strategies are compatible with different types of low-quality images. Taking MSE as a hard constraint to supervise variance may cause unstable training and performance decrement. For example, for images with distortion, CDM shows superior while CETCO undergoes inferior performance. If the deviation between the two is forcibly constrained, the accurate CDM prediction may become inaccurate. Therefore, soft-coupling loss (SCL) is proposed to solve this issue. Under the slacked constraint of SCL, the inaccurate prediction of CDM in blurred images is penalized, thus supervising and enhancing the offsets regression quality in the CDM branch. Similarly, in [13], mutual regression is also considered for robust table structure parsing. However, in our SrcLPD, the CETCO module can be regarded as a removable auxiliary training branch that is flexible and concise.

Specifically, for each corner point p_i, the SCL is formulated as

$$\mathcal{L}_{SCL_i} = \begin{cases} \alpha \cdot d_i \cdot \left[\exp\left(\frac{d_i}{\beta}\right) - 1\right], & d_i \le \beta \\ d_i, & \text{otherwise} \end{cases} \tag{1}$$

where α and β represent modulation factor and gradient change tolerance. The differential of $\mathcal{L}_{SCL_i}$ with respect to d_i is

$$\frac{\partial \mathcal{L}_{SCL_i}}{\partial d_i} = \begin{cases} \frac{\alpha}{\beta} \cdot d_i \cdot \exp(\frac{d_i}{\beta}) + \alpha \cdot \left(\exp\left(\frac{d_i}{\beta}\right) - 1\right), & d_i \le \beta \\ 1, & \text{otherwise} \end{cases} \tag{2}$$

The Manhattan distance d_i is utilized to measure the difference between the two types of predictions. Specifically, $d_i = |x_{c_i} - x_{v_i}| + |y_{c_i} - y_{v_i}|$, where (x_{c_i}, y_{c_i}), (x_{v_i}, y_{v_i}) denote the i-th corner point coordinates predicted by CETCO and CDM, respectively.

Without loss of generality, Eq. 1, Eq. 2, as well as MSE type hard constraint, can be visualized in Fig. 3. We can perceive from Fig. 3 that the backpropagation is moderated by slight gradient since the close predictions from dual branches, and prediction difference beyond the tolerance parameters β obtain constant gradient.

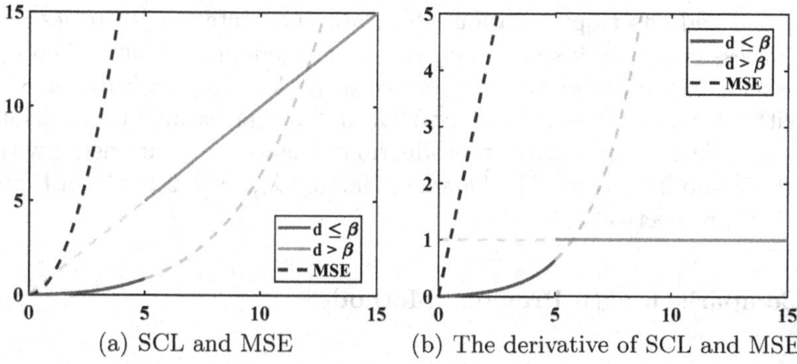

(a) SCL and MSE (b) The derivative of SCL and MSE

Fig. 3. Loss curves and corresponding gradient with $\beta = 5$, $\alpha = 0.1$.

Optimization. The overall loss of SrcLPD includes $\mathcal{L}_{co}^{h}$ for corner heatmaps, $\mathcal{L}_{co}^{o}$ for corner offsets, $\mathcal{L}_{ce}^{h}$ for center heatmap, $\mathcal{L}_{ce}^{o}$ for center-to-corner offsets, and $\mathcal{L}_{SCL}$ for four corner soft-coupling losses. Following [30], focal loss [11] is utilized to instantiate $\mathcal{L}_{co}^{h}$, $\mathcal{L}_{ce}^{h}$ and L1 loss is considered as $\mathcal{L}_{co}^{o}$, $\mathcal{L}_{ce}^{o}$. Totally, the optimization of SrcLPD is supervised by

$$\mathcal{L} = \lambda_1 \mathcal{L}_{co}^{h} + \lambda_2 \mathcal{L}_{ce}^{h} + \lambda_3 \mathcal{L}_{co}^{o} + \lambda_4 \mathcal{L}_{ce}^{o} + \lambda_5 \sum_{i=1}^{4} \mathcal{L}_{SCL_i} \tag{3}$$

4 Experiments

4.1 Datasets and Evaluation Metrics

CCPD. The Chinese City Parking Dataset (CCPD) [27] is adopted for SrcLPD training and evaluating. The latest version downloaded from github repo[1] is compatible with our motivation since more challenging low-quality images are added. The train, validation and test samples are 100,000, 99,996, 141,982, respectively.

IoU and TIoU. Intersection over union (IoU) is a common evaluation protocol. A detection is true positive if the prediction BBox has IoU > 0.7 with respect to ground truth. Furthermore, a more reasonable protocol called tightness-aware IoU (TIoU) [12] proposed for scene text detection is also considered for comprehensive performance analysis in this paper. The evaluation threshold of TIoU in the experiment is set to 0.5.

4.2 Implementation Details

The proposed SrcLPD is implemented by PyTorch. Most training details are consistent with [30], such as input images resolution, the backbone, the loss function

[1] https://github.com/detectRecog/CCPD.

for heatmap and offset optimization, etc. Data augmentation strategies including brightness variation, histogram equalization, random rotation and cropping, and perspective transformation are achieved by package imgaug[2]. Adam [8] optimizer with 30 epochs, 36 batch size, and 2×10^{-4} initial learning rate is adopted. Seed of 42 is set for experiments reproduction. The learning rate decay with 0.1 factor at 25, and 28 epochs. The balancing factors $\lambda_i, i = 1, 2, 3, 4, 5$ in Eq. 3 are $\{4, 4, 2, 1, 1\}$, respectively.

4.3 Comparison with Previous Methods

Table 1. Comparisons of various license plate detectors on various evaluation subsets of CCPD [27]. Take Rotate 10k as an example, Rotate means the license plate images show rotation appearance and 10k indicates this subset contains about 10,000 images.

Method	Input Size	Base 100k	Rotate 10k	Tilt 30k	Blur 20k	DB 10k	FN 20k	Weather 10k	Challenge 50k
Faster-RCNN [18]	–	–	94.4	88.2	81.6	66.7	76.5	–	89.8
YOLOv3 [17]	320	–	96.7	89.2	82.2	71.3	82.4	–	91.5
WPOD-Net [20]	–	99.1	97.9	96.0	–	86.1	87.7	95.4	92.9
Nguyen et al. [14]	800	97.5	97.2	97	–	**95.6**	89.2	97.7	**98.6**
Qin et al. [15]	512	99.6	98.5	95,9	–	92.4	94.5	99.1	94.1
CenterNet [30]	512	**100.0**	**98.8**	**97.9**	96.9	93.1	95.7	**99.9**	97.9
SrcLPD (ours)	512	**100.0**	98.7	**97.9**	**97.6**	94.2	**96.0**	**99.9**	98.1

As shown in Table 1, SrcLPD is compared with 7 advanced LPD methods, among which CenterNet [30] is regarded as the baseline model. The CenterNet is retrained on the CCPD dataset and shows outstanding performance. For a fair comparison, same image augmentation strategies with SrcLPD is deployed for the training of CenterNet. The proposed SrcLPD achieves state-of-the-art (SOTA) performance on 2 subsets. And SrcLPD exhibits the same or slightly fluctuating results on all subsets compared to the baseline. On the Blur 20k and FN 20k subsets, SrcLPD demonstrates the expected improvement, surpassing previous results by 0.7% and 0.3%. The proposed smooth regression constraint (SRC) and soft-coupling loss (SCL) integrate the advantage of both corner point and center point regression approach, making SrcLPD robust to blur and shooting distance disturbances. On the Rotate 10k and tilt 30k subsets, SrcLPD only suffers a negligible 0.1% degradation or totally consistent performance compared with SOTA. Therefore, the efficiency of the proposed SrcLPD for license plate detection in low-quality images has been verified. The detection visualization of randomly selected sample of each subset is shown in Fig. 4. We display the corner heatmap and the predicted quadrangle BBox in detail.

[2] https://github.com/aleju/imgaug.

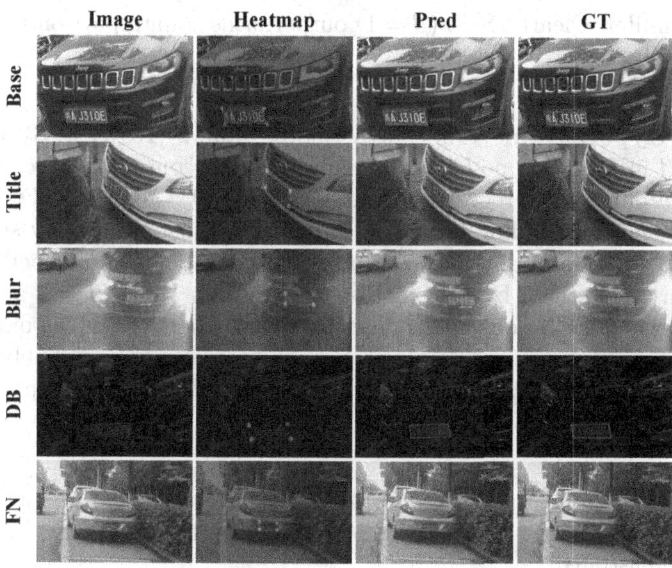

Fig. 4. Visualization results. From left to right are the input image, the corner heatmap, the quadrangle BBox, and GT. The examples shown are randomly selected from the corresponding subset.

4.4 Ablation Studies

All ablation studies using the following experimental settings are conducted. 1) 10k images randomly selected in the CCPD base train are adapted for training. 2) 4 complete hard cases subsets (Tilt, Blur, FN, and Challenge) are adopted for evaluation. 3) The threshold for the IoU and TIoU metrics are 0.7 and 0.5, respectively.

Smooth Regression Constraint. As shown in Table 2 (**Row 1, 2**), the direct application of detection heads at the corner-point and center-point in the model pipeline to construct an SRC without coupling loss significantly damages baseline performance. The detection metrics of all subsets are reduced. We attribute this degradation to the conflict between the optimization processes of different license plate detection heads.

Soft-Coupling Loss. As shown in Table 2 (**Row, 1, 3**), we first consider the smooth L1 function, which is formally similar to SCL (Eq. 1), to instantiate coupling loss. Smooth L1 still assigns large gradient when the results of the two regression heads are similar, which makes the optimization process oscillate and leads to sub-optimal results. In Table 2 (**Row 4**), SCL with segment point $\beta = 1$ is studied. Compared with smooth L1, SCL brings soft gradient to the left of the point β, which stabilizes the optimization when the results of different regression heads are similar. Clearly, SCL ($\beta = 1$) outperforms counterpart on the FN and Challenge subsets.

Effect of Hyperparameters. Furthermore, Table 2 (**Row, 4–8**) shows the impact of β in Eq. 1 on model performance. The discontinuity β regulates the degree of coupling of two regression strategies, i.e., CDM and CETCO, by changing the gradient of coupling loss. The coupling loss instantiated by smooth L1 shows that a small value of the discontinuity will degrade the model performance, which is further verified by the experimental results. When $\beta = 1$ and $\beta = 3$, the superiority of SCL does not appear. However, when $\beta = 5$, the outstanding detection performance is achieved. Furthermore, it also has relatively superior performance when $\beta = 8$. With the further increase of β, the model performance gradually degrades.

Table 2. Ablation studies on the proposed SRC and the coupling loss. IoU > 0.7 and TIoU > 0.5 are considered as evaluation protocols. The effect of hyper-parameter β in Eq. 1 is also illustrated.

Row	SRC	Coupling loss		Tilt 30k		Blur 20k		FN 20k		Challenge 50k	
				IoU	TIoU	IoU	TIoU	IoU	TIoU	IoU	TIoU
1	×	×		97.2	79.0	**96.8**	77.2	94.3	73.7	97.2	76.8
2	✓	×		96.9	77.9	96.2	76.9	93.2	73.6	96.8	76.7
3	✓	Smooth L1		97.3	78.7	96.1	75.5	92.3	72.6	96.0	74.9
4		**SCL (ours)**	$\beta = 1$	97.2	78.8	95.9	76.0	93.7	73.4	96.8	76.0
5			$\beta = 3$	97.1	79.1	96.1	76.5	92.3	73.2	96.3	76.0
6			$\beta = 5$	**97.5**	79.2	96.6	77.1	**94.8**	**75.1**	**97.4**	**77.5**
7			$\beta = 8$	**97.5**	**79.3**	96.5	**77.5**	93.4	73.8	96.6	77.1
8			$\beta = 12$	96.9	77.8	95.2	75.1	89.7	71.6	95.0	74.7

5 Conclusion

In this paper, we proposed SrcLPD, which integrates two bounding box regression strategies for license plate detection. Our SrcLPD is robust to blurring and distortion, therefore showing superior license plate detection performance in low-quality images. Rather than trivially adding detection heads to existing license plate detectors, a novel soft-coupling loss (SCL) is designed to better model and instantiate the proposed smooth regression constraint (SRC) approach. Experiments demonstrate that SCL is a significant factor in explicitly exploiting the advantages of both regression methods and stabilizing the training process. In the future, robust license plate detection models in more types of low-quality images will be investigated.

Acknowledgment. This work was supported by the National Natural Science Foundation of China (U1909209), the National Key Research and Development Program of China (2021YFE0100100, 2021YFE0205400), the Research Funding of Education of Zhejiang Province (Y202249784), and the Fundamental Research Funds for the Provincial Universities of Zhejiang (GK219909299001-023).

References

1. Fan, X., Zhao, W.: Improving robustness of license plates automatic recognition in natural scenes. IEEE Trans. Intell. Transp. Syst. **23**(10), 18845–18854 (2022)
2. Ge, Z., Liu, S., Wang, F., Li, Z., Sun, J.: Yolox: Exceeding yolo series in 2021. arXiv preprint arXiv:2107.08430 (2021)
3. Glorot, X., Bordes, A., Bengio, Y.: Deep sparse rectifier neural networks. In: Proceedings of the Fourteenth International Conference on Artificial Intelligence and Statistics, pp. 315–323. JMLR Workshop and Conference Proceedings (2011)
4. He, K., Zhang, X., Ren, S., Sun, J.: Deep residual learning for image recognition. In: 2016 IEEE Conference on Computer Vision and Pattern Recognition (CVPR), pp. 770–778 (2016). https://doi.org/10.1109/CVPR.2016.90
5. Hsu, G.S., Chen, J.C., Chung, Y.Z.: Application-oriented license plate recognition. IEEE Trans. Veh. Technol. **62**(2), 552–561 (2012)
6. Ioffe, S., Szegedy, C.: Batch normalization: accelerating deep network training by reducing internal covariate shift. In: International Conference on Machine Learning, pp. 448–456. PMLR (2015)
7. Jiang, Y., et al.: An efficient and unified recognition method for multiple license plates in unconstrained scenarios. IEEE Trans. Intell. Transp. Syst. **24**, 5376–5389 (2023)
8. Kingma, D.P., Ba, J.: Adam: a method for stochastic optimization. arXiv preprint arXiv:1412.6980 (2014)
9. Law, H., Deng, J.: CornerNet: detecting objects as paired keypoints. In: Proceedings of the European conference on computer vision (ECCV), pp. 734–750 (2018)
10. Lin, T.Y., Dollár, P., Girshick, R., He, K., Hariharan, B., Belongie, S.: Feature pyramid networks for object detection. In: Proceedings of the IEEE Conference on Computer Vision and Pattern Recognition, pp. 2117–2125 (2017)
11. Lin, T.Y., Goyal, P., Girshick, R., He, K., Dollár, P.: Focal loss for dense object detection. In: Proceedings of the IEEE International Conference on Computer Vision, pp. 2980–2988 (2017)
12. Liu, Y., Jin, L., Xie, Z., Luo, C., Zhang, S., Xie, L.: Tightness-aware evaluation protocol for scene text detection. In: Proceedings of the IEEE/CVF Conference on Computer Vision and Pattern Recognition, pp. 9612–9620 (2019)
13. Long, R., et al.: Parsing table structures in the wild. In: Proceedings of the IEEE/CVF International Conference on Computer Vision, pp. 944–952 (2021)
14. Nguyen, D.L., Putro, M.D., Vo, X.T., Jo, K.H.: Triple detector based on feature pyramid network for license plate detection and recognition system in unusual conditions. In: 2021 IEEE 30th International Symposium on Industrial Electronics (ISIE), pp. 1–6. IEEE (2021)
15. Qin, S., Liu, S.: Towards end-to-end car license plate location and recognition in unconstrained scenarios. Neural Comput. Appl. **34**(24), 21551–21566 (2022)
16. Redmon, J., Divvala, S., Girshick, R., Farhadi, A.: You only look once: unified, real-time object detection. In: Proceedings of the IEEE Conference on Computer Vision and Pattern Recognition, pp. 779–788 (2016)

17. Redmon, J., Farhadi, A.: Yolov3: an incremental improvement. arXiv preprint arXiv:1804.02767 (2018)
18. Ren, S., He, K., Girshick, R., Sun, J.: Faster R-CNN: towards real-time object detection with region proposal networks. In: Advances in Neural Information Processing Systems, vol. 28 (2015)
19. Silva, S.M., Jung, C.R.: A flexible approach for automatic license plate recognition in unconstrained scenarios. IEEE Trans. Intell. Transp. Syst. **23**(6), 5693–5703 (2021)
20. Silva, S.M., Jung, C.R.: License plate detection and recognition in unconstrained scenarios. In: Proceedings of the European conference on computer vision (ECCV), pp. 580–596 (2018)
21. Wang, Q., Lu, X., Zhang, C., Yuan, Y., Li, X.: LSV-IP: large-scale video-based license plate detection and recognition. IEEE Trans. Pattern Anal. Mach. Intell. **45**(1), 752–767 (2022)
22. Wang, W., et al.: Shape robust text detection with progressive scale expansion network. In: Proceedings of the IEEE/CVF Conference on Computer Vision and Pattern Recognition, pp. 9336–9345 (2019)
23. Wang, W., et al.: Pan++: towards efficient and accurate end-to-end spotting of arbitrarily-shaped text. IEEE Trans. Pattern Anal. Mach. Intell. **44**(9), 5349–5367 (2021)
24. Wang, W., et al.: Efficient and accurate arbitrary-shaped text detection with pixel aggregation network. In: Proceedings of the IEEE/CVF International Conference on Computer Vision (ICCV), October 2019
25. Wang, Y., Bian, Z.P., Zhou, Y., Chau, L.P.: Rethinking and designing a high-performing automatic license plate recognition approach. IEEE Trans. Intell. Transp. Syst. **23**(7), 8868–8880 (2021)
26. Xie, L., Ahmad, T., Jin, L., Liu, Y., Zhang, S.: A new CNN-based method for multi-directional car license plate detection. IEEE Trans. Intell. Transp. Syst. **19**(2), 507–517 (2018)
27. Xu, Z., et al.: Towards end-to-end license plate detection and recognition: a large dataset and baseline. In: Proceedings of the European Conference on Computer Vision (ECCV), pp. 255–271 (2018)
28. Yuan, Y., Zou, W., Zhao, Y., Wang, X., Hu, X., Komodakis, N.: A robust and efficient approach to license plate detection. IEEE Trans. Image Process. **26**(3), 1102–1114 (2016)
29. Zhang, W., Mao, Y., Han, Y.: SLPNet: towards end-to-end car license plate detection and recognition using lightweight CNN. In: Peng, Y., et al. (eds.) PRCV 2020. LNCS, vol. 12306, pp. 290–302. Springer, Cham (2020). https://doi.org/10.1007/978-3-030-60639-8_25
30. Zhou, X., Wang, D., Krähenbühl, P.: Objects as points. arXiv preprint arXiv:1904.07850 (2019)

A Feature Refinement Patch Embedding-Based Recognition Method for Printed Tibetan Cursive Script

Cai Rang Dang Zhi[1,2,3], Heming Huang[1,2,3](✉), Yonghong Fan[1,2,3], and Dongke Song[1,2,3]

[1] School of Computer Science and Technology, Qinghai Normal University, Xining 810008, China
tb-ad@qq.com

[2] State Key Laboratory of Tibetan Intelligent Information Processing and Application, Xining 810008, China

[3] Key Laboratory of Tibetan Information Processing, Ministry of Education, Xining 810008, China

Abstract. Recognition of Tibetan cursive scripts has important applications in the field of automated Tibetan office software and ancient document conservation. However, there are few studies on recognition of Tibetan cursive scripts. This paper proposes a printed Tibetan cursive script recognition method based on feature refinement patch embedding. Firstly, the feature refinement patch embedding module (FRPE) serializes the line text image of feature sequences. Secondly, a global modeling of feature vectors is carried out by using a single transformer encoder. Finally, the output of the recognition result is decoded by using a fully connected layer. Experimental results show that, compared with the baseline model, the proposed method improves the accuracy by 9.52% on the dataset CSTPD, a database containing six Tibetan cursive fonts. Moreover, it achieves an average accuracy rate of 92.5% on the dataset CSTPD. Similarly, it also works better than the baseline model on Tibetan text recognition synthetic data for natural scene images.

Keywords: Tibetan recognition · cursive scripts · feature refinement patch embedding · Transformer

1 Introduction

The recognition of Tibetan cursive scripts is crucial for automated Tibetan office software and ancient document conservation. It enables efficient processing of Tibetan language documents by converting handwritten or cursive texts into digital format [1]. This enhances productivity and efficiency in Tibetan office environments. Additionally, recognizing cursive scripts is important for preserving ancient Tibetan documents, which hold immense historical and cultural value.

Q. Liu et al. (Eds.): PRCV 2023, LNCS 14426, pp. 383–399, 2024.
https://doi.org/10.1007/978-981-99-8432-9_31

By accurately digitizing these scripts, researchers and conservators can create digital replicas, ensuring long-term preservation and accessibility. Overall, further advancements in Tibetan cursive script recognition are essential for enhancing office software and preserving Tibetan culture and history. However, to the serious overlap and conglutination of strokes, character recognition faces difficulties such as segmentation, feature extraction, context dependence, and Scarcity of data. Therefore, it is necessary to conduct complete and systematic research on the recognition of cursive scripts.

To mitigate the influence of segmentation errors on the final recognition results, some researchers have opted for text line recognition as their focus. By treating the entire line of text as a single recognition object, they aim to overcome the challenges posed by the overlapping and connected strokes in handwritten or cursive scripts. RenQingDongZhu uses the CRNN networks [2] to directly recognize printed Tibetan and ancient woodcut text lines and get a satisfactory result. However, the module BiLSTM of CRNN [3] needs a huge amount of computation, and it reduces the efficiency greatly.

To solve the above problems, this paper proposes a printed Tibetan cursive script recognition method based on feature refinement patch embedding. In the proposed recognition approach, the first step involves standardizing the image size of the long text line to 64 × 320 pixels. Following this, the unified text line is divided into patches of equal size. Each patch is then transformed [4] into a one-dimensional vector using the feature refinement module. These vectors are further processed by a single Transformer encoder to encode the input information. Lastly, the output from the previous steps is decoded using a fully-connected layer to generate recognition results at the character level for Tibetan script. This decoding process involves mapping the encoded information to specific characters in the Tibetan script, enabling the recognition system to produce accurate results. The advantages of this method are as follows: it does not require character segmentation; it is not limited by the width of the text line and can directly output the textual content of the entire text line; and there is no need for an explicit language model in the model framework, and inference is fast. The average recognition accuracy of the proposed model reaches 92.5% on the dataset CSTPD which include 6 printed Tibetan cursive scripts and the framework of the proposed model is more concise than that of CRNN, as shown in Fig. 1.

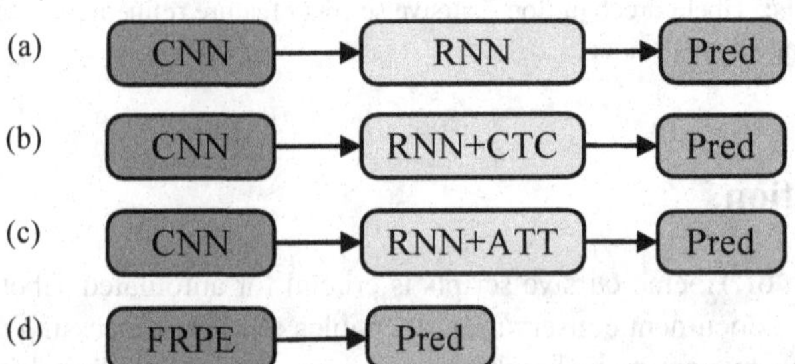

Fig. 1. Generic text recognition network framework. (a) Traditional CRNN model. (b) (c) Adding CTC and attention decoding methods to the traditional CRNN model framework. (d) The framework of the proposed method.

The contributions of this work are as follows:

(1) A transformer framework based on feature refinement patches is designed to improve the effect of recognition.
(2) A label-slicing algorithm suiting for images of cursive scripts Tibetan printed is proposed.
(3) The pre-trained model of printed Tibetan cursive scripts, code, and the test set of CSTPD are shared on Gitcode[1].

2 Related Work

In recent years, the research on text recognition has made great progress. In this section, the related works are reviewed, and they are grouped into two categories namely the traditional CNN-based recognition methods and the Transformer-based recognition methods.

2.1 CNN-Based Text Recognition

Since the great success of CNNs in image classification [5], their applications in computer vision have become increasingly widespread, including the recognition of printed Tibetan script.

The process of early text recognition methods are as follows. At first, given cropped images of text instances and apply CNNs to encode the images into the feature space; then, decoders are applied to extract characters from these features. The whole recognition step can be divided into two stages: text regions are predicted by a detector [6], and the text recognizer recognizes all characters in the cropped image instances [7]. For example, in [8], based on baseline location information and the center of gravity of the connected domain, Chen proposes a Ding-cutting method for Tibetan characters, and a CNN is used for final recognition. To enhance the feature extraction ability of convolution in CNN, Wang uses CBAM hybrid attention for the results of the pooling layer [9], and it effectively improves the recognition effect.

When a neural network is used for the recognition of printed Tibetan cursive script [10], the objects are line-level image text units instead of letters or characters [11]. Since the strokes of the Tibetan cursive scripts are seriously interlaced and adhered, for example, the strokes of current character are interlaced even with the preceding seven characters, as shown in Fig. 2. Therefore, character segmentation is a challenging problem to the recognition of printed Tibetan cursive script.

CNN is used for feature extraction of line level text images. This framework extracts and encodes features from images for text recognition [12]. A recurrent neural network (RNN) is then used to capture the sequence dependencies, and it convert the line-level text images into an image sequence prediction task. The final features are decoded into entire character sequences.

[1] https://gitcode.net/weixin_40632216/wm_ocr.

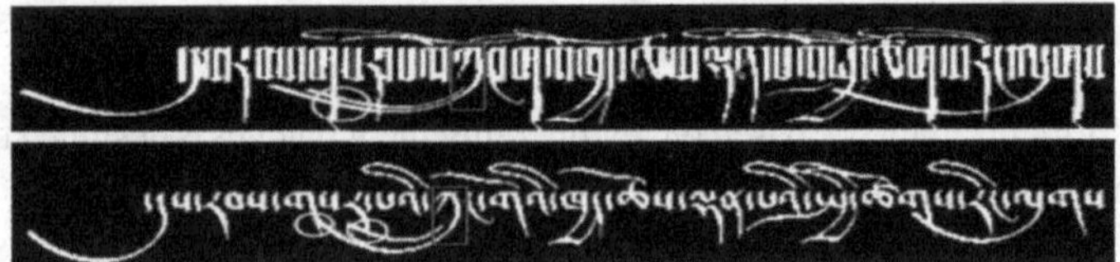

Fig. 2. Sample of interlaced strokes in Tibetan cursive scripts. The red rectangle indicates the current character while the yellow circles indicate the positions where the stroke of the current character intersects with that of the preceding characters. (Color figure online)

2.2 Transformer-Based Text Recognition

Transformer [13] has been widely used in text recognition models due to its strong ability of capturing long-term dependencies.

Dosovitskiy et al. attempt to apply Transformer directly to image recognition tasks by proposing ViT (Vision Transformer) [14]. Firstly, an image is segmented into multiple patches; secondly, the linear embedding sequence of of each patch is used as the input of a transformer; finally, a multilayer perceptron is used for classification. Inspired by ViT, Li et al. propose the TrOCR [15]. Firstly, a pre-trained image transformer is used as the encoder; secondly, an encoder-decoder model based on Transformer is designed with a pre-trained text transformer as the decoder; finally, pre-training is performed on a large amount of synthetic data. Experiments show that the best results are achieved in such tasks as printed character recognition, handwrittencharacrer recognition, and scene text recognition tasks. To be able to fully utilize the multiscale features in convolution of a Transformer, Zhang et al. propose a multiscale deformable self-attention-based scene text detection and recognition method [16].

Transformer has achieved good results in many recognition tasks, and to further reduce the sequence length and simulate the down sampling function in CNN, Liu et al. propose Swin transformer [17]. In Swin transformer, a 4×4 convolution is used for 4-fold downsampling, and the length of each patch is converted to a one-dimensional sequence with length $4 \times 4 \times C$, where C is the number of image channels. Swin Transformer effectively solves the problem of sequence overlength;furthermore, the offset window is proposed to reduce the computational effort, and self-attention is performed in the local non-overlapping window. But the computational complexity of Swin transformer is higher. The above studies finally prove that Swin transformer exceeds CNN in image classification, target detection, and segmentation tasks. Swin Transformer has great potential as a visual feature extraction backbone network. Therefore, Huang et al. propose an end-to-end scene text detection and recognition method called Swin Textspotter by using Swin Transformer as a backbone network [18]. Besides, a mixture of CNN and Transformer are used by some researchers [19]. For example, Fu et al. propose t a hybrid of CNN and Transformer to extract visual features from scene text images [20], and then a contextual attention module consisting of a variant of the transformer, is used as part of the decoding process. Similarly, Raisiet,et al. [21] apply a two-dimensional learnable sinusoidal position coding that allows Transformer encoders to focus more on spatial dependencies.

How to strike a balance between accuracy and efficiency is an important issue, as ViT becomes an increasingly common approach in text recognition tasks. Ateinza simplifies the encoder-decoder framework for text recognition by using ViT as an encoder and a non-autoregressive decoder [22]. ViT consists of a Transformer encoder where the word embedding layer is replaced by a linear layer.

When text detection and text recognition are trained as a whole, the requirement of labeling cost is higher. To reduce the annotation cost, Yair et al. train Transformer by using the Hungarian matching loss, which can be trained in a weakly supervised manner and achieve the same effect of fully supervised learning [23].

3 Methodology

For character recognition, two modules are necessary: one for feature extraction and another for sequence modeling, and often they are often complex in design. In this paper, a simple single Transformer encoder feature extractor is used to unify the sequence conversion and model tasks, as shown in Fig. 3.

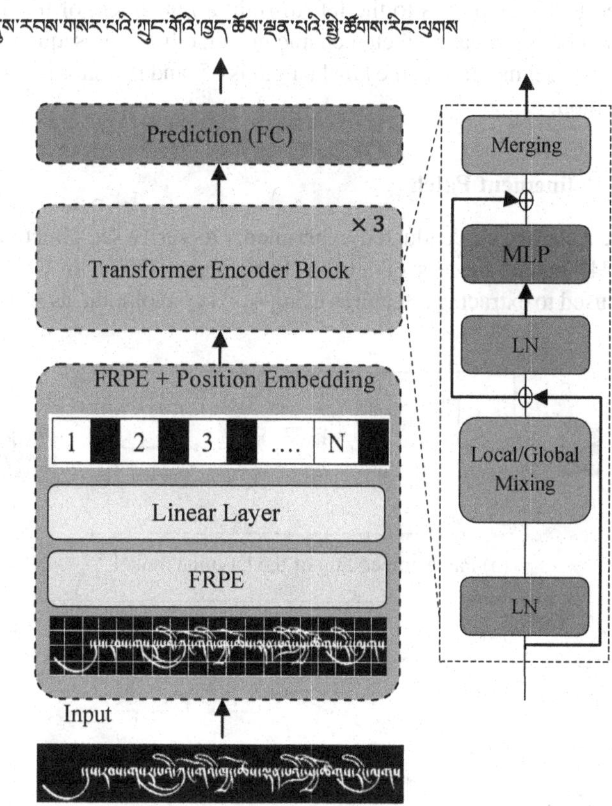

Fig. 3. A Network Framework for the Recognition of Printed Tibetan Cursive Script Based on Feature Refinement Patch Embedding.

3.1 Label Text Split

By default, Tibetan text is segmented at letter level, however, it is not conducive to label alignment. Because it destroys syllable, the basic unit of Tibetan text, see the second row of Table 1.

Table 1. Different label split methods

Label Text	་དུས་རབས་གསར་པའི་ཀྲུང་གོའི་ཁྱད་ཆོས་ལྡན་པའི་སྤྱི་ཚོགས་རིང་ལུགས	Text length
Letters	་ད ུ ས ་ ར བ ས ་ ག ས ར ་ པ འ ི ་ ཀ ྲ ུ ང ་ ག ོ འ ི ་ ཁ ྱ ད ་ ཆ ོ ས ་ ལ ྡ ན ་ པ འ ི ་ ས ྤ ྱ ི ་ ཚ ོ ག ས ་ ར ི ང ་ ལ ུ ག ས	61
Characters	་དུ ས ་ ར བ ས ་ ག ས ར ་ པ འི ་ ཀྲུ ང ་ གོ འི ་ ཁྱ ད ་ ཆོ ས ་ ལྡ ན ་ པ འི ་ སྤྱི ་ ཚོ ག ས ་ རི ང ་ ལུ ག ས	45

For example, in letter level, one complete character དུ is separated to ད and ུ, and there are many such errors. In character level, however, structure of Tibetan syllable has not been destroyed and it conforms to the left-to-right arrangement of the Tibetan text. A blank is inserted between each Tibetan character to facilitate subsequent segmentation. For character level segmentation, the label length is 45, and it is shorter than that of letter level segmentation.

3.2 Feature Refinement Patch

In their study, Asher et al. conducted experiments to verify the effectiveness of patch embedding [24] in the success of Vision Transformer (ViT). In SVTR [25], patch embedding is used to extract text features using 4×4 convolution, as shown in Fig. 4(a).

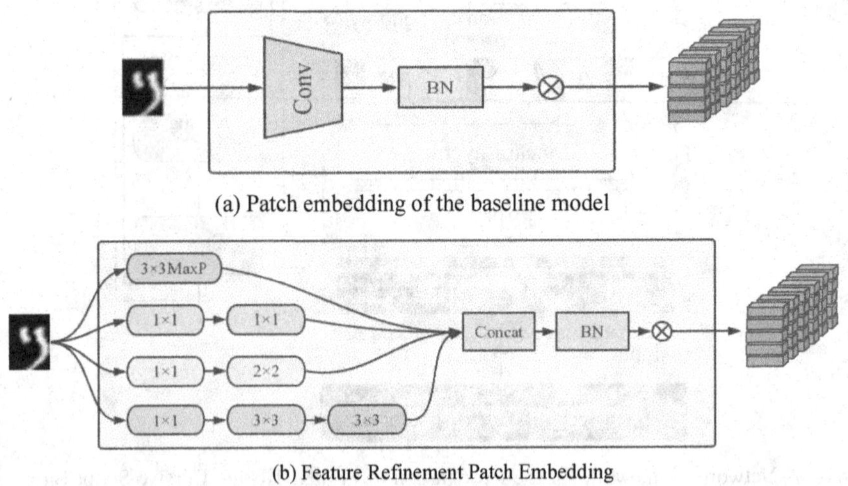

(a) Patch embedding of the baseline model

(b) Feature Refinement Patch Embedding

Fig. 4. Patch embedding.

Figure 4(a) depicts the patch embedding technique employed in the baseline model. Initially, the patch embedding module takes the input text image and utilizes a convolutional neural network (CNN) to extract relevant features, resulting in a three-dimensional feature map. Afterwards, a convolutional layer and a non-linear activation function are employed to process this feature map, generating a two-dimensional feature matrix. This matrix can be interpreted as a collection of small blocks that effectively represent the spatial dimensions of the text image.

In Tibetan cursive script, strokes often intertwine with each other, and using a smaller convolution window allows for the extraction of more intricate features. When the convolution window is larger, it becomes more challenging to accurately identify the intersecting strokes. Hence, we propose feature refinement patch embedding, which chooses smaller convolution window sizes to efficiently capture the desired finer features, as shown in Fig. 4(b).

To better characterize the detailed features, three different convolutional kernels are used at first to extract features, and then, all features extracted by the different convolutional kernels are fused. In addition, small-sized convolutions instead of large-sized convolutions can reduce network parameters, increase network depth, and expand the receptive field. The size of the original image is $W \times H \times C$, which is converted into D $\times \frac{W}{4} \times \frac{H}{4}$ dimensional patch vectors by the patch feature refinement embedding module. The FRPE module is performed as follows:

Firstly, the original image with an input size of 64×320 is directly subjected to a pooling operation, where the size of the convolution kernel is 3×3 and the stride is 2. Direct pooling operations can generate features that is different from conventional convolutions, and the output feature map size is $C \times \frac{w}{2} \times \frac{h}{2}$, where C is the number of channels and its value is 3.

Secondly, the structure of the 2nd and 3rd branches is like the Conv + Conv operation, and the differences are the size of the convolution kernel and the number of convolution kernels; the feature information of different depths can be extracted by the above two branches. The outputs of the 2nd and 3rd branches are $384 \times \frac{w}{2} \times \frac{h}{2}$ and $256 \times \frac{w}{2} \times \frac{h}{2}$, respectively.

Again, a convolution operation is added to the 4th branch to enhance the semantic information, and the feature map output is $256 \times \frac{w}{2} \times \frac{h}{2}$.

Finally, the sampling results of the four different convolutional branches are fused by the join operation, and the obtained feature map is $899 \times \frac{w}{2} \times \frac{h}{2}$. Since a 2-layer patch feature refinement embedding module is used, the size of the final feature is $1795 \times \frac{w}{4} \times \frac{h}{4}$.

3.3 Transformer Encoder Block

The basic architecture of the Transformer Encoder Block is a visual transformer [16], which takes the original image patches and uses them directly for image feature extraction by linear projection, as shown in Fig. 5. A Transformer Encoder Block consists of four components: layer normalization (LN), attention, multi-layer perceptron (MLP), and merging.

Firstly, the input of the Transformer Encoder Block is an RGB image segmented by the feature refinement module into non-overlapping patches $x \in R^{W \times H \times C}$. Secondly,

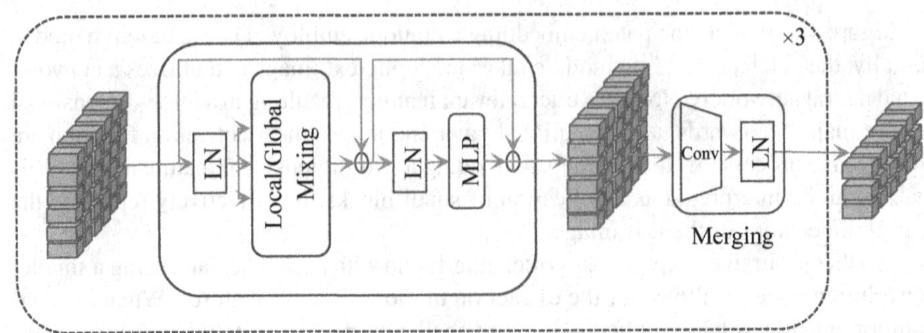

Fig. 5. Transformer encoder block.

the image is transformed into a 2D sequence of patches with $X_p \in R^{N\times(P^2C)}$, where P^2 is the resolution of each image patch, and C is the feature of X_p number of channels; thus, a 2-dimensional image is represented as a sequence with $N \frac{WH}{P^2}$ markers and is used as a valid input sequence of the transformer block. Finally, markers of X_p are linearly transcribed into a D-dimensional patch embedding. Thus, the input to the Transformer encoder is:

$$Z_0 = \left[X_{class}; X_p^1E; X_p^2E; \ldots.; X_p^NE\right] + E_{pos}, \quad (1)$$

where $X_{class} \in R^{1\times D}$ is the class embedding, $E \in R^{D\times(P^2C)}$ is the linear projection matrix, and $E_{pos} \in R^{(N+1)\times D}$ is the location embedding.

The output of the Transformer encoder is:

$$Z_l\prime = MSA(LN(Z_{l-1})) + Z_{l-1}, \quad (2)$$

where $l = 1, 2, \ldots, L, L$ is the depth of the encoder block.

The output of MLP is:

$$Z_l = MLP\left(LN\left(Z_l'\right)\right) + Z_{l-1}'. \quad (3)$$

Character predictions are formed by combining a series of linear predictions:

$$y_i = Linear\left(Z_L'\right), \quad (4)$$

where $i = 1, \ldots, S, S$ refers to the sum of the maximum text length and two start position markers.

The transformer encoder block can enhance the correlation between textual and non-textual components through local and global mixing and use different self-attention mechanisms to sense contextual information.

There are two different feature fusion methods: local mixing and global mixing. Where local mixing refers to mixing different channel features at the same scale to obtain a richer feature representation. Specifically, it is achieved by concatenating or weighted average of different channel features. Global mixing refers to fusing features of different scales to improve the receptive field of the model and capture global image information. Specifically, it is achieved by concatenating or weighted average of features of different scales. The two feature fusion methods can be combined to further improve the performance of the model.

3.4 Merging and SubSample

To reduce computational cost and remove redundant features, a 3×3 convolution kernel is used in the Transformer encoder to perform down-sampling with a step of 2 in the height dimension and a step of 1 in the width dimension. Thus, the dimension of features reduces from $h \times w \times d$ to $\frac{h}{2} \times w \times d$.

The merging operation halves the height while maintaining a constant width. Compressing the height dimension creates a multiscale representation of each character without affecting the patch layout in the width dimension.

In the last stage, the merging operation is replaced by the subsample operation. At first, the pooling step in the height dimension is set to 1 and an MLP is connected; and then, recognition is performed by a nonlinear activation layer followed by a dropout layer.

4 Experiments

4.1 Datasets and Evaluation Metrics

There is no publicly available database for the recognition of printed Tibetan cursive scripts. Therefore, we have established a database containing six Tibetan cursive fonts (CSTPD). Database CSTPD is divided into two groups, 12903162 for training and 21540 for testing. There are 6 types of cursive scripts, namely, Betsu, Drutsa, Tsutong, Tsuring, Tsumachu, and Chuyig. Table 2 show us some samples of six types of Tibetan cursive fonts.

Table 2. Some samples of six printed Tibetan cursive scripts of CSTPD

Name of Tibetan cursive scripts	Sample
Betsu	
Drutsa	
Tsutong	
Tsuring	
Tsumachu	
Chuyig	

To the Tibetan text recognition synthetic data for natural scene images, 2000000 for training and 20000 for testing. It is provided by the Research and application demonstration of multimodal network audio-visual recognition technology for regional dialects and ethnic language project team. Some of its data instances are shown in Fig. 6.

During the experiment, accuracy, as shown in Formula (5), and character error rates, as shown in Formula (6), are used to evaluate the performance of the model on the

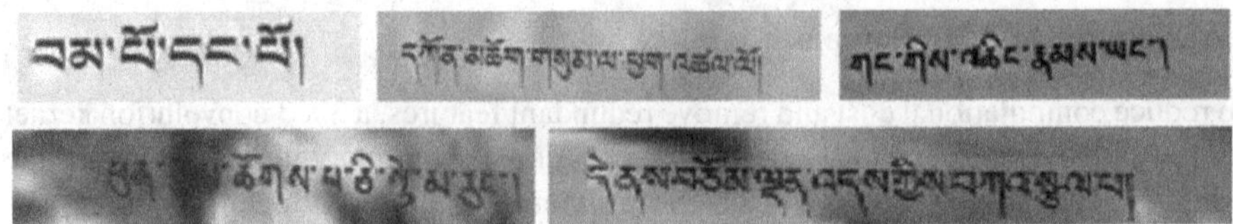

Fig. 6. Some samples of Tibetan text recognition synthetic data for natural scene images.

CSTPD dataset.

$$\text{Accuracy} = \frac{\text{Number of correctly matched samples}}{\text{Total number of samples}} \times 100(\%), \quad (5)$$

$$\text{Character Error Rate} = \frac{\sum_{l=1}^{L} \frac{S+D+I}{N}}{L}, \quad (6)$$

where L represents the total number of samples, N represents the number of characters in a sample, and S, D, and I represent the number of character substitutions, deletions, and insertions, respectively.

The depth of the Transformer encoder block is set to 3, embedded_ dim is set to 768, and the maximum text length is set to 50. Other parameters are shown in Table 3.

Table 3. Model parameters

Learning rate	Optimizer	Activation function	Text length	Embed_dim
0.0005	AdamW	GELU	50	768

4.2 Comparative Recognition Experiment

A comparative experiment is conducted between the proposed method and the SVTR method on the CSTPD, and the results are shown in Fig. 7. The recognition accuracy of the FRPE method reaches 92%, and it is 9.52% higher than that of the baseline model SVTR.

Figure 8 shows us the comparative experiment of the baseline model and the proposed model on 6 cursive scripts. It can be seen that: The accuracies of baseline model SVTR are over 90% to three types of fonts: Betsu, Drutsa, and Tsumachu. The accuracies of the baseline model to Chuyig and Tsutong have reached over 80%. However, the accuracy of the baseline model on Tsuring is poor, only 45%. The accuracies of model FRPE are over 90% on such five font types as Betsu, Drutsa, Tsumac hu, Chuyig, and Tsutong. And its accuracy on Tsuring is 88%. Compared with the baseline model SVTR, the proposed method FRPE improves the recognition accuracy of Tsuring to 95.55%.

The experiments demonstrate that the superior performance of our method, which yield an average recognition accuracy of 92.5% to all 6 cursive scripts. Our findings

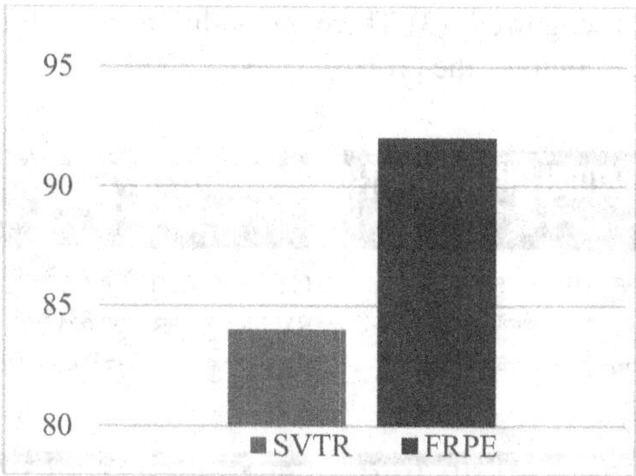

Fig. 7. Comparison experiment with baseline model.

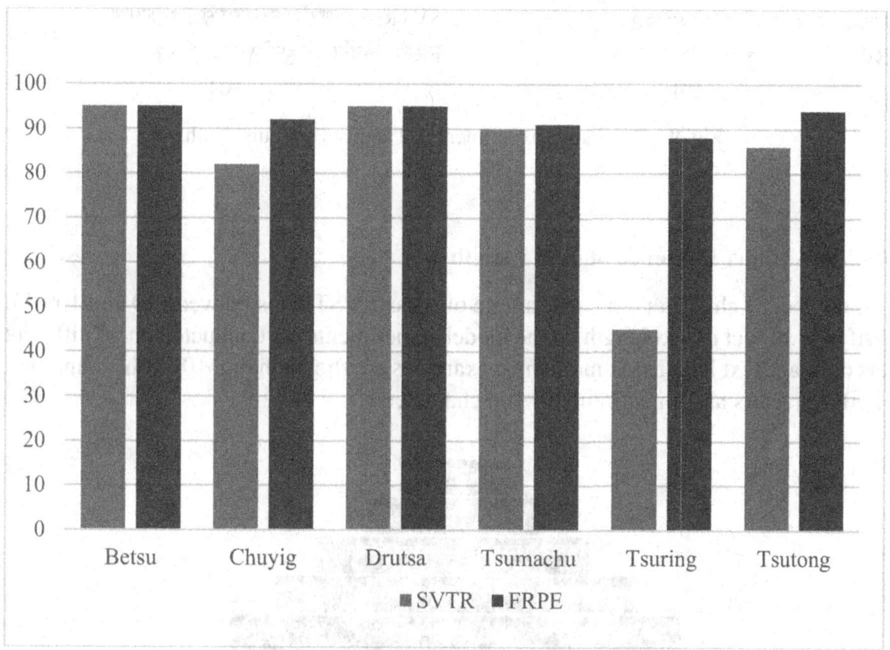

Fig. 8. Comparative experiment of 6 cursive scripts.

highlight the effectiveness of the proposed approach in accurately recognizing text in challenging cursive scripts.

To further validate the effectiveness of this method, a qualitative experiment is conducted, and the results are shown in Fig. 9. Figure 9 presents some examples where the model FRPE outperforms the baseline model SVTR. For example, as shown in the first row and first column of Fig. 9 (a), the model SVTR has three errors in recognition of Tsuring script: (1) Syllable segmentation error between syllables ཀུགས and བོད; (2)

The character རེis misrecognized; (3) There are additional syllable delimiter "·"in the recognition process. In contrast, the proposed model FRPE avoid these errors.

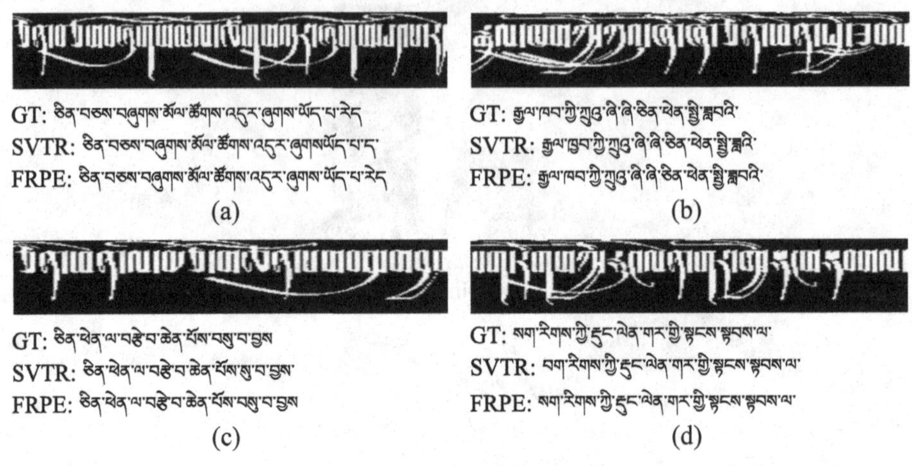

Fig. 9. Qualitative experiment. GT indicates ground truth.

4.3 The Model Influence of Text Length

The number of characters in each sample of dataset CSTPD is between 10 and 100. To verify the impact of text length on the model, experiments are conducted on 10 different sets of image text lengths. Some sample examples are shown in Fig. 10, with a minimum of 10 characters and a maximum of 100 characters.

Fig. 10. Ten different lengths of image text test samples.

In the investigation presented in Fig. 11, we aim to examine the impact of text length on the rate of errors in character recognition. Notably, the analysis reveals those characters with a length of 10 exhibit the highest error rate, amounting to an alarming 0.2275. However, as the length of characters increases, a discernible trend emerges wherein the error rate gradually diminishes. Remarkably, of particular significance is the notable reduction in character recognition errors as the character length surpasses 30. This finding underscores the significance of text length in influencing the accuracy of character recognition systems.

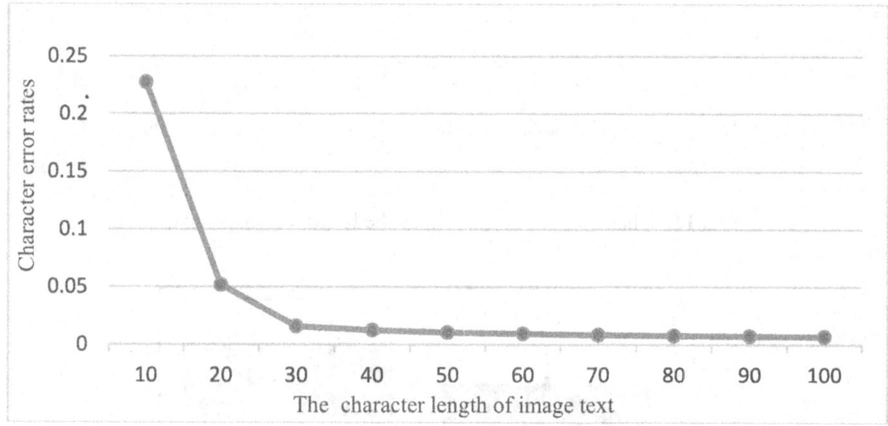

Fig. 11. Characters error rate versus the length of image text.

However, the character error rate remains nearly the same for character lengths greater than 30. This shows that when the length of the character increases to a certain length, the character length of the image will no longer affect the recognition effect.

To further ensure the fairness of the experiment, the number of samples with different text lengths will be kept consistent. The total number of samples is 3120. The experimental results are consistent with the previous experimental results.

Figure 12 shows that when the number of test samples is consistent, there is a decrease in the character recognition error rate, compared with the experimental results in Fig. 11. However, as the length of characters increased, the decreasing trend of character recognition error rate is basically consistent with that in Fig. 11.

4.4 Performance of the Model on Complex Tibetan Image Text Datasets

To further validate the effectiveness of the improved model, a comparison test is conducted on a Tibetan text recognition synthetic data for natural scene images. The baseline model SVTR and FRPE have the same number of rounds, and the model is validated once every 2000 iterations. The whole process is visualized by box-and-line and violin plots as shown in Fig. 13.

Figure 13 shows us that, compared to the model SVTR, FRPE has a taller and narrower box and a shorter whisker length, which means that the predictions of the model FRPE are closer to the actual distribution. The violin plots of the two models are shown in Fig. 14.

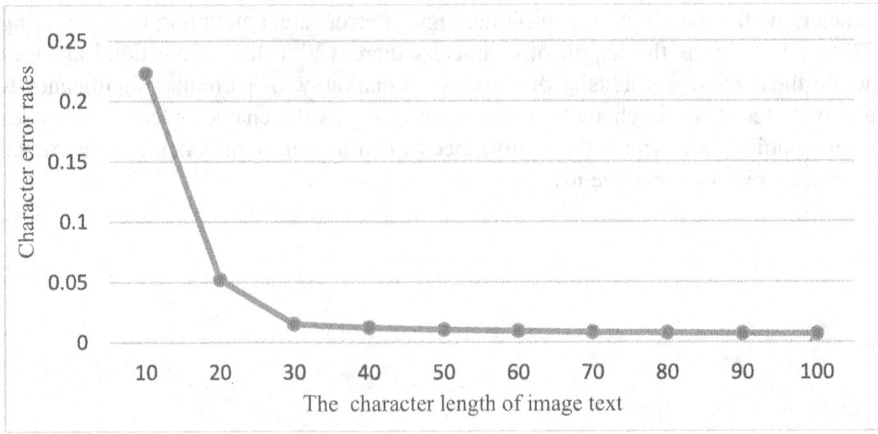

Fig. 12. Characters error rate versus the length of image text.

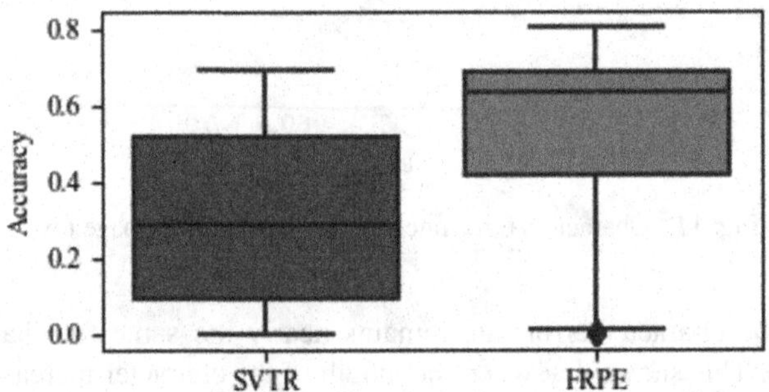

Fig. 13. Box Plots. The boxes indicate the interquartile range of the data, the horizontal line indicates the median, and the whiskers indicate the overall range of the data.

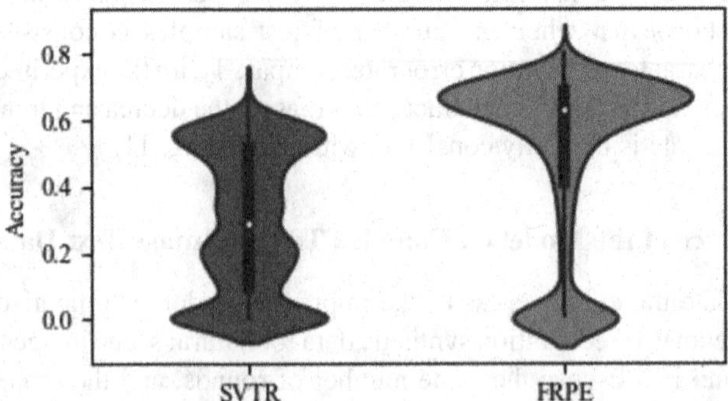

Fig. 14. Violin Plots. The white dot represents the median, and the range of the black box is from the lower quartile to the upper quartile. The thin black lines represent the whiskers. (Color figure online)

Figure 14 shows us the model FRPE has a higher and wider violin, which means that the predicted results of model FRPE are closer to the distribution of actual test scores. In contrast, model SVTR has a lower and narrower violin, which may mean that the predicted results of model SVTR deviate more from the distribution of actual test scores.

Through the above comparison, it can be concluded that, comparing to model SVTR, model FRPE is better at predicting complex Tibetan text images because its prediction results are closer to the distribution of actual test scores. This indicates that model FRPE has better prediction accuracy and reliability.

5 Conclusion

In this research paper, we introduce a novel approach for recognizing characters of printed Tibetan cursive scripts. Our method is based on feature refinement patch embedding, which aims to enhance the recognition performance of Tibetan cursive scripts, with a specific focus on Tsuring cursive script. To effectively capture the intricate stroke details of Tibetan cursive script, we employ feature refinement facet embedding to extract discriminative features. These features are then fed into a single Transformer encoder for sequence modeling. Finally, a fully connected layer is utilized to generate the prediction results. Through comprehensive comparative experiments, we demonstrate the superior performance of our proposed model, FRPE, in recognizing printed Tibetan cursive scripts. This research adheres to the writing style commonly found in scientific literature.

Acknowledgement. This work is supported by the National Science Foundation of China under Grant No. 62066039, Natural Science Foundation of Qinghai Province under Grant No. 2022-ZJ-925, and Research and Application Demonstration of Multimodal Network Audio-Visual Recognition Technology for Regional Dialects and Ethnic Language No. 20231201, and the "111" Project Grant under No. D20035.

References

1. Zhang, G.: Scene Tibetan Detection and Recognition System for Mobile Device. Northwest University for Nationalities (2022)
2. RenQingDongZhu. Tibetan Ancient Book Woodcut Character Recognition Based on Deep Learning. Tibet University (2021)
3. Shi, B., Bai, X., Yao, C.: An end-to-end trainable neural network for image-based sequence recognition and its application to scene text recognition. IEEE Trans. Pattern Anal. Mach. Intell. **39**(11), 2298–2304 (2017)
4. Raisi, Z., Naiel, M.A., Younes, G., et al.: Transformer-based text detection in the wild. In: 2021 IEEE/CVF Conference on Computer Vision and Pattern Recognition Workshops (CVPRW), pp. 3162–3171. IEEE, Nashville, TN, USA (2021)
5. Krizhevsky, A., Sutskever, I., Hinton, G.E.: ImageNet classification with deep convolutional neural networks. Commun. ACM **60**(6), 84–90 (2017)
6. Yang, W., Zou, B.: A character flow framework for multi-oriented scene text detection. J. Comput. Sci. Technol. **36**, 465–477 (2021)

7. Ronen, R., Tsiper, S., Anschel, O., Lavi, I., Markovitz, A., Manmatha, R.: GLASS: global to local attention for scene-text spotting. In: Avidan, S., Brostow, G., Cissé, M., Farinella, G.M., Hassner, T. (eds.) Computer Vision – ECCV 2022. ECCV 2022. LNCS, vol .13688, pp 249–266 Springer, Cham (2022). https://doi.org/10.1007/978-3-031-19815-1_15
8. Chen, Y.: Design and Implementation of Printed Tibetan Language Recognition Software on Android Platform. Northwest University for Nationalities (2020)
9. Wang, Y.: Design and Implementation of a Tibetan Image Text Recognition System Based on Android. Qinghai Normal University (2022)
10. Zhao, D.: Research on offline handwritten Wumei Tibetan language recognition technology based on BP network. Qinghai Normal University (2009)
11. San, Z.: Research On Multi Font Tibetan Printed Font Recognition Based on Neural Network. Qinghai Normal University (2021)
12. Chen, X., Jin, L., Zhu, Y., Luo, C., Wang, T.: Text recognition in the wild: a survey. ACM Comput. Surv. **54**(2), 1–35 (2021)
13. Vaswani, A., et al.: Attention is all you need. In: Proceedings of the 31st International Conference on Neural Information Processing Systems December, pp. 6000–6010. Curran Associates Inc, Red Hook, NY, USA (2017)
14. Dosovitskiy, A., Beyer, L., Kolesnikov, A., et al.: An image is worth 16x16 words: transformers for image recognition at scale [M/OL]. arXiv, http://arxiv.org/abs/2010. 11929, Accessed 02 2022
15. Li, M., Lv, T., Chen, J., et al.: TrOCR: transformer-based optical character recognition with pre-trained models [M/OL]. arXiv, http://arxiv.org/abs/2109.10282. Accessed 02 Oct 2022
16. Zhang, X., Su, Y., Tripathi, S., et al.: Text spotting transformers. In: 2022 IEEE/CVF Conference on Computer Vision and Pattern Recognition (CVPR), pp. 9509–9518. IEEE, New Orleans, LA, USA (2022)
17. Liu, Z., Lin, Y., Cao, Y., et al.: Swin transformer: hierarchical vision transformer using shifted windows. In: 2021 IEEE/CVF International Conference on Computer Vision (ICCV), pp. 9992–10002. IEEE, Montreal, QC, Canada (2021)
18. Huang, M., Liu, Y., Peng, Z., et al.: SwinTextSpotter: scene text spotting via better synergy between text detection and text recognition. In: 2022 IEEE/CVF Conference on Computer Vision and Pattern Recognition (CVPR), pp. 4583–4593. IEEE, New Orleans, LA, USA (2022)
19. Fang, S., Xie, H., Wang, Y., et al.: Read like humans: autonomous, bidirectional and iterative language modeling for scene text recognition. In: 2021 IEEE/CVF Conference on Computer Vision and Pattern Recognition (CVPR), pp. 7094–7103. IEEE, Nashville, TN, USA (2021)
20. Fu, Z., Xie, H., Jin, G., et al.: Look back again: dual parallel attention network for accurate and robust scene text recognition. In: Proceedings of the 2021 International Conference on Multimedia Retrieval, pp. 638–644. Association for Computing Machinery, New York, NY, USA (2021)
21. Raisi, Z., Naiel, M.A., Younes, G., et al.: 2lspe: 2d learnable sinusoidal positional encoding using transformer for scene text recognition. In: 18th Conference on Robots and Vision, pp. 119–126. IEEE, Burnaby, BC, Canada (2021)
22. Atienza, R.: Vision transformer for fast and efficient scene text recognition. In: Lladós, J., Lopresti, D., Uchida, S. (eds.) Document Analysis and Recognition – ICDAR 2021. ICDAR 2021. Lecture Notes in Computer Science, vol. 12821, pp. 319–334. Springer, Cham (2021). https://doi.org/10.1007/978-3-030-86549-8_21
23. Kittenplon, Y., Lavi, I., Fogel, S., et al.: Towards weakly-supervised text spotting using a multi-task transformer. In: 2022 IEEE/CVF Conference on Computer Vision and Pattern Recognition (CVPR), pp. 4594–4603. IEEE, New Orleans, LA, USA (2022)

24. Trockman, A., Zico Kolter, J.: Patches are all you need? [M/OL]. arXiv, https://arxiv.org/abs/2201.09792, 2 Jan 2022
25. Du, Y., Chen, Z., Jia, C., et al.: SVTR: Scene text recognition with a single visual model. In: 31st International Joint Conference on Artificial Intelligence, pp. 12593–12602. Morgan Kaufmann, Vienna, Austria (2022)

End-to-End Optical Music Recognition with Attention Mechanism and Memory Units Optimization

Ruichen He[1] and Junfeng Yao[1,2,3](✉)

[1] Center for Digital Media Computing, School of Film, School of Informatics, Xiamen University, Xiamen 361005, China
yao0010@xmu.edu.cn

[2] Key Laboratory of Digital Protection and Intelligent Processing of Intangible Cultural Heritage of Fujian and Taiwan, Ministry of Culture and Tourism, Xiamen, China

[3] Institute of Artificial Intelligence, Xiamen University, Xiamen 361005, China

Abstract. Optical Music Recognition (OMR) is a research field aimed at exploring how computers can read sheet music in music documents. In this paper, we propose an end-to-end OMR model based on memory units optimization and attention mechanisms, named ATTML. Firstly, we replace the original LSTM memory unit with a better Mogrifier LSTM memory unit, which enables the input and hidden states to interact fully and obtain better context-related expressions. Meanwhile, the decoder part is augmented with the ECA attention mechanism, enabling the model to better focus on salient features and patterns present in the input data. We use the existing excellent music datasets, PrIMuS, Doremi, and Deepscores, for joint training. Ablation experiments were conducted in our study with the incorporation of diverse attention mechanisms and memory optimization units. Furthermore, we used the musical score density metric, SnSl, to measure the superiority of our model over others, as well as its performance specifically in dense musical scores. Comparative and ablation experiment results show that the proposed method outperforms previous state-of-the-art methods in terms of accuracy and robustness.

Keywords: Optical Music Recognition · Attention mechanism · Memory units optimization · SnSl

1 Introduction

Optical Music Recognition(OMR) aims to explore computational methods for interpreting music notation from visual representations such as images and scans. Over the years, Optical Music Recognition (OMR) technology has undergone significant advancements. However, despite these developments, there still exist several obstacles in the field, and the current state-of-the-art often fails to provide

Q. Liu et al. (Eds.): PRCV 2023, LNCS 14426, pp. 400–411, 2024.
https://doi.org/10.1007/978-981-99-8432-9_32

optimal solutions [1]. In particular, the complex nature of musical notation, combined with factors such as image degradation and variability in notation styles, means that extracting accurate information from musical scores can still be a daunting challenge.

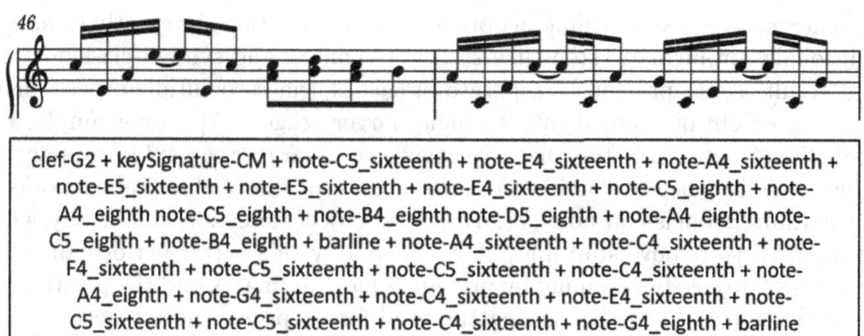

Fig. 1. Polyphonic musical score image and its corresponding semantic information.

Digitizing various real-life objects has become one of the hottest research directions in artificial intelligence, and with the popularization of computer technology, digitizing real objects into electronic data is of great significance for people to operate the electronic versions of these objects on computers. Currently, Optical Character Recognition(OCR) is widely used and has brought great help to people's lives. One of the problems with the current neural network structure for music score recognition is that the depth of neural network structures used for the task has remained relatively shallow [2,3]. In comparison, OCR text recognition technology had already started using a CTC loss method to solve text recognition problems back in 2015. After several years of research, the structure of music score recognition models has been continuously optimized, for example, by adding convolutional layers and optimizing encoding methods to make progress in recognizing polyphonic music scores [4]. Additionally, many excellent datasets have been introduced in the music score recognition field, such as DOREMI, DeepScores, and PrIMuS. However, the performance of the current music score recognition models on these datasets is still not satisfactory, and there is a need to improve the neural network structure, incorporate lightweight modules, optimize the LSTM mechanism, and improve the model's recognition rate and robustness.

Optical music recognition (OMR) presents different challenges compared to other image recognition tasks. Documents can contain various graphical content such as maps, charts, tables, comics, engineering drawings, and patents [1,5]. For instance, comics narrate a story by combining text and visual elements, making the recovery of the correct reading order difficult. The hardest aspects of OMR involve the two-dimensional nature of graphical content, long-distance spatial relationships, and the flexibility of individual elements in terms of their scales and rotations. These difficulties are more prevalent in the recognition of graphical content than in text recognition.

2 Related Work

The emergence of deep learning in optical music recognition has greatly improved traditionally problematic steps like staff-line removal and symbol classification. As a result, some previously separate stages of the recognition process have become obsolete or merged into a single, larger stage [6,7]. For example, the music object detection stage used to involve a separate segmentation stage to remove staff lines before classification. However, convolutional neural networks in deep learning models can now perform music object detection holistically without the need to remove staff lines. This not only yields better performance but also allows for one-step training by providing pairs of images and the positions of the music objects, thus eliminating the need for preprocessing altogether [8,9]. This has led to some steps becoming obsolete or merging into a larger stage. However, deep learning models using convolutional neural networks can now perform music object detection holistically without the need for staff-line removal [10,11]. This approach not only improves performance but also eliminates the need for preprocessing by training the models with pairs of images and corresponding positions of the music objects.

The use of an end-to-end neural network for Optical Music Recognition (OMR) was first proposed by Jorge Calvo-Zaragoza et al. This was achieved by utilizing an end-to-end neural model, trained on the PrIMuS dataset, which contains over 80,000 real score images of single-staff Western musical notation, used for training and evaluating neural methods. The model combines the capabilities of Convolutional Neural Networks (CNNs) and Recurrent Neural Networks (RNNs), with CNNs processing input images and RNNs handling the sequential nature of the problem [12,13]. Traditional methods typically require splitting into multiple processing steps, while the end-to-end neural network approach can perform music object detection and recognition tasks holistically. The authors also introduce semantically encoded representations, as shown in Fig. 1. Laying the foundation for future research in encoder-decoder methods [14]. This work proposes an innovative approach to solving OMR problems using an end-to-end neural network, but is limited in the encoder section and can only be applied to single-voice music transcription.

The recognition task of polyphonic music scores has not been well addressed in the overall development of OMR research, with most methods focused on recognizing single-voice or monophonic music scores. Sachinda Edirisooriya et al. proposed two new decoder models, flagdecoder and rnndecoder, based on the image encoder and encoder proposed in past end-to-end OMR research. The first approach treats it as a multitask binary classification, simultaneously detecting the presence of each pitch and staff symbol, and if the symbol is a note, detecting the rhythm [15–17]. The second approach treats it as a multi-sequence detection, using a vertical RNN decoder on symbols present at each vertical image slice, allowing the model to output a predetermined number of symbol predictions at a single horizontal position in the image. The authors' method achieves good results in the task of recognizing polyphonic music scores, however, they acknowledge that there are many areas for optimization in their new decoder approach, and the recognition of music scores outside of the MuseScore engraver in terms of datasets has a relatively general performance.

3 Method

3.1 Architecture Overview

This article proposes a method whose architecture overview is depicted in Fig. 2. The method employs a neural network architecture consisting of two major parts: the convolutional neural network layer and the recurrent neural network layer. The input music score image undergoes multiple convolutional layers in the CNN part to extract relevant features, producing a corresponding sequence of features [18]. The sequence of features is then fed into an ML (Mogrifier LSTM) network for classification, while the hyperparameter of the Mogrifier layer is set to 5. The intermediate and final results of the classification are then input into an EA (ECA Attention) layer to obtain the final output.

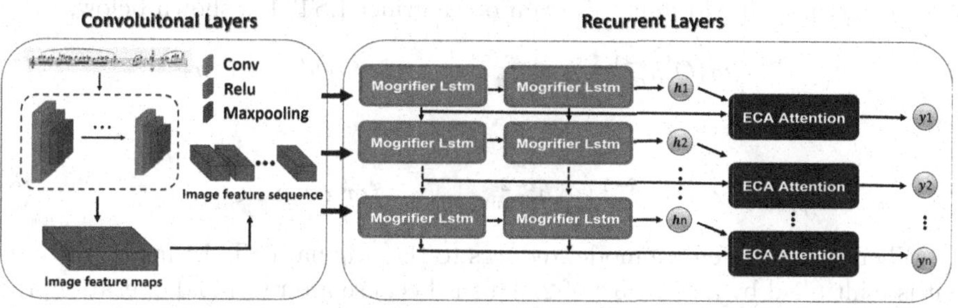

Fig. 2. ATTML Structure.

3.2 Memory Unit Optimization Component

Using LSTM is generally considered a way to alleviate the problems of vanishing gradients and information loss by better modeling long-range semantics [19]. However, it is noted that in LSTM, the current input is independent of the previous state and only interacts with gates, lacking interaction before that, which may lead to a loss of contextual information. As shown in Fig. 3, Mogrifier LSTM is a novel LSTM architecture that enhances the representational and generalization abilities of LSTM through controlled nonlinear transformations of the update and reset gates [20]. Mogrifier LSTM works differently by using

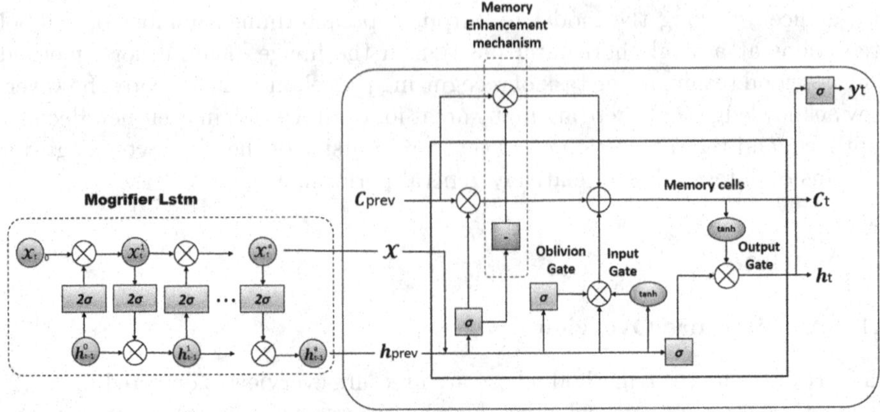

Fig. 3. Structure of memory optimizaiton units.

the idea of "hidden state cycling" from the previous time step, which inputs the neural network's output at the current time point into a different model to be processed and then returns it to this network for further processing.

To distinguish the interactions between X and the hidden state H, matrices Q and R are set, and a hyperparameter i is defined to control how X and H should interact. The formula diagram of Mogrifier LSTM is shown below:

$$x^i = 2\sigma(Q^i h_{prev}^{i-1}) \odot x^{i-2}, \qquad for\ odd\ i \in [1...r] \tag{1}$$

$$h_{prev}^i = 2\sigma(R^i x^{i-1}) \odot h_{prev}^{i-2}, \qquad for\ even\ i \in [1...r] \tag{2}$$

When $i = 0$, the entire model reduces to the original LSTM.Finally, the output is multiplied by a constant of 2. This is because after sigmoid operation, the output values are distributed in the range of (0, 1), and repeatedly multiplying by them will make the values smaller and smaller. Therefore, multiplying by a value of 2 ensures the stability of the numerical values.

3.3 ATTention-Based OMR Decoder

In music score recognition task, the attention module we used is the ECA module. It has been empirically proven that the ECA module is more effective than the SE and CBAM attention mechanisms in music score recognition tasks. It can capture elements such as notes, beats, and key signatures in the input features while ensuring computational efficiency, optimizing network performance, and improving the accuracy of the model in various tasks [21]. In traditional attention mechanisms, the positional information in the input sequence is usually focused on, but the importance of channels in the input is ignored. However, the ECA attention mechanism can pay attention to the importance of each channel in the input sequence, and obtain the weight of each channel through adaptive weighting, thereby capturing important patterns in input features. As shown in Fig. 4, starting from a single music score image, the corresponding sequence of image features is extracted. The current feature layer U is of the dimension $\{C, H, W\}$, where C is the number of channels in the music score image, and H and W are the height and width of the image. We need to perform average pooling on the $\{H, W\}$ dimensions of the feature map, and the pooled feature map size changes from $\{C, H, W\}$ to $\{C, 1, 1\}$, which can be understood as a corresponding number for each channel. In the SE attention mechanism module, the vector after global pooling will be passed through an MLP network layer. The specific result of the MLP network layer is an FC layer, a Relu layer, followed by a fully connected FC layer, and finally a Sigmoid layer, to obtain the weights of the corresponding channels. The formula for attention mechanism is as follows:

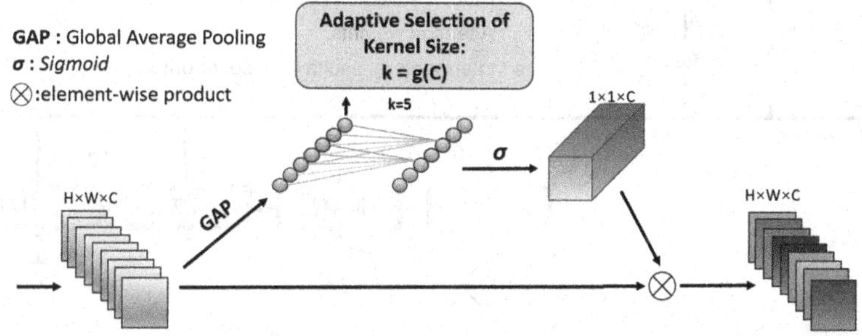

Fig. 4. Attention mechanism structure in the model.

$$C = \phi(k) = 2^{(y*k-b)} \tag{3}$$

$$k = \left| \frac{log_2(C)}{\gamma} + \frac{b}{\gamma} \right| odd \tag{4}$$

In this formula, k represents the size of the convolutional kernel, and C represents the number of channels. The constraint 'odd' means that k can only take odd values. The parameters γ and b are both set to 2 and 1 in the paper, respectively, and are used to adjust the ratio between the number of channels C and the size of the convolutional kernel k.

In the AFFML decoder, the process of generating attention weights and the interaction process of variables in each optimization module proposed above is illustrated in Fig. 5. In the first part, the input x is processed by the Mogrifier Lstm unit to obtain the corresponding h (temporal state sequence), which is mapped from x_t to h_t as follows:

$$h_t = f_1(h_t - 1, x_t) \tag{5}$$

In the second section, we build an attention mechanism using the deterministic ECA attention model. For a sequence $H = (h_1, h_2, ..., h_n)$ that is the output of the Mogrifier LSTM unit, we use it as input for the attention module.

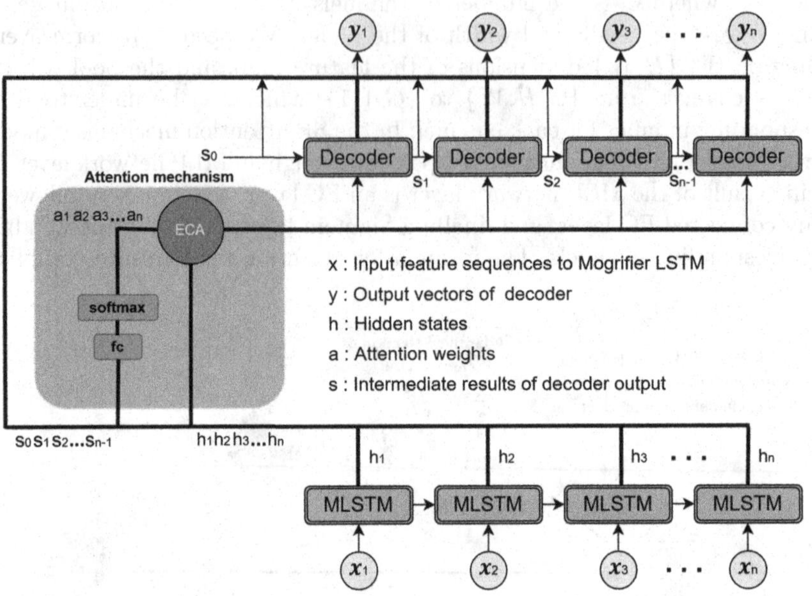

Fig. 5. The RNN structure in the ATTML that we propose.

4 Experiment

4.1 Dataset

PrIMuS. [12] PrIMuS dataset contains 87,678 real-music incipits, was created using an export from the RISM database. Each incipit is provided with the original sheet music image, the corresponding MusicXML file, and its corresponding symbolic semantic representation.

Deepscores. [22] DeepScores contains high-quality music score images that have been segmented into 300,000 music scores containing symbols of different shapes and sizes. The dataset has undergone detailed statistical analysis and has been compared with computer vision datasets such as PASCAL VOC, SUN, SVHN, ImageNet, MS-COCO, and other OMR datasets to optimize the database.

DoReMi. [23] DoReMi allows harmonization with two existing datasets, DeepScores and MUSCIMA++. DoReMi was generated using a music notation software and includes over 6400 printed sheet music images with accompanying metadata useful in OMR research. DoReMi provides OMR metadata, MIDI, MEI, MusicXML, and PNG files, each aiding a different stage of OMR.

4.2 Experiment Set up

The ATTML-NET proposed in this papers is trained and tested using the PyTorch framework. Our training was supervised training under the semantic training label files. Our input consisted of segmented standard single-staff sheet music, along with standard MusicXML files and corresponding semantic sequence label files. The batch size for training was set to 8, and we trained ATTML-Net for 100 epochs to achieve optimal results. We set the learning rate to 0.0001, and the gamma value in ECA was set to 2, while the weighted value was set to 1. Additionally, the hyperparameter i in MogrifierLstm was set to 5. With this configuration, we performed joint training using the three datasets mentioned above.

4.3 Comparisons with the OMR Methods

In music score recognition task, the current existing methods adopt the encoder-decoder architecture, using ResNet-50-like network structure as the feature extractor in the optical music recognition model. As shown in Fig. 6, We compared the performance of different models on recognizing monophonic score, polyphonic score, and dense polyphonic score to validate the performance of different models under different symbol density. As the current music score recognition methods still use rather simple CRNN structure, we compare some OCR models structure in OMR application with the existing methods while comparing with the original OMR methods to verify its effectiveness. As shown in Table 1, We mainly compare the effects of different decoders on the recognition results in this task.

Table 1. The performances of the ATTML models compared with other OMR Methods.

Dataset model	PrIMuS		Deepscores		DoReMi	
	mAP	Rank-1	mAP	Rank-1	mAP	Rank-1
Baseline [12]	89.92	90.63	78.36	79.91	74.39	77.77
ResNet50 + FlagDecoder [24]	85.91	87.36	81.73	83.94	72.26	73.36
ResNet50 + RnnDecoder [6]	86.72	89.63	79.66	82.93	77.36	79.69
ASTER [25]	86.86	88.73	83.72	84.93	79.63	81.74
ABCNet [26]	89.73	91.86	87.16	89.73	84.66	85.93
ATTML	89.28	92.23	89.61	91.97	87.93	89.21

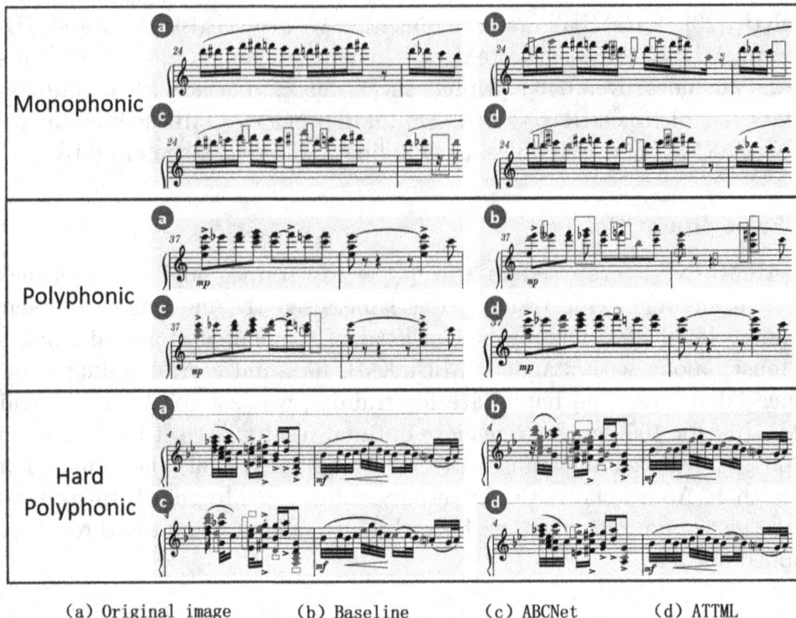

Fig. 6. The output of the music score images corresponding to the semantic representations obtained by different methods.

4.4 Ablation Study

To demonstrate the effectiveness of the proposed approach, we conducted ablative analysis by removing certain components from the system and observing the resulting changes in performance. Table 2 illustrates the effectiveness of the modules we incorporated in the decoder, including the Mogrifier Lstm module and the ECA attention mechanism module. We also added SE attention and Cbam attention mechanisms for comparison to investigate the impact of different attention mechanisms on music recognition accuracy. The figures demonstrate the effectiveness of the modules we proposed.

Table 2. Ablation study of proposed methods on PrIMuS, Deepscores and DoReMi.

SE [25]	CBAM [27]	AFF [28]	ECA [21]	MLSTM [20]	AP
✓					83.3
✓				✓	85.6(+2.3)
	✓			✓	86.5(+3.2)
		✓		✓	87.6(+4.3)
			✓		87.1(+3.8)
			✓	✓	89.7(+6.4)

We conducted experiments to test the effects of individually incorporating attention modules and combining attention modules with memory optimization units. These experiments were conducted on three curated music datasets. we performed experiments based on the ratio of single-note pitch density and symbol looseness level.

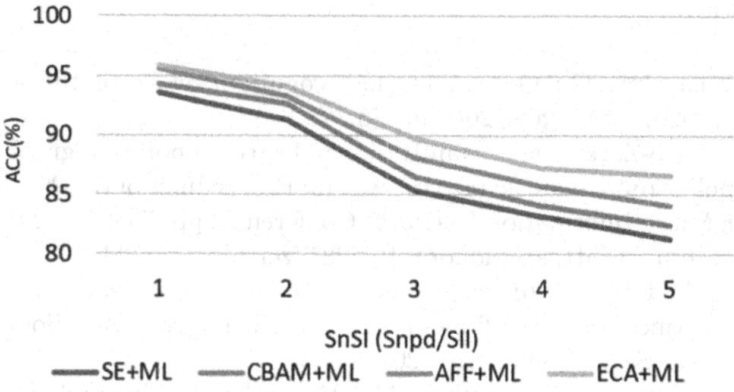

Fig. 7. The performance of each model under different levels of music notation density.

As shown in Fig. 7, The results demonstrate that our proposed model achieved good accuracy across all tested scenarios. SnSl is an index used to assess the overall density of a musical score, where Snpd measures the density of notes on an individual staff, while Sll quantifies the level of spacing between symbols.

5 Conclusion

In this paper, we propose an end-to-end stereo music score recognition method based on attention mechanism and memory unit optimization. We capture the importance of each channel in the input sequence using the ECA attention mechanism and obtain the weight for each channel through adaptive weighting, thus capturing important patterns in the input features. Simultaneously, we optimize the memory module using the Mogrifier LSTM memory optimization unit by leveraging the previous and subsequent hidden states h. We conduct comprehensive experiments on three publicly available high-quality music score datasets, and the experimental results demonstrate that our proposed ATTML method achieves the state-of-the-art performance in the field of stereo music score recognition. Our work provides a novel approach for music score recognition and offers a new direction for future research in stereo music score recognition, which has high practical value.

Acknowledgements. The paper is supported by the Natural Science Foundation of China (No. 62072388), Collaborative Project fund of Fuzhou-Xiamen-Quanzhou Innovation Zone(No.3502ZCQXT202001), the industry guidance project foundation of science technology bureau of Fujian province in 2020(No.2020 H0047), and Fujian Sunshine Charity Foundation.

References

1. Shatri, E., Fazekas, G.: Optical music recognition: state of the art and major challenges (2020). arXiv:abs/2006.07885
2. Pacha, A., Calvo-Zaragoza, J., Jan Hajič, J.: Learning notation graph construction for full- pipeline optical music recognition. In: Proceedings of the 20th International Society for Music Information Retrieval Conference, pp. 75–82. ISMIR, Delft, The Netherlands (2019). https://doi.org/10.5281/zenodo.3527744
3. Dorfer, M., Arzt, A., Widmer, G.: Learning audio-sheet music correspondences for score identification and offline alignment. In: International Society for Music Information Retrieval Conference (2017)
4. Moss, F.C., Köster, M., Femminis, M., Métrailler, C., Bavaud, F.: Digitizing a 19th-century music theory debate for computational analysis, vol. 2989, pp. 12. 159–170. CEUR Workshop Proceedings (2021). http://infoscience.epfl.ch/record/289818
5. Géraud, T.: A morphological method for music score staff removal. In: 2014 IEEE International Conference on Image Processing (ICIP), pp. 2599–2603 (2014). https://doi.org/10.1109/ICIP.2014.7025526
6. Edirisooriya, S., Dong, H.W., McAuley, J., Berg-Kirkpatrick, T.: An empirical evaluation of end-to-end polyphonic optical music recognition (2021)
7. Dosovitskiy, A., et al.: FlowNet: learning optical flow with convolutional networks. In: Proceedings of the IEEE International Conference on Computer Vision (ICCV) (2015)
8. Pacha, A., Eidenberger, H.: Towards a universal music symbol classifier. In: 2017 14th IAPR International Conference on Document Analysis and Recognition (ICDAR), vol. 02, pp. 35–36 (2017). https://doi.org/10.1109/ICDAR.2017.265
9. Raphael, C., Wang, J.: New approaches to optical music recognition. In: Proceedings of the 12th International Society for Music Information Retrieval Conference, pp. 305–310. ISMIR, Miami, United States (2011). https://doi.org/10.5281/zenodo.1414856
10. Kaliakatsos-Papakostas, M.A., Epitropakis, M.G., Vrahatis, M.N.: Musical composer identification through probabilistic and feedforward neural networks. In: Di Chio, C., Brabazon, A., Di Caro, G.A., Ebner, M., Farooq, M., Fink, A., Grahl, J., Greenfield, G., Machado, P., O'Neill, M., Tarantino, E., Urquhart, N. (eds.) EvoApplications 2010. LNCS, vol. 6025, pp. 411–420. Springer, Heidelberg (2010). https://doi.org/10.1007/978-3-642-12242-2_42
11. Ríos-Vila, A., Calvo-Zaragoza, J., Iñesta, J.M.: Exploring the two-dimensional nature of music notation for score recognition with end-to-end approaches. In: 2020 17th International Conference on Frontiers in Handwriting Recognition (ICFHR), pp. 193–198 (2020). https://doi.org/10.1109/ICFHR2020.2020.00044
12. Jorge, C.Z., David, R.: End-to-end neural optical music recognition of monophonic scores. Appl. Sci. **8**(4), 606 (2018)

13. Alfaro-Contreras, M., Calvo-Zaragoza, J., Iñesta, J.M.: Approaching end-to-end optical music recognition for homophonic scores. In: Morales, A., Fierrez, J., Sánchez, J.S., Ribeiro, B. (eds.) IbPRIA 2019. LNCS, vol. 11868, pp. 147–158. Springer, Cham (2019). https://doi.org/10.1007/978-3-030-31321-0_13
14. Hankinson, A., Roland, P., Fujinaga, I.: The music encoding initiative as a document- encoding framework. In: Proceedings of the 12th International Society for Music Information Retrieval Conference, pp. 293–298. ISMIR, Miami, United States (2011). https://doi.org/10.5281/zenodo.1417609
15. Kim, S., Lee, H., Park, S., Lee, J., Choi, K.: Deep composer classification using symbolic representation (2020)
16. Pinheiro, P.H.O., Collobert, R.: Recurrent convolutional neural networks for scene parsing (2013). arXiv:abs/1306.2795
17. Tsai, T.J., Ji, K.: Composer style classification of piano sheet music images using language model pretraining (2020). arXiv:abs/2007.14587
18. Ríos-Vila, A., Iñesta, J.M., Calvo-Zaragoza, J.: On the use of transformers for end-to-end optical music recognition. In: Pinho, A.J., Georgieva, P., Teixeira, L.F., Sánchez, J.A. (eds.) Pattern Recognition and Image Analysis, pp. 470–481. Springer International Publishing, Cham (2022). https://doi.org/10.1007/978-3-031-04881-4_37
19. Merity, S., Keskar, N.S., Socher, R.: Regularizing and optimizing LSTM language models (2017). arXiv:abs/1708.02182
20. Melis, G., Kociský, T., Blunsom, P.: Mogrifier LSTM (2019). arXiv:abs/1909.01792
21. Wang, Q., Wu, B., Zhu, P., Li, P., Zuo, W., Hu, Q.: ECA-Net: efficient channel attention for deep convolutional neural networks. In: 2020 IEEE/CVF Conference on Computer Vision and Pattern Recognition (CVPR), pp. 11531–11539 (2020). https://doi.org/10.1109/CVPR42600.2020.01155
22. Tuggener, L., Elezi, I., Schmidhuber, J., Pelillo, M., Stadelmann, T.: Deepscores-a dataset for segmentation, detection and classification of tiny objects. In: 2018 24th International Conference on Pattern Recognition (ICPR), pp. 3704–3709 (2018). https://doi.org/10.1109/ICPR.2018.8545307
23. Shatri, E., Fazekas, G.: DoReMi: first glance at a universal OMR dataset (2021). arXiv:bs/2107.07786
24. Liu, A., et al.: Residual recurrent CRNN for end-to-end optical music recognition on monophonic scores. In: Proceedings of the 2021 Workshop on Multi-Modal Pre-Training for Multimedia Understanding, MMPT '21, pp. 23–27. Association for Computing Machinery, New York (2021). https://doi.org/10.1145/3463945.3469056
25. Hu, J., Shen, L., Sun, G.: Squeeze-and-excitation networks. In: 2018 IEEE/CVF Conference on Computer Vision and Pattern Recognition, pp. 7132–7141 (2018). https://doi.org/10.1109/CVPR.2018.00745
26. Liu, Y., Chen, H., Shen, C., He, T., Jin, L., Wang, L.: ABCNet: real-time scene text spotting with adaptive Bezier-curve network. In: 2020 IEEE/CVF Conference on Computer Vision and Pattern Recognition (CVPR), pp. 9806–9815 (2020). https://doi.org/10.1109/CVPR42600.2020.00983
27. Woo, S., Park, J., Lee, J.Y., Kweon, I.S.: CBAM: convolutional block attention module. In: Ferrari, V., Hebert, M., Sminchisescu, C., Weiss, Y. (eds.) Computer Vision - ECCV 2018. Lecture Notes in Computer Science(), vol. 11211, pp. 3–19. Springer, Cham (2018). https://doi.org/10.1007/978-3-030-01234-2_1
28. Dai, Y., Gieseke, F., Oehmcke, S., Wu, Y., Barnard, K.: Attentional feature fusion. In: 2021 IEEE Winter Conference on Applications of Computer Vision (WACV), pp. 3559–3568 (2020)

Tripartite Architecture License Plate Recognition Based on Transformer

Ran Xia[1], Wei Song[1,2,3(✉)], Xiangchun Liu[1], and Xiaobing Zhao[1,2,3]

[1] School of Information Engineering, Minzu University of China, Beijing 100081, China
songwei@muc.edu.cn

[2] National Language Resource Monitoring and Research Center of Minority Languages, Minzu University of China, Beijing 100081, China

[3] Key Laboratory of Ethnic Language Intelligent Analysis and Security Governance of MOE, Minzu University of China, Beijing 100081, China

Abstract. Under natural conditions, license plate recognition is easily affected by factors such as lighting and shooting angles. Given the diverse types of Chinese license plates and the intricate structure of Chinese characters compared to Latin characters, accurate recognition of Chinese license plates poses a significant challenge. To address this issue, we introduce a novel Chinese License Plate Transformer (CLPT). In CLPT, license plate images pass through a Transformer encoder, and the resulting Tokens are divided into four categories via an Auto Token Classify (ATC) mechanism. These categories include province, main, suffix, and noise. The first three categories serve to predict the respective parts of the license plate - the province, main body, and suffix. In our tests, we employed YOLOv8-pose as the license plate detector, which excels in detecting both bounding boxes and key points, aiding in the correction of perspective transformation in distorted license plates. Experimental results on the CCPD, CLPD, and CBLPRD datasets demonstrate the superior performance of our method in recognizing both single-row and double-row license plates. We achieved an accuracy rate of 99.6%, 99.5%, and 89.3% on the CCPD Tilt, Rotate, and Challenge subsets, respectively. In addition, our method attained an accuracy of 87.7% in the CLPD and 99.9% in the CBLPRD, maintaining an impressive 99.5% accuracy even for yellow double-row license plates in the CBLPRD.

Keywords: License Plate Recognization · License Plate Detection · Transformer

1 Introduction

Automatic License Plate Recognition (ALPR) is a computational system that automatically detects and recognizes license plates from images or videos using computer vision and machine learning technologies. Compared to pure Latin character license plates, Chinese license plate recognition proposes additional

Q. Liu et al. (Eds.): PRCV 2023, LNCS 14426, pp. 412–423, 2024.
https://doi.org/10.1007/978-981-99-8432-9_33

challenges. Chinese license plates have two structures: single line and double line, and the algorithm's adaptability to non-single-line license plates needs to be further considered. Chinese license plates include characters that represent 34 provincial-level administrative regions, which increases the types of characters to be recognized, and certain license plates feature distinct suffixes, such as '挂' , '学' , '警' , etc. Compared to Latin characters, Chinese characters have a high degree of glyph complexity and similarity, making recognition more challenging. These characters are also more susceptible to misidentification due to factors such as lighting conditions, blur, and shooting angles.

In this paper, we propose a novel Chinese License Plate Transformer (CLPT) for the recognition of Chinese license plates. This system is inspired by the transformer model's [11] remarkable performance and adaptability in various vision tasks. Our method has the following four insights:

1. We propose a Tripartite Architecture(TA) that deconstructs all Chinese license plates into a province-main-suffix format. This separation not only comprehensively encompasses all types of Chinese license plates but also enables the model to leverage the inherent structural characteristics of license plates. Consequently, this approach significantly enhances the accuracy of license plate recognition.
2. We propose an Auto Token Classify(ATC) mechanism, designed to complement the TA architecture. This mechanism adaptively categorizes all output tokens from the transformer into several groups, aligning with specific subtasks, including province classification, recognition of the Latin character main body, and suffix classification.
3. Compared to conventional algorithms that directly utilize YOLO for license plate detection, the extension of YOLOv8, known as YOLOv8-pose, offers the capability to predict the four key points of a license plate additionally, without imposing a significant computational burden. Harnessing these key points for perspective transformation correction enhances the model's proficiency in recognizing license plates under distorted viewing angles.
4. Compared to traditional Recurrent Neural Networks (RNNs), the utilization of a Transformer architecture does not inherently constrain recognition content to a predefined left-to-right single-line sequence. As a consequence, Transformer models demonstrate notable advantages in the recognition of dual-line license plates.

2 Related Work

2.1 License Plate Recognition

Raj et al. [7] segmented the characters on the license plate for OCR recognition. However, this method is dependent on the segmentation model and therefore has an error accumulation issue. Xu et al. [15] proposed RPnet, an end-to-end license plate recognition system that finally uses seven classifiers to predict the characters on the license plate separately. However, this method can only identify

7-digit blue plates rigidly. GONG Y et al. [3] proposed predicting the rotation angle θ in the LPD part to correct the rotated license plate and then use CTC to predict characters of variable length. However, this method struggles with non-planar rotated license plates; Wang et al. [13] proposed a shared-parameter classification head for the CCPD dataset, which segments the prediction of blue license plates into province, city alphabet, and a sequence of five Latin characters. However, this method presents challenges when attempting to apply it universally to other types of license plates.

2.2 Transformer

Transformers have shown good performance in the field of natural language processing. ViT [2] (Vision Transformer) applied Transformers to the visual field and achieved excellent results. However, due to the large architecture and slow inference speed of ViT, it is limited in its application in the license plate recognition task. With the birth of lightweight Transformers, such as MobileViT [6], Deit [10], etc., we propose a new solution to this problem. Wu et al. [14] proposed TinyViT, a new family of tiny and efficient vision transformers, pretrained on large-scale datasets with their proposed fast distillation framework. While ensuring lightweight and high efficiency, TinyViT possesses a hierarchical structure that can better handle the detailed features in Chinese characters. We use the lightest TinyViT-5M as the pre-training encoder, divide the output results into three sub-tasks of province and suffix classification, and Latin character body sequence recognition through the ATC mechanism. This not only achieves excellent license plate recognition performance, but also provides a new way of thinking for using Transformer models for license plate recognition.

3 Proposed Method

3.1 License Plate Detection

Specifically in license plate detection tasks, the bbox-based YOLO algorithm may face accuracy issues due to possible rotation and distortion of the license plate, as its rectangular representation struggles to capture the detailed characteristics of these distortions.

We adopted the YOLO-Pose algorithm, an extension of the traditional YOLO, which includes key point prediction. This feature provides a significant advantage over regular YOLO algorithms in addressing rotated license plates. By modifying the detection head, our model simultaneously predicts the bbox and the four corner points of the license plate. This key point information enables us to perform a perspective transformation, effectively correcting for rotation and distortion without significantly adding complexity to our approach.

3.2 License Plate Recognition

Tripartite Architecture(TA). As shown in Fig. 1, our Tripartite Architecture (TA) partitions the license plate into three components: province, main body, and suffix. This strategy, which leverages the structural information of the license plate, positions the province and suffix at fixed points, each containing specific Chinese characters.

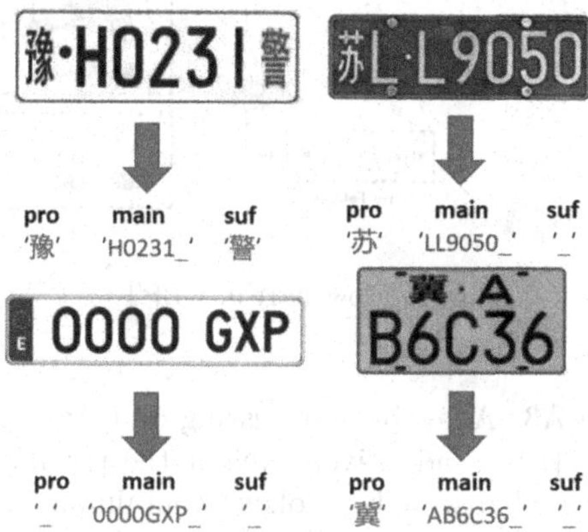

Fig. 1. partitioning of Different Types of License Plates

This design allows the main body to focus solely on Latin characters, reducing the prediction task complexity. To further enhance flexibility for various license plate structures, we incorporate the '_' character. Within the province and suffix, it represents an absent character, while it signals sequence termination in the main body. This design handles different license plate structures and lengths effectively, showcasing strong generalization capabilities.

CLPT. As shown in Fig. 2, the network mainly consists of a Transformer encoder with a pyramid structure, an ATC module, and post-processing corresponding to the token. The 224×224 image is first encoded into a series of tokens that enter the Transformer Encoder after the Patch Embedding process. The Transformer Encoder of TinyViT consists of downsampling three times and Transformer Block, forming a hierarchical structure. Specifically, downsampling in the Transformer Encoder uses MBConv, and the Swin structure is used in the Transformer Block to perform self-attention on tokens within the window. This process gradually downsamples the original encoded 56×56 tokens to 7×7 tokens. After the encoding is complete, each of the 49 tokens of 320 dimensions contains features of the corresponding patch after sufficient self-attention interaction. At this time, the ATC module performs soft grouping on these 49 tokens. For noise tokens, we do not do any subsequent processing; for province and suffix tokens, we perform a global average pooling on these tokens, followed by a fully connected layer, to classify the province character or suffix character. For the variable-length Latin character sequence, we need to select n key tokens from it for sequence prediction. n represents the maximum length of the middle character sequence, which includes a special symbol '_' to indicate the end of the

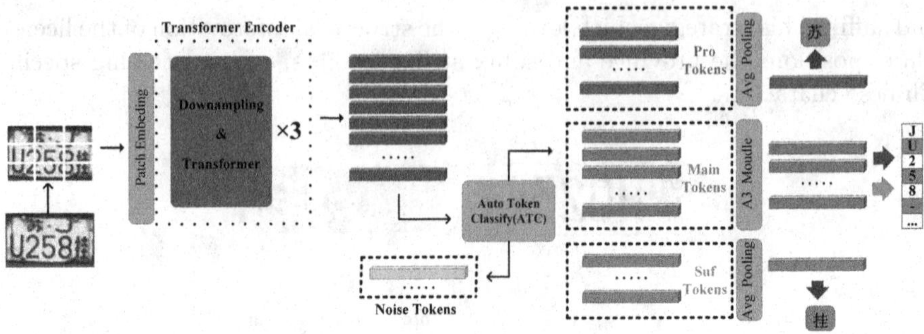

Fig. 2. Flowchart of CLPT

sequence. We use the A3 (Adaptive Addressing and Aggregation) module proposed by Wang et al. [12] to adaptively weight the results through the spatial attention mechanism and fuse n key tokens containing character information from these 49 tokens. We connect a fully connected layer after key tokens to get the prediction of the main part. All three prediction processes use cross-entropy as the loss function, and the sum is added in a certain ratio to obtain the final loss function formula as follows:

$$\text{Loss} = \lambda_1 \cdot \text{Loss}_{\text{pro}} + \lambda_2 \cdot \text{Loss}_{\text{main}} + \lambda_3 \cdot \text{Loss}_{\text{suf}} \tag{1}$$

where λ is the weight of different losses. In our method, λ_1, λ_2, and λ_3 are set to 0.35, 0.5, and 0.15, respectively.

Auto Token Classify(ATC) Mechanism. Once the image has passed through the encoder, each token obtains its corresponding information. We use the ATC mechanism to softly classify the output tokens, allowing tokens containing specific corresponding information to complete different tasks. Specifically, some tokens contain provincial information, some contain information of the Latin character main body, and some contain suffix information. In addition, we added a Noise Token category to store tokens that primarily contain noise (Fig. 3).

For a token classifier with four categories (province, main body, suffix, and noise), we first map the input x to the scores of the four categories. This can be expressed as:

$$\mathbf{s}(x) = \mathbf{W}_2\text{ReLU}(\text{LayerNorm}(\mathbf{W}_1 x + \mathbf{b}_1)) + \mathbf{b}_2 \tag{2}$$

where $\mathbf{W}_1$, $\mathbf{b}_1$, $\mathbf{W}_2$, and $\mathbf{b}_2$ are the weight and bias parameters of the network.

Then, we use the softmax function to transform the output of the classifier into a probability distribution:

$$\mathbf{p}(x) = \text{softmax}(\mathbf{s}(x)) \tag{3}$$

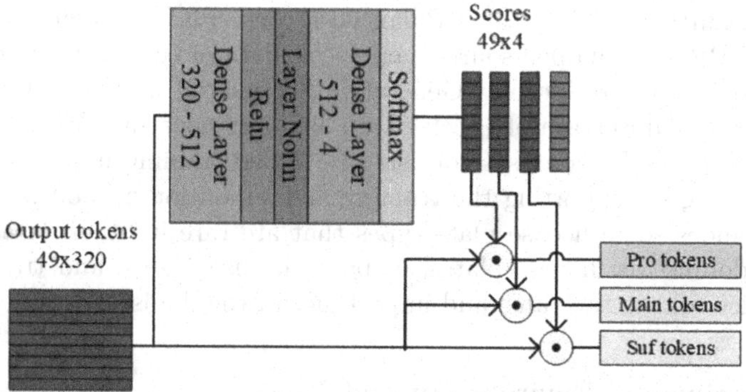

Fig. 3. $\odot$ represents the element-wise multiplication operation. The gray part in the "scores" indicates the proportion of Noise Tokens (Color figure online)

Finally, the ATC computes the result tensor corresponding to each category, i.e., province, Latin character main body, and suffix, excluding the noise token. This process involves multiplying the input vector element-wise by the probability of each category:

$$\mathbf{r}_i = x \odot \mathbf{p}_i(x) \quad \text{for } i \in 0, 1, 2 \tag{4}$$

where $\odot$ represents the element-wise multiplication operation, and $\mathbf{p}_i(x)$ is the output of the softmax function, corresponding to the probability of category i. Finally, we obtain the result tensor $\mathbf{r}_i$ corresponding to each category (province, main body, suffix), which encodes the information related to each category in the input vector x.

4 Experiments

4.1 Datasets

CCPD Dataset. The Chinese City Parking Dataset (CCPD) [15] is a large license plate recognition dataset comprising about 290k images from various parking lots in China. The dataset includes the following subsets: CCPD-base (200k), CCPD-db (20k), CCPD-fn (20k), CCPD-rotate (10k), CCPD-tilt (10k), CCPD-weather (10k), CCPD-challenge (10k). Half of the CCPD-base subset is used for training, while the remaining subsets are utilized for testing.

CLPD Dataset. The Comprehensive License Plate Dataset (CLPD) is a richly-annotated dataset containing 1200 images of various types of vehicles, covering all 31 provinces in mainland China, with diverse shooting conditions and regional codes. Notably, the dataset includes both seven-letter and eight-letter license plates, presenting increased recognition complexity, and making it a significant tool for our experimental setup.

CBLPRD Dataset. The "China-Balanced-License-Plate-Recognition-Dataset-330k" (CBLPRD) is an open-source, large-scale dataset containing 330k Chinese license plate images produced by Generative Adversarial Networks (GANs). The images in this dataset are of excellent quality and cover a variety of Chinese license plate types.The dataset consists of 300,000 training images and 10,000 validation images, supporting the training and validation of models. In particular, it includes some license plate types that are rare in other datasets, such as yellow double-row license plates, embassy license plates, and tractor green plates, which add to the value and importance of the dataset.

4.2 Experimental Environment and Tools

Our network was run on a computer with a 24G RTX 3090 graphics card and an 11th generation Intel Core i7-11700K processor. We implemented the deep learning algorithm based on Pytorch. For yolov8-pose, we used the Adam optimizer, set the batch size to 128, set the learning rate to 0.01, and used mosaic enhancement and random perspective transformation. For CLPT, we used the Adadelta optimizer, set the batch size to 128, set the learning rate to 1, and did not use any data augmentation.

5 Results

5.1 CCPD

For license plate detection, we utilized the method proposed by [15], focusing solely on precision. A prediction is deemed correct if the Intersection over Union (IoU) between the predicted bounding box and the ground truth exceeds 0.7. As presented in Table 1, YOLOv8 outperforms the other methods across all subsets, particularly achieving a 7.0% and 5.8% boost in the Rotate and Challenge subsets, respectively.

Table 1. Comparison of the average precision (percentage) of license plate detection in different subsets. AP represents the average accuracy of the entire dataset.

Method	AP	Base	DB	FN	Rotate	Tilt	Weather	Challenge
Faster-RCNN [9]	92.9	98.1	92.1	83.7	91.8	89.4	81.8	83.9
TE2E [4]	94.2	98.5	91.7	83.8	95.1	94.5	83.6	93.1
RPnet [15]	94.5	**99.3**	89.5	85.3	94.7	93.2	84.1	92.8
YOLOv4 [1]	95.1	96.8	93.7	93.1	93.5	94.7	96.6	85.5
YOLOv3 [8]	96.0	97.1	97.2	93.3	91.6	94.6	97.9	90.5
YOLOv8	**99.0**	**99.3**	**99.1**	**98.8**	**98.6**	**99.2**	**99.7**	**96.3**

For combined license plate detection and recognition, a positive sample is confirmed when the IoU between the bounding box and ground truth surpasses 0.6 and all characters are predicted correctly. We tested using both bounding box results (without rotation correction) and keypoint results (with distortion correction). As depicted in Table 2, aside from the DB subset, our method yielded the best results.

Table 2. The table compares license plate detection and recognition accuracy across various subsets, distinguishing between using bounding boxes ([bbox]) and corrected keypoints ([keypoint]), both detected by YOLOv8.

Methods	Base	DB	FN	Rotate	Tilt	Weather	Challenge
Ren et al. [9]	92.8	97.2	94.4	90.9	82.9	87.3	76.3
Liu et al. [5]	95.2	98.3	96.6	95.9	88.4	91.5	83.8
Xu et al. [15]	95.5	98.5	96.9	94.3	90.8	92.5	85.1
Zhang et al. . [16]	93.0	99.1	96.3	97.3	95.1	96.4	83.2
Zhou et al. [17]	97.5	**99.2**	98.1	98.5	90.3	95.2	86.2
ours[bbox]	**99.8**	**99.2**	**98.8**	98.5	98.8	**98.3**	**89.3**
ours[kepoint]	**99.8**	98.9	**98.8**	**99.5**	**99.6**	98.1	89.0

Keypoint correction notably enhances the accuracy in handling rotated license plates (Rotate and Tilt subsets) improving the results by 1.6% and 0.2% respectively, however, it shows a slight decrease of 0.1% on the FN and Weather subsets, and 0.5% on the DB subset. Conversely, predicting with bounding boxes results in higher accuracy when the license plates are brighter or darker (DB subset) (Fig. 4).

Fig. 4. Results display in CCPD dataset

5.2 CLPD

In the context of automatic license plate recognition, the generalization capability of a model holds significant importance as an evaluation metric. Following the methodology proposed in reference [16], we exclusively employ the base subset of the CCPD dataset for training, while utilizing the CLPD dataset, which encompasses license plate samples from various diverse scenarios, as our test set. By adopting this approach, the results obtained from the CLPD dataset effectively showcase the model's ability to generalize to other datasets. This evaluation technique offers a comprehensive validation of the model's performance across different scenarios and conditions. We consider only those license plates as positive samples which have been entirely and correctly identified, given the fact that only completely accurate predictions bear practical meaning in license plate recognition. The experimental results on the CLPD dataset are presented in Table 3. With the base subset of CCPD as our training set, without any addition of synthetic license plates, our method achieves a Top1 accuracy of 83.4% on the CLPD. Further, with the inclusion of the CBLPRD dataset as additional augmentation data, we manage to reach a Top1 accuracy of 87.7%.

Table 3. Comparison of License Plate Recognition Accuracy on CLPD Dataset

Method	Accuracy
Xu et al. [15]	66.5
Zhang et al. (real data only) [16]	70.8
Zhang et al. (real + synthetic data) [16]	76.8
Zou et al. [17]	78.7
ours(real data only)	**83.4**
ours(real+synthetic data)	**87.7**

5.3 CBLPRD

RNNs carry an inherent assumption that characters are arranged in a sequence from left to right and are contained in a single line. For instance, the classical Convolutional Recurrent Neural Network (CRNN) compresses the original image height to one during the CNN process, thereby inputting it into the subsequent RNN. Although LPRNet does not utilize an RNN, it primarily extracts horizontal features using a 1×13 convolutional kernel and then transforms the feature map into a sequential format via height-wise pooling. These methods show limitations when faced with double-row license plates. A bidirectional RNN can alleviate this issue to some extent.

However, in contrast, our proposed Transformer-based model is not limited by these assumptions. Its feature vectors, extracted by the adaptive addressing

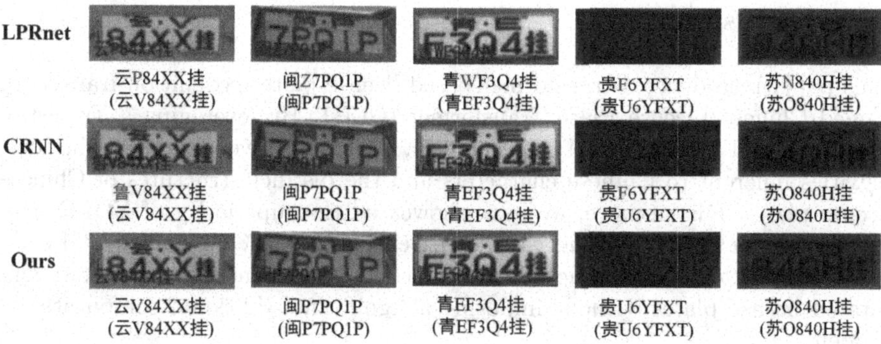

Fig. 5. The figure shows the recognition results of different algorithms on two-way license plates. The words in parentheses are the ground truth.

module, can adapt more effectively to various license plate structures. In Fig. 5, we observe that for LPRNet and CRNN, all misclassifications occur in the first row of characters in dual-row license plates. This underscores the difficulty these algorithms face in accurately identifying license plates with dual-row structures. In contrast, our proposed method demonstrates commendable proficiency in the precise recognition of such dual-row license plates. We conducted experiments on the CBLPRD dataset and, in addition, listed and tested the accuracy of yellow double-row license plates in the validation set to observe the algorithm's ability to recognize double-row license plates. The experimental results shown in Table 4 confirmed our hypothesis: on the validation set, the accuracy of LPRNet is 84.3%, but it cannot recognize double-row license plates; the bidirectional RNN (BiLSTM) in CRNN alleviates this problem to a certain extent, but when processing double-row license plates, the accuracy still significantly decreases by 8.9%. Our Transformer model has an accuracy of 99.9% on the validation set, and when dealing with yellow double-row license plates, the accuracy remains as high as 99.5%.

Table 4. Performance Comparison of License Plate Recognition Algorithms on Validation Set, with Additional Emphasis on Yellow Double-row Plates Accuracy

Algorithm	Validation Set	Yellow Double-row Plates
LPRNet	84.3	0.0 (−84.3)
CRNN	97.7	88.8(−8.9)
CLPT(ours)	**99.9**	**99.5** (−0.4)

These experimental results clearly show that, compared to RNN-based models, our Transformer model demonstrates exceptionally high adaptability and robustness when recognizing complex license plate formats (such as double-row license plates).

6 Conclusion

This paper introduces a transformer-based license plate recognition framework named Chinese License Plate Transformer (CLPT).By leveraging a Tripartite Architecture(TA) and the ATC mechanism, CLPT effectively manages the complexities inherent to Chinese characters and the distinct structures of Chinese license plates. Furthermore, we have showcased the superiority of YOLOv8 in license plate detection and suggested an extension of YOLOv8, named YOLOv8-Pose. This extension enhances the detection performance for distorted and rotated license plates without imposing a significant additional computational burden.

References

1. Bochkovskiy, A., Wang, C.Y., Liao, H.Y.M.: YOLOV4: optimal speed and accuracy of object detection. arXiv preprint: arXiv:2004.10934 (2020)
2. Dosovitskiy, A., et al.: An image is worth 16x16 words: transformers for image recognition at scale. arXiv preprint: arXiv:2010.11929 (2020)
3. Gong, Y., et al.: Unified Chinese license plate detection and recognition with high efficiency. J. Vis. Commun. Image Represent. **86**, 103541 (2022)
4. Li, H., Wang, P., Shen, C.: Toward end-to-end car license plate detection and recognition with deep neural networks. IEEE Trans. Intell. Transp. Syst. **20**(3), 1126–1136 (2018)
5. Liu, W., et al.: SSD: single shot multibox detector. In: Leibe, B., Matas, J., Sebe, N., Welling, M. (eds.) ECCV 2016. LNCS, vol. 9905, pp. 21–37. Springer, Cham (2016). https://doi.org/10.1007/978-3-319-46448-0_2
6. Mehta, S., Rastegari, M.: MobileViT: light-weight, general-purpose, and mobile-friendly vision transformer. arXiv preprint: arXiv:2110.02178 (2021)
7. Raj, S., Gupta, Y., Malhotra, R.: License plate recognition system using yolov5 and CNN. In: 2022 8th International Conference on Advanced Computing and Communication Systems (ICACCS), vol. 1, pp. 372–377. IEEE (2022)
8. Redmon, J., Farhadi, A.: YOLOV3: an incremental improvement. arXiv preprint: arXiv:1804.02767 (2018)
9. Ren, S., He, K., Girshick, R., Sun, J.: Faster R-CNN: towards real-time object detection with region proposal networks. In: Advances in Neural Information Processing Systems, vol. 28 (2015)
10. Touvron, H., Cord, M., Douze, M., Massa, F., Sablayrolles, A., Jégou, H.: Training data-efficient image transformers & distillation through attention. In: International Conference on Machine Learning, pp. 10347–10357. PMLR (2021)
11. Vaswani, A., et al.: Attention is all you need. In: Advances in Neural Information Processing Systems, vol. 30 (2017)
12. Wang, P., Da, C., Yao, C.: Multi-granularity prediction for scene text recognition. In: Avidan, S., Brostow, G., Cisse, M., Farinella, G.M., Hassner, T. (eds.) Computer Vision - ECCV 2022. Lecture Notes in Computer Science, vol. 13688, pp. 339–355. Springer, Cham (2022). https://doi.org/10.1007/978-3-031-19815-1_20
13. Wang, Y., Bian, Z.P., Zhou, Y., Chau, L.P.: Rethinking and designing a high-performing automatic license plate recognition approach. IEEE Trans. Intell. Transp. Syst. **23**(7), 8868–8880 (2021)

14. Wu, K., et al.: TinyViT: fast pretraining distillation for small vision transformers. In: Avidan, S., Brostow, G., Cisse, M., Farinella, G.M., Hassner, T. (eds.) Computer Vision - ECCV 2022. Lecture Notes in Computer Science, vol. 13681, pp. 68–85. Springer, Cham (2022)
15. Xu, Z., et al.: Towards end-to-end license plate detection and recognition: a large dataset and baseline. In: Ferrari, V., Hebert, M., Sminchisescu, C., Weiss, Y. (eds.) ECCV 2018. LNCS, vol. 11217, pp. 261–277. Springer, Cham (2018). https://doi.org/10.1007/978-3-030-01261-8_16
16. Zhang, L., Wang, P., Li, H., Li, Z., Shen, C., Zhang, Y.: A robust attentional framework for license plate recognition in the wild. IEEE Trans. Intell. Transp. Syst. **22**(11), 6967–6976 (2020)
17. Zou, Y., et al.: License plate detection and recognition based on YOLOV3 and ILPRNET. SIViP **16**(2), 473–480 (2022)

Fundamental Theory of Computer Vision

Focus the Overlapping Problem on Few-Shot Object Detection via Multiple Predictions

Mandan Guan[1], Wenqing Yu[1], Yurong Guo[1], Keyan Huang[2], Jiaxun Zhang[2], Kongming Liang[1(✉)], and Zhanyu Ma[1]

[1] Beijing University of Posts and Telecommunications, Beijing 100876, China
liangkongming@bupt.edu.cn

[2] Beijing Institute of Spacecraft System Engineering, Beijing 100094, China

Abstract. The overlapping problem is one of the main challenges of few-shot object detection (FSOD). It refers to the situation where some parts of the target are obscured by other objects or occluders. Since the support set is insufficient to provide enough samples with overlapping objects, it is difficult to get the correct bounding boxes due to the missing information. In this paper, we aim to detect highly overlapping instances in few-shot scenes and present a proposal-based method by combining both basic training and fine-tuning stages. Specifically, we predict a set of bounding boxes instead of a single bounding box for each proposal after proposal generation. Then, we introduce a new NMS strategy to prevent the erroneous removal of our desired bounding box by traditional NMS. Simultaneously, we introduce a new loss to distinguish the obtained bounding boxes to avoid the convergence of bounding boxes during the final training process. While benchmarking on MS-COCO and Pascal VOC, our method is confirmed to be efficient and obtains good generalizability. Our proposed method served as a good stimulus for future study in the field of FSOD.

Keywords: Few-shot object detection · Overlapping detection · Non-maximum suppression

1 Introduction

Object detection has achieved tremendous gains in the rapid development of artificial intelligence, with a vast amount of data. However, not enough data or annotations are available in reality [14,25]. Therefore, FSOD has gained considerable attention in recent years. Two lines of work address the challenging problem and lead to good performances. To teach a meta-learner to aid in the transfer of knowledge from base classes, meta-learning based techniques first develop a stage-wise and periodic meta-training paradigm [11,15,16]. The other involves fine-tuning based detectors rowing over meta-learning using the balanced dataset introduced in TFA [20,23]. The overlapping problem is common

Q. Liu et al. (Eds.): PRCV 2023, LNCS 14426, pp. 427–439, 2024.
https://doi.org/10.1007/978-981-99-8432-9_34

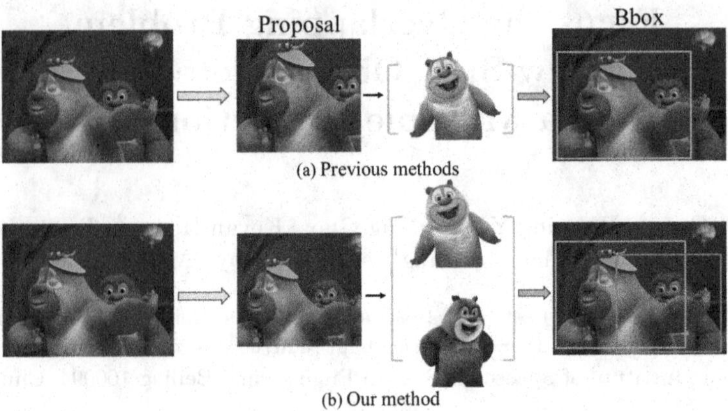

Fig. 1. This illustrates a common detection task with overlapping objects affecting Big Bear and Little Bear. For (a), previous methods employ two proposals to delineate the objects, but they are prone to be discarded during post-processing, leading to missed detections. (b) proposes a single proposal to predict two corresponding instances, significantly mitigating the likelihood of missed detections.

and difficult in FSOD, which has received little attention. The support set is too small to give enough samples that have overlapping objects to train the model. So, incorrect pickup or even missed detection of target objects in the query set occurs.

As shown in Fig. 1 (a) is a common occurrence. Existing methods usually predict only one instance per proposal. However, it tends to prioritize easier-to-detect targets, neglecting more challenging ones that are affected by overlapping.

In this paper, we propose a method FSMP to address the problem that highly overlapped objects existing in the field of FSOD by predicting multiple instances for each proposal. We predict a set of instances as shown in Fig. 1 (b). This increases the likelihood of obtaining the correct bounding boxes. Specifically, in the case shown in Fig. 1, method (b) predicts two bounding boxes for one proposal. Therefore, assuming we have only one proposal in total, there is only one bounding box in method (a). While there are two bounding boxes in method (b). The probability of incorrectly suppressing the bounding box is greatly reduced when predicting more proposals. More concisely, we predict a set of bounding boxes from one proposal instead of a single bounding box. Moreover, traditional Non-maximum Suppression (NMS) used in post-processing can mistakenly suppress correctly retained proposals, further exacerbating the issue. To solve this, we introduce diffNMS. It skips NMS for instances from the same proposal, addressing incorrect suppression issues. To expand the gap between these bounding boxes that might be trained as the same, distLoss is introduced. It encourages characteristic convergence during training. Overall, our approach improves the accuracy of detection with overlapped objects with limited samples, specifically addressing challenges related to overlapping.

Our method is both efficient and straightforward. In summary, our contributions are as follows: (i)We utilize a set of predicted bounding boxes instead of a single bounding box, effectively addressing overlapping problems in FSOD. (ii)To prevent traditional NMS from erroneously filtering out the required bounding boxes, we enhance the post-processing operation by skipping NMS when the predicted instances originate from the same proposal. (iii)To differentiate the obtained bounding boxes from the same proposal, we introduce distLoss. Finally, we validate our method on MS-COCO [17] and Pascal VOC [5] and observe significant improvements. Furthermore, our method exhibits consistent accuracy even on nearly no-overlapping scenarios, demonstrating its generalizability.

2 Related Work

2.1 Few-Shot Object Detection

It aims to identify the target using a limited number of annotated samples. Several recent works [6–9, 12, 21, 23, 26] have proposed different approaches in this area. A simple two-stage fine-tuning approach (TFA) [23] has also been introduced, where a Faster R-CNN model is initially trained on the base class. Another recent work, FSCE [21], addresses class confusion between novel classes by training Region of Interest (RoI) features using a separate supervised contrast learning head. LVC [13] proposes two novel strategies to enhance detection accuracy by referencing pseudo-labels and adjusting bounding boxes. [8] introduces graph neural networks to establish relationships between novel class and base class.

2.2 Object Detection in Crowded Scenes

Earlier research primarily focused on the pedestrian re-identification problem, showing significant interest in pedestrian detection methods such as detecting by parts and improving hand-crafted rules in training target [27]. There is also a section that employs sliding window strategies to address the issue. Additionally, several studies have proposed novel loss functions [24]. Subsequent studies aimed to extract features using deep neural networks (DNNs) and then utilize an enhanced decision forest for further processing. However, it should be noted that the trade-off for these benefits is a longer detection time.

2.3 Non-maximum Suppression

Non-maximum Suppression(NMS) constitutes a vital post-processing component in the procedure of object detection. Once **Intersection over Union (IoU)** surpasses a predetermined threshold, the respective bounding box is promptly discarded, thereby resulting in a diminished recall. Researchers have put forth several methodologies to enhance NMS. For instance, soft NMS [1] and softer

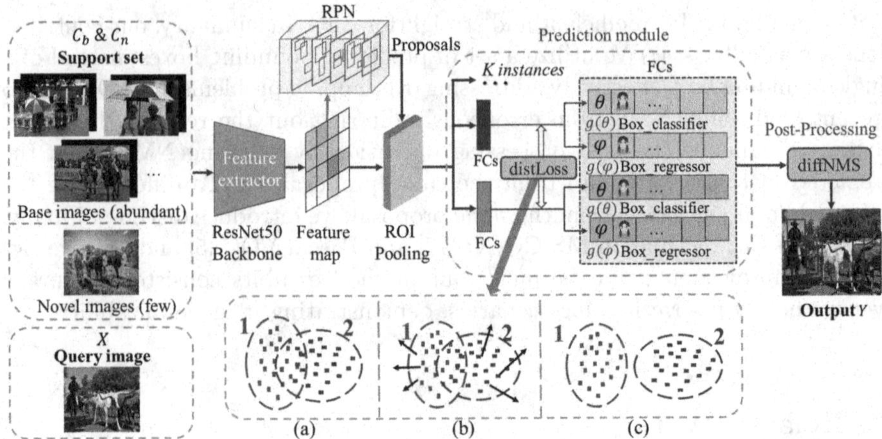

Fig. 2. The overall framework. We train the weights with samples from the base class and then fine-tune the entire model by adding a few samples of the novel class. For the proposals from RPN, we predict K bounding boxes during the proposal prediction stage. Then we add distLoss to distance two bounding boxes from the same proposal. As the figure below shows, (a) is the two bounding boxes from the same proposal. (b) is the training procedure where we distance the two. For (c), we have successfully separated the single features. Finally, diffNMS is added as the post-processing procedure.

NMS [2] adopt a decay mechanism to reduce the score of neighboring bounding boxes instead of instantaneously discarding them.

In summary, the overlapping problem in FSOD is both common and challenging, with limited practical solutions available. To address this issue, we propose various strategies, focusing on category expansion to generic categories. Different from [3], we aim to establish the relationship between the novel class and base class. By adopting this approach, we contribute to solving the overlapping problem in few-shot scenarios, an area that has received little attention thus far.

3 Methodology

In this section, we begin our method FSMP by providing an overview of our methods which is shown in Fig. 2.

3.1 Problem Definition

In this paper, we employ the same problem setting as in TFA [23]. Assuming we are given a dataset, $\mathcal{D} = \{(x, y), x \in \mathcal{X}, y \in \mathcal{Y}\}$, x represents the input image, and $y = \{(c_i, l_i), i = 1, \ldots, N\}$ denotes the categories $c \in C_b \cup C_n$ and bounding box coordinates l of the N object instances in the image x and two annotation sets, where C_b denotes the set of base class and C_n denotes the set of novel class. P denotes the set for proposals. When working with synthetic few-shot datasets such as COCO, the training set for novel classes is balanced, ensuring that there are the same number of annotated objects for each class (referred to as k-shot). In reality, we have seen that there is frequently a high degree of overlap between targets in the samples. The ability to discern individual objects from heavily overlapping targets is severely hampered. For each proposal, we construct a set of corresponding related ground-truth sets $\mathrm{G}(b_i)$ rather than forecasting a single one,

$$\mathrm{G}\left(b_i\right)=\left\{g_j \in \mathcal{G} \mid \operatorname{IoU}\left(b_i, g_j\right) \geq \theta\right\} \tag{1}$$

where b is the bounding box, $\mathrm{G}(b_i)$ represents the set of all ground truth bounding boxes g, and θ denotes the threshold for IOU.

3.2 Few-Shot Object Detection

Firstly, we conduct pre-training of the model on the base class. Then we fine-tune the model using k samples from the novel class, introducing modifications to its classification scores and prediction offsets.

Base Model Training. We exclusively train the features on the base class C_b during the initial stage, and the joint loss is,

$$\mathrm{L}=L_{rpn}+L_{cls}+L_{loc} \tag{2}$$

where L_{rpn} is used on the output of the RPN to tell the difference between the center and the background of the anchors. L_{cls} is a cross-entropy loss for the box classifier, while L_{loc} is the loss for the box regressor. The pre-trained model gains the representation of the important attributes by training on a sizable dataset. We use the pre-trained model at first, initializing it with C_b. We then use the pre-trained model to extract the feature maps. Then we use RPN and ROI Pooling to train the model and get the weights.

Few-shot Fine-Tuning. In the following stage, we create a few-shot samples. k samples from C_b and C_n are chosen at random, and initialization weights are then allocated at random to the networks in charge of forecasting the novel class. Only the last layers that box classification and box regression are adjusted; the feature extractor F as a whole is unchanged. We improve FSOD performance by combining basic training and fine-tuning stages.

3.3 Multiple Predictions

Since the few-shot samples cannot provide enough annotations that have overlapping objects, it is difficult to train the model. And this will result in an incorrect pickup or even missed detection of target objects in the query set. We modify the traditional proposal-based prediction approach to particularly target the issue of overlapping at various heights in order to overcome this challenge. After generating proposals, we go on to forecast a set of instances for each proposal.

We attempt to predict K bounding boxes by using K convolutional layers. The fact that any proposal can predict K cases that have a common ground truth is noteworthy. This inspiration has led us to concentrate our efforts on improving three crucial areas:

Prediction Sets. Present-day proposal-based methods are centered on the paradigm of instance prediction. In our method, a set of corresponding instances denoted (d_i, m_i), is predicted for every proposal. Here, d_i is the confidence score for each prediction class, and m_i is the offset of the frame position coordinates. The sets of predictions $S(b_i)$ can be:

$$\mathrm{S}(b_i) = \left\{ \left(\mathbf{d}_i^{(1)}, \mathbf{m}_i^{(1)}\right), \left(\mathbf{d}_i^{(2)}, \mathbf{m}_i^{(2)}\right), \ldots, \left(\mathbf{d}_i^{(k)}, \mathbf{m}_i^{(k)}\right) \right\} \tag{3}$$

where d_i^k and m_i^k is the k_{th} prediction. Our method predicts a collection of instance score sets for each proposal, which may present the following challenges: When presented with a proposal containing more than K instances, we select the highest confidence scores. The remaining instances are then effectively accounted for by additional prediction boxes generated by the proposals. In cases where a bounding box contains only a single object, we employ a strategy by introducing a fictional box. This incorporation of a fake box ensures that the confidence score associated with this generated box is deliberately set to an astonishingly low value, regardless of the object category. Consequently, the presence of this low-confidence bounding box does not compromise the ultimate detection outcome during post-processing operations.

NMS Improvement. The conventional post-processing method is incapable of effectively managing the additional regions generated by our method. The pseudo-code of NMS used in the post-processing step is as Algorithm 1. Typically, these regions are filtered out due to their lesser scores or higher IOU values. Consequently, the framework for multi-instance prediction utilized in our earlier method is ineffectual. The suppression of predicted instances originating from the same proposal is the primary reason for the inadequacy of traditional NMS, as determined by meticulous analysis. We develop a gating that bypasses NMS for instances originating from the same proposal. When instances are derived from distinct proposals, however, we use NMS as usual.

Loss for Restriction. As stated previously, a set of bounding boxes that correspond to the same ground truth are predicted. During training, the performance of the model is evaluated using a loss function that quantifies the difference between the predicted and observed outcomes. So, a new loss function distLoss is introduced to pull back the distance between these bounding boxes. It can be like:

$$\mathrm{distLoss}(\alpha) = \tfrac{1}{K} \sum_{i=1}^{K} |d_{i+1} - d_i| \tag{4}$$

$$\mathrm{distLoss}(\beta) = \tfrac{1}{K} \sum_{i=1}^{K} |m_{i+1} - m_i| \tag{5}$$

Algorithm 1: Pseudo-code of NMS used in the post-processing step.

```
Input: B = {b_1, ..., b_N}, P = {p_1, ..., p_N}, M = {m_1, ..., m_N}, θ
       B is the set of detection bounding boxes
       P is the proposals getting after RPN
       M is the set of corresponding detection box offset scores
       θ is the NMS threshold
Q ← {};
while B ≠ empty do
    a ← arg max M ;
    A ← b_a ;
    Q ← Q ∪ A B ← B − A ;
    for b_i ∈ B   &&   b_i ∈ p_j do
        if iou(A, b_i) ≥ θ then
            B ← B − b_i; M ← M − m_i;
        end
    end
end
return Q, M
```

where α can be sets of parameters in a box classifier network, and d is the classification score for Lls, the loss of the classification network. β is the parameter of the box regressor network, and m is the delta for the location to represent the loss Dls in a box regressor network. We let the model place the two boxes as far apart as feasible. This method ensures that different bounding boxes do not converge to virtually identical properties.

4 Experiments

In this section, we describe our experimental procedure in detail, including the parameters and specifics of the experimental apparatus. Using MS-COCO [17] for both training and testing and Pascal VOC [5] for testing, we undertake thorough comparisons with prior methods on established FSOD benchmarks. We assess the efficacy using standard object detection metrics, Average Precision (AP). We employ the *detectron2* framework to implement the efficient Faster-RCNN [20] network.

4.1 Experiment Settings

Datasets. Our method is based on MS-COCO [17] and Pascal VOC [5]. We adhere to the data segmentation and training examples provided by [12]. We specify 60 categories as a base class that does not intersect with PASCAL VOC. The remaining 20 categories are used as novel classes with two distinct settings: $k = 10$ and $k = 30$, where k represents the number of novel classes introduced.

Detailed Settings. We use standard ResNet-50 and ResNet-101 pre-trained on ImageNet [4] as the backbone. All of ours use Faster-RCNN [20] with FPN as standard. All of our experiments are based on CUDA 10.1, and we use two GPUs to analyze data in parallel while setting the batch-size to 8. We perform 120, 000 iterations in total for each. We choose SGD as the optimizer, which has a momentum of 0.9 and a weight decay of 0.001. To make training easier, we set the learning rate to 0.001.

4.2 Results on the Dataset

We know that COCO contains various overlapping image types and that a single training and testing procedure may not fully capture the detection accuracy. We carried out many sets of trials with identical hyperparameters in order to address this problem and average it. We replicated the experiments using the current experimental setup in order to create a baseline and documented the associated performance. The success of our methods can be assessed compared to previous results. The following source has more information on this comparison: compared

Table 1. Performance of FSOD on the MS-COCO benchmark. We discuss performance on the 20 novel MS-COCO classes in the FSOD settings. The best outcomes have each been bolded. Performance on the base classes is also reported. FSMP(Ours) outperforms all prior methods on the average rank metric and performs best compared to the state-of-the-art.

Shot	Methods	Backbone	Avg. Rank	Metrics					
				nAP	nAP50	nAP75	bAP	bAP50	bAP75
10-shot	TFA w/cos [23]	ResNet-101	6.3	10.0	19.1	9.3	32.4	50.6	35.7
	FSCE [21]	ResNet-101	5.5	11.9	-	10.5	-	-	-
	DCNet [10]	ResNet-101	3.0	12.8	23.4	11.2	-	-	-
	QA-FewDet [8]	ResNet-101	6.7	11.6	23.9	9.8	-	-	-
	Meta Faster-RCNN [9]	Resnet-101	3.3	12.7	25.7	10.8	-	-	-
	DMNet [19]	Resnet-101	9.0	10.0	17.4	10.4	-	-	-
	KR-FSD [22]	Resnet-101	9.0	10.2	21.5	8.7	-	-	-
	N-PME [18]	Resnet-101	8.3	10.6	21.1	9.4	-	-	-
	LVC [13]	ResNet-50	8.7	11.9	22.0	10.6	30.9	49.6	36.3
	FSMP(Ours)	ResNet-101	1.0	**13.00**	**29.10**	**11.31**	**33.97**	**51.13**	**37.25**
	FSMP(Ours)	ResNet-50	2.8	12.75	28.15	10.59	33.09	50.05	36.70
30-shot	TFA w/cos [23]	ResNet-101	7.0	13.7	24.9	13.4	34.2	52.3	38.0
	FSCE [21]	ResNet-101	7.0	16.4	-	16.2	-	-	-
	DCNet [10]	ResNet-101	3.0	18.6	32.6	17.5	-	-	-
	QA-FewDet [8]	ResNet-101	6.3	16.5	31.9	15.5	-	-	-
	Meta Faster-RCNN [9]	Resnet-101	6.0	16.6	31.8	15.8	-	-	-
	DMNet [19]	Resnet-101	4.7	17.1	29.7	17.7	-	-	-
	KR-FSD [22]	Resnet-101	9.3	14.1	28.6	13.2	-	-	-
	N-PME [18]	Resnet-101	9.0	14.1	26.5	13.6	-	-	-
	LVC [13]	ResNet-50	2.8	17.3	30.8	16.9	31.7	56.2	39.8
	FSMP(Ours)	ResNet-101	1.0	**19.87**	**36.02**	**17.82**	**35.89**	**57.01**	**42.32**
	FSMP(Ours)	ResNet-50	2.7	19.00	34.87	16.31	35.07	55.32	40.91

Table 2. Few-shot detection performance across the two splits on the VOC benchmark. In most cases our method works better than the baseline, proving the generalizability.

Method/Shot	Backbone	Novel Split 1					Novel Split 2				
		1	2	3	5	10	1	2	3	5	10
TFA w/ cos [23]	ResNet-101	39.8	36.1	44.7	55.7	56.0	23.5	26.9	34.1	35.1	39.1
DCNet [10]		33.9	37.4	43.7	51.1	59.6	23.2	24.8	30.6	36.7	46.6
FSCE [21]		**44.2**	43.8	51.4	**61.9**	**63.4**	27.3	29.5	**43.5**	**44.2**	45.2
CGDP+FSCN [3]	ResNet-50	40.7	45.1	46.5	57.4	62.4	27.3	31.4	40.8	42.7	46.3
LVC [13]		37.7	42.0	50.3	57.0	58.0	19.8	22.8	35.6	42.7	44.2
Ours (FSMP)		41.2	**46.0**	**52.6**	60.1	62.3	**29.3**	**31.7**	39.2	42.0	**48.1**

to the existing method TFA w/cos [23], FSCE [21], DCNet [10], QA-FewDet [8], Meta Faster-RCNN [9], DMNet [19], KR-FSD [22], N-PME [18] and our baseline method [13], we got good results on different shots as well as indicators, almost all of which were raised by 1-4 percentage points, shown in Table 1. In all the training counts, our method outperforms the preceding method and ranks first over others. This improvement demonstrates that our procedure is effective. We report the average AP, AP50, and AP75 of the base classes and 20 novel classes on COCO in Table 1. AP75 means the matching threshold is 0.75, a more strict metric than AP50. We find our method performs better on 30-shot than that on 10-shot generally. Again, we consistently outperform previous methods across all shots on both novel AP and novel AP75. We achieve around 1 point improvement in AP over the best-performing method. In addition, we compared the efficacy of the existing model structure on the Pascal VOC dataset, as shown in Table 2. The results were the best on some shots, others not. However, it is good for transfer learning to demonstrate generalizability.

4.3 Ablation Study and Visualization

Importance of Designs. We carried out the ablation study to look into our design decisions. We first show how well multi-instance prediction, NMS improvement, and loss improvement work. Our experiments were built with ResNet-50, 8 batch size, and 60,000 iterations. To assess the quality of the model, we employed two evaluation metrics nAP and bAP. This is the experimental setting used to determine the efficacy of our three techniques, as indicated in Table 3. From the experiments, we find if we just add MI technique, the detection effect is not much improved. The reason is that when we predict the process of multiple instances, it may make the probability of false and wrong detection greater. Therefore, we need to introduce additional strategies to strengthen our approach. As mentioned before, we introduce DN and DL. Through the experimental data, we can see that each strategy is effective and the stacking of multiple strategies will result in the strongest effect. For lightweight, we set the value of the predicted number of instances K to 2 in the previous. To better determine the effect played by K

Table 3. Ablative performance on COCO 30-shot task by gradually applying the proposed components to the baseline. **MI**: multiple instances. **DN**: diffNMS. **DL**: distLoss.

	Components			Metrics	
	MI	DN	DL	nAP	bAP
LVC [13]				10.6	30.8
	✓			10.5	31.2
	✓	✓		12.0	33.5
FSMP(Ours)	✓	✓	✓	12.1	34.7

Table 4. Performance on 30-shot task for loss selection, including four ways: no loss for restriction, loss for deltas, loss for classification, and loss for both. **Lls**: logits_loss. **Dls**: deltas_loss.

	Metrics			
	nAP	bAP	nAP50	bAP50
baseline	18.1	33.5	32.9	54.3
+Lls	18.2	34.0	32.8	54.9
+Dls	18.0	33.0	32.2	55.0
+Lls +Dls	18.9	34.7	33.4	55.1

in this experiment, we conduct the ablation experiment on the value of K for the setting of 15000 iteration on the base classes. The results of the experiment are shown in Fig. 3. We find that as the value of K increases, the accuracy for each metric is growing. However, when K rises to 5 and above, the accuracy does not change much.

Increments by Loss. We have mentioned before that each proposal predicts a set of different bounding boxes, and they share the same ground truth. Meanwhile, we have previously shown that the bounding boxes from the same proposal may be trained to converge. As a result, we must set the loss to pull the distance. We set up an ablation study to examine the effects of various losses on the experiments since the prediction procedure takes into account two factors, as indicated in Table 4. We first just add no loss and get the results. Then, we add *logits_loss*(*Lls*) separately to test the accuracy. We find that when we just add *Lls*, it has some accuracy improvements. When we just add *deltas_loss*(*Dls*), it may get some drop. And when we combine both for some weight initialization, it works the best. The reason it drops is that for overlapping objects the distance

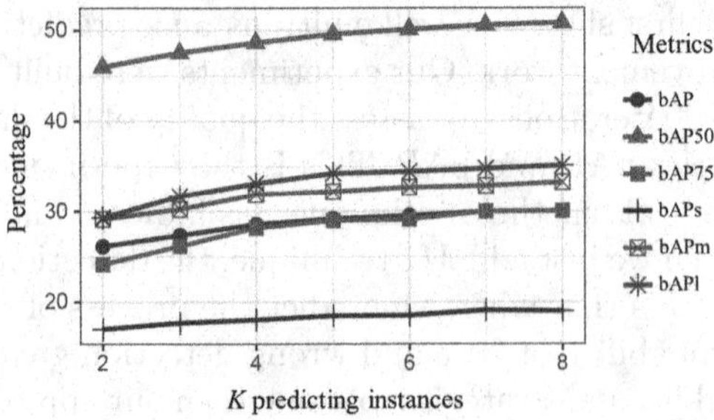

Fig. 3. For the selection of K, this experiment was performed on the base class, where the horizontal coordinate indicates the value of K and the vertical coordinate represents the percentage.

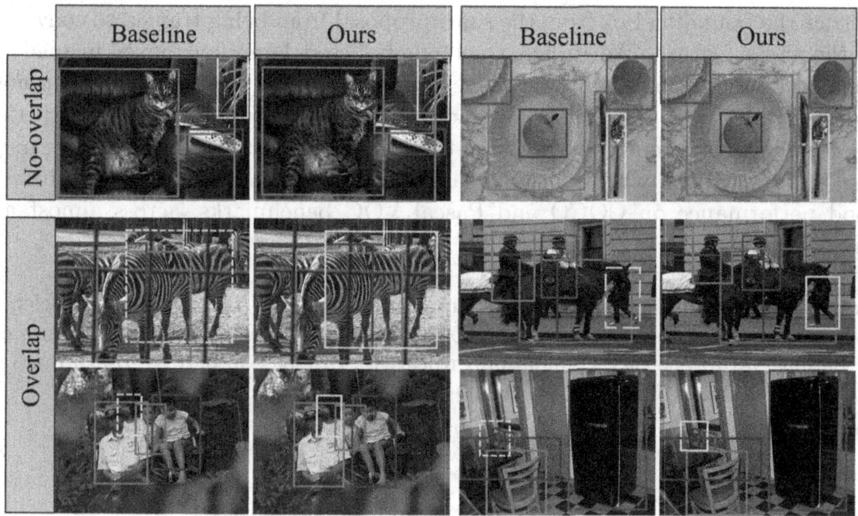

Fig. 4. Visual comparison between our strategy with the baseline. The occurrences that our approach can be used to detect but was not picked up by the baseline are represented by the dotted lines. We arranged the effects into three rows of two instances each. No overlapping scenes appear in the first row of images, and there are overlaps in the next two lines. From the visualization, we find our detection results can be on par with the baseline in no-overlap scenarios. Moreover, in scenes with overlap, we can detect objects that were not detected before.

between indicators is so small that it may cause confusion when just add alone. But if we add both, there are more restrictions for the reduction and it performs better. The most appropriate weight we find is 0.6, 0.3.

Visualization. We illustrate the impact of our model visually in Fig. 4. The images are arranged in three rows. No overlapping scenes appear in the first row of images, and there are overlaps in the next two lines. For each line, two examples are listed. The left side represents the test results of the baseline, and the right side represents the results of ours. The occurrences that our approach can be used to detect but was not picked up by the baseline are represented by the dotted lines. Finally, we find that our method can perform few-shot detection successfully with good generalization even without the presence of overlapping objects to all extents of overlap.

5 Conclusion

We propose a new way to predict multi-instances for the overlapping problem in FSOD. We also introduce two other strategies to help solve this problem: we utilize a diffNMS strategy to prevent post-processing operations from incorrectly suppressing the bounding box, and then we add a new loss function distLoss to restrict the bounding box from the same proposal from being trained to converge to the same feature. We utilize a set of predicted bounding boxes instead of a single bounding box, which can increase the accuracy to detect overlapping objects. But it will also introduce more interfering information, although the negative impact is much smaller than the positive impact. In the future we may explore how to deal with interfering information. Overall, our method achieves good performance on COCO and Pascal VOC benchmarks, across almost all numbers of shots.

Acknowledgements. This work was supported in part by National Natural Science Foundation of China (NSFC) No. 62106022, 62225601, U19B2036, U22B2038, in part by Beijing Natural Science Foundation Project No. Z200002, in part by Youth Innovative Research Team of BUPT No. 2023QNTD02.

References

1. Anderson, C.H., Burt, P.J., van der Wal, G.S.: Change detection and tracking using pyramid transform techniques. In: Other Conferences (1985)
2. Bodla, N., Singh, B., Chellappa, R., Davis, L.S.: Soft-NMS - improving object detection with one line of code. In: ICCV, pp. 5562–5570 (2017)
3. Chu, X., Zheng, A., Zhang, X., Sun, J.: Detection in crowded scenes: one proposal, multiple predictions. In: CVPR, pp. 12211–12220 (2020)
4. Deng, J., Dong, W., Socher, R., Li, L-J., Li, K., Fei-Fei, L.: ImageNet: a large-scale hierarchical image database. In: CVPR, pp. 248–255 (2009)
5. Everingham, M., Van Gool, L., Williams, C.K.I., Winn, J., Zisserman, A.: The pascal visual object classes (VOC) challenge. In: IJCV, pp. 303–338 (2010)
6. Fan, Q., Zhuo, W., Tai, Y.-W.: Few-shot object detection with attention-RPN and multi-relation detector. In: CVPR, pp. 4012–4021 (2019)
7. Fan, Z., Ma, Y., Li, Z., Sun, J.: Generalized few-shot object detection without forgetting. In: CVPR, pp. 4525–4534 (2021)
8. Han, G., He, Y., Huang, S., Ma, J., Chang, S.-F.: Query adaptive few-shot object detection with heterogeneous graph convolutional networks. In: ICCV, pp. 3243–3252 (2021)
9. Han, G., Huang, S., Ma, J., He, Y., Chang, S-F.: Meta faster R-CNN: towards accurate few-shot object detection with attentive feature alignment. In: AAAI, pp. 780–789 (2022)
10. Hu, H., Bai, S., Li, A., Cui, J., Wang, L.: Dense relation distillation with context-aware aggregation for few-shot object detection. In: CVPR (2021)
11. Huang, S., Zeng, X., Wu, S., Yu, Z., Azzam, M., Wong, H.-S.: Behavior regularized prototypical networks for semi-supervised few-shot image classification. PR **112**, 107765 (2021)
12. Kang, B., Liu, Z., Wang, X., Yu, F., Feng, J., Darrell, T.: Few-shot object detection via feature reweighting. In: ICCV, pp. 8419–8428 (2018)
13. Kaul, P., Xie, W., Zisserman, A.: Label, verify, correct: a simple few shot object detection method. In: CVPR, pp. 14217–14227 (2021)
14. Liang, K., Guo, Y., Chang, H., Chen, X.: Visual relationship detection with deep structural ranking. In: AAAI (2018)
15. Lin, T.-Y., Dollár, P., Girshick, R.B., He, K., Hariharan, B., Belongie, S.J.: Feature pyramid networks for object detection. In: CVPR, pp. 936–944 (2016)

16. Lin, T.-Y., Goyal, P., Girshick, R.B., He, K., Dollár, P.: Focal loss for dense object detection. In: TPAMI, pp. 318–327 (2017)
17. Lin, T.Y., et al.: Microsoft COCO: common objects in context. In: Fleet, D., Pajdla, T., Schiele, B., Tuytelaars, T. (eds.) ECCV 2014. LNCS, vol. 8693, pp. 740–755. Springer, Cham (2014). https://doi.org/10.1007/978-3-319-10602-1_48
18. Liu, W., Wang, C., Yu, S., Tao, C., Wang, J., Wu, J.: Novel instance mining with pseudo-margin evaluation for few-shot object detection. In: ICASSP, pp. 2250–2254 (2022)
19. Lu, Y., Chen, X., Wu, Z., Yu, J.: Decoupled metric network for single-stage few-shot object detection. IEEE Trans. Cybern. **53**, 514–525 (2022)
20. Ren, S., He, K., Girshick, R.B., Sun, J.: Faster R-CNN: towards real-time object detection with region proposal networks. In: TPAMI, pp. 1137–1149 (2015)
21. Sun, B., Li, B., Cai, S., Yuan, Y., Zhang, C.: FSCE: few-shot object detection via contrastive proposal encoding. In: CVPR, pp. 7348–7358 (2021)
22. Wang, J., Chen, D.: Few-shot object detection method based on knowledge reasoning. Electronics **11**, 1327 (2022)
23. Wang, X., Huang, T.E., Darrell, T., Gonzalez, J., Yu, F.: Frustratingly simple few-shot object detection. ArXiv e-prints (2020)
24. Wang, X., Xiao, T., Jiang, Y., Shao, S., Sun, J., Shen, C.: Repulsion loss: detecting pedestrians in a crowd. In: CVPR, pp. 7774–7783 (2017)
25. Wang, Y., Yao, Q., Kwok, J.T., Ni, L.M.: Generalizing from a few examples: a survey on few-shot learning. ArXiv e-prints (2019)
26. Zhang, Y., Chu J., Leng L., Miao, J.: Mask-refined R-CNN: a network for refining object details in instance segmentation. In: MDPI, p. 1010 (2020)
27. Zheng, A., Zhang, Y., Zhang, X., Qi, X., Sun, J.: Progressive end-to-end object detection in crowded scenes. In: CVPR, pp. 847–856 (2022)

Target-Aware Bi-Transformer for Few-Shot Segmentation

Xiangling Wang, Xiaoliu Luo, and Taiping Zhang(✉)

Chongqing University, Chongqing, China
{202114131112,20160602026t,tpzhang}@cqu.edu.cn

Abstract. Traditional semantic segmentation tasks require a large number of labels and are difficult to identify unlearned categories. Few-shot semantic segmentation (FSS) aims to use limited labeled support images to identify the segmentation of new classes of objects, which is very practical in the real world. Previous researches were primarily based on prototypes or correlations. Due to colors, textures, and styles are similar in the same image, we argue that the query image can be regarded as its own support image. In this paper, we proposed the Target-aware Bi-Transformer Network (TBTNet) to equivalent treat of support images and query image. A vigorous Target-aware Transformer Layer (TTL) also be designed to distill correlations and force the model to focus on foreground information. It treats the hypercorrelation as a feature, resulting a significant reduction in the number of feature channels. Benefit from this characteristic, our model is the lightest up to now with only 0.4M learnable parameters. Furthermore, TBTNet converges in only 10% to 25% of the training epochs compared to traditional methods. The excellent performance on standard FSS benchmarks of PASCAL-5^i and COCO-20^i proves the efficiency of our method. Extensive ablation studies were also carried out to evaluate the effectiveness of Bi-Transformer architecture and TTL.

Keywords: Semantic segmentation · Few-shot learning · Transformer

1 Introduction

Semantic segmentation aims to assign each pixel of an image to a certain class, which is one of the cornerstones of computer vision tasks. With the development of the deep convolution network [12,15], it has made considerable progress. However, the training process requires an enormous amount of labeled data, which is a labor-intensive task. So semi- and weakly-supervised segmentation [11,24,27] is invented to reduce the dependence on expensive labels. But all the above methods can only recognize the classes in the train episode. To address this limitation, Few-shot segmentation (FSS) task was proposed.

Supported by Chongqing University.

Q. Liu et al. (Eds.): PRCV 2023, LNCS 14426, pp. 440–452, 2024.
https://doi.org/10.1007/978-981-99-8432-9_35

There are many FSS approaches that have been proposed in recent years [10,22,26]. The typical method follows the meta-learning paradigm [2], which is easy to overfit due to insufficient data for training. The FSS model is supposed to predict the segmentation of query images based on the condition of support images and corresponding annotations. Nowadays the prevalent approaches are based on prototype [9,23,26] and pixel-wise correlation [10,19,28]. Prototype-based methods want to obtain the prototype of the target from support images of high-level features and then utilize the prototype to segment query images. Pixel-wise-based methods take the similarity of each pixel between the query and support images as features to train the model. We can regard similarity as a class-agnostic feature, so the model rarely overfit.

Objects in different pictures, even if they belong to the same category, may have vastly different features, especially for those parts that are not easily distinguishable. Self-similarity may alleviate this phenomenon, but all the above methods ignored it.

In the paper, we proposed the Target-aware Bi-Transformer Network (TBT-Net), which can integrate two types of similarity. As shown in Fig. 1, we first construct an intermediate prediction based on cross-similarity. According to [28], using high-level feature similarity can make a preliminary prediction, which highlights the most recognizable area for the class. After that we utilize self-similarity to refine the segmentation of query images. It is because that self-similarity contains the structure information of an image, which can expand the part segmentation to the whole.

We also adopt a pyramid structure to implement our model. Our Target-aware Bi-Transformer Module (TBTM) can aggregate two affinity matrices at the same layer, guided by previous layer information, and then transfer the refined similarity to the next layer. That is because high-level intermediate prediction can roughly locate the target, but the boundary of an object is hard to distinguish due to low resolution. Increasing the segmentation resolution progressively with the expansion of affinity matrix can make the boundary more accurate.

In order to make the model only concentrates to the categories we are interested in, we propose a Target-aware Transformer Module (TTM), which consists of two Target-aware Transformer Layers (TTL). Inheriting the virtue of transformer, each pixel of the query image has a global receptive field for the support image. Guided by the mask, TTL only focus on the target in the foreground. Since our model takes affinity matrices as inputs, which are low-dimensional features, the learned parameters are much less than the vanilla FSS models [10,13,19,28]. Besides, the training time is shorter, and the computing complexity is lower than other methods due to the fewer parameters. Although our model is small, it can also achieve state-of-the-art performance. All in all, our contributions are:

- Bi-Transformer architecture is proposed for few-shot segmentation work, which can take the advantage of self-similarity to boost performance.

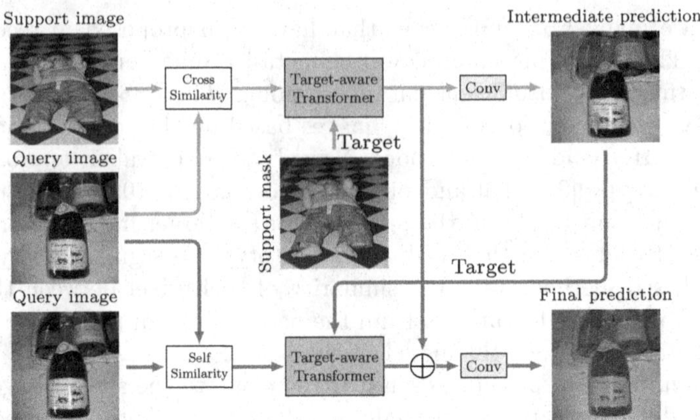

Fig. 1. Illustration of the Bi-Transformer architecture for few-shot segmentation.

- We propose a novel Target-aware Transformer, which can efficiently extract the target's hypercorrelation information under the guidance of the mask.
- Our TBTNet only has 0.4 M learnable parameters, it is the lightest FSS model to date.
- Our model can converge quickly and achieve SOTA performance.

2 Relate Works

2.1 Few-Shot Semantic Segmentation

Few-shot semantic segmentation is a branch of semantic segmentation that aims to assign each pixel to a particular class with only a few examples. It was first proposed by [2], which adopts a meta-learning paradigm to propagate support branch annotation to query branch. Soon afterward, Jake *et al.* [9] imports the idea of prototype into FSS, which extracts the prototype from the support set of a certain class and segments the query image by the prototype. Recently, Liu *et al.* [26] try to alleviate the intra-class variations, they generate an intermediate prototype from both query and support images.

Although the prototype-based method has had great success in FSS, it disregards a lot of pixel structure information, which hinders the performance of this approach. PFNet [28] uses the affinity matrix between query and support high-level features to obtain a prior segmentation with a parameters-free method. The idea of hypercorrelation was introduced by [10], which is a 4D tensor transformed from affinity matrices and squeezed with 4D convolution. Followed [10], ASNet [5] replaced 4D convolution with an attention mechanism based on the transformer to compress hypercorrelation.

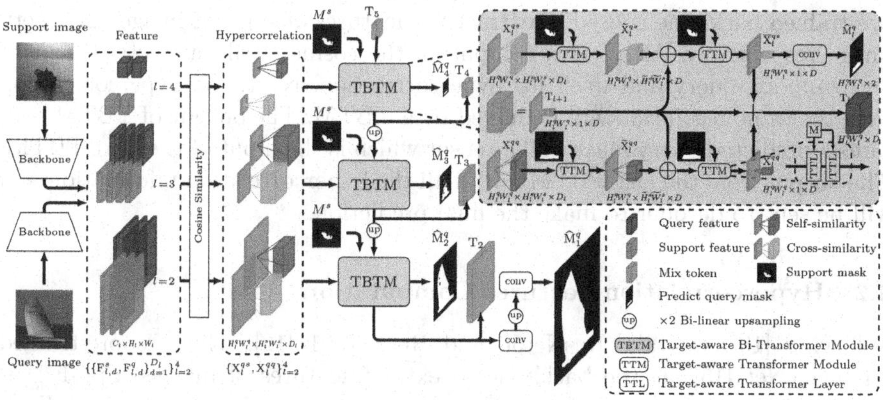

Fig. 2. Overall network architecture. Our TBTNet consists of four main sub-modules: feature extraction, similarity computation, TBTM pyramidal encoder, and a simple convolution decoder. For more details please refer to Sect. 4.

2.2 Vision Transformer

Ashish *et al.* [3] first proposed transformer in the Nature Language Processing (NLP) field, which is the standard architecture now. After that, Vit [6] introduced the transformer to Computer Vision (CV) and achieved great success. Recently, many transformer-based methods have been proposed in FSS. CyCTR [8] screens out reliable support features as query tokens to implement cross attention with the query image. DCAMA [25] aggregates mask by the attention between query and support features.

3 Problem Setting

There are two sets of data D_{train} and D_{test}. The former is used to train the FSS model, and the last one is for testing, to evaluate the accuracy of the model. Each set contains many episodes $E = \{I^q, M^q, I^s, M^s\}$ where I^s and I^q represent support image and query image, M^s and M^q denote the corresponding binary mask of the certain category. For the k-shot scenario, $E = \{I^q, M^q, I_1^s, M_1^s, I_2^s, M_2^s, ..., I_K^s, M_K^s\}$. The categories of D_{train} and D_{test} are disjoint, which means $C_{train} \cap C_{test} = \varnothing$, where C_{train} and C_{test} are the classes of D_{train} and D_{test}. During the training stage, we randomly sample episodes E from D_{train} to learn a network that can predict M^q by $\{I^q, I_1^s, M_1^s, I_2^s, M_2^s, ..., I_K^s, M_K^s\}$. At the inference stage, our model samples episodes from D_{test} and predicts the novel class target segmentation M^q.

4 Method

4.1 Overview

As shown in Fig. 2, our Target-aware Bi-Transformer Network (TBTNet) consists of three Target-aware Bi-Transformer Modules and two decoders. Firstly, a pre-trained backbone is used to extract the features of support image and query image respectively. After that, computing the cosine similarity between query and support, query and itself for hypercorrelation. Next, the hypercorrelation with the same resolution will be mixed by a TBTM. The output of TBTM contains a predicted query mask and a tensor which is the input of the next TBTM. The output from the last TBTM mixed all the hypercorrelation information and will be sent to decoder to make the final prediction.

4.2 Hypercorrelation Features Computation

Following [10], we take ResNet50 and ResNet101 [12], which is pre-trained on ImageNet [15], as the backbone to extract features of images. $\mathrm{F}^s_{l,d}, \mathrm{F}^q_{l,d} \in \mathbb{R}^{C_l \times H_l \times W_l}$ are the features of support and query images $I^s, I^q \in \mathbb{R}^{3 \times H \times W}$ respectively.

$$\{\{\mathrm{F}^*_{l,d}\}^{D_l}_{d=1}\}^4_{l=2} = \mathrm{ResNet}(I^*), \tag{1}$$

where l denotes the output layer of ResNet, D_l means the number of block at layer l and $* \in \{s, q\}$.

Cross- and Self- similarity. Since the features extracted from the backbone contain rich semantic information, we compute the cosine between features and get cross-similarity $\mathrm{A}^{qs}_{l,d} \in \mathbb{R}^{H^q_l W^q_l \times H^s_l W^s_l}$:

$$\mathrm{A}^{qs}_{l,d}(p^q, p^s) = \mathrm{ReLU}\left(\frac{\mathrm{F}^q_{l,d}(p^q)^T\, \mathrm{F}^s_{l,d}(p^s)}{\|\mathrm{F}^s_{l,d}(p^q)\| \|\mathrm{F}^q_{l,d}(p^s)\|}\right), \tag{2}$$

where p^* is 2D positions and $\mathrm{F}^*_{l,d}(p^*) \in \mathbb{R}^{C_l \times 1}$. We compute the self-similarity $\mathrm{A}^{qq}_{l,d}$ in the same way as cross-similarity, only replacing $\mathrm{F}^s_{l,d}$ with $\mathrm{F}^q_{l,d}$.

Hypercorrelation. To obtain cross-hypercorrelation X^{qs}_l and self-hypercorrelation $\mathrm{X}^{qq}_l \in \mathbb{R}^{H^q_l W^q_l \times H^s_l W^s_l \times D_l}$, we stack all the affinity matrix at the same layer.

$$\mathrm{X}^{q*}_l = \mathrm{Stack}(\{\mathrm{A}^{q*}_{l,d}\}^{D_l}_{d=1}), \tag{3}$$

4.3 Target-Aware Bi-Transformer Module

Previous approaches only use cross-similarity to predict the segmentation of query image, often leading to incomplete result. It greatly limits the capability of the model. In contrast, self-similarity contains the structural information inherent in the image, which helps make the prediction more complete. Therefore, we designed the TBTM, which first passes the cross-hypercorrelation through two Target-aware Transformer Modules (TTM) and a Convolution Block to obtain the intermediate prediction, then, under the guidance of the prediction, refines the self-hypercorrelation through the other two TTMs. As shown in Fig. 2, TTM aims to reduce the support spatial sizes progressively and change the channels of the hypercorrelation.

$$\widetilde{X}_l^{qs} = \mathrm{TTM}(X_l^{qs}, M^s), \tag{4}$$

$$\overline{X}_l^{qs} = \mathrm{TTM}(\widetilde{X}_l^{qs} \oplus T_{l+1}, M^s), \tag{5}$$

where $M^s \in \{0,1\}^{H \times W}$ is the binary segmentation map of support image. $\widetilde{X}_l^{qs} \in \mathbb{R}^{H_l^q W_l^q \times \widetilde{H}_l^s \widetilde{W}_l^s \times D}$ and $\overline{X}_l^{qs} \in \mathbb{R}^{H_l^q W_l^q \times 1 \times D}$ are the output of TTM. Noting that both $\widetilde{H}_l^s$ and $\widetilde{W}_l^s$ are smaller than H_l^s and W_l^s respectively. T_{l+1} denotes the MixToken from previous layer which has mixed the self- and cross-similarity information. It is initialized to 0 and $T_{l+1} \in \mathbb{R}^{H_l^q W_l^q \times 1 \times D}$. We utilized broadcasted element-wise addition to sum $\widetilde{X}_l^{qs}$ and T_{l+1} because of their shapes are different. $\overline{X}_l^{qs}$ will be sent to a convolution block to compute $\widehat{M}_l^q$:

$$\widehat{M}_l^q = \mathrm{ReLU}(\mathrm{Conv}(\mathrm{ReLU}(\mathrm{Conv}(\overline{X}_l^{qs})))), \tag{6}$$

where $\widehat{M}_l^q$ denotes the predicted segmentation of query image at layer l, $\widehat{M}_l^q \in \mathbb{R}^{2 \times H_l^q \times W_l^q}$. The convolution block consists of two times alternating convolution layers and ReLU activation functions. We can get a binary segmentation map $M_l^q \in \{0,1\}^{H_l^q \times W_l^q}$ easily from $\widehat{M}_l^q$:

$$M_l^q(x,y) = \begin{cases} 0 \text{ if } \widehat{M}_l^q(0,x,y) > \widehat{M}_l^q(1,x,y) \\ 1 \qquad \text{otherwise} \end{cases}, \tag{7}$$

where $x \in \{0,1,...,W_l^q - 1\}, y \in \{0,1,...,H_l^q - 1\}$ indicate the 2D coordinates of features. And then, we can treat M_l^q as a pseudo mask and deal with the self-similarity X_l^{qq} as same as X_l^{qs}:

$$\widetilde{X}_l^{qq} = \mathrm{TTM}(X_l^{qq}, M_l^q), \tag{8}$$

$$\overline{X}_l^{qq} = \mathrm{TTM}(\widetilde{X}_l^{qq} \oplus T_{l+1}, M_l^q), \tag{9}$$

Finally, we used a residual structure to update T

$$T_l = T_{l+1} + \overline{X}_l^{qs} + \overline{X}_l^{qq}, \tag{10}$$

Up to now, we have gotten all the output of TBTM, $\widehat{M}_l^q$ and T_l. Noting that we upsampled T_l before taking it as the input of the next TBTM to make sure the spatial sizes match.

4.4 Target-Aware Transformer Module

The traditional Transformer has global attention, it is difficult to make the model only focus on specific categories due to the support images contain multiple objects. Therefore, we propose the Target-aware Transformer Module (TTM) to make the model only calculates the hypercorrelation of the target in the mask. TTM consists of multiple Target-aware Transformer Layers(TTL), and the structure of TTL is illustrated in Fig. 3. In order to gradually reduce support spatial sizes, *i.e.* $H^s W^s$, we replaced the linear layer with a convolution layer to project input X_{in} into X_Q, X_K, X_V, and a shortcut term X_{SC}:

$$X_\bigstar = \text{Conv}_\bigstar(\text{Drop}(X_{in})), \tag{11}$$

where $\bigstar \in \{Q, K, V, SC\}$, $X_{in} \in \mathbb{R}^{H^q W^q \times H^s W^s \times D_{in}}$, and Drop means randomly setting elements 0 with rate β. We only perform drop operation on self-hypercorrelation branch, *i.e.* $X_{in} = X_l^{qq}$. Taking the mask as a filter so that only foreground information is retained in the X_V. We regarded query spatial sizes, *i.e.* $H^q W^q$, as batchsize and carried out Batch Matrix Multiplication(BMM) to compute:

$$\dot{X}_{out} = \text{Soft}\max(X_Q {X_K}^T)(X_V \odot \widetilde{M}^s), \tag{12}$$

where:

$$\widetilde{M}^s = \text{DownSample}(M^s) \in \{0,1\}^{\ddot{H}^s \times \ddot{W}^s}, \tag{13}$$

and $\odot$ means broadcasted dot product. Two multi-layer perception and normalization layers follow to calculate the final output $X_{out} \in \mathbb{R}^{H^q W^q \times \dot{H}^s \dot{W}^s \times D_{out}}$:

$$\ddot{X}_{out} = \text{Norm}(\text{MLP}(\dot{X}_{out}) + \dot{X}_{out} + X_{SC}), \tag{14}$$

$$X_{out} = \text{Norm}(\text{MLP}(\ddot{X}_{out}) + \ddot{X}_{out}), \tag{15}$$

Now we have reduced support spatial sizes from $H^s W^s$ to $\dot{H}^s \dot{W}^s$ meanwhile changed channels from D_{in} to D_{out}.

4.5 Segmentation Decoder

The structures of both decoders are the same as Convolution Block in the TBTM. It is simple but efficient to obtain the final prediction $\widehat{M}_1^q \in \mathbb{R}^{2 \times H_1^q \times W_1^q}$.

The model parameters are optimized by the cross-entropy loss between a series of predictions $\{\widehat{M}_l^q\}_{l=1}^4$ and the ground-truth $M^q \in \{0,1\}^{H \times W}$ overall pixel locations. Noting that we unsampled all the predictions to the same size with M_q by bilinear interpolation before computing loss. We also set a hyperparameter α to adjust the weighs of $\{\mathcal{L}_l = \text{CE}(\widehat{M}_l^q, M^q)\}_{l=1}^4$:

$$\mathcal{L}_{total} = (1 - 3 \times \alpha)\mathcal{L}_1 + \alpha \sum_{l=2}^{4} \mathcal{L}_l, \tag{16}$$

where CE denotes cross-entropy and $\alpha = 0.1$ in all the experiments.

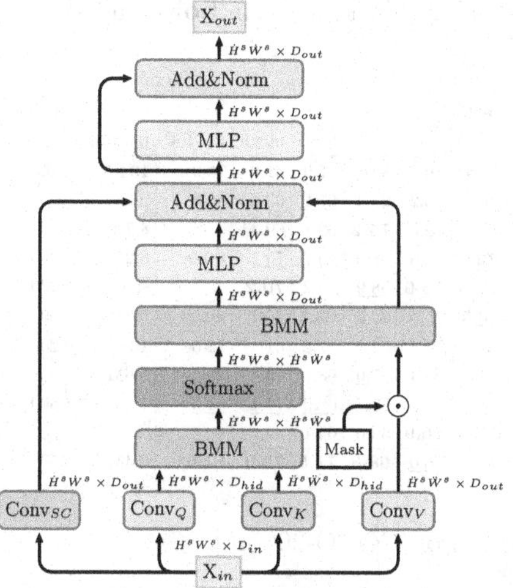

Fig. 3. Illustration of the proposed Target-aware Transformer Layer's calculation process.

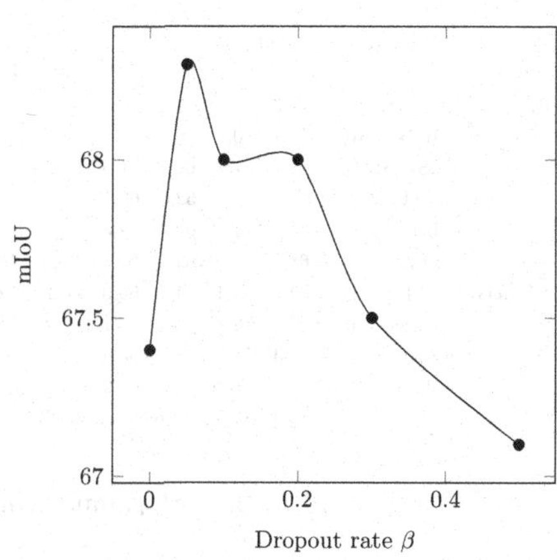

Fig. 4. Ablation study on the dropout rate.

5 Experiments

In this section, we conducted extensive experiments on PASCAL-5^i [2] and COCO-20^i [16] datasets which are prevalent in the few-shot segmentation field. And we use mIoU and FB-IoU as metrics to compare our results with recently excellent methods. Finally, we analyze the influence of each proposed module through extensive ablation experiments. All experiments are implemented on PyTorch [1]. Following HSNet [10], we use Adam [14] as the optimizer to update model parameters and the learning rate is set to 0.001. The batch size is set to 8 for all experiments. Both query and support images' spatial sizes are set to 400×400 without any data augmentation. Borrowed from ASNet [5], we set $H_2^q, W_2^q = 50$, $H_2^s, W_2^s, H_3^s, W_3^s, H_3^q, W_3^q = 25$ and $H_4^q, W_4^q, H_4^s, W_4^s = 13$. Different from other methods [5,10,19], our train epoch is only set to 50 for PASCAL-5^i and 20 for COCO, which is much less than others.

5.1 Datasets

PASCAL-5^i includes PASCAL VOC 2012 [7] and extended annotations from SDS [4] datasets, which contain 20 object categories of images. All the images are evenly divided into 4 folds $i = \{0, ..., 3\}$, each fold contains 5 classes images $C_{test}^i = \{5\times i, ..., 5\times i+4\}$ for testing and the rest 15 classes $C_{train}^i = \{0, ..., 19\} - C_{test}^i$ for training. Following [28], we randomly sampled 1000 support-query pairs for testing.

COCO-20^i [16] is based on MSCOCO [20], which is much more difficult than PASCAL-5^i. We divided it into 4 folds as same as PASCAL-5^i, but each fold contains 60 and 20 categories images for training and testing respectively.

Table 1. Performance comparison on PASCAL-5^i [2]. Best results in **bold**, and second best are underlined.

Backbone network	Methods	1-shot						5-shot						learnable params	train epoch
		5^0	5^1	5^2	5^3	mean	FB-IoU	5^0	5^1	5^2	5^3	mean	FB-IoU		
ResNet50	PFENet [28]	61.7	69.5	55.4	56.3	60.8	73.3	63.1	70.7	55.8	57.9	61.9	73.9	10.8M	200
	HSNet [10]	64.3	70.7	60.3	60.5	64.0	76.7	70.3	73.2	67.4	67.1	69.5	80.6	2.6M	–
	SSP [18]	60.5	67.8	**66.4**	51.0	61.4	–	67.5	72.3	**75.2**	62.1	69.3	–	8.7M	–
	VAT [19]	67.6	72.0	62.3	60.1	65.5	77.8	**72.4**	73.6	68.6	65.7	70.1	**80.9**	3.2M	300
	IPRNet [17]	65.2	**72.9**	63.3	61.3	65.7	–	70.2	**75.6**	68.9	66.2	**70.2**	–	–	200
	Ours	**68.7**	72.0	62.4	**62.6**	**66.4**	**77.9**	70.6	75.0	66.6	**68.1**	70.1	80.1	0.3M	50
ResNet101	PFENet [28]	60.5	69.4	54.4	55.9	60.1	72.9	62.8	70.4	54.9	57.6	61.4	73.5	10.8M	200
	HSNet [10]	67.3	72.3	62.0	63.1	66.2	77.6	71.8	74.4	67.0	68.3	70.4	80.6	2.6M	–
	ASNet [5]	69.0	73.1	62.0	63.6	66.9	78.0	73.1	75.6	65.7	69.9	71.1	81.0	1.3M	500
	IPMT [26]	**71.6**	**73.5**	58.0	61.2	66.1	–	**75.3**	**76.9**	59.6	65.1	69.2	–	–	200
	Ours	70.2	73.3	**63.6**	**66.1**	**68.3**	**79.0**	72.2	76.0	**68.3**	**71.5**	**72.0**	**81.6**	0.4M	50

Table 2. Performance comparison on COCO-20^i [16].

Backbone network	Methods	1-shot						5-shot						learnable params	train epoch
		5^0	5^1	5^2	5^3	mean	FB-IoU	5^0	5^1	5^2	5^3	mean	FB-IoU		
ResNet50	PFENet [28]	36.5	38.6	34.5	33.8	35.8	–	36.5	43.3	37.8	38.4	39.0	–	10.8M	50
	CMNet [21]	**48.7**	33.3	26.8	31.2	35.0	–	**49.5**	35.6	31.8	33.1	37.5	–	–	50
	IPMT [26]	41.4	45.1	**45.6**	40.0	43.0	–	43.5	49.7	48.7	**47.9**	47.5	–	–	50
	VAT [19]	39.0	43.8	42.6	39.7	41.3	68.8	44.1	51.1	50.2	46.1	47.9	72.4	3.3M	–
	Ours	39.8	**46.9**	44.6	**43.8**	**43.8**	**70.6**	45.6	**54.7**	**51.5**	47.2	**49.7**	**72.7**	0.3M	20
ResNet101	PFENet [28]	34.3	33.0	32.3	30.1	32.4	–	38.5	38.6	38.2	34.3	37.4	–	10.8M	50
	HSNet [10]	37.2	44.1	42.4	41.3	41.2	69.1	45.9	53.0	51.8	47.1	49.5	72.4	2.6M	–
	ASNet [5]	**41.8**	45.4	43.2	41.9	43.1	69.4	**48.0**	52.1	49.7	48.2	49.5	72.7	1.3M	–
	IPMT [26]	40.5	45.7	44.8	39.3	42.6	–	45.1	50.3	49.3	46.8	47.9	–	–	50
	Ours	40.2	**47.5**	**46.6**	**45.3**	**44.9**	**71.2**	46.2	**55.5**	**52.7**	**49.4**	**50.9**	**73.3**	0.4M	20

5.2 Comparison with State-of-the-Arts

As shown in Tables 1, 2, we compared the performance of TBTNet and recently excellent approaches on PASCAL-5^i [2] and COCO-20^i [16] respectively. Extensive experiments indicate that our model can achieve higher accuracy and shorter train time with fewer parameters. TBTNet outperformed all other models on PASCAL-5^i whether took ResNet50 or ResNet101 as the backbone. It achieved the best or second-best results on each fold, especially over ASNet 2.5 mIoU on fold 3 with ResNet101. TBTNet exceeds the previous SOTA model ASNet 1.4 mIoU and achieved a new record. As for the numbers of learnable parameters, our TBTNet only has 0.4 M which is 3.7% of PFENet's and 30.8% of ASNet's. Due to the small number of parameters, our model is easy to train and only needs 50 epochs to converge which is 10% of ASNet's and 25% of others. To the best of our knowledge, TBTNet is the model with the shortest training period to date.

On the more difficult datasets COCO-20^i, TBTNet also achieved remarkable performance. Our model got the best score on folds 1, 2, 3 and mIoU, no matter whether in 1-shot or 5-shot conditions and two backbones. It manifests that TBTNet can generalize well, with almost no bias towards categories. Under

1-shot configuration, TBTNet outperformed ASNet by 1.8 mIoU when taking ResNet101 as backbone. As on the PASCAL dataset, our training period was only 40% of the others.

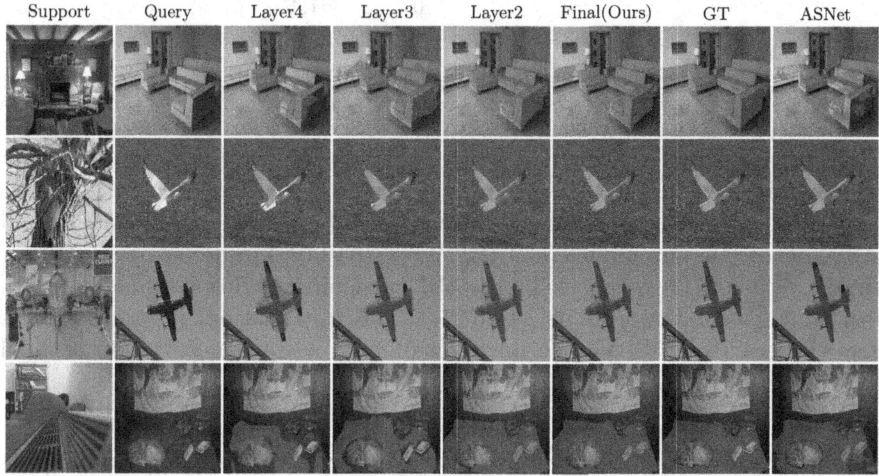

Fig. 5. Qualitative comparison between our proposed TBTNet and ASNet. From left to right: support image, query image, intermediate prediction of TBTNet at layer 4, 3, 2, finale prediction, ground truth and the prediction of ASNet.

In Fig. 5, we visualized the inference procedure of TBTNet and compared the predictions with ASNet, which is one of the SOTA models. We observed that segmentation can be optimized gradually with increasing resolution layer by layer. Our model can use self-similarity to make the segmentation more complete.

5.3 Ablation Study

All ablation study experiments are carried out on PASCAL-5^i [2] with the ResNet101 backbone and the 1-shot setting.

Effectiveness of Bi-Transformer Architecture. We conduct an ablation study by modifying the structure of TBTM to evaluate the influence of self-similarity. As shown in the Table 3, "Bi-T" means whether utilize self-similarity branch in the TBTM. In other words, the unselected "Bi-T" represents $\mathrm{T}_l = \mathrm{T}_{l+1}+\overline{\mathrm{X}}_l^{qs}$, and vice versa $\mathrm{T}_l = \mathrm{T}_{l+1}+\overline{\mathrm{X}}_l^{qs}+\overline{\mathrm{X}}_l^{qq}$ in Eq. (10). Experiments indicate that self-similarity branch can lead to 3.1% increase in mIoU. The improvement can prove that Bi-Transformer architecture is very efficacious for FSS.

Table 3. Ablation study on the Bi-Transformer architecture and our proposed TTL.

Bi-T	TTL	ASL	Dropout rate	mIoU
	✓		0	65.4(+0.0)
✓	✓		0	67.4(+2.0)
✓	✓		0.05	68.3(+0.0)
✓		✓	0.05	67.7(-0.6)

Effectiveness of TTL. To explore the strength of the proposed TTL, we compare it with Attention Squeeze Layer (ASL) [5]. In the Table 3, "TTL" and "ASL" denote the sub-module in TTM. We set β of both experiments as 0.05 for fairness. When TTL is replaced with ASL, a significant drop can be observed, with mIoU descending from 68.3 to 67.7. It indicates that our proposed TTL is more efficient than ASL, which may benefit from a more multivariate residual structure in Eq. (14).

Ablation Study on the Dropout Rate. We conducted a series of experiments to find the optimal parameter β, and all the results are shown in Fig. 4. The mIoU reaches its peak at 68.3 when β is 0.05. As β increases, mIoU rises and then falls. It is because appropriate β can effectively prevent overfitting, and enhance the generalization ability of the model, whereas an excessive β will lead to the loss of too much information, thus hindering performance.

6 Conclusion

In this paper, we introduce Bi-Transformer architecture to few-shot segmentation. To utilize self-similarity information efficiently, we proposed TBTM to integrate it with cross-similarity. A novel TTL is also been proposed to compact the similarity information which is a variant of the transformer. Our TBTNet is a lightweight and fast convergence model. Its effectiveness has been demonstrated by its outstanding performance on the standard benchmarks for few-shot segmentation. We hope that our research will shed light on other domains where similarity analysis is required.

References

1. Adam, P., et al.: PyTorch: an imperative style, high-performance deep learning library. arXiv:1912.01703 (2019)
2. Amirreza, S., Shray, B., Liu, Z., Irfan, E., Byron, B.: One-shot learning for semantic segmentation. arXiv:1709.03410 (2017)
3. Ashish, V., et al.: Attention is all you need. arXiv:1706.03762 (2017)

4. Hariharan, B., Arbeláez, P., Girshick, R., Malik, J.: Simultaneous detection and segmentation. In: Fleet, D., Pajdla, T., Schiele, B., Tuytelaars, T. (eds.) ECCV 2014. LNCS, vol. 8695, pp. 297–312. Springer, Cham (2014). https://doi.org/10.1007/978-3-319-10584-0_20
5. Dahyun, K., Minsu, C.: Integrative few-shot learning for classification and segmentation. In: 2022 IEEE/CVF Conference on Computer Vision and Pattern Recognition (CVPR), pp. 9969–9980 (2022)
6. Dosovitskiy, A., et al.: An image is worth 16×16 words: transformers for image recognition at scale. arXiv:2010.11929 (2020)
7. Everingham, M., Eslami, S., Gool, L., Williams, C.K.I., Winn, J., Andrew, Z.: The pascal visual object classes challenge: a retrospective. Int. J. Comput. Vision **111**, 98–136 (2014)
8. Gengwei, Z., Guoliang, K., Yunchao, W., Yi, Y.: Few-shot segmentation via cycle-consistent transformer (2021)
9. Jake, S., Kevin, S., Zemel, R.: Prototypical networks for few-shot learning (2017)
10. Juhong, M., Dahyun, K., Minsu, C.: Hypercorrelation squeeze for few-shot segmentation. In: 2021 IEEE/CVF International Conference on Computer Vision (ICCV), pp. 6921–6932 (2021)
11. Jungbeom, L., Joon, O.S., Sangdoo, Y., Junsuk, C., Eunji, K., Sung-Hoon, Y.: Weakly supervised semantic segmentation using out-of-distribution data. In: 2022 IEEE/CVF Conference on Computer Vision and Pattern Recognition (CVPR), pp. 16876–16885 (2022)
12. Kaiming, H., Zhang, X., Shaoqing, R., Jian, S.: Deep residual learning for image recognition. In: 2016 IEEE Conference on Computer Vision and Pattern Recognition (CVPR), pp. 770–778 (2015)
13. Kaixin, W., Liew, J., Yingtian, Z., Daquan, Z., Jiashi, F.: Panet: few-shot image semantic segmentation with prototype alignment. In: 2019 IEEE/CVF International Conference on Computer Vision (ICCV), pp. 9196–9205 (2019)
14. Kingma, D.P., Jimmy, B.: Adam: a method for stochastic optimization. CoRR (2014)
15. Krizhevsky, A., Ilya, S., Hinton, G.E.: Imagenet classification with deep convolutional neural networks. Commun. ACM **60**, 84–90 (2012)
16. Minh, N.K.D., Todorovic, S.: Feature weighting and boosting for few-shot segmentation. In: 2019 IEEE/CVF International Conference on Computer Vision (ICCV), pp. 622–631 (2019)
17. Okazawa, A.: Interclass prototype relation for few-shot segmentation. ArXiv (2022)
18. Qi, F., Wenjie, P., Yu-Wing, T., Chi-Keung, T.: Self-support few-shot semantic segmentation (2022)
19. Sunghwan, H., Seokju, C., Jisu, N., Stephen, L., Wook, K.S.: Cost aggregation with 4D convolutional swin transformer for few-shot segmentation. arXiv:2207.10866 (2022)
20. Lin, T.Y., et al.: Microsoft COCO: common objects in context. In: Fleet, D., Pajdla, T., Schiele, B., Tuytelaars, T. (eds.) ECCV 2014. LNCS, vol. 8693, pp. 740–755. Springer, Cham (2014). https://doi.org/10.1007/978-3-319-10602-1_48
21. Weide, L., Chi, Z., Henghui, D., Tzu-Yi, H., Guosheng, L.: Few-shot segmentation with optimal transport matching and message flow. arXiv:2108.08518 (2021)
22. Xiaolin, Z., Yunchao, W., Yi, Y., Thomas, H.: SG-One: similarity guidance network for one-shot semantic segmentation. IEEE Trans. Cybern. **50**, 3855–3865 (2018)
23. Xiaoliu, L., Zhao, D., Taiping, Z.: Intermediate prototype network for few-shot segmentation. Signal Process. **203**, 108811 (2022)

24. Xingjia, P., et al.: Unveiling the potential of structure preserving for weakly supervised object localization. In: 2021 IEEE/CVF Conference on Computer Vision and Pattern Recognition (CVPR), pp. 11637–11646 (2021)
25. Xinyu, S., et al.: Dense cross-query-and-support attention weighted mask aggregation for few-shot segmentation. arXiv:2207.08549 (2022)
26. Yuanwei, L., Nian, L., Xiwen, Y., Junwei, H.: Intermediate prototype mining transformer for few-shot semantic segmentation. arXiv:2210.06780 (2022)
27. Yuchao, W., et al.: Semi-supervised semantic segmentation using unreliable pseudo-labels. In: 2022 IEEE/CVF Conference on Computer Vision and Pattern Recognition (CVPR), pp. 4238–4247 (2022)
28. Zhuotao, T., Hengshuang, Z., Michelle, S., Zhicheng, Y., Ruiyu, L., Jiaya, J.: Prior guided feature enrichment network for few-shot segmentation. IEEE Trans. Pattern Anal. Mach. Intell. **44**, 1050–1065 (2020)

Convex Hull Collaborative Representation Learning on Grassmann Manifold with L_1 Norm Regularization

Yao Guan[1], Wenzhu Yan[1(✉)], and Yanmeng Li[2]

[1] School of Computer and Electronic Information, Nanjing Normal University, Nanjing, China
{guanyao,yanwenzhu}@nnu.edu.cn

[2] School of Internet of Things, Nanjing University of Posts and Telecommunications, Nanjing, China
liyanmeng@njupt.edu.cn

Abstract. Collaborative representation learning mechanism has recently attracted great interest in computer vision and pattern recognition. Previous image set classification methods mainly focus on deriving collaborative representation models on Euclidean space. However, the underlying manifold geometry structure of the image set is not well considered. In this paper, we propose a novel manifold convex hull collaborative representation framework with L_1 norm regularization from geometry-aware perspective. Our model achieves the goal of inheriting the highly expressive representation capability of the Grassmann manifold, while also maintaining the flexible nature of the convex hull model. Notably, the collaborative representation mechanism emphasizes the exploration of connections between different convex hulls on Grassmann manifold. Besides, we regularize the collaborative representation coefficients by using the L_1 norm, which exhibits superior noise robustness and satisfies the data reconstruction requirements. Extensive experiments and comprehensive comparisons demonstrate the effectiveness of our method over other image set classification methods.

Keywords: Image set classification · Manifold learning · L_1 norm · Collaborative representation

1 Introduction

In traditional visual recognition task, objects of interest are trained and recognized based on single image. However, with the rapid development of multimedia network technology, a lot of videos and images of a person or an object are conveniently available. Therefore, image set based classification [3,6,14,22] has attracted increasing interest in computer vision and machine learning. Compared with traditional single image classification, the set oriented classification strategy can effectively improve the robustness and recognition rate since each set

Q. Liu et al. (Eds.): PRCV 2023, LNCS 14426, pp. 453–465, 2024.
https://doi.org/10.1007/978-981-99-8432-9_36

contains a large number images covering more information for the target subject. However, large intra-class variations caused by differences in position, posture and illumination, also bring great challenges to effectively model an image set.

Existing image set based classification methods can be mainly classified into two categories: parametric methods and non-parametric methods. In the past decades, non-parametric methods have attracted increasing attentions due to their stable and promising performance. Different from the distribution function in parametric methods, non-parametric methods utilize more flexible models to represent image sets, such as manifold [12,23], affine or convex hull [1,27], sparse coding [24,27]. To our knowledge, adopting affine/convex hull to model image sets can reveal unseen appearance implicitly. Works conducted on affine/convex hulls have achieved significant progress. For example, based on the fundamental work of affine/convex hull in [1], SANP in [9] exploited the sparse property of affine hull to jointly optimize the nearest points between two image sets. To reduce the computation complexity, RNP [24] utilized a regularized hull to acquire a more concise form with L_2 norm constraint. Recently, Cevikalp et al. [2] expanded the scope by incorporating a more comprehensive distance description using binary hierarchy for both linear and kernelized affine/convex hulls, resulting in faster learning process. To better explore the the geometry structure for image sets, researchers represent an image set as a manifold point in the domain of non-Euclidean geometry [11,12,23]. For example, the methods in [12,16] model image sets as points on Grassmann manifold and adopt the Projection Metric (PM) to compute the projection distance which is connected with the true geodesic distance on the Grassmann manifold with a fixed scale of $\sqrt{2}$. Furthermore, motivated by the principles of sparse representation coding in Euclidean space, extensive research efforts have been devoted to exploring similar issues on the Grassmann manifold [8,23,26]. Wei et al. created the Grassmann locality-aware group sparse coding model (GLGSC) [23] to preserve locality information and exploit the relationships between sets to simultaneously capture inter- and intra-set variations. Given the high capacity for deep learning to acquire non-linear representation, several recent studies have developed deep Riemannian manifold learning frameworks [13,17]. However, the backpropagation computing of manifold is time consuming.

As far as we know, very few studies have specifically focused on exploring the representation ability of convex hull on non-linear manifold, which motivates our research in this area. In this paper, we aim to extend the convex model from Euclidean space onto the Grassmann manifold equipped with better robustness ability. The main contributions of this paper include: (1) To better leverage the geometric structures presented in the set data, our approach involves combining the flexibility of the convex hull model with the expressive power of the Grassmann manifold. By doing so, we seek to enhance the ability of convex hull collaborative representation on nonlinear manifold to better represent the probe image sets. (2) To enhance the robustness of the model against the noise data, we impose the L_1-norm regularization on convex hull representation coefficients in our manifold convex hull collaborative representation model to achieve sparsity. (3) Extensive experimental results conducted on several real-world set-based datasets clearly demonstrate the competitive performance of our method in terms of classification accuracy and robustness.

We organize our paper as follows: Section 2 first presents the related works. The proposed method and the valuated experiments are exhibited in Sect. 3 and Sect. 4, respectively. Then, The conclusions are provided in Sect. 5.

2 Preliminaries

2.1 Sparse/Collaborative Representation Learning

Given a sample $\mathbf{y}$ and a dictionary $\mathbf{D}$, the framework of representation learning can be formulated as follows :

$$min_{\mathbf{a}} \quad \Pi(\mathbf{a}) \quad s.t. \|\mathbf{y} - \mathbf{Da}\| \le \varepsilon, \tag{1}$$

where $\Pi(\mathbf{a})$ is usually some kind of constraint of $\mathbf{a}$, ε is a small constant. Sparse representation classification (SRC) introduced the $\|\mathbf{a}\|_1$ in [25] to encode $\mathbf{y}$ over a learned dictionary $\mathbf{D}$. To fully study the role of collaboration between the different classes, we can further adopt the following L_2-norm regularization with a low computational burden:

$$\mathbf{a} = \min_{\mathbf{a}} \|y - \mathbf{Xa}\| + \lambda\|\mathbf{a}\|_2, \tag{2}$$

where $\mathbf{X}$ can be considered as a dictionary of all samples. Inspired by the success of CRC, Zhu et al. [27] proposed image set based collaborative representation model by representing each set as a convex hull. The distance between set $\mathbf{Y}$ and sets $\mathbf{X}$ can be defined as:

$$min_{\mathbf{a},\mathbf{b}} \|\mathbf{Ya} - \mathbf{Xb}\| + \lambda_1\|\mathbf{a}\|_2 + \lambda_2\|\mathbf{b}\|_2 \quad s.t. \sum a_i = 1, \sum b_j = 1 \tag{3}$$

where $\mathbf{a}$ and $\mathbf{b}$ are the hull representation coefficients. In (3), the hull $\mathbf{Ya}$ of the probe set $\mathbf{Y}$ is collaboratively represented over the gallery sets $\mathbf{Xb}$.

2.2 Grassmann Manifold

The space of $d \times p$ $(0 < p < d)$ matrices with orthonormal columns is a Riemann manifold known as a Stiefel manifold $St(p,d)$, *i,e.*, $St(p,d)=\{\mathbf{X} \in \mathbb{R}: \mathbf{X}^T\mathbf{X}=\mathbf{I}_p\}$, where $\mathbf{I}_p$ denotes the identity matrix of size $p \times p$. Grassmann manifold $Gr(p,d)$ can be defined as a quotient manifold of $St(p,d)$ with the equivalence relation $\sim$ being: $\mathbf{X}_1 \sim \mathbf{X}_2$ if and only if $\text{Span}(\mathbf{X}_1) = \text{Span}(\mathbf{X}_2)$, where $\text{Span}(\mathbf{X})$ denotes the subspace spanned by columns of $\mathbf{X} \in St(p,d)$.

Supposing that $\bar{\mathbf{Y}}$ and $\bar{\mathbf{X}}_j$ are points on Grassmann manifold Gr(p,d). The manifold operators for $\bar{\mathbf{Y}}$ and $\bar{\mathbf{X}}_j$ contain subtraction, summation and multiplication. We view the geodesic distance between two manifold points on Grassmann manifold. Since matrix subtraction, summation and scalar multiplication are not defined appropriately, it's very hard to directly solve complex operations on manifold. Fortunately, as depicted in [8], Grassmann manifold can be embedded into the space of the symmetric matrices via mapping

$$\Gamma : Gr(p,d) \rightarrow Sym(d), \quad \Gamma(\mathbf{X}) = \mathbf{X}\mathbf{X}^T. \tag{4}$$

The above projection distance is proven to be the true geodesic distance on the Grassmann manifold at a fixed scale. Mathematically, the mapping $\Gamma(\mathbf{G}) = \mathbf{G}\mathbf{G}^T$ leads to an efficient learning scheme to simplify the designed models on Grassmann manifold.

3 Method

In this section, we propose a new geometry-aware collaborative representation model based on manifold convex hull learning with L_1 norm regularization to deal with image set classification task.

3.1 Problem Setup

Let $\mathbf{X}_k = \{\mathbf{x}_k^j\}_{1 \le j \le n_k} \in R^{d \times n_k}$ be a gallery set with n_k images and each image is a d-dimensional sample vector. Note that x_k^j denotes the j-th image of the k-th gallery set. Then, a collection of gallery sets can be denoted as $\mathbf{X} = [\mathbf{X}_1, \ldots, \mathbf{X}_k, \ldots, \mathbf{X}_K]$. Similarly, let the probe set with p_k images be $\mathbf{Y}_k = \{\mathbf{y}_k^i\}_{1 \le i \le p_k} \in R^{d \times p_k}$. Here, $\mathbf{y}_k^i$ denotes the i-th image of the k-th probe set. Given a probe set $\mathbf{Y}$, our goal is to find the corresponding class label of the whole set.

3.2 Multiple Subspaces Extraction

Due to arbitrary poses, partial occlusions, illumination variations and object deformations, the image set usually has complicated structure that a single subspace could not convey such intra-set variations. A better way is to segment the image set into several clusters of linear model. In this paper, we follow [18] and use Hierarchical Divisive Clustering (HDC) algorithm to extract m Maximal Linear Patchs (MLPs) from an image set $\mathbf{X}_k$. i.e., $\mathbf{X}_k = \{\mathbf{X}_k^{(1)}, \mathbf{X}_k^{(2)}, \ldots, \mathbf{X}_k^{(m)}\}$. After extracting MLPs from an image set, we can use an orthonormal basic matrix $\bar{\mathbf{X}}_k^{(i)}$ to denote the MLP:

$$\mathbf{X}_k^{(i)}\mathbf{X}_k^{(i)T} = \bar{\mathbf{X}}_k^{(i)} \Lambda_k^{(i)} \bar{\mathbf{X}}_k^{(i)T}, \quad \bar{\mathbf{X}}_k^{(i)} \in R^{d \times q}, \tag{5}$$

where $\bar{\mathbf{X}}_k^{(i)}$ and $\Lambda_k^{(i)}$ are the eigenvector matrix of the q largest eigenvalues and diagonal matrix of the q largest eigenvalues, respectively. Then, each image set can be represented by several local linear subspaces spanned by orthonormal basis matrices. Specifically, the gallery sets $\bar{\mathbf{X}} = [\bar{\mathbf{X}}_1^{(1)}, \bar{\mathbf{X}}_1^{(2)}, \ldots, \bar{\mathbf{X}}_1^{(m)}, \ldots, \bar{\mathbf{X}}_K^{(1)}, \bar{\mathbf{X}}_K^{(2)}, \ldots, \bar{\mathbf{X}}_K^{(m)}]$ and the probe set $\bar{\mathbf{Y}} = [\bar{\mathbf{Y}}^{(1)}, \bar{\mathbf{Y}}^{(2)}, \ldots, \bar{\mathbf{Y}}^{(m)}]$.

3.3 The L_1 Norm Regularized Convex Hull Collaborative Representation Model

We first define the convex hull of points on Grassmann manifold and derive the the geometric similarity between two convex hulls as follows.

Definition 1. *Given a set of Grassmann manifold points* $\mathcal{G} = \{\bar{\mathbf{Y}}_i\}, i = 1, ..., m$, *the convex hull of* $\mathcal{G}$ *can be considered as a convex model that contains all the convex combinations of* $\bar{\mathbf{Y}}_i, i = 1, ..., m$:

$$\Omega_{\mathcal{G}} = \biguplus_{i=1}^{m} a_i \otimes \bar{\mathbf{Y}}_i \;\; s.t. \sum_{i=1}^{m} a_i = 1, a_i \geq 0, \tag{6}$$

where α_i is the weight of the i-th Grassmann manifold point $\bar{\mathbf{Y}}_i$. The operators $\biguplus$ and $\otimes$ are summation and multiplication over Grassmann manifold, respectively. In the previous works [1,27], convex hull and collaborative representation learning are explored simply on Euclidean space, which ignores the nonlinear structural information hidden in image sets. Motivated by the collaborative representation learning mechanism, we aim to fully capture the correlation relationship among manifold convex hulls of gallery sets and formulate the manifold convex hull collaborative representation model regularized by L_1-norm as follows:

$$\begin{aligned} &min_{\mathbf{a},\mathbf{b}} \left\| \biguplus_{i=1}^{m} a_i \otimes \bar{\mathbf{Y}}_i \ominus \biguplus_{j=1}^{n} b_j \otimes \bar{\mathbf{X}}_j \right\|_{geod}^{F} + \lambda_1 \|\mathbf{a}\|_1 + \lambda_2 \|\mathbf{b}\|_1 \\ &s.t. \sum a_i = 1, \sum b_j = 1, \end{aligned} \tag{7}$$

where $\bar{\mathbf{Y}}_i$ and $\bar{\mathbf{X}}_j$ are points on Grassmann manifold Gr(p,d). We let $\sum a_i = 1$ to avoid the trival solution $\mathbf{a} = \mathbf{b} = \mathbf{0}$. Notably, L_1 norm regularization will lead to sparser solution for the representation coefficients, which can improve the classification performance if the features are not informative enough, i.e., is more robustness to outliers and noises.

Then, based on the project metric in (4), we have

$$\begin{aligned} &J = min_{\mathbf{a},\mathbf{b}} \left\| \sum_{i=1}^{m} a_i \bar{\mathbf{Y}}_i \bar{\mathbf{Y}}_i^T - \sum_{j=1}^{n} b_j \bar{\mathbf{X}}_j \bar{\mathbf{X}}_j^T \right\|_F^2 + \lambda_1 \|\mathbf{a}\|_1 + \lambda_2 \|\mathbf{b}\|_1 \\ &s.t. \sum a_i = 1, \end{aligned} \tag{8}$$

where λ_1 and λ_2 are positive constants to balance the representation residual and the regularizer, the coefficient vectors $\mathbf{a} = [a_1, a_2, \ldots, a_m]^T$ and $\mathbf{b} = [b_1, b_2, \ldots, b_n]^T$. The above model achieves the goal of establishing manifold convex hull collaborative representation on nonlinear manifold, which inherits the highly expressive representation capability of the Grassmann manifold, while also maintaining the flexible nature of the convex hull model.

3.4 Optimization

For simplicity, we first define similarity between gallery subspace $\bar{\mathbf{X}}$ and the probe subspace $\bar{\mathbf{Y}}$ by an $n\times m$ dimensional matrix and the ij-th entry of this matrix is:

$$[H(\bar{\mathbf{X}},\bar{\mathbf{Y}})]_{i,j}=Tr(\bar{\mathbf{X}}_i^T\bar{\mathbf{Y}}_j\bar{\mathbf{Y}}_j^T\bar{\mathbf{X}}_i)=\left\|\bar{\mathbf{X}}_i^T\bar{\mathbf{Y}}_j\right\|_F^2, \tag{9}$$

similarly, the similarity matrix on probe subspaces $\bar{\mathbf{Y}}$ and the similarity matrix on gallery subspaces $\bar{\mathbf{X}}$ can be defined by $m\times m$ and $n\times n$ symmetric matrix, respectively:

$$[H(\bar{\mathbf{X}})]_{i,j}=\left\|\bar{\mathbf{X}}_i^T\bar{\mathbf{X}}_j\right\|_F^2, [H(\bar{\mathbf{Y}})]_{i,j}==\left\|\bar{\mathbf{Y}}_i^T\bar{\mathbf{Y}}_j\right\|_F^2. \tag{10}$$

With the definition of Eq. (9) and Eq. (10), the first term in Eq. (8) can be rewritten as:

$$min_{\mathbf{a},\mathbf{b}}\{\mathbf{a}^TH(\bar{\mathbf{Y}})\mathbf{a}-2\mathbf{b}^TH(\bar{\mathbf{X}},\bar{\mathbf{Y}})\mathbf{a}+\mathbf{b}^TH(\bar{\mathbf{X}})\mathbf{b}\}. \tag{11}$$

Let $\mathbf{U}\Sigma\mathbf{U}^T$ be the singular value decomposition (SVD) of the symmetric matrix H($\bar{\mathbf{X}}$) and H($\bar{\mathbf{Y}}$). Specifically, we have:

$$\begin{aligned}H(\bar{\mathbf{Y}})&=\mathbf{U}_y\mathbf{V}_y\mathbf{U}_y^T=\mathbf{U}_y\mathbf{V}_y^{\frac{1}{2}}\mathbf{V}_y^{\frac{1}{2}}\mathbf{U}_y^T=\mathbf{P}^T\mathbf{P},\\ H(\bar{\mathbf{X}})&=\mathbf{U}_x\mathbf{V}_x\mathbf{U}_x^T=\mathbf{U}_x\mathbf{V}_x^{\frac{1}{2}}\mathbf{V}_x^{\frac{1}{2}}\mathbf{U}_x^T=\mathbf{Q}^T\mathbf{Q}.\end{aligned} \tag{12}$$

Mathematically, for Eq. (8), J is equivalent to the following optimization problem:

$$\begin{aligned}&min_{\mathbf{a},\mathbf{b}}\|\mathbf{Pa}-\mathbf{kQb}\|_2^2+\mathbf{b}^T\mathbf{Mb}+\lambda_1\|\mathbf{a}\|_1+\lambda_2\|\mathbf{b}\|_1\\ &s.t.\sum a_i=1,\end{aligned} \tag{13}$$

where $\mathbf{k}^T$=$(\mathbf{Q}^T)^{-1}H(\bar{\mathbf{X}},\bar{\mathbf{Y}})\mathbf{P}^{-1}$ and $\mathbf{M}$=$\mathbf{Q}^T(\mathbf{I}-\mathbf{k}^T\mathbf{k})\mathbf{Q}$ and $\mathbf{I}$ is a diagonal matrix whose elements are 1.

Then, for Eq. (13), we derive the Lagrangian function as follows:

$$\begin{aligned}L(\mathbf{a},\mathbf{b},\lambda)=&\|\mathbf{Pa}-\mathbf{kQb}\|_2^2+\mathbf{b}^T\mathbf{Mb}+\lambda_1\|\mathbf{a}\|_1+\lambda_2\|\mathbf{b}\|_1\\ &+\langle\lambda,\mathbf{ea}-1\rangle+\frac{c}{2}\|\mathbf{ea}-1\|_2^2,\end{aligned} \tag{14}$$

where λ is the Lagrange multiplier, $c>0$ is the penalty parameter, $\mathbf{e}$ is a row vector whose elements are 1 and $\langle\cdot,\cdot\rangle$ is the inner product. Notably, we use the alternating minimization method, which is very efficient to solve multiple variable optimization problems [7]. Specifically, **a** and **b** are optimized alternatively with the other one fixed as follows:

Step 1: fix **b**, update **a**,

$$\begin{aligned}\mathbf{a}^{(t+1)} &= argmin_{\mathbf{a}} L(\mathbf{a}, \mathbf{b}^{(t)}, \lambda^{(t)}) \\ &= argmin_{\mathbf{a}} \left\| \mathbf{Pa} - \mathbf{kQb}^{(t)} \right\|_2^2 + \lambda_1 \left\|\mathbf{a}\right\|_1 + \left\langle \lambda^{(t)}, \mathbf{ea} - 1 \right\rangle + \frac{c}{2}\|\mathbf{ea} - 1\|_2^2 \\ &= argmin_{\mathbf{a}} \left\| \begin{bmatrix} \mathbf{P} \\ \sqrt{\frac{c}{2}}\mathbf{e} \end{bmatrix} \mathbf{a} - \begin{bmatrix} \mathbf{kQb}^{(t)} \\ \sqrt{\frac{c}{2}}(1 - \frac{\lambda^{(t)}}{c}) \end{bmatrix} \right\|_2^2 + \lambda_1 \|\mathbf{a}\|_1. \end{aligned} \tag{15}$$

The problem in Eq. (15) can be easily solved by some representative l_1-minimization approaches such as LARS [5].

Step 2: fix **a**, update **b**,

$$\begin{aligned}\mathbf{b}^{(t+1)} &= argmin_{\mathbf{b}} L(\mathbf{a}^{(t+1)}, \mathbf{b}, \lambda^{(t)}) \\ &= argmin_{\mathbf{b}} \left\| \mathbf{Pa}^{(t+1)} - \mathbf{kQb} \right\|_2^2 + \mathbf{b}^T\mathbf{Mb} + \lambda_2 \|\mathbf{b}\|_1 \\ &= argmin_{\mathbf{b}} \left\| \begin{bmatrix} \mathbf{kQ} \\ \mathbf{M}^{\frac{1}{2}} \end{bmatrix} \mathbf{b} - \begin{bmatrix} \mathbf{Pa}^{(t+1)} \\ \mathbf{0} \end{bmatrix} \right\|_2^2 + \lambda_2 \|\mathbf{b}\|_1. \end{aligned} \tag{16}$$

Once $\mathbf{a}^{(t+1)}$ and $\mathbf{b}^{(t+1)}$ are obtained, λ is updated as follows:

$$\lambda^{(t+1)} = \lambda^{(t)} + c(\mathbf{ea}^{(t+1)} - 1). \tag{17}$$

Algorithm 1: The proposed method

Input: probe set $\mathbf{Y}$, gallery sets $\mathbf{X}=[\mathbf{X}_1, \ldots, \mathbf{X}_k, \ldots, \mathbf{X}_K]$, λ_1, λ_2
Output: the label of probe set $\mathbf{Y}$

```
Using HDC and SVD to extract subspace from X,Y to X̄, Ȳ.
Calculate matrix H(X̄,Ȳ), H(X̄) and H(Ȳ) by Eq.(8) and Eq.(9).
Initialize b^(0), λ^(0) and t=0.
while t < max_num do
    Update a by Eq.(15);
    Update b by Eq.(16);
    Update λ by Eq.(17);
    t=t+1;
end
Identity(Y)=argmin_i ||Ȳa − X̄_i b_i||_2^2
```

The algorithm of our method is summarized in Algorithm 1. Eventually, we use the representation residual of hull $\bar{\mathbf{Y}}\mathbf{a}$ by each $\bar{\mathbf{X}}_i$ to determine the class label of $\mathbf{Y}$. The classifier in the proposed method is:

$$Identity(\mathbf{Y}) = argmin_i \left\| \bar{\mathbf{Y}}\mathbf{a} - \bar{\mathbf{X}}_i\mathbf{b}_i \right\|_2^2. \tag{18}$$

4 Experiment

4.1 Datasets and Settings

To demonstrate the performance of the proposed model, we evaluate four typical visual classification tasks: object categorization, gesture recognition, handwritten digits recognition and face recognition.

For the object categorization task, we utilize the ETH-80 dataset which contains 80 sets belonging to 8 categories. Each category contains 10 sub-categories with 41 images, captured from different angles. We randomly assigned 5 sets to serve as gallery sets and the other 5 sets to act as probe sets per category.

For the gesture recognition task, we consider the Cambridge Gesture dataset which contains 900 sequences from 9 different gestures. More specifically, each gesture consists of 100 image sequences performed by 2 subjects under 5 illuminations and 10 arbitrary motions.

For the handwritten digits recognition task, we utilize the MNIST datasets [4]. This dataset contains 70,000 images samples which can be divided into 300 sets. We randomly selected 5 sets for training and the other 5 sets for testing per class.

For the face recognition task, we utilize the Extended Yale database B and the YouTube Celebrities dataset. The YaleB consists of 16,128 images from 28 individuals under nine different poses and 64 illumination conditions, making it a large-scale and challenging database with significant illumination variations. We build 28 image sets depending on different poses of each subject, where each image set contains approximately 20–64 frames. The YTC dataset is a challenging and large-scale dataset consisting of 1,910 video sequences of 47 celebrities collected from YouTube. The frames in these videos are highly compressed and low-resolution, resulting in significant variations in pose and illumination. For YaleB and YTC datasets, we randomly select 3 image sets for training and 6 image sets for testing per class. Furthermore, we also utilize the LFW-a dataset to evaluate our method's performance with deep learned features. The data in LFW-a consists of deeply extracted features from the LFW [10] dataset by using ResNet. we randomly select 4 sets for training and the other 4 sets for testing per class.

4.2 Comparative Methods

We compare the performance of our proposed method with several existing image set classification methods, including, Manifold-Manifold Distance (MMD) [20], Manifold Discriminant Analysis (MDA) [18], Multiple Riemannian Manifold-Valued Descriptors Based Multi-Kernel Metric Learning(MRMML) [15], Projection Metric Learning (PML) [12], Grassmann Locality-Aware Group Sparse Coding model (GLGSC) [23], Affine Hull based Image Set Distance (AHISD) and kernel AHISD [1], Convex Hull based Image Set Distance (CHISD) and kernel CHISD [1], Regularized Nearest Points (RNP) [24], Image Set-Based Collaborative Representation and Classification (ISCRC) [27], Covariance Discriminative

Learning (CDL) [19], and Prototype Discriminative Learning (PDL) [21]. Note that the important parameters of these methods are set according to the original literatures. For our method, we set $\lambda_1 = 0.001$ and $\lambda_2 = 0.001$ on all datasets. The results are listed in Table 1 on all datasets.

4.3 Results and Analysis

From the Table 1, we can obtain that our method achieves better classification performance on most of the evaluated datasets. Specifically, affine/convex hull based methods (AHISD and CHISD) have relatively lower scores, particularly on gesture with scores of 25.78% and 26.44%, respectively. Compared with the affine/convex hull models, the methods based on Riemann manifold motivated from geometry structure perform well. For example, CDL on SPD manifold achieves 95.75% on ETH-80, and the best score of 82.91% on gesture dataset, the Grassmann manifold based models (PML and GLGSC) also achieve excellent classification performance on all datasets, which means that the nonlinear manifold based methods are conductive to deal with the image set classification problem. Notably, as the GLGSC aims to exploit the relationships among image sets based on spare representation on Grassmann manifold, the classification performance is superior to the sparse coding methods (RNP, ISCRC) in Euclidean space. For the proposed model, it inherits the highly expressive representation capability of the Grassmann manifold, while also maintaining the flexible nature of the convex hull model. The results in Table 1 shows that our method holds high recognition accuracies in different classification tasks, such as 96.50% and 70.24% on ETH-80 and YTC datasets, which definitively demonstrates the effectiveness of establishing manifold convex hull collaborative representation on nonlinear manifold.

4.4 Evaluations on Noisy Datasets

In practical applications, image sets usually contain noisy data, which makes this classification task more difficult. To validate the robustness of our method to noisy data , we conduct experiments on the two databases (ETH-80, YaleB), with several representative methods. We implement noise by adding 10 percent images of an image set from another class to each image set. The original data is denoted as "Clean", the noise only in the gallery sets is denoted as "N_G", the noise only in the probe sets is denoted as "N_P" and the noise in both gallery and probe sets is denoted as "N_GP". The experimental results are shown in Fig. 1. From Fig. 1, it can be seen that our proposed method shows better robustness in the four cases, "Clear", "N_G", "N_P" and "N_GP". Besides, we have observed that after adding noise, the accuracy only slightly decreased while performances of the other methods largely decreased. This shows that our method can effectively suppress the impact of noise.

Table 1. Average recognition rates (%) of different methods on the six datasets.

Methods	ETH-80	Cambridge	Mnist	YaleB	YTC	LFW-a
MMD	79.25 ± 0.07	49.22 ± 0.21	99.80 ± 0.00	40.95 ± 0.04	65.85	99.60
MDA	76.25 ± 0.21	24.66 ± 0.16	**100.00 ± 0.00**	61.84 ± 0.14	69.60	53.14
AHISD	71.50 ± 0.16	25.78 ± 0.13	99.60 ± 0.00	58.39 ± 0.06	66.42	**100.00**
KAHISD	79.00 ± 0.23	26.44 ± 0.11	**100.00 ± 0.00**	62.98 ± 0.09	–	–
CHISD	75.50 ± 0.08	25.78 ± 0.10	**100.00 ± 0.00**	61.31 ± 0.14	66.81	**100.00**
KCHISD	79.25 ± 0.17	26.67 ± 0.13	**100.00 ± 0.00**	63.75 ± 0.13	–	–
CDL	95.75 ±0.13	**82.91 ± 0.05**	**100.00 ± 0.00**	89.40 ± 0.05	69.38	99.59
RNP	70.50 ± 0.19	28.78 ± 0.15	**100.00 ± 0.00**	64.64 ± 0.06	66.77	**100.00**
ISCRC	72.25 ± 0.33	26.78 ± 0.11	82.40 ± 0.35	74.82 ± 0.08	69.72	41.53
PML	87.75 ± 0.23	79.89 ± 0.25	**100.00 ± 0.00**	74.70 ± 0.16	67.48	99.76
PDL	77.75 ± 0.16	32.86 ± 0.18	99.80 ± 0.00	**92.20 ± 0.07**	70.04	**100.00**
GLGSC	91.25 ± 0.18	73.56 ± 0.28	**100.00 ± 0.00**	84.64 ± 0.11	66.10	99.11
MRMML	94.25 ± 0.02	84.66 ± 0.02	**100.00 ± 0.00**	50.59 ± 0.08	36.20	–
Proposed	**96.50 ± 0.08**	**80.56 ± 0.17**	**100.00 ± 0.00**	**91.66 ± 0.03**	**70.24**	**100.00**

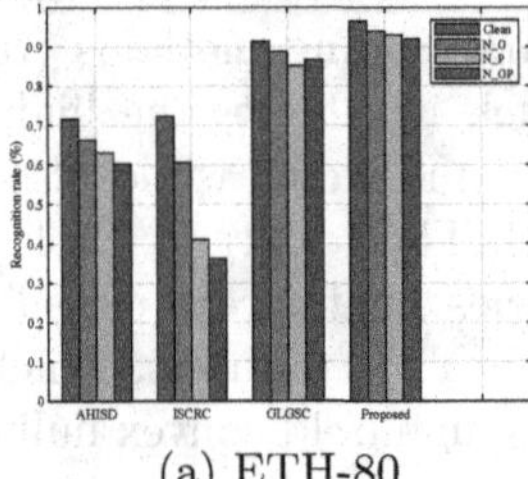

(a) ETH-80

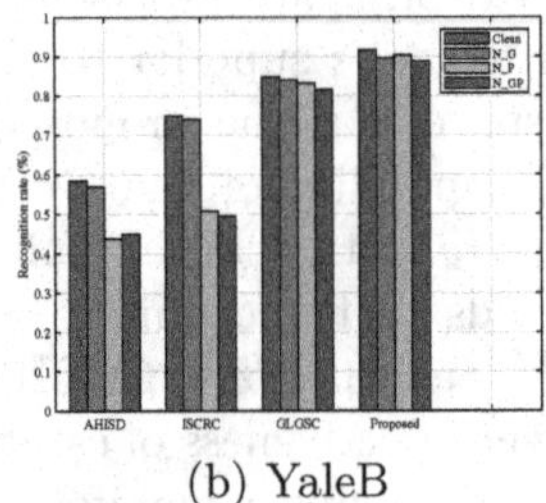

(b) YaleB

Fig. 1. The average recognition rates of different methods on different noisy datasets. Clean: the original data; N_G: the data with noise only in the gallery sets; N_P: the data with noise only in the probe sets; N_GP: the data with noise in both gallery and probe sets.

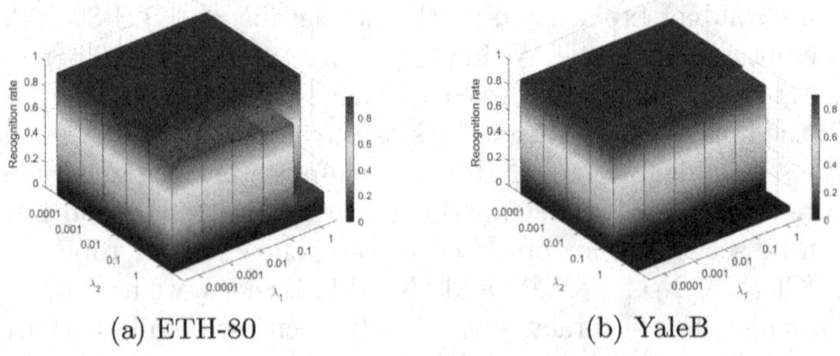

(a) ETH-80 (b) YaleB

Fig. 2. Recognition accuracies on ETH-80 and YaleB with different λ_1 and λ_2.

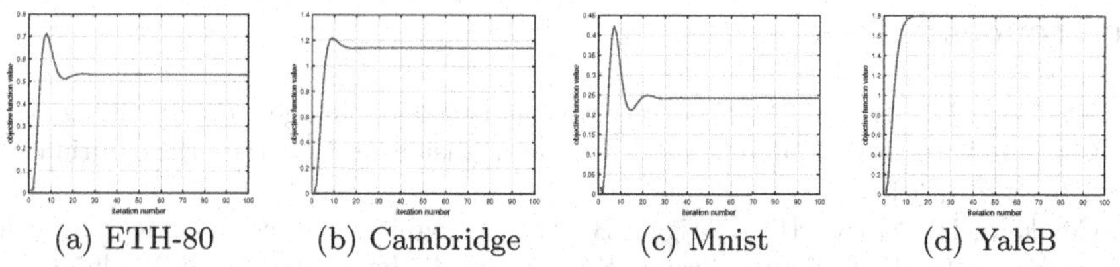

(a) ETH-80 (b) Cambridge (c) Mnist (d) YaleB

Fig. 3. Convergence of the proposed method.

4.5 Parameter Sensitivity Analysis

To verify whether the proposed method is sensitive to parameters λ_1 and λ_2, in this section, we discuss the recognition accuracy with different parameter values. The relationships between recognition accuracy and different parameter values on ETH-80 and YaleB are shown in Fig. 2, respectively. From Fig. 2, it can be obtain that the proposed model is less sensitive to λ_1 and λ_2 on ETH-80 and YaleB datasets with a relatively smaller value ($\leq$0.01). Thus, we set $\lambda_1 = 0.001$ and $\lambda_2 = 0.001$ on all datasets in our experiments.

4.6 Convergence Analysis

In this section, we conduct experiments to validate the convergence of our method on four benchmark datasets. The convergence curves of the objective function value versus the iteration number are plotted in Fig. 3. From the results, we can obtain that our method can converge within as few as 20 to 30 iterations.

5 Conclusion

In this paper, to provide a proper modeling of the set data, we propose a new manifold convex hull collaborative representation method. Definitely, Our model can successfully leverage the highly expressive representational capability of the Grassmann manifold while preserve the flexible nature of the convex hull model, thus, it exhibits a strong feature extraction ability. Moreover, an efficient alternate optimization algorithm is further designed to solve the problem. Eventually, the experimental results on benchmark datasets show the efficiency to the classification accuracy and robustness to noise. In the future work, we will adopt the kernel mapping to exploit the nonlinear problem.

Acknowledgment. This work is supported by the National Natural Science Foundation of China under Grant No. (62106107, 62276138), and by the Natural Science Foundation of Jiangsu Province, China (Youth Fund Project) under Grant No. (BK20210560) and by the Natural Science Research of Jiangsu Higher Education Institutions of China under Grant No. (21KJB520011), and by the Start Foundation of Nanjing University of Posts and Telecommunications under Grant No. (NY222024).

References

1. Cevikalp, H., Triggs, B.: Face recognition based on image sets. In: 2010 IEEE Computer Society Conference on Computer Vision and Pattern Recognition, pp. 2567–2573 (2010)
2. Cevikalp, H., Yavuz, H.S., Triggs, B.: Face recognition based on videos by using convex hulls. IEEE Trans. Circuits Syst. Video Technol. **30**(12), 4481–4495 (2019)
3. Chen, Z., Xu, T., Wu, X.J., Wang, R., Kittler, J.: Hybrid Riemannian graph-embedding metric learning for image set classification. IEEE Trans. Big Data **9**(1), 75–92 (2023)
4. Deng, L.: The MNIST database of handwritten digit images for machine learning research [best of the web]. IEEE Signal Process. Mag. **29**(6), 141–142 (2012)
5. Efron, B., Hastie, T., Johnstone, I., Tibshirani, R.: Least angle regression (2004)
6. Gao, X., Feng, Z., Wei, D., Niu, S., Zhao, H., Dong, J.: Class-specific representation based distance metric learning for image set classification. Knowl.-Based Syst. **254**, 109667 (2022)
7. Gunawardana, A., Byrne, W.: Convergence theorems for generalized alternating minimization procedures. J. Mach. Learn. Res. **6**, 2049–2073 (2005)
8. Harandi, M., Sanderson, C., Shen, C., Lovell, B.: Dictionary learning and sparse coding on grassmann manifolds: an extrinsic solution. In: 2013 IEEE International Conference on Computer Vision, pp. 3120–3127 (2013)
9. Hu, Y., Mian, A.S., Owens, R.: Face recognition using sparse approximated nearest points between image sets. IEEE Trans. Pattern Anal. Mach. Intell. **34**(10), 1992–2004 (2012)
10. Huang, G.B., Mattar, M., Berg, T., Learned-Miller, E.: Labeled faces in the wild: a database for studying face recognition in unconstrained environments. Technical report (2007)
11. Huang, Z., Van Gool, L.: A Riemannian network for SPD matrix learning. In: Thirty-First AAAI Conference on Artificial Intelligence (2017)
12. Huang, Z., Wang, R., Shan, S., Chen, X.: Projection metric learning on Grassmann manifold with application to video based face recognition. In: 2015 IEEE Conference on Computer Vision and Pattern Recognition (CVPR), pp. 140–149 (2015)
13. Huang, Z., Wu, J., Van Gool, L.: Building deep networks on Grassmann manifolds. In: Proceedings of the AAAI Conference on Artificial Intelligence, vol. 32 (2018)
14. Sun, Y., Wang, X., Peng, D., Ren, Z., Shen, X.: Hierarchical hashing learning for image set classification. IEEE Trans. Image Process. **32**, 1732–1744 (2023)
15. Wang, R., Wu, X.J., Chen, K.X., Kittler, J.: Multiple Riemannian manifold-valued descriptors based image set classification with multi-kernel metric learning. IEEE Trans. Big Data **8**(3), 753–769 (2022)
16. Wang, R., Wu, X.J., Kittler, J.: Graph embedding multi-kernel metric learning for image set classification with Grassmannian manifold-valued features. IEEE Trans. Multimedia **23**, 228–242 (2021)
17. Wang, R., Wu, X.J., Kittler, J.: SymNet: a simple symmetric positive definite manifold deep learning method for image set classification. IEEE Trans. Neural Netw. Learn. Syst. **33**(5), 2208–2222 (2021)
18. Wang, R., Chen, X.: Manifold discriminant analysis (2009)
19. Wang, R., Guo, H., Davis, L.S., Dai, Q.: Covariance discriminative learning: a natural and efficient approach to image set classification. In: 2012 IEEE Conference on Computer Vision and Pattern Recognition, pp. 2496–2503 (2012)

20. Wang, R., Shan, S., Chen, X., Gao, W.: Manifold-manifold distance with application to face recognition based on image set. In: 2008 IEEE Conference on Computer Vision and Pattern Recognition, pp. 1–8 (2008)
21. Wang, W., Wang, R., Shan, S., Chen, X.: Prototype discriminative learning for image set classification. IEEE Signal Process. Lett. **24**(9), 1318–1322 (2017)
22. Wei, D., Shen, X., Sun, Q., Gao, X.: Discrete metric learning for fast image set classification. IEEE Trans. Image Process. **31**, 6471–6486 (2022)
23. Wei, D., Shen, X., Sun, Q., Gao, X., Yan, W.: Locality-aware group sparse coding on Grassmann manifolds for image set classification. Neurocomputing **385**, 197–210 (2020)
24. Yang, M., Zhu, P., Van Gool, L., Zhang, L.: Face recognition based on regularized nearest points between image sets. In: 2013 10th IEEE International Conference and Workshops on Automatic Face and Gesture Recognition (FG), pp. 1–7 (2013)
25. Zhang, L., Yang, M., Feng, X.: Sparse representation or collaborative representation: which helps face recognition? In: 2011 International Conference on Computer Vision, pp. 471–478. IEEE (2011)
26. Zhang, S., Wei, D., Yan, W., Sun, Q.: Probabilistic collaborative representation on Grassmann manifold for image set classification. Neural Comput. Appl. **33**(7), 2483–2496 (2021)
27. Zhu, P., Zuo, W., Zhang, L., Shiu, S.C.K., Zhang, D.: Image set-based collaborative representation for face recognition. IEEE Trans. Inf. Forensics Secur. **9**(7), 1120–1132 (2014)

FUFusion: Fuzzy Sets Theory for Infrared and Visible Image Fusion

Yuchan Jie[1], Yong Chen[2], Xiaosong Li[1(✉)], Peng Yi[3], Haishu Tan[1], and Xiaoqi Cheng[2]

[1] School of Physics and Optoelectronic Engineering, Foshan University, Foshan 528225, China
lixiaosong@buaa.edu.cn

[2] School of Mechatronic Engineering and Automation, Foshan University, Foshan 528225, China

[3] Jiangsu Shuguang Photoelectric Co., Ltd, Yangzhou 225009, China

Abstract. Infrared and visible image fusion combines the high resolution, rich structure of visible images with the remarkable information of infrared images for a wide range of applications in tasks such as target detection and segmentation. However, the representation and retention of significant and non-significant structures in fused images is still a big challenge. To address this issue, we proposed a novel infrared and visible image fusion approach based on the theory of fuzzy sets. First, we proposed a novel filter that integrates the SV-bitonic filter into a least squares model. By leveraging both global and local image features, this new filter achieved edge preservation and smoothness while effectively decomposing the image into structure and base layers. Moreover, for the fusion of the structure layers, according to the salient structural characteristics, we proposed a feature extraction method based on fuzzy inference system. Additionally, the intuitionistic fuzzy sets similarity measure was utilized to extract details from the residual structure layer. Extensive experiments demonstrate that the proposed model outperforms the state-of-the-art methods on publicly available datasets. The code is available at https://github.com/JEI981214/FUFusion_PRCV.

Keywords: Infrared and Visible Image Fusion · SV-bitonic filter · Fuzzy Inference System

1 Introduction

Infrared and visible image fusion (IVIF) has received significant attention in the field of image processing as a pixel-level fusion task, and has typical applications in target detection [1], target tracking [2], etc. Visible images are consistent with the human visual system and can provide texture and details with high spatial resolution; however, they are susceptible to complex weather conditions, such as low light and fog. Infrared images are obtained by sensing radiation in the infrared spectrum and can separate the target from the background. They can be resistant to adverse weather interference but have low resolution and contain little detailed information. Therefore, IVIF can

Q. Liu et al. (Eds.): PRCV 2023, LNCS 14426, pp. 466–478, 2024.
https://doi.org/10.1007/978-981-99-8432-9_37

simultaneously highlight salient targets in infrared images and structural details in visible images, improving the comprehensiveness of the image description of the same scene and facilitating subsequent vision tasks.

IVIF can be roughly divided into three categories: deep learning methods (DLMs), sparse representation methods (SRMs), and transformed domain methods (TDMs).

Wu et al. [3] were the first to introduce DLM into IVIF [4]. In their work, a deep Boltzmann machine was utilized to determine the choice of coefficients for sub-images. Since then, deep learning strategies have been gradually combined with IVIF. Regarding the methods employed, a range of DLMs, generative adversarial networks [5], convolutional neural networks (CNN) [6], and transformers [7] have been developed. CNN-based IVIF uses well-designed network framework and loss function to extract features. To strengthen visual awareness and incorporate useful information under dim lighting conditions, DIVFusion [8] utilized scene-illumination separation network and color fidelity loss function to complete the fusion task. Even though these methods yield favorable fusion outcomes, owing to the lack of ground truth, their fusion performance is limited. It is also worth noting that these DLMs pay less attention to the image decomposition.

Conventional IVIF depends on handcrafted feature extraction and fusion rules to preserve the intrinsic characteristics of the source image. The goal of SRM is to study a redundant dictionary and then describe the image features with sparse representation (SR) coefficients. Yang et al. [9] developed a model of detail injection and rule of visual significance SR to retain detail and salient information. Li et al. [10] combined the merits of spatial domain approaches and SR to complete the fusion of noisy images. Nonetheless, an SRM only considers the underlying features of an image and pays less attention to the visually salient information of the infrared image, which leads to loss of brightness in the fused image. In addition, an SRM only considers the underlying features and pays less attention to the salient information of the infrared images.

TDM is more effective in IVIF, and the key to its performance is the design of decomposition and fusion rules. To reduce the computational complexity associated with transforming the original image to the frequency domain, MDLatLRR [11] extracts the details and background features of the image at multiple levels by learning a projection matrix. To preserve salient radiating targets of infrared images, EPGF-IFD [13] decomposes the image into six distinct feature layers using edge-preserving guided filtering based on image gradient mapping. The methods [11–13] are effective in identifying significant radiating targets in infrared images but have low decision accuracy for blurred and indeterminate edges.

To address the limitations of the existing approaches, this study proposed a fuzzy sets theory-based method for IVIF (FUFusion). First, a novel filter combining an SV-bitonic filter (SVB), and a least squares (LS) model (SVB-LS filter) is proposed, which considers both global and structural features of source images and realizes a fast and efficient "structure-based" decomposition. To extract the texture and edge information in the structural layer, a significant structural feature extraction algorithm based on a fuzzy inference system is developed. Through expert knowledge, fuzzy and uncertain structural information is processed, and significant edges can be effectively reasoned and decided. Meanwhile, considering the details in the remaining structural layer, the intuitionistic fuzzy sets similarity measure is introduced to extract and measure the structural fuzzy features using its stronger uncertainty and fuzziness expression capability. Finally, basic layers containing rich energy and contrast are fused by energy-based rules. The final fused image can be obtained by combining the pre-fused salient structural layer, residual structural layer, and basic layer. The main contributions of this work are as follows:

(1) We propose a new method for IVIF. In this work, a new SVB-LS filter is proposed to utilize the SV-bitonic filter's high adaptability to structurally complex images, while also considering the global features of the image. It can well complete a computationally efficient and edge-preserving "structure-base" decomposition.
(2) Based on the edge detection technique of fuzzy logic inference, we propose an improved structured layer notable feature extraction algorithm. It effectively avoids artifacts at the edges of fused images, and is robust to feature extraction from images with lighting variations and high contrast.
(3) To the best of our knowledge, this is the first time that the intuitive fuzzy set similarity measurement is introduced into IVIF, exploiting its powerful ability to represent complex ambiguities and uncertainties to fully extract detailed information from the remaining structural layers.

2 Proposed Method

In this section, an infrared and visible image fusion method is proposed and discussed in detail. The schematic of the model is displayed in Fig. 1.

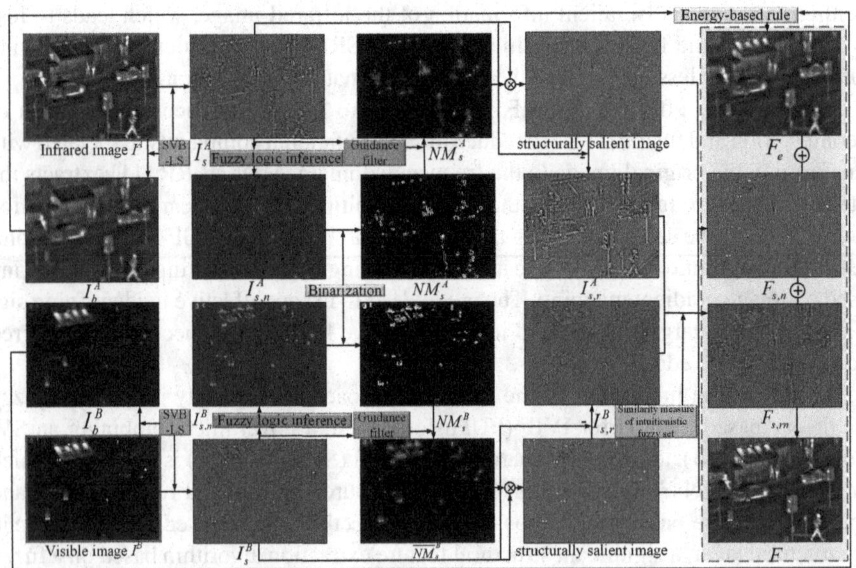

Fig. 1. Overview of the proposed FUFusion model.

2.1 SVB-LS Model

Image decomposition is essential for extracting structural features. The proposed SVB-LS model employs the SV-bitonic filter [14] to smoothen image gradients. Further, it embeds them into the LS model that can consider the global and local characteristics of images. The SV-bitonic filter uses an elliptical mask set with different orientations and aspect ratios and flexibly defines two radial parameters of the ellipse. Importantly, the presence of complex structures in infrared and visible images cannot be neglected. The introduction of the SV-bitonic filter makes the SVB-LS model more adaptable to complex structures. Moreover, the LS filter has high efficiency. Therefore, the SVB-LS filter can maintain its high computational efficiency and edge-preserving smoothing quality during "structure-base" decomposition.

We propose the SVB-LS model for the following primary reasons. The LS algorithm is a model that penalizes gradients close to zero, resulting in the inability to retain large gradients (Fig. 2(b)). The structural-change-based SV-bitonic filter possesses a structural morphological operation with a local adaptive mask, and filtering can effectively preserve the image edges (Fig. 2(c)). In order to consider the global consistency of the image and effectively preserve the details of the local structure simultaneously, the following optimization model can be employed.

$$\min_{I_b^A} \sum_p ((I_{b,p}^A - I_p^A)^2 + \lambda \sum_{*\in\{x,y\}} (\nabla I_{b,*,p}^A - (f_t^{SVB}(\nabla I_*^A))_p)^2) \tag{1}$$

Here, $* \in \{x, y\}$, p represents the position of the pixel point. As the output, I_b^A is the base layer of I^A. $f_t^{SVB}(\nabla I_*^A)$ represents the x- and y-directional gradients of I^A filtered by the SV-bitonic filter [14]. If λ is sufficiently large, ∇I_*^A will be close to $f_t^{SVB}(\nabla I_*^A)$; according to a previous study, $\lambda = 1024$ [15]. The result of the SVB-LS filter is displayed in Fig. 2 (d), which retains some of the global information and more structural features than do Fig. 2 (b) and (c).

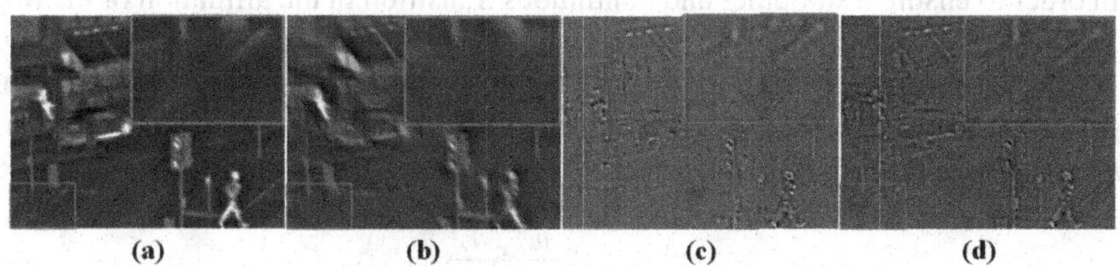

Fig. 2. Structural layer comparison: (a) original image, and different structural layers obtained from (b) LS, (c) SV-bitonic, and (d) SVB-LS filters.

I_b^A can be obtained using the fast Fourier and inverse fast Fourier transforms, as follows:

$$I_b^A = F^{-1}\left(\frac{F(I^A) + \lambda \sum_{*\in\{x,y\}} \overline{F(\delta_*)} \cdot F(f_t^{SVB}(\nabla I_*^A))}{F(1) + \lambda \sum_{*\in\{x,y\}} \overline{F(\delta_*)} \cdot F(\delta_*)}\right) \tag{2}$$

where $F(\cdot)$ and $F^{-1}(\cdot)$ denote the fast Fourier and inverse fast Fourier transform operators, respectively; $\overline{F(\cdot)}$ is the conjugate of $F(\cdot)$; and $F(1)$ represents the fast Fourier transform of the Delta function. Similarly, we can determine I_b^B.

In summary, the proposed SVB-LS model can be implemented according to the following two steps: (I) gradient filtering of the input image in the x and y directions using an SV-bitonic filter and (II) combining Eq. (2) with the solution of Eq. (1) to obtain the output. I_b^A can be subtracted from I^A to obtain the corresponding structural layer I_s^A. Similarly, we can obtain I_s^B.

2.2 Significant Feature Fusion

Generally, rich blurring features are present at the edges of infrared and visible images. The application of fuzzy theory in detecting significant features in the structural layer can accurately represent the uncertainty of image grayscale and geometry. This enhances the fusion results to align more closely with the visual characteristics perceived by the human eye. The proposed edge detection method based on a fuzzy inference system is inspired by a previous study [16]. In our method, we consider I_s^A and I_s^B as inputs, which are processed by the system based on certain rules. Subsequently, the probability of a pixel being identified as an edge pixel point is used as the system output. Finally, the edge of the image is obtained by threshold determination. In addition, Mamdani was chosen as the fuzzy inference type for the fuzzy inference system.

First, the horizontal and vertical gradients of I_s^A are calculated as inputs to the fuzzy inference system.

$$I_{s,gx}^A = I_{s,g}^A(i+1,j) - I_{s,g}^A(i,j) \tag{3}$$

$$I_{s,gy}^A = I_{s,g}^A(i,j+1) - I_{s,g}^A(i,j) \tag{4}$$

where $I_{s,gx}^A$ and $I_{s,gy}^A$ represent the horizontal and vertical gradients of I_s^A, respectively.

In order to ensure a smoother and continuous transition in the affiliation of the fuzzy sets, before being used as inputs to the fuzzy inference system, each pixel point gradient is mapped between [0,1] using a 0-mean Gaussian affiliation function. Moreover, the weight of the fuzzy inference can be controlled by adjusting the standard deviation to enhance the edge detection performance.

$$\overset{\smile}{I}{}_{s,gx}^{A} = e^{\frac{-(I_{s,gx}^A - \rho)^2}{2\sigma^2}} \tag{5}$$

where $\overset{\smile}{I}{}_{s,gx}^{A}$ is the gradient obtained by 0-mean Gaussian subordination function, $\sigma = 0.1$ is the standard deviation, $\rho = 0$ is the mean value. Similarly, we can obtain $\overset{\smile}{I}{}_{s,gx}^{A}$.

A triangular subordination function is defined to intuitively describe the fuzzy system. The analytic formula of this function is generally expressed as:

$$\mu_{FS}(x) = \begin{cases} \frac{x-\alpha}{\beta-\alpha} & \alpha < x \& x < \beta \\ \frac{\gamma-x}{\gamma-\beta} & \beta < x \& x \le \gamma \\ 0 & x \le \alpha | \gamma \le x \end{cases} \tag{6}$$

Here, $x = \{I^A_{s,gx}(i,j), I^A_{s,gy}(i,j)\}$ is the input, | and & is the "OR" and "AND" operations, respectively. To calculate the affiliation parameter and maintain consistency in the fuzzification of $I^A_{s,gx}$ and $I^A_{s,gy}$, their minimum, average, and maximum values are acquired from the two connection matrices to obtain α, γ, and β, respectively.

The inference rules of the designed fuzzy inference system are defined as follows: (I) R_1 = "If $\breve{I}^A_{s,gx}$ and $\breve{I}^A_{s,gy}$ are zero, then $I^A_{s,n}$ is black." and (II) R_2 = "If $\breve{I}^A_{s,gx}$ or $\breve{I}^A_{s,gy}$ is not zero, then $I^A_{s,n}$ is white." In other words, if the difference between the horizontal and vertical pixels is small, the output is black; otherwise, it is white. The "AND" operation takes the smaller of the fuzzy values and can be expressed as $min(\breve{I}^A_{s,gx}, \breve{I}^A_{s,gy})$. The "OR" operation is employed to achieve the larger fuzzy value and can be expressed as $max(1 - \breve{I}^A_{s,gx}, 1 - \breve{I}^A_{s,gy})$.

The centroid method, which is insensitive to the shape of the affiliation function of the fuzzy output, is used here to determine the affiliation-weighted average centroid of the fuzzy output. The output of the fuzzy inference system for detecting the significant features of I^A_s is $I^A_{s,n}$. Similarly, $I^B_{s,n}$ can be obtained.

Salient feature map binarization can generate a decision map for selecting the salient feature pixels of the source images more rationally.

$$NM^A_s = \psi\left(I^A_{s,n}\right) \tag{7}$$

where $\psi(\cdot)$ is the binarization operation. Similarly, NM^B_s can be obtained. To preserve significant edges, we post-processed the feature decision map using a guided filter [17].

$$\overline{NM}^A_s = G_{r,\theta}(NM^A_s, I^A_s) \tag{8}$$

We can similarly obtain $\overline{NM}^B_s$.

The pre-fused structurally notable image $F_{s,n}$ can be represented as follows:

$$F_{s,n} = \overline{NM}^A_s \times I^A_s + \overline{NM}^B_s \times I^B_s \tag{9}$$

2.3 Insignificant Feature Fusion

The pre-fused significant layers obtained using the fuzzy inference system may still lose a few critical details in I^A_s and I^B_s. Therefore, an intuitionistic fuzzy set (IFS)-based similarity measure is introduced to adequately extract and measure the fuzzy features of infrared and visible images.

The remaining structural layers are first fuzzified using the triangular affiliation function described in Eq. (6), and $x = \{I^A_{s,r}(i,j), I^B_{s,r}(i,j)\}$ serves as the input here.

IFS theory describes three characteristic parameters: degrees of affiliation, non-affiliation, and hesitation. Among them, the degree of hesitation represents the degree of uncertainty between the degrees of affiliation and non-affiliation. Therefore, IFS theory can deal with non-deterministic situations more accurately and has a high information representation capability. Further, it can better simulate human subjective preferences and reasoning processes in line with human cognitive styles. To fully describe the uncertainty and fuzziness of the features in residual structural layers, we further transformed the fuzzy set affiliation matrix into the IFS affiliation matrix, which can be represented by Eqs. (10)–(12).

$$\mu_{IFS}(x) = [\mu_{FS}(x)]^{\eta} \tag{10}$$

$$\upsilon_{IFS}(x) = 1 - [\mu_{FS}(x)]^{1/\eta} \tag{11}$$

$$\varpi_{IFS}(x) = [\mu_{FS}(x)]^{1/\eta} - [\mu_{FS}(x)]^{\eta} \tag{12}$$

where $\eta = 2$, μ_{IFS} is the IFS affiliation, υ_{IFS} is the IFS unaffiliated degree, and ϖ_{IFS} represents the IFS hesitancy function.

Based on this background, to accurately determine the remaining structural layer pixels of the fused image, we introduced a target scheme according to the similarity measurement of IFS. The primary idea lies in the assumption that there is an explicit IFS C and that $\mu_C = 1, \upsilon_C = 0, and \varpi_C = 0$ at this point. The similarity of the two unknown elements IFS-A and IFS-B was measured and determined separately using the clear intuitionistic fuzzy sets of IFS-C. Similarity of A to C and B to C in intuitionistic fuzzy sets can be expressed as follows:

$$\begin{aligned} s^{B,C}(i,j) = 1 - \frac{1}{2k}\sum_{i=1}^{k}\Bigg(&\left|\frac{(\mu^C - \mu^B_{IFS}(x)[(\mu^B_{IFS}(x))^2 + (\mu^C)^2]^{1/2}}{\mu^C + \mu^B_{IFS}(x)}\right| \\ &+\left|\frac{(\upsilon^B_{IFS}(x) - \upsilon^C))[(1-\upsilon^B_{IFS}(x))^2 + (1-\upsilon^C)^2]^{1/2}}{2 - \upsilon^B_{IFS}(x) - \upsilon^C}\right|\Bigg) \end{aligned} \tag{13}$$

Similarly, we can obtain $s^{A,C}$. $s^{B,C}$ and $s^{A,C}$ are the similarities of intuitionistic fuzzy sets of B and C and those of A and C, respectively.

A single pixel often correlates with the neighboring pixels. Therefore, a convolution operation on the similarity matrices of the infrared and visible images can fully consider the neighborhood information of the pixels, which can be represented as follows:

$$sc^{AC}(i,j) = \sum\nolimits_p^2 \sum\nolimits_q^2 \chi(p,q) s^{A,C}(i,j) \tag{14}$$

where χ is the given convolutional kernel, and sc is the convolutional similarity matrix. Similarly, we can obtain sc^{BC}.

Finally, the fusion of the remaining structural layers can be represented as follows:

$$F_{s,r} = \begin{cases} I^A_{s,r}(i,j) & sc^{AC}(i,j) > sc^{BC}(i,j) \\ [I^A_{s,r}(i,j) + I^B_{s,r}(i,j)]/2 & sc^{AC}(i,j) = sc^{BC}(i,j) \\ I^B_{s,r}(i,j) & sc^{AC}(i,j) < sc^{BC}(i,j) \end{cases} \tag{15}$$

Here, $F_{s,r}$ is the pre-fused remaining structural layer, and $I^A_{s,r}$ and $I^B_{s,r}$ are the remaining structural layers of I^A and I^B, respectively. sc^{AC} denotes the convolutional similarity matrix of intuitionistic fuzzy sets A and C. Similarly one can obtain sc^{BC}.

2.4 Fusion of Base Layers

The base layer presents approximate features, and the pixel amplitude is sufficient to represent the amount of energy in source images. Therefore, we used an energy-based rule to determine the pre-fused base layer F_b:

$$F_b(i,j) = \begin{cases} I_b^A(i,j), \ I_b^A(i,j) > I_b^B(i,j) \\ I_b^B(i,j), \ I_b^A(i,j) \le I_b^B(i,j) \end{cases} \tag{16}$$

where I_b^A and I_b^B denote the base layers of I^A and I^B, respectively.

The final fused image is obtained by summing the pre-fused structural notable layer $F_{s,n}$, the pre-fused structural residual layer $F_{s,rn}$, and the pre-fused base layer F_b.

3 Experimental Analysis

3.1 Experimental Settings

We demonstrated the effectiveness of FUFusion primarily through extensive experiments using the RoadScene[1] dataset with 211 aligned sets of images and the M3FD[2] dataset with 300 sets of images. We randomly selected 60 sets of RoadScene and M3DF datasets for parameter analysis and ablation analysis. Furthermore, we compared FUFusion with seven advanced approaches: DIVFusion [8], MFEIF [18], SDDGAN [19], SEAFusion [20], TEMF [21], GANMcC [22], and LatLRR [23].

Six objective evaluation metrics, namely, $\mathrm{Q^{MI}}$ [24], $\mathrm{Q^{NCIE}}$ [25], $\mathrm{Q^{G}}$ [26], $\mathrm{Q^{M}}$ [27], SSIM [28], and VIF [29], were employed to objectively evaluate the performance of FUFusion in terms of the information content, features, similarity to the structure of the source image, and human eye perception. Furthermore, $\mathrm{Q^{MI}}$ measures the dependency between variables, $\mathrm{Q^{NCIE}}$ measures the non-linear correlation between discrete variables, $\mathrm{Q^{G}}$ measures the amount of edge information supplied from input to output, $\mathrm{Q^{M}}$ determines the retention of features in the fused image, SSIM measures the structural similarity between images, and VIF measures the conformity of the fused image to human eye perception.

The experiments were conducted using an NVIDIA GeForce GTX 1070 and 2400 MHz Intel Core i7-7700HQ.

3.2 Parameter Analysis

In the proposed model, the threshold r of $f_t^{SVB}(\cdot)$ in Eq. (1) is the key to preserve edges of structure layers. We selected the best r from the subjective and objective aspects.

From Fig. 3, the structure of the enlarged car logo is blurred when $r < 5$, indicating that the optimal parameter is no less than 5. We can Observe from Table 1 that the values of $\mathrm{Q^{MI}}$ and VIF are decreasing as r increases, representing a decrease in the perception of the fused results. Combined subjective and objective assessment, $r = 5$.

Table 1. Mean values of metrics on the selected images (bold: the best; blue: sub-optimal).

r	5	6	7	8
Q^{MI}	**0.5855**	0.5708	0.5579	0.5468
VIF	**0.3232**	0.3196	0.3179	0.3152

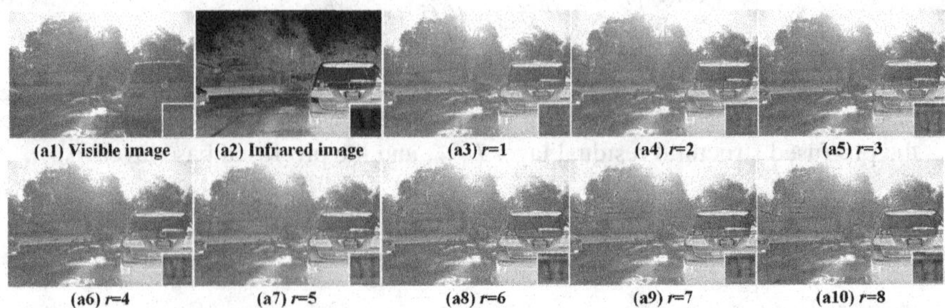

(a1) Visible image (a2) Infrared image (a3) r=1 (a4) r=2 (a5) r=3 (a6) r=4 (a7) r=5 (a8) r=6 (a9) r=7 (a10) r=8

Fig. 3. The fused images produced by different *t*.

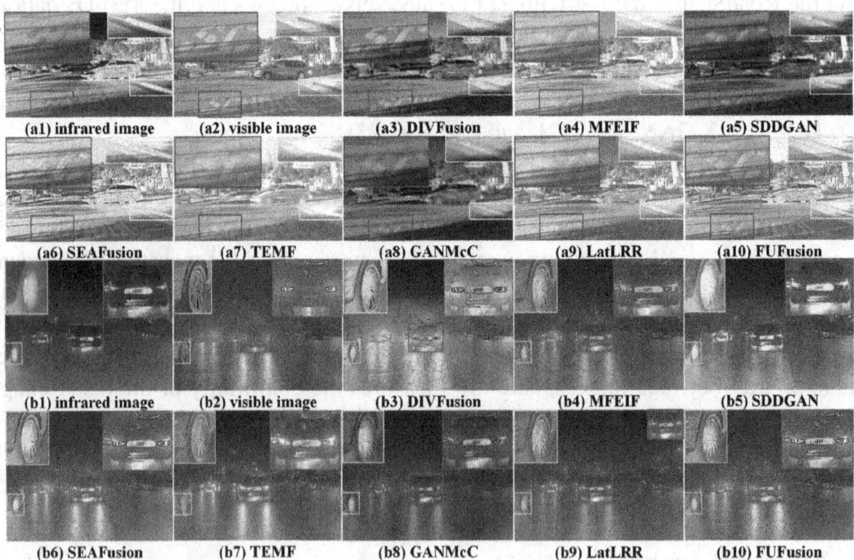

(a1) infrared image (a2) visible image (a3) DIVFusion (a4) MFEIF (a5) SDDGAN (a6) SEAFusion (a7) TEMF (a8) GANMcC (a9) LatLRR (a10) FUFusion (b1) infrared image (b2) visible image (b3) DIVFusion (b4) MFEIF (b5) SDDGAN (b6) SEAFusion (b7) TEMF (b8) GANMcC (b9) LatLRR (b10) FUFusion

Fig. 4. Subjective evaluation using the RoadScene and M3FD datasets.

Table 2. Objective evaluation of fused images produced by different filters.

Filters	Q^{MI}	Q^{NCIE}	Q^{G}	Q^{M}	SSIM	VIF
SVB filter	**0.6357**	**0.8146**	0.5515	0.9830	0.5271	0.3243
LS filter	0.4286	0.8089	0.4168	0.6934	0.5258	0.2194
SVB-LS filter	0.6186	0.8140	**0.5586**	**1.0256**	**0.5280**	**0.3244**

3.3 Ablation Experiment

In the proposed FUFusion, we developed a new SVB-LS filter to realize "structure-base" decomposition. To validate its effectiveness, we performed a comparison experiment where images are decomposed by the SVB filter, LS filter, and the SVB-LS filter, respectively. From Table 2, four metrics of the fused images based on the SVB-LS filter are ranked first, which proved the effectiveness of the SVB-LS filter.

3.4 Subjective Comparison

Results of subjective evaluation using the road dataset are shown in Fig. 4. In general, SEAFusion and FUFusion possess a high degree of clarity. As shown in the enlarged areas in Fig. 4, SDDGAN, TEMF, and GANMcC lose significant edge information including landmarks, kerbs, and texture of tires. MEFIF, DIVFusion, and LatLRR retain some structure and edges but have slightly less contrast than does FUFusion. By contrast, FUFusion provides more edge and texture information and has superior retention of luminance in visible light images.

3.5 Objective Comparison

Quantitative comparisons of FUFusion and seven previously mentioned methods are summarized in Table 3. As outlined in Table 3, MFEIF had the best VIF metric values, suggesting that MFEIF is more consistent with human visual perception. The Q^{MI}, Q^{NCIE}, Q^{G}, and Q^{M} of SEAFusion were ranked second, representing a higher degree of retention of content and features in source images. Conversely, DIVFusion, SDDGAN, TEMF, GANMcC, and LatLRR were slightly inferior. The SSIM of GANMcC ranks first, indicating that GANMcC can preserve more structural similarity from source images. FUFusion ranked first in terms of the Q^{MI}, Q^{NCIE}, Q^{G}, and Q^{M} metrics, and VIF was among the top three. Hence, we can conclude that FUFusion has less distortion and higher fidelity of the original image, with adequate performance in preserving the structure, details, and contrast of source images.

[1] https://github.com/hanna-xu/RoadScene.

[2] https://github.com/JinyuanLiu-CV/TarDAL.

Table 3. Mean values of metrics on the RoadScene and M3FD datasets.

Methods	Q^{MI}	Q^{NCIE}	Q^{G}	Q^{M}	SSIM	VIF
DIVFusion [8]	0.3659	0.8063	0.2989	0.3649	0.3942	0.1964
MFEIF [18]	0.4341	0.8075	0.3877	0.4323	0.6015	**0.3458**
SDDGAN [19]	0.4223	0.8077	0.3237	0.3800	0.5848	0.2702
SEAFusion [20]	0.4593	0.8083	0.4645	0.5408	0.5536	0.2832
TEMF [21]	0.4030	0.8067	0.3226	0.3799	0.5609	0.2635
GANMcC [22]	0.3652	0.8061	0.3881	0.4342	**0.6269**	0.3164
LatLRR [23]	0.3725	0.8064	0.2743	0.3374	0.5799	0.1895
FUFusion	**0.6460**	**0.8152**	**0.5697**	**1.2661**	0.5543	0.3391

4 Conclusion

Herein, we proposed a novel fuzzy-set-theory-based fusion method, called FUFusion, for IVIF. First, we developed a new SVB-LS filter that fully considers global information and local structural features to achieve image decomposition with high computational efficiency. Second, an algorithm based on a fuzzy inference system was proposed to extract significant structural features. Notably, this algorithm has high robustness to edge structures with large illumination changes and can avoid artifacts. Based on this background, an intuitive fuzzy set similarity measurement with representation of fuzzy and uncertain information was introduced to extract detailed and texture information of the remaining structural layers. Finally, a simple and effective energy-based method was employed to extract contrast and illumination information. Quantitative and qualitative evaluations indicated that the performance of FUFusion exceeds those of seven advanced fusion methods.

Importantly, images captured in real scenes may suffer from unaligned, inconsistent resolution interference due to parallax. Therefore, in future, FUFusion will be optimized for improved adaptability to real scenes. Simultaneously, the relevant subsequent tasks will be fully considered to broaden its scope of practical applications.

Acknowledgements. This research was supported by the National Natural Science Foundation of China (Grant No. 62201149).

References

1. Zhou, H., Sun, M., Ren, X., Wang, X.Y.: Visible-thermal image object detection via the combination of illumination conditions and temperature information. Remote Sens. **13**, 3656 (2021)
2. Chandrakanth, V., Murthy, V., Channappayya, S.S.: Siamese cross-domain tracker design for seamless tracking of targets in RGB and thermal videos. IEEE Trans. Artif. Intelli. **4**, 161–172 (2022)
3. Wu, W., Qiu, Z.M., Zhao, M., Huang, Q.H., Lei, Y.: Visible and infrared image fusion using NSST and deep Boltzmann machine. Optik **157**, 334–342 (2018)

4. Zhang, X., Demiris, Y.: Visible and infrared image fusion using deep learning. IEEE Trans. Pattern Anal. Mach. Intelli. **45**, 10535–10554 (2023)
5. Gao, Y., Ma, S.W., Liu, J.J.: DCDR-GAN: a densely connected disentangled representation generative adversarial network for infrared and visible image fusion. IEEE Trans. Circ. Syst. Video Technol. **33**, 549–561 (2023)
6. Zhao, Z.X., Xu, S., Zhang, J.S., Liang, C.Y., Zhang, C.X., Liu, J.M.: Efficient and model-based infrared and visible image fusion via algorithm unrolling. IEEE Trans. Circ. Syst. Video Technol. **32**, 1186–1196 (2022)
7. Chang, Z.H., Feng, Z.X., Yang, S.Y., Gao, Q.W.: AFT: adaptive fusion transformer for visible and infrared images. IEEE Trans. Image Process. **32**, 2077–2092 (2023)
8. Tang, L.F., Xiang, X.Y., Zhang, H., Gong, M.Q., Ma, J.Y.: DIVFusion: darkness-free infrared and visible image fusion. Inform. Fusion **91**, 477–493 (2023)
9. Yang, Y., Zhang, Y., Huang, S., Zuo, Y., Sun, J.: Infrared and visible image fusion using visual saliency sparse representation and detail injection model. IEEE Trans. Instrum. Meas. **70**, 1–15 (2021)
10. Li, X., Zhou, F., Tan, H.: Joint image fusion and denoising via three-layer decomposition and sparse representation. Know.-Based Syst. **224**, 107087 (2021)
11. Li, H., Wu, X.J., Kittler, J.: MDLatLRR: a novel decomposition method for infrared and visible image fusion. IEEE Trans. Image Process. **29**, 4733–4746 (2020)
12. Mo, Y., Kang, X.D., Duan, P.H., Sun, B., Li, S.T.: Attribute filter based infrared and visible image fusion. Inform. Fusion **75**, 41–54 (2021)
13. Ren, L., Pan, Z., Cao, J., Zhang, H., Wang, H.: Infrared and visible image fusion based on edge-preserving guided filter and infrared feature decomposition. Signal Process. **186**, 108108 (2021)
14. Treece, G.: Morphology-based noise reduction: structural variation and thresholding in the bitonic filter. IEEE Trans. Image Process. **186**, 336–350 (2020)
15. Liu, W., Zhang, P.P., Chen, X.G., Shen, C.H., Huang, X.L., Yang, J.: Embedding bilateral filter in least squares for efficient edge-preserving image smoothing. IEEE Trans. Circuits Syst. Video Technol. **30**, 23–35 (2020)
16. Alshennawy, A.A., Aly, A.A.: Edge detection in digital images using fuzzy logic technique. Int. J. Comput. Inform. Eng. **3**, 540–548 (2009)
17. Li, S., Kang, X., Hu, J.: Image fusion with guided filtering. IEEE Trans. Image Process. **22**, 2864–2875 (2013)
18. Liu, J.Y., Fan, X., Jiang, J., Liu, R.S., Luo, Z.X.: Learning a deep multi-scale feature ensemble and an edge-attention guidance for image fusion. IEEE Trans. Circuits Syst. Video Technol. **32**, 105–119 (2022)
19. Zhou, H.B., Wu, W., Zhang, Y.D., Ma, J.Y., Ling, H.B.: Semantic-supervised infrared and visible image fusion via a dual-discriminator generative adversarial network. IEEE Trans. Multimedia **25**, 635–648 (2023)
20. Tang, L.F., Yuan, J.T., Ma, J.Y.: Image fusion in the loop of high-level vision tasks: a semantic-aware real-time infrared and visible image fusion network. Inform. Fusion **82**, 28–42 (2022)
21. Chen, J., Li, X.J., Luo, L.B., Mei, X.G., Ma, J.Y.: Infrared and visible image fusion based on target-enhanced multiscale transform decomposition. Inf. Sci. **508**, 64–78 (2020)
22. Ma, J., Zhang, H., Shao, Z., Liang, P., Han, X.: GANMcC: A generative adversarial network with multiclassification constraints for infrared and visible image fusion. IEEE Trans. Instrum. Meas. **70**, 1–14 (2021)
23. Li, H., Wu, X.-J.: Infrared and visible image fusion using Latent low-rank representation. arXiv preprint arXiv:1804.08992 (2022)
24. Qu, G., Zhang, D., Yan, P.: Information measure for performance of image fusion. Electron. lett. **38**, 313–315 (2002)

25. Wang, Q., Shen, Y., Jin, J.: Performance evaluation of image fusion techniques. Image Fusion Algorithms Appl. **19**, 469–492 (2008)
26. Xydeas, C.S., Petrovic, V.: Objective image fusion performance measure. Electron. Lett. **36**, 308–309 (2000)
27. Wang, P.-W., Liu, B.: A novel image fusion metric based on multi-scale analysis. In: Proceedings of the 2008 9th International Conference on Signal Processing, pp. 965–968 (2008)
28. Wang, Z., Bovik, A.C., Sheikh, H.R., Simoncelli, E.P.: Image quality assessment: from error measurement to structural similarity. IEEE Trans. Image Process. **13**, 600–613 (2004)
29. Han, Y., Cai, Y.Z., Cao, Y., Xu, X.M.: A new image fusion performance metric based on visual information fidelity. Inform. Fusion **14**, 127–135 (2013)

Progressive Frequency-Aware Network for Laparoscopic Image Desmoking

Jiale Zhang, Wenfeng Huang, Xiangyun Liao(✉), and Qiong Wang(✉)

Guangdong Provincial Key Laboratory of Computer Vision and Virtual Reality Technology, Shenzhen Institute of Advanced Technology, Chinese Academy of Sciences, Beijing, China
{xy.liao,wangqiong}@siat.ac.cn

Abstract. Laparoscopic surgery offers minimally invasive procedures with better patient outcomes, but smoke presence challenges visibility and safety. Existing learning-based methods demand large datasets and high computational resources. We propose the Progressive Frequency-Aware Network (PFAN), a lightweight GAN framework for laparoscopic image desmoking, combining the strengths of CNN and Transformer for progressive information extraction in the frequency domain. PFAN features CNN-based Multi-scale Bottleneck-Inverting (MBI) Blocks for capturing local high-frequency information and Locally-Enhanced Axial Attention Transformers (LAT) for efficiently handling global low-frequency information. PFAN efficiently desmokes laparoscopic images even with limited training data. Our method outperforms state-of-the-art approaches in PSNR, SSIM, CIEDE2000, and visual quality on the Cholec80 dataset and retains only 629K parameters. Our code and models are made publicly available at: https://github.com/jlzcode/PFAN.

Keywords: Medical Image Analysis · Vision Transformer · CNN

1 Introduction

Laparoscopic surgery provides benefits such as smaller incisions, reduced postoperative pain, and lower infection rates [14]. The laparoscope, equipped with a miniature camera and light source, allows visualization of surgical activities on a monitor. However, visibility can be hindered by smoke from laser ablation and cauterization. Reduced visibility negatively impacts diagnoses, decision-making, and patient health during intraoperative imaging and image-guided surgery, and hampers computer vision algorithms in laparoscopic tasks such as depth estimation, surgical reconstruction and lesion identification. Although smoke evacuation equipment is commonly used, its high cost and impracticality make image processing-based approaches a more attractive alternative [27]. However, traditional image processing algorithms have limitations in efficacy and can cause visual distortions. Approaches based on atmospheric scattering models inaccurately treat smoke as a homogeneous scattering medium, potentially

Q. Liu et al. (Eds.): PRCV 2023, LNCS 14426, pp. 479–492, 2024.
https://doi.org/10.1007/978-981-99-8432-9_38

leading to tissue misidentification and surgical accidents. End-to-end deep learning approaches show promise, but acquiring large training datasets is difficult and time-consuming, especially for medical applications. Moreover, most deep learning-based models have large parameter counts, making them unsuitable for resource-constrained medical devices. Laparoscopic models must be adaptable to various smoke concentrations and brightness levels, applicable across different surgical environments, lightweight, and effective with limited datasets.

In this study, we propose the Progressive Frequency-Aware Net (PFAN), an efficient, lightweight model built within the generative adversarial networks (GANs) framework for laparoscopic smoke removal. We address smoke removal by focusing on the image frequency domain, integrating high-frequency and low-frequency features to translate smoke-filled images into clear and no-artifacts smoke-free images. With only 629K parameters, PFAN demonstrates remarkable laparoscopic image desmoking results. In summary, the contributions of this work include:

(1) Our proposed PFAN model effectively combines CNN and ViT to take into account frequency domain features of laparoscopic images. PFAN employs the MBI (CNN-based) and LAT (ViT-based) components to sequentially extract high and low-frequency features from the images. This approach establishes a robust feature extraction framework by leveraging the CNNs' local high-frequency feature extraction capabilities and the Transformers' global low-frequency feature extraction strengths.
(2) Our work introduces two innovations to the PFAN model: the Multi-scale Bottleneck-Inverting (MBI) Block, which extracts local high-frequency features using a multi-scale inverted bottleneck structure, and the Locally-Enhanced Axial Attention Transformer (LAT), which efficiently processes global low-frequency information with Squeeze-Enhanced Axial Attention and Locally-Enhanced Feed Forward Layer.
(3) The lightweight PFAN model effectively removes smoke from laparoscopic images with a favorable performance-to-complexity balance. It is suitable for resource-constrained devices. Evaluation results indicate its superiority over state-of-the-art methods, highlighting its effectiveness in removing surgical smoke from laparoscopic images.

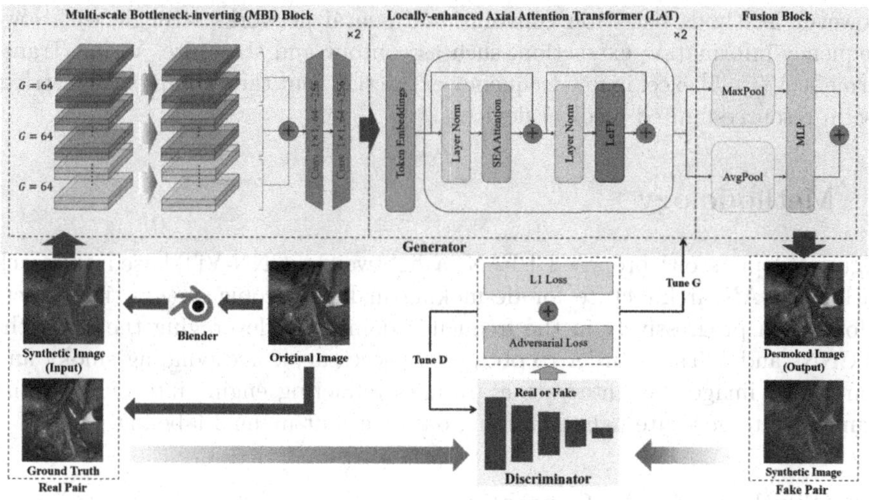

Fig. 1. The flowchart of PFAN illustrates a framework consisting of a generator network (G) and a discriminator network (D). Within this proposed approach, the generator G incorporates Multi-scale Bottleneck-Inverting (MBI) Blocks and Locally-Enhanced Axial Attention Transformer (LAT) Blocks.

2 Related Work

2.1 Traditional Theory-Based Desmoking Methods

Traditional desmoking techniques include image restoration and enhancement. Restoration methods, like the dark channel prior (DCP) by He et al. [7], use atmospheric degradation and depth information, but face limitations in laparoscopic imaging. Enhancement techniques, such as Retinex algorithm [22], wavelet-based algorithms [24], improving local contrast, increasing visibility and interpretability [18]. Wang et al. [29] created a variational desmoking approach, but it relies on assumptions regarding smoke's heterogeneous nature and varying depths.

2.2 Deep Learning-Based Desmoking Methods

Deep learning advances have fostered diverse frameworks for laparoscopic image smoke removal. Sabri et al. [2] employed synthetic surgical smoke images with different smoke densities and utilized CNNs to remove smoke in a supervised setting, while DehazeNet [23], AOD-Net [16] and DM^{2}F-Net [4] relied on atmospheric scattering models, inappropriate for surgical environments. GANs [6], using game-theoretic approaches, generate realistic images. Techniques like Pix2Pix [13] employ conditional GANs for domain mapping. In medical imaging, GANs have been effective in PET-CT translation and PET image denoising [1].

However, methods based on convolutional neural networks struggle with low-frequency information extraction, such as contour and structure. Vision Transformers (ViT) [5] excel in low-frequency extraction, but their complexity restricts use in resource-limited medical devices.

3 Methodology

Figure 1 depicts our proposed PFAN, a lightweight CNN-ViT-based approach within a GAN architecture for desmoking in laparoscopic images, it extracts information progressively in the frequency domain by leveraging the strengths of CNNs and ViTs. In order to obtain the necessary corresponding smoky and non-smoky images, we integrate a graphics rendering engine into our learning framework to generate paired training data without manual labeling.

3.1 Synthetic Smoke Generation

We employ the Blender engine to generate smoke image pairs for model training, offering two advantages over physically-based haze formation models [23] and Perlin noise functions [3]. First, laparoscopic surgical smoke is localized and depth-independent, making traditional haze models unsuitable. Second, modern rendering engines provide realistic and diverse smoke shapes and densities using well-established, physically-based built-in models. With Blender's render engine, denoted by ϕ, we generate the smoke evolution image sequence, S_{Smoke}, by adjusting parameters such as smoke source density, intensity, temperature, location (S_d, S_i, S_t, S_l), and light location and intensity (L_l, L_i):

$$S_{Smoke} = \phi(S_d, S_i, S_t, S_l, L_l, L_i) \tag{1}$$

Let I_{Smoke} represent one frame of the smoke image sequence. To create a synthetic smoke evolution image sequence (I_{Syn}) within the surgical scene, we overlay a randomly generated frame of smoke evolution image sequence (I_{Smoke}) onto each smoke-free laparoscopic image ($I_{Smoke-free}$). The following formula represents this process:

$$I_{Syn} = I_{Smoke-free} + I_{Smoke} \tag{2}$$

The synthesized laparoscopic image sequence shows the evolution process of smoke. In the first frame of the synthesized image sequence, smoke is only present at a specific location within the image, simulating the situation of burning lesion areas in laparoscopic surgery. As time progresses, it disperses from the burning point outwards according to random density, temperature, and intensity parameters. The synthesis of an extensive range of realistic images depicting simulated surgical smoke is made possible through the utilization of a robust rendering engine. By incorporating variations in smoke, such as location, intensity, density, and luminosity, over-fitting is prevented in the network's training.

3.2 Multi-scale Bottleneck-Inverting (MBI) Block

The MBI Block is designed to efficiently extract high-frequency features, drawing inspiration from various well-established neural networks [11,12,20,26]. Here, we denote input smoke images as $\mathcal{X}_{Smoke} \in \mathbb{R}^{H \times W \times 3}$, and the set of high-frequency information extracted by each MBI Block can be defined as $\{\mathcal{X}_{HF} = \mathcal{X}_{HF_1}, ..., \mathcal{X}_{HF_k}\}$. Within each MBI Block, group convolution is represented as GConv, and the multi-scale feature can be obtained as:

$$\mathcal{X}_{MS} = GConv_{i,g}(\mathcal{X}_{Smoke}) + GConv_{j,g}(\mathcal{X}_{Smoke}) + GConv_{k,g}(\mathcal{X}_{Smoke}) \quad (3)$$

Here, i, j, and k represent the size of the receptive field, which were set to 3, 7, and 11, respectively. We choose GELU [9] instead of RELU as the activation function following each convolution layer, given its smoother properties and proven higher performance. The parameter g indicates that, during group convolution, input features will be divided into g groups. In this paper, this value is set to 64, which matches the feature channels, resulting in a significant reduction of parameters by 1/64 in comparison to standard convolution. Next, we merge the multi-scale feature $\mathcal{X}_{MS}$ and expand it to a high-dimensional representation using point-wise convolution. Following this, features are projected back to a low-dimensional representation through point-wise convolution, represented as

$$\mathcal{X}_{HF} = PwConv_{high \to low}(PwConv_{low \to high}(\mathcal{X}_{MS})) \quad (4)$$

3.3 Locally-Enhanced Axial Attention Transformer (LAT) Block

Applying ViT models to desmoke laparoscopic images faces challenges. ViT's multi-head self-attention layer applies global attention, neglecting differing frequencies and local high-frequency information. Additionally, ViT's computational cost increases quadratically with token count, limiting its use with high-resolution feature maps. To overcome these issues, we introduce the Locally-Enhanced Axial Attention Transformer (LAT) Block. It combines streamlined squeeze Axial attention for global low-frequency semantics and a convolution-based enhancement branch for local high-frequency information. The LAT Block captures long-range dependencies and global low-frequency information with low parameter counts.

Given the features at MBI Block outputs, $\mathcal{X}_{MBI}$, LAT first reshapes the input into patch sequences using $H \times H$ non-overlapping windows. And then Squeeze-Enhanced Axial Attention computes attention maps ($\mathcal{X}_{Sea}$) for each local window. To further process the information, LAT replaces the multi-layer perceptron (MLP) layers in a typical ViT with a Locally-Enhanced Feed Forward Layer. Additional skip connections enable residual learning and produce $\mathcal{X}_{LAT}$.

$$\mathcal{X}_{Sea} = SEA(LN(\mathcal{X}_{MBI})) + \mathcal{X}_{MBI}, \quad \mathcal{X}_{LAT} = LEFF(LN(\mathcal{X}_{Sea})) + \mathcal{X}_{Sea} \quad (5)$$

Here, $\mathcal{X}_{Sea}$ and $\mathcal{X}_{LAT}$ correspond to the outputs of the Squeeze-Enhanced Axial Attention and LEFF modules, respectively. LN denotes layer normalization [15]. We discuss Squeeze-Enhanced Axial Attention and LEFF in detail in subsequent sections.

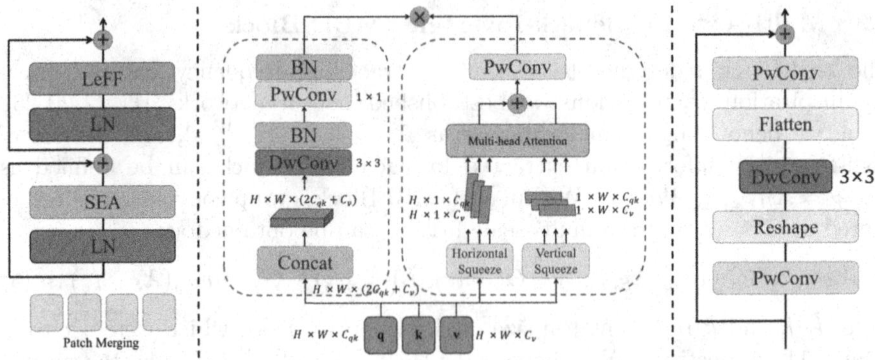

Fig. 2. Left: the schematic illustration of the proposed Locally-Enhanced Axial Attention Transformer Block. Middle: Squeeze-Enhanced Axial Attention Layer. Right: Locally-Enhanced feed-forward network.

Squeeze-Enhanced Axial Attention (SEA). The Squeeze-Enhanced Axial Attention utilized in the Locally-Enhanced Axial Attention Transformer (LAT) is designed to extract global information in a succinct way. Initially, we compute $\mathbf{q}$, $\mathbf{k}$, and $\mathbf{v}$ by $\mathbf{q} = W_q * \mathcal{X}, \mathbf{k} = W_k * \mathcal{X}, \mathbf{v} = W_v * \mathcal{X}$, where $\mathcal{X} \in \mathbb{R}^{H \times W \times C}$. $W_q, W_k \in \mathbb{R}^{C_{qk} \times C}$ and $W_v \in \mathbb{R}^{C_v \times C}$ are learnable weights. Then, a horizontal squeeze $\mathbf{q}_h$ is executed by averaging the query feature map along the horizontal direction and a vertical squeeze $\mathbf{q}_v$ is applied in the vertical direction.

$$\mathbf{q}_h = \frac{1}{W}\left(\mathbf{q}^{(C_{qk},H,W)}\mathbb{1}_W\right)^{\rightarrow(H,C_{qk})}, \quad \mathbf{q}_h = \frac{1}{H}\left(\mathbf{q}^{(C_{qk},W,H)}\mathbb{1}_H\right)^{\rightarrow(W,C_{qk})} \tag{6}$$

The notation $\mathbf{z}^{\rightarrow(\cdot)}$ represents the permutation of tensor $\mathbf{z}$'s dimensions, and $\mathbb{1}_m \in \mathbb{R}_m$ is a vector with all elements equal to 1. The squeeze operation on $\mathbf{q}$ is also applied to $\mathbf{k}$ and $\mathbf{v}$, resulting in $\mathbf{q}_h, \mathbf{k}_h, \mathbf{v}_h \in \mathbb{R}^{H \times C_{qk}}, \mathbf{q}_v, \mathbf{k}_v, \mathbf{v}_v \in \mathbb{R}^{W \times C_{qk}}$. The squeeze operation consolidates global information along a single axis, thereby significantly improving the subsequent global semantic extraction process, as demonstrated by the following equation.

$$\mathbf{y}_{(i,j)} = \sum_{p=1}^{H} softmax_p\left(\mathbf{q}_h^{i\mathsf{T}}\mathbf{k}_h^p\right)\mathbf{v}_h^p + \sum_{p=1}^{W} softmax_p\left(\mathbf{q}_v^{j\mathsf{T}}\mathbf{k}_v^p\right)\mathbf{v}_v^p \tag{7}$$

As can be seen, in Squeeze-Enhanced Axial Attention, each position of the feature map only propagates information in two squeezed axial features, while in traditional self-attention (as in the following equation), each position of the feature map calculates self-attention with all positions.

$$\mathbf{y}_{(i,j)} = \sum_{p \in \mathcal{G}_{(i,j)}} softmax_p\left(\mathbf{q}_{(i,j)}^{\mathsf{T}}\mathbf{k}_p\right)\mathbf{v}_p \tag{8}$$

The traditional global self-attention is as above, where $\mathcal{G}(i,j)$ means all positions on the feature map at location (i,j). When a conventional attention module is applied to a feature map with dimensions $H \times W \times C$ the time complexity becomes $O(H^2W^2(C_{qk}+C_v))$, resulting in low efficiency. However, with SEA, the time complexity for squeezing q, k, v is $O((H+W)(2C_{qk}+C_v))$ and the attention operation takes $O((H^2+W^2)(C_{qk}+C_v))$ time. Consequently, our squeeze Axial attention successfully lowers the time complexity to $O(HW)$, ensuring a more efficient and faster process.

Locally-Enhanced Feed-Forward Network (LEFF). Adjacent pixels play a crucial role in image desmoking, as demonstrated in [29], which highlights their essential contribution to image dehazing and denoising. However, previous research [27] has highlighted the limited ability of the Feed-Forward Network (FFN) within the standard Transformer to effectively utilize local context. To address this limitation, we introduce a depth-wise convolutional block to LAT, inspired by recent studies [17]. As depicted in Fig. 2 (Right), we begin by applying a linear projection layer to each token to augment its feature dimension. Subsequently, we reshape the tokens into 2D feature maps and implement a 3×3 depth-wise convolution to capture local information. Afterward, we flatten the features back into tokens and reduce the channels using another linear layer to align with the input channel dimensions. *LeakyReLU* serves as the activation function following each linear or convolution layer.

Fusion Block. We employ Channel Attention [31] as the Fusion Block in our approach to enhance the cross-channel feature fusion capabilities. The Channel Attention mechanism models inter-dependencies between channels of features, enabling adaptive adjustment of feature responses across different channels, and assigning corresponding weights. Embedding channel attention can facilitate adaptive enhancement and fusion of convolution and corresponding Transformer features in the LAT module. The attention map, $\mathcal{X}_{CA}$, can be calculated using the function, where σ represents the *Sigmoid* function.

$$\mathcal{X}_{CA} = \sigma\left(LEFF(AvgPool(\mathcal{X}_{LAT})) + LEFF(MaxPool(\mathcal{X}_{LAT}))\right) \quad (9)$$

Afterward, the low-frequency information $\mathcal{X}_{LF}$ is acquired as described in (10).

$$\mathcal{X}_{LF} = \mathcal{X}_{LAT} \cdot \mathcal{X}_{CA} \quad (10)$$

To achieve the smoke-free result, $\mathcal{X}_{Smoke-free}$, the low-frequency information of the original input smoke image $\mathcal{X}_{LF}$ is combined with the high-frequency information $\mathcal{X}_{HF}$, which is the output of MBI blocks.

$$\mathcal{X}_{Smoke-free} = \mathcal{X}_{HF} + \mathcal{X}_{LF} \quad (11)$$

4 Experiment

4.1 Data Collections

We used images from the Cholec80 dataset [28], consisting of 80 cholecystectomy surgery videos by 13 surgeons. We sampled 1,500 images at 20-second intervals from these videos, selecting 660 representative smoke-free images. As detailed in Sect. 3.1, we added synthetic random smoke, yielding 660 image pairs, divided in an 8:1:2 ratio for training, validation, and testing. Synthetic smoky images were generated according to Sect. 3.1. Importantly, each dataset contained distinct videos, ensuring no overlap.

4.2 Implementation Details

Our experiments utilized six NVIDIA RTX 2080Ti GPUs. Initially, we trained the Discriminator (PatchGAN) for one epoch to provide a rough smoke mask, followed by iterative training of the Discriminator and Generator while freezing the PatchGAN's parameters during the Generator's training. We employed an Adam solver with a learning rate of 0.0002, momentum parameters $\beta 1 = 0.5$ and $\beta 2 = 0.999$, and a batch size of 6. Consistent with prior research, random cropping was used for generating training and validation patches.

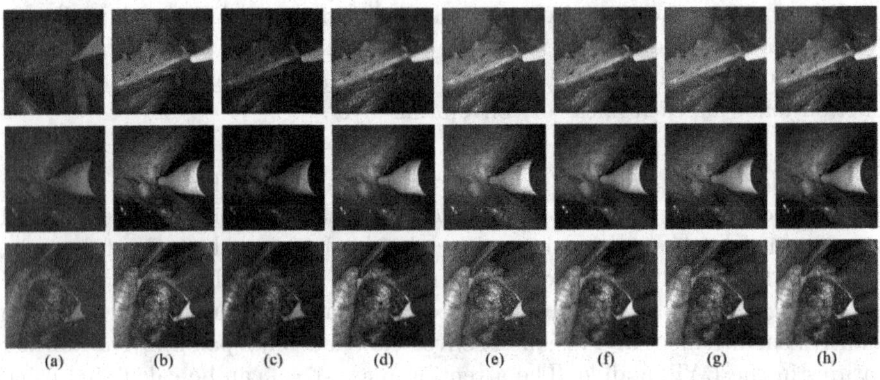

Fig. 3. Comparison experiments between SOTAs. (a) Input (b) Ground Truth, (c) Dark Channel Prior(DCP) [7] (d) CycleGAN + ResNet, (e) CycleGAN + U-Net, (f) Pix2Pix + ResNet, (g) Pix2Pix + U-Net, and (h) Ours.

5 Result

In our quantitative evaluations, we assess desmoking performance by comparing smoke-free images to their desmoked counterparts using the following metrics: number of Parameters, Peak Signal-to-Noise Ratio (PSNR), Structural Similarity Index (SSIM) [10], and CIEDE2000 [21] (which represents color reconstruction accuracy for the human visual system). We compare the proposed method with eight state-of-the-art desmoking and dehazing methods including both a traditional image processing approach (DCP [7]) and the most recent deep learning-based methods (original CycleGAN [19] (with U-Net [25]), CycleGAN with ResNet [8] (6Blocks), CycleGAN with ResNet (9Blocks), original

Pix2Pix [13] (with U-Net), Pix2Pix with ResNet (6Blocks), Pix2Pix with ResNet (9Blocks), Pix2Pix with Uformer [30]).

Table 1. Quantitative results. The best and second-best results are highlighted and underlined, respective

Model		Parameters↓	PSNR↑	SSIM↑	CIEDE2000↓
DCP		/	27.6250	0.5528	35.9952
CycleGAN	U-Net	54414K	28.7449	0.7621	10.3298
CycleGAN	ResNet6	7841K	29.0250	0.7826	9.5821
CycleGAN	ResNet9	11383K	29.0926	0.7802	9.2868
Pix2Pix	U-Net	54414K	29.2967	0.7073	8.8060
Pix2Pix	ResNet6	7841K	29.8249	0.8358	6.9364
Pix2Pix	ResNet9	11383K	29.8721	0.8417	<u>6.7046</u>
Pix2Pix	Uformer	85605K	29.7030	0.8026	8.0602
Ablation Models					
w/o Multi-scale		613K	<u>29.9970</u>	0.8692	6.9362
w/o Fusion Block		629K	29.4425	0.7814	8.1200
w/o MBI		540K	29.7599	<u>0.9029</u>	6.9149
w/o LAT		90K	28.8936	0.7857	10.1284
Ours		629K	**30.4873**	**0.9061**	**5.4988**

Table 1 demonstrates our model's superior performance compared to alternative methods based on synthetic datasets. The highest PSNR and SSIM values, and the lowest CIEDE2000 value, emphasize our approach's effectiveness in smoke removal tasks. Figure 3 presents a subjective evaluation of desmoking results, emphasizing previous approaches' limitations in adequately removing smoke. Non-deep learning methods often produce low-brightness, color-shifted images due to DCP's unsuitability for surgical applications with complex lighting and varied smoke. Although deep learning techniques better restore brightness, CycleGAN and Pix2Pix cannot fully eliminate smoke, as evidenced by residual smoke in some image portions (Fig. 3). These methods also result in unclear tissue contours due to CNN-based models' restricted global low-frequency feature extraction. In contrast, our methodology yields cleaner images with enhanced brightness, sharp details, and distinct edges.

5.1 Evaluation Under Different Smoke Densities

Smoke impairs image information, often irreversibly, depending on thickness. To evaluate networks' desmoking performance at varying densities, we analyzed light, medium, and heavy smoke levels. We generated test sets for each density level with fixed starting positions and temperatures. Figure 4 displays rendered

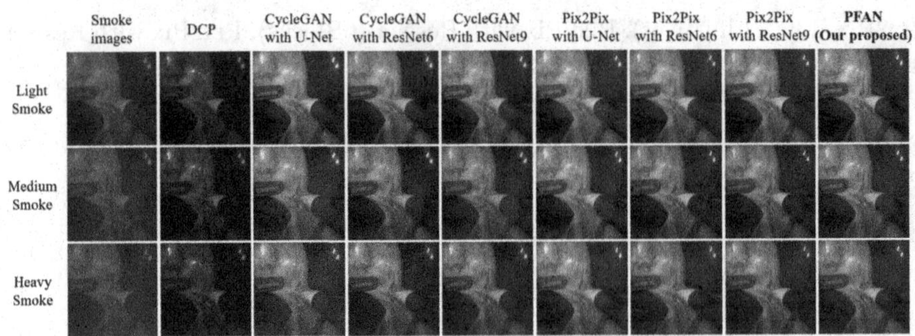

Fig. 4. Qualitative comparison between SOTAs under different smoke densities

smoke images (I_{Syn}) and desmoked results from five methods and our approach. DCP struggles to restore dark-red tissue colors, whereas deep learning-based techniques perform better using context. Pix2Pix produces similar results but falters for some images, introducing artificial reflections. Our method achieves clean results with minor saturation deviations, even under dense smoke conditions. Table 2 compares our approach to five alternatives, consistently yielding the highest SSIM and PSNR while reducing CIEDE2000, outperforming other established methods.

Table 2. Quantitative comparison between SOTAs under different smoke densities

Smoke Density		Light Smoke			Medium Smoke			Heavy Smoke		
Model		PSNR↑	SSIM↑	CIEDE2000↓	PSNR↑	SSIM↑	CIEDE2000↓	PSNR↑	SSIM↑	CIEDE2000↓
DCP		27.6611	0.6215	30.1270	27.6811	0.5887	32.9143	27.6944	0.5807	33.8072
CycleGAN	U-Net	29.0426	0.7778	8.5370	28.9490	0.7607	10.7167	28.8837	0.7639	10.7521
CycleGAN	ResNet6	29.0713	0.7958	8.2868	28.7621	0.7741	11.7635	28.7647	0.7755	11.6661
CycleGAN	ResNet9	29.3232	0.8002	7.8017	28.7466	0.7650	11.9671	28.9379	0.7711	10.8202
Pix2Pix	U-Net	29.2652	0.7270	8.9004	29.4071	0.7119	9.1812	29.4474	0.7199	8.9037
Pix2Pix	ResNet6	29.9776	0.8404	6.6498	30.1833	0.8288	6.8033	30.2138	0.8344	6.2970
Pix2Pix	ResNet9	29.9492	0.8484	6.6610	30.1498	0.8372	**6.7079**	30.3287	0.8434	6.7079
Ours		**30.1209**	**0.8856**	**6.5182**	**30.2740**	**0.8704**	6.8001	**30.5223**	**0.8762**	**6.1147**

5.2 Ablation Studies

We design a series of ablation experiments to analyze the effectiveness of each of the modules we propose. The ablation results are reported in Table 1.

Effectiveness of the MBI Block: The goal of the MBI Block is to effectively capture multi-scale, high-frequency details. Figure 5 demonstrates that removing the MBI Block results in remaining smoke and blurry edges and textures in some image portions. This limitation in high-frequency detail extraction makes it challenging to obtain satisfactory desmoking outcomes. In Table 1, our PFAN outperforms the model without the MBI Block in terms of PSNR, SSIM, and CIEDE2000 metrics. This comparison highlights the critical role of MBI Blocks in achieving superior results.

Effectiveness of the LAT Block: The ViT-based LAT Blocks aim to extract global low-frequency information. Figure 5 shows that the model without LAT Blocks achieves a visually similar desmoking effect to the Ground Truth (GT); however, the color appears dull and exhibits noticeable distortion compared to the original smoke-free image. The higher CIEDE2000 value indicates insufficient low-frequency feature extraction. Furthermore, the lower PSNR and SSIM values demonstrate the effectiveness of the LAT module.

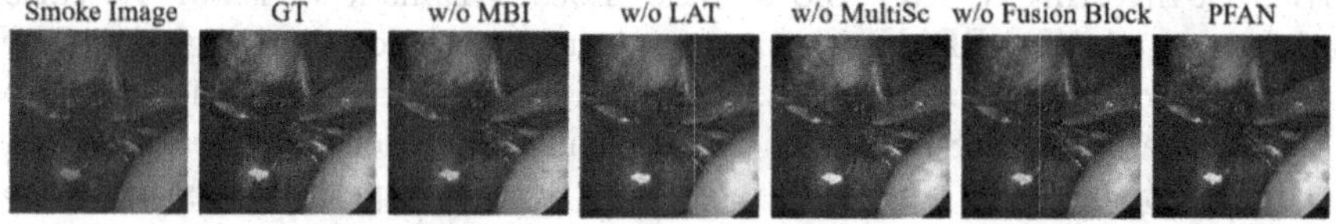

Fig. 5. Qualitative results of ablation experiments.

Effectiveness of the Multi-scale MBI: Our approach employs group convolution with varying receptive fields in the MBI Block, facilitating multi-scale high-frequency information extraction. We conducted an ablation study, replacing multi-scale convolutions in the MBI Block with only 3×3 group convolutions. Figure 5 reveals substantial improvements in smoke removal, but the tissue in the central scalpel area appears blurred. Table 1 demonstrates the "w/o Multi-scale" model achieves comparable performance to PFAN in terms of CIEDE2000 and PSNR; however, the SSIM value is significantly inferior, highlighting the importance of Multi-scale group convolutions in the MBI Block.

Effectiveness of Fusion Block: The Fusion Block in our proposed method leverages channel attention for adaptive discriminative fusion between image Transformer features and convolutional features, enhancing the network's learning capability. Importantly, omitting channel attention leads to the most significant decline in SSIM value among the four ablation experiments. Additionally, noticeable differences in both PSNR and CIEDE2000 emerge compared to the PFAN results, underscoring channel attention's crucial role in PFAN.

6 Limitations

Our method has a few limitations. It overlooks external factors such as water vapor and pure white gauze, which can degrade image quality and then impede desmoking performance. Future iterations should incorporate these elements into training and testing to ensure clinical applicability. Moreover, our proposed single-frame desmoking method may introduce temporal discontinuity in video desmoking tasks due to smoke density fluctuations. Thus, based on our current method, further investigation into spatial-temporal convolution techniques is necessary for enhancing laparoscopic video desmoking.

7 Conclusion

In conclusion, we proposed a groundbreaking deep learning method PFAN for laparoscopic image desmoking. By incorporating the lightweight and efficient CNN-ViT-based approach with the innovative CNN-based Multi-scale Bottleneck-Inverting (MBI) Blocks and Locally-Enhanced Axial Attention Transformers (LAT), PFAN effectively captures both low and high-frequency information for desmoking analysis, even with a limited dataset. The evaluation on the synthetic Cholec80 dataset, with various smoke-dense images, showcases the superiority of PFAN compared to existing SOTAs in performance and visual effects. Additionally, PFAN maintains a lightweight design, making it a feasible and desirable choice for implementation in medical equipment. Our desmoking method enables advanced applications. It enhances surgical safety by providing real-time desmoked images, serving as a valuable reference during ablation procedures. Furthermore, beyond aiding surgeons directly, the technology can also improve the robustness of various vision-based surgical assistance systems when used as a preprocessing step.

References

1. Armanious, K., Jiang, C., Fischer, M., Thomas, K., Nikolaou, K.: MedGAN: medical image translation using GANs, pp. 1–16 (2016)
2. Bolkar, S., Wang, C., Cheikh, F.A., Yildirim, S.: Deep smoke removal from minimally invasive surgery videos. In: Proceedings - International Conference on Image Processing, ICIP, pp. 3403–3407 (2018). https://doi.org/10.1109/ICIP.2018.8451815
3. Bolkar, S., Wang, C., Cheikh, F.A., Yildirim, S., Bolkar, S., Wang, C., Cheikh, F.A., Yildirim, S.: 2018 25th IEEE International Conference on Image Processing (ICIP), pp. 3403–3407 (2018)
4. Deng, Z., et al.: Deep multi-model fusion for single-image dehazing. In: Proceedings of the IEEE/CVF International Conference on Computer Vision, pp. 2453–2462 (2019)
5. Dosovitskiy, A., et al.: An image is worth 16 × 16 words: transformers for image recognition at scale (2020). https://arxiv.org/abs/2010.11929

6. Goodfellow, I., et al.: Generative adversarial networks. Commun. ACM **63**(11), 139–144 (2020). https://doi.org/10.1145/3422622
7. He, K., Sun, J., Tang, X.: Single image haze removal using dark channel prior. In: 2009 IEEE Conference on Computer Vision and Pattern Recognition, CVPR 2009 (January 2011), pp. 1956–1963 (2009). https://doi.org/10.1109/CVPRW.2009.5206515
8. He, K., Zhang, X., Ren, S., Sun, J.: Deep residual learning for image recognition. In: Proceedings of the IEEE Conference on Computer Vision and Pattern Recognition, pp. 770–778 (2016)
9. Hendrycks, D., Gimpel, K.: Gaussian Error Linear Units (GELUs), pp. 1–10 (2016). https://arxiv.org/abs/1606.08415
10. Horé, A., Ziou, D.: Image quality metrics: PSNR vs. SSIM. In: Proceedings - International Conference on Pattern Recognition, pp. 2366–2369 (2010). https://doi.org/10.1109/ICPR.2010.579
11. Huang, W., Liao, X., Qian, Y., Jia, W.: Learning hierarchical semantic information for efficient low-light image enhancement. In: 2023 International Joint Conference on Neural Networks (IJCNN), pp. 1–8 (2023). https://doi.org/10.1109/IJCNN54540.2023.10190996
12. Huang, W., Liao, X., Zhu, L., Wei, M., Wang, Q.: Single-image super-resolution neural network via hybrid multi-scale features. Mathematics **10**(4), 1–26 (2022). https://doi.org/10.3390/math10040653
13. Isola, P., Efros, A.A., Ai, B., Berkeley, U.C.: Image-to-Image translation with conditional adversarial networks
14. Jaschinski, T., Mosch, C.G., Eikermann, M., Neugebauer, E.A., Sauerland, S.: Laparoscopic versus open surgery for suspected appendicitis. Cochrane Database Syst. Rev. **2018**(11) (2018). https://doi.org/10.1002/14651858.CD001546.pub4
15. Kotwal, A., Bhalodia, R., Awate, S.P.: Joint desmoking and denoising of laparoscopy images. In: Proceedings - International Symposium on Biomedical Imaging 2016-June, pp. 1050–1054 (2016). https://doi.org/10.1109/ISBI.2016.7493446
16. Li, B., Peng, X., Wang, Z., Xu, J., Feng, D.: AOD-Net: All-in-One dehazing network, pp. 4780–4788 (2017). https://doi.org/10.1109/ICCV.2017.511
17. Li, Y., Zhang, K., Cao, J., Timofte, R., Van Gool, L.: LocalViT: bringing locality to vision transformers (2021). https://arxiv.org/abs/2104.05707
18. Li, Y., Miao, Q., Liu, R., Song, J., Quan, Y., Huang, Y.: Neurocomputing a multi-scale fusion scheme based on haze-relevant features for single image dehazing. Neurocomputing **283**, 73–86 (2018). https://doi.org/10.1016/j.neucom.2017.12.046
19. Liu, W., Hou, X., Duan, J., Qiu, G.: End-to-End single image fog removal using enhanced cycle consistent adversarial networks **29**(1), 7819–7833 (2020). https://doi.org/10.1109/TIP.2020.3007844
20. Liu, Z., Mao, H., Wu, C.Y., Feichtenhofer, C., Darrell, T., Xie, S.: [ConvNeXt_CVPR22] A ConvNet for the 2020s. In: CVPR, pp. 11976–11986 (2022). https://arxiv.org/abs/2201.03545
21. Luo, M.R., Cui, G., Rigg, B.: The development of the CIE 2000 colour-difference formula: Ciede 2000. Color Research & Application: Endorsed by Inter-Society Color Council, The Colour Group (Great Britain), Canadian Society for Color, Color Science Association of Japan, Dutch Society for the Study of Color, The Swedish Colour Centre Foundation, Colour Society of Australia, Centre Français de la Couleur **26**(5), 340–350 (2001)
22. Rahman, Z.u., Jobson, D.J., Woodell, G.A., Science, C.: Retinex Processing for Automatic Image Enhancement

23. Removal, I.H., Cai, B., Xu, X., Jia, K., Qing, C.: DehazeNet: an end-to-end system for single. IEEE Trans. Image Process. **25**(11), 5187–5198 (2016). https://doi.org/10.1109/TIP.2016.2598681
24. Rong, Z., Jun, W.L.: Improved wavelet transform algorithm for single image dehazing. Optik **125**(13), 3064–3066 (2014). https://doi.org/10.1016/j.ijleo.2013.12.077
25. Ronneberger, O., Fischer, P., Brox, T.: U-Net: convolutional networks for biomedical image segmentation. In: Navab, N., Hornegger, J., Wells, W.M., Frangi, A.F. (eds.) MICCAI 2015. LNCS, vol. 9351, pp. 234–241. Springer, Cham (2015). https://doi.org/10.1007/978-3-319-24574-4_28
26. Szegedy, C., et al.: Going deeper with convolutions. In: Proceedings of the IEEE Conference on Computer Vision and Pattern Recognition, pp. 1–9 (2015)
27. Tchaka, K., Pawar, V.M., Stoyanov, D.: Chromaticity based smoke removal in endoscopic images. In: Medical Imaging 2017: Image Processing **10133**(February 2017), 101331M (2017). https://doi.org/10.1117/12.2254622
28. Twinanda, A.P., Shehata, S., Mutter, D., Marescaux, J., De Mathelin, M., Padoy, N.: EndoNet: a deep architecture for recognition tasks on laparoscopic videos. IEEE Trans. Med. Imaging **36**(1), 86–97 (2017). https://doi.org/10.1109/TMI.2016.2593957
29. Wang, C., Alaya Cheikh, F., Kaaniche, M., Beghdadi, A., Elle, O.J.: Variational based smoke removal in laparoscopic images. Biomed. Eng. Online **17**(1), 1–18 (2018). https://doi.org/10.1186/s12938-018-0590-5
30. Wang, Z., Cun, X., Bao, J., Zhou, W., Liu, J., Li, H.: Uformer: a general u-shaped transformer for image restoration. In: Proceedings of the IEEE Conference on Computer Vision and Pattern Recognition (CVPR) (2022)
31. Woo, S., Park, J., Lee, J.-Y., Kweon, I.S.: CBAM: convolutional block attention module. In: Ferrari, V., Hebert, M., Sminchisescu, C., Weiss, Y. (eds.) ECCV 2018. LNCS, vol. 11211, pp. 3–19. Springer, Cham (2018). https://doi.org/10.1007/978-3-030-01234-2_1

A Pixel-Level Segmentation Method for Water Surface Reflection Detection

Qiwen Wu, Xiang Zheng(✉), Jianhua Wang, Haozhu Wang, and Wenbo Che

Key Laboratory of Marine Technology and Control Engineering, Ministry of Transport (Shanghai Maritime University), Shanghai 201306, China
xiangzheng@shmtu.edu.cn

Abstract. Water surface reflections pose challenges to unmanned surface vehicles or robots during target detection and tracking tasks, leading to issues such as the loss of tracked targets and false target detection. Current methods for water surface reflection detection primarily rely on image thresholding, saturation, edge detection techniques, which have poor performance in segmentation, as they are more suitable for handling simper image scenarios and are insufficient for the detection of water surface images characterized by complex background information, intricate edge details, and the inclusion of abundant contextual elements from both shores. To bridge the gap, we propose a novel model named WRS-Net for achieving pixel-wise water reflection segmentation, which leverages an encoder-decoder architecture and incorporates two novel modules, namely Multi-scale Fusion Attention Module (MSA) and Interactive Convergence Attention Module (ICA). In addition, a water surface reflection dataset for sematic segmentation is constructed. The MSA extracts detailed local reflection features from shallow networks at various resolutions. These features are subsequently fused with high-level semantic information captured by deeper networks, effectively reducing feature loss and enhancing comprehensive extraction of both shallow features and high-level semantic information. Additionally, the ICA consolidates the preservation of local reflection details while simultaneously considering the global distribution of the reflected elements, by encapsulating the outputs of the MSA, the multiple feature maps of various scales, with the outputs of the decoder. The experiment results demonstrate enhanced performance of the proposed method in contour feature extraction and effective reflection segmentation capabilities. Specifically, the proposed method achieves mIoU, mPA, and average accuracy of 94.60%, 97.70%, and 97.96%, respectively, on the water reflection semantic segmentation dataset.

Keywords: water surface reflection detection · semantic segmentation · attention network · feature fusion · deep learning · neural networks

1 Introduction

Water reflections are phenomena caused by the reflection and refraction of light on the water's surface, frequently interfering with the image's content and structure and diminishing its clarity. Reflections frequently influence the detection of visual images,

Q. Liu et al. (Eds.): PRCV 2023, LNCS 14426, pp. 493–505, 2024.
https://doi.org/10.1007/978-981-99-8432-9_39

interfere with the detection of targets, and cause issues such as target loss, false detection [1]. In vision-based water surface target tracking by unmanned surface vehicles (USVs), the water surface reflection poses great challenges for the feature extraction of the target, particularly small targets. Surface reflections are also the main interference in edge detection, binarization, extraction of shorelines for surface image processing. Therefore, effective water reflection detection is necessary. Currently, the primary solution to the water surface image reflection problem relies on traditional image segmentation techniques. Wang [2], Huang et al. [3] adopted adaptive histogram threshold region segmentation, and segmentation based on saturation, brightness, and edge detection [4–6] to identify water surface reflection.

The previously mentioned image segmentation techniques are incapable of extracting semantic information, are susceptible to environmental factors such as lighting and weather and are only applicable to the processing of uncomplicated images. Traditional image segmentation techniques are no longer applicable to water reflection images due to their irregular shapes, sensitivity to environmental changes, high degree of similarity to the actual scene, complex target background, and rich information about the reflections on both sides of the water.

The semantic segmentation method based on deep neural networks can classify images at the pixel level, and has been applied for image processing [7], autonomous driving [8], and face recognition [9]. It utilizes its learning and feature extraction capabilities to acquire complex semantic information in images, thereby enhance the performance of detection.

To address the issues mentioned above for water reflection detection and motivated by the advantages of semantic segmentation methods, a new model called WRS-Net, a semantic segmentation model with an encoder-decoder structure, is proposed for pixel-wise detection of reflection features in water surface images. The key contributions are summarized as follows: (1) A novel semantic segmentation model named WRS-Net with two distinctive components, the MSA module and the ICA module, is proposed for comprehensive identification of reflection features in water surface image. Multiple MSA modules are cascaded with dual-input technique to extract multi-scale reflection features in order to achieve improved contextual feature connectivity and mitigate feature loss. The ICA modules consolidate the preservation of local feature details by combining the outputs of the MSA modules with the outputs of the decoder layers. (2) The proposed segmentation model is verified on the water surface reflection dataset consisting images under various weather conditions. The experiment results demonstrate the effectiveness of the proposed model in detecting surface reflections, with improved overall classification accuracy compared to other existing models. This research can aid unmanned surface vessels in vision-based autonomous navigation subjected to surface reflection interferences.

2 Related Work

Deep learning-based image semantic segmentation began with Shelhamer et al.'s [10] Fully Convolutional Networks (FCN) proposal. Olaf et al. [11] proposed a U-net based on a symmetric codec structure by adding the feature output of the encoder to the feature output of the decoder and fusing the two in a hopping connection to reduce feature loss while effectively enhancing the network's segmentation effect on image edge details. Seg-Net, which was proposed by V. Badrinarayanan [12], employs an average pooling layer to supplement the decoding structure with features from the encoding structure.

To acquire multi-region contextual information, Zhao [13] proposed PSP-Net (Pyramid Scene Parsing Network) based on the pyramid pooling structure method. Chen et al. utilized hole convolution to enlarge the perceptual field and subsequently proposed Deeplabv1 [14], v2 [15], v3 [16], v3+ [17], which derives multi-scale feature information through the convolution of cavities with varying expansion rates, effectively enhancing the segmentation effect. Yang [18] proposed the Dense-ASPP network by combining the ASPP structure with Dense-Net's [19] concept of dense connectivity. On the basis of U-net, Zhou [20] conducted additional research and proposed the U-net++ network with multi-branch feature fusion, further combining contextual features to increase the feature utilization rate. Ding [21] proposed the Asymmetric Convolution Network (ACNet), this method improves the overall segmentation accuracy. The CG-Net (Context Guided Network) was devised by Wu [22], which outperformed other segmentation methods. Zhu [23] investigated the information distribution of the low-level network in depth and made full use of the shallow-level network with rich texture features to propose the semantic segmentation network STL-Net (Statistical Texture Learning Network), which improves feature utilization effectively.

The attention mechanism has been extensively utilized in segmentation tasks to concentrate on local critical information while accounting for feature differences. Chen [24] proposed a semantic segmentation method that makes use of the attention mechanism in multi-scale inputs. Zhao [25] proposed a point-wise spatial attention network (PSA-Net) to enhance prediction accuracy via mapping distribution information between pixel points. Fu et al. [26] devised a position attention and channel attention module to spatial and channel dimensions' semantic relationships were modelled to accomplish partitioned feature segmentation, respectively. Niu [27] utilized the channel attention module to design the HMA-Net (Hybrid Multiple Attention Network), which demonstrated superior performance for aerial image segmentation. Huang [28] proposed CC-Net (Criss-Cross Attention for Semantic Segmentation), a network that employs a novel cross-attention module to derive inter-feature dependencies by gathering contextual information of all pixels on the cross-path.

3 Methods

This paper draws inspiration from the traditional semantic segmentation network Deeplabv3+ and leverages the ASPP structure, as the foundation for the proposed the WRS-Net. The general structure design of WRS-Net is depicted in Fig. 1. It employs an encoding-decoding structure, taking a single image as input and generating the resultant reflection detection map of an end-to-end manner.

MSA modules are added in the encoding step, which are cascaded by employing a dual input technique to extract multi-scale reflection information features. This method effectively combines shallow texture features and deep semantic features to address the issue of insufficient contextual feature connection, effectively reduce local feature loss, and better distinguish reflections with high similarity to physical objects on both sides of the river.

In the decoding stage, the ICA module is designed, which combines the output containing feature information encoded by the MSA module with the output of each layer of the decoder, consolidating the local reflection detail features while taking into account the global reflection distribution features. This model is advantageous for suppressing interference from non-target pixel points and decreasing the likelihood of false recognition. The encoder and decoder are linked by an ASPP structure, which expands the network's overall perceptual field by convolving cavities with varying expansion rates, thereby reducing internal data loss and facilitating the acquisition of edge-inverted features.

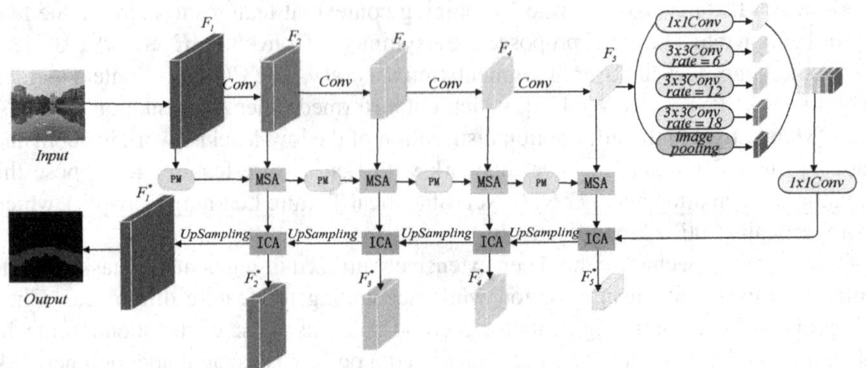

Fig. 1. WRS-Net structure diagram

3.1 Patching Merging (PM)

PM [29] modules, as illustrated in Fig. 2, are employed to accomplish the effect of down-sampling feature extraction through hierarchical design while conserving some computing power. First, four small feature maps of the same size are sampled in increments of 2 in each row and column direction to acquire a feature tensor, which is then expanded by a Layer-Norm into a feature map that is one-fourth the size of the input and contains four times as many channels. For further feature fusion, the output feature map can be fused with a feature map of the same resolution.

3.2 Multi-Scale Fusion Attention Module (MSA)

Due to the high similarity between the local reflection and the physical objects on both banks, the simple down-sampling convolution level to obtain information of different scale features ignores the difference between different scale features and is unable to distinguish the target pixels with high similarity.

As depicted in Fig. 3, the MSA module developed in this paper is capable of extracting superficial network features with varying resolutions and fusing them with deeper

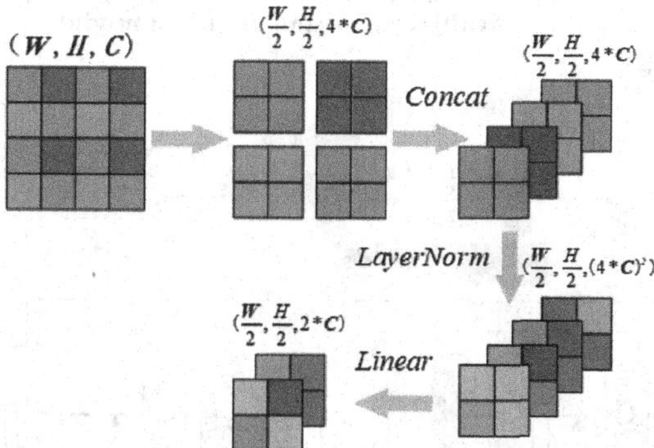

Fig. 2. PM module structure diagram

network features. The shallow feature map can acquire the small reflection information of the local area of the image of water surface reflection, whereas the deep feature map can capture the image's high-level semantic information. The combination of the two can effectively reduce feature loss, thereby substantially improving the segmentation effect for an image of water surface reflection with a complex background and abundant information.

The module has two inputs and the first input feature map is twice as large as the second input. In particular, the first input passes through a PM Module to obtain a feature map of the same size as the second output, and then a concat operation is performed with the second input to obtain a fused feature at both scales. The features pass through a Residual Module to improve feature utilization, and then undergo a channel fusion process with the refined features of the first input passing through an Attention Module, further fusing the feature information of the two scales and enhancing the preservation of the reflection's detailed features. Finally, a 1 * 1 post-channel convolutional transform is used as the input for subsequent feature fusion.

3.3 Interactive Convergence Attention Module (ICA)

The typical encoder-decoder structure, which acquires image features by convolving multiple down-samples and mapping detection results by successive up-samples, is straightforward and practical, allowing for rapid target detection. This structure, however, disregards the cascade relationship between the encoder and decoder and lacks direct layer-to-layer feature correlation.

This paper presents the design for the ICA module, which is depicted in Fig. 4. The module combines the output containing feature information encoded at multiple scales by the MSA module with the output of each layer of the decoder, consolidating the local cepstral detail features while taking the global cepstral distribution features into account and facilitating contextual feature acquisition. The model selects useful information of the input features by learning the attention map as a feature index, extracts good features while suppressing bad noise information, and then adds and multiplies the features after channel fusion with the features calculated by refinement pixel by pixel, effectively

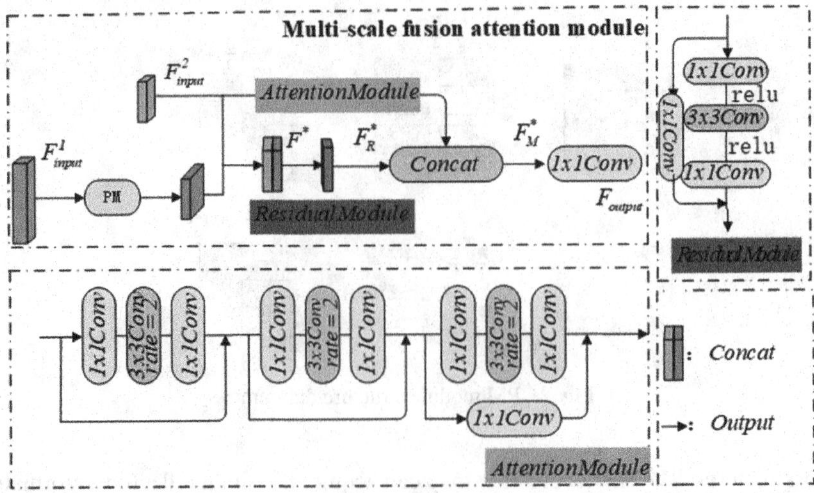

Fig. 3. MSA module structure diagram

connecting the contextual features and obtaining refined refinement output features, which has a greater effect on the extraction of detailed edge information from images of water reflection.

The inputs of the ICA are the output F1 of the coding layer and the output F2 of the corresponding decoding layer of the coding layer. One branch of F1 is spliced with the F2 channel to produce F3, while the other branch is used as the input to the attention module, which then calculates the attention feature index matrix F4. The output of the module after a convolution operation is obtained by multiplying feature F5 by F4 pixel by pixel to obtain feature F6, which is then added to feature F5 pixel by pixel to yield the output of the module.

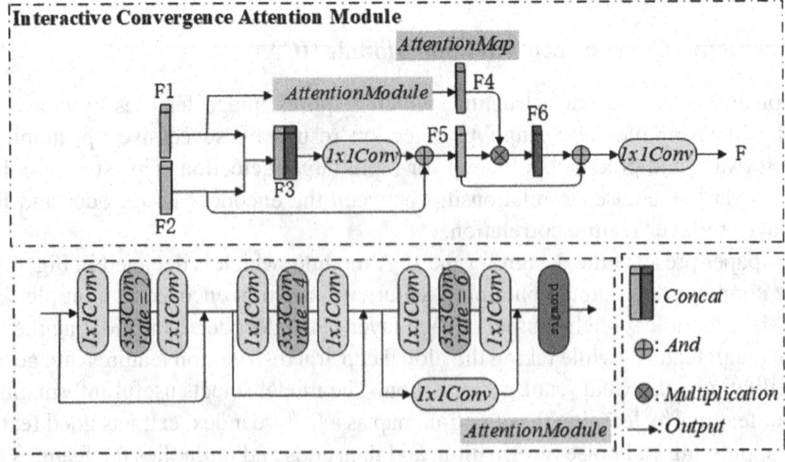

Fig. 4. ICA module structure diagram

The primary method of calculation is as follows:

$$F = \mathbb{C}\{(1 + A(F1) * [\mathbb{C}(Cat(F1, F2)) + F2]\} \tag{1}$$

$$F3 = Cat(F1, F2) \tag{2}$$

$$F4 = A(F1) \tag{3}$$

$$F5 = \mathbb{C}(Cat(F1, F2)) + F2 \tag{4}$$

$$F6 = (1 + A(F1)) * [\mathbb{C}(Cat(F1, F2)) + F2] \tag{5}$$

Cat denotes the channel feature stitching operation; $Cat(F1, F2)$ denotes the channel stitching of the channel numbers of feature maps F1 and F2; $\mathbb{C}$ denotes the 1×1 convolution operation; *denotes a pixel-by-pixel multiplication operation; +denotes a pixel-by-pixel addition operation; *A* denotes the index of the feature weights obtained by the attention module. The attention mechanism consists of three residual blocks, each of which contains a 1×1 convolution and a 3×3 dilated convolution; the dilation rates of the three null convolutions are 2, 4, and 6 respectively; and the output of the residual blocks is used as input to the Sigmoid function to obtain the feature index map, which is computed as follows:

$$\sigma(x, y, c) = \frac{1}{1 + \exp(-E(x, y, c))} \tag{6}$$

where x is the horizontal coordinate of the respective feature point on the feature map, y is the vertical coordinate of the respective feature point, c is the number of channels of the respective feature map, E is the feature point (x, y), the eigenvalue of the feature point with channel number c, and σ is the output weight of the feature value after the Sigmoid function.

4 Data and Evaluation Metrics

Data Sources: There are three primary sources for the water surface reflection semantic segmentation dataset developed in this paper: (1) USV-Inland [1], the world's first water surface unmanned dataset, was announced in March 2021 by Ouka Smart Hublot and Tsinghua University. (2) water surface related video images captured by the laboratory boat-mounted camera. (3) water surface reflection videos or images collected from the Internet.

In a realistic scenario, the USV-Inland dataset is the first multi-sensor, multi-weather inland river unmanned vessel water surface dataset. A total of 27 raw data segments were acquired in various scenarios involving inland rivers. The data collection environment is diverse and rich, and the data contains images collected under a variety of natural conditions, such as sunny days, intense light, overcast days, and cloudy days, serving as the primary data source for the water surface reflection dataset we created.

By supplementing the water surface reflection dataset with reflection image data of different waters, water clarity, and lighting conditions, the image data collected by a laboratory unmanned surface boat in several inland waterways in Wu-Zhong District, Suzhou, Jiangsu Province, can provide a data enhancement effect. The majority of Internet-collected water surface reflection images are symmetrical, clearer, and more vividly hued; therefore, the addition of these data increases the sample size and further improves the method's ability to generalize. A total of 1668 images of water surface reflection were acquired from the three data sources listed above, and some of these images are displayed in Fig. 5. We enriched some of the images with difficult-to-learn complex backgrounds by primarily folding, inverting, rotating, cutting, and sewing. The augmented dataset consists of 2055 images of water reflections and their respective annotated images. The dataset was divided into a training set and a validation set, with 1750 training set images and 305 validation set images.

Fig. 5. Water reflection image

Metrics: The network segmentation performance is evaluated and compared using the three most representative semantic segmentation network performance evaluation metrics, namely accuracy (Precision), mean intersection and merge ratio (mIoU), and mean pixel accuracy (mPA).

5 Experiments

This experimental environment is Ubuntu 18.04 with Pytorch 1.10 [30], CUDA 11.2 and an Nvidia RTX2080ti GPU for training. The Adam optimizer was utilized for optimization, with momentum set to 0.9, batch-size set to 4, learning rate set to 1e−3, epoch set to 100, and learning rate decay by cosine annealing (cos) beginning at the 50th epoch.

5.1 Experimental Analysis

To verify the effectiveness of WRS-Net for segmentation of water reflection images, we did extensive experiments on the training set of the reflection dataset, tested on the validation set, and compared the performance with other segmentation networks,

Table 1. Comparison of experimental results of different segmentation methods

Methods	mIoU	mPA	Precision
U-Net	82.28%	86.87%	89.15%
Seg-Net	86.20%	91.17%	94.25%
PSP-Net	88.17%	92.54%	94.85%
DenseASPP	91.31	93.58	95.76
Deeplabv3+ (mobilenetv2)	92.10%	95.55%	96.58%
BiSe-Net(xception)	92.27%	95.86%	96.64%
CGNet	92.72%	96.35%	96.33%
WRS-Net	**94.60%**	**97.70%**	**97.96%**

including U-Net, Seg-Net, PSP-Net, Deeplabv3+, Dense-ASPP, BiSe-Net, CG-Net, the results of the different network segmentation performance metrics are shown in Table 1.

From Table 1, it can be deduced that WRS-Net can obtain an accuracy of 97.96%, demonstrating a significant advantage over other segmentation networks. Comparing with BiSe-Net (xception) and CGNet, which have better segmentation performance, the mIoU and mPA values of this paper's method are improved by 2.33% and 1.88%, and by 1.84% and 1.35%, respectively; the improvement of WRS-Net's segmentation performance over U-Net network is 1.35%. WRS-Net demonstrated the greatest performance improvement over U-Net, with mIoU, mPA, and accuracy increasing by 12.32%, 10.83%, and 8.81%, respectively.

Comparing the segmentation evaluation index values of various networks demonstrates that WRS-Net performs better at segmenting water surface reflections. The primary reason is that the MRS module performs multi-scale feature fusion of features with different resolutions and then uses the fused feature information as a supplement to the high-level semantic information, thereby effectively enhancing the utilization of feature information; The ICA module connects the encoder and decoder, passing the feature information encoded by the backbone network to the decoder for decoding operations and effectively connecting the contextual features and yielding refined output features.

The visual reflection segmentation maps of each network for the test maps are depicted in Figs. 6 and 7. From the yellow boxes marked on each detection result map, it is clear that the segmentation of the water reflection images by U-net is relatively imprecise and that the performance of the target edge segmentation is inadequate; The PSP-net network is superior to the U-Net at segmenting the edge details of reflections, but less effective at classifying pixels in continuous regions; BiSe-Net and CG-Net are superior at segmentation, with CG-Net outperforming BiSe-Net at classifying edge pixels.

Compared to CG-Net, the method presented in this paper demonstrates further optimization for the segmentation of target edges and fine targets, effectively classifying reflection pixels and non-reflection pixels, segmenting the contour edges of reflections more precisely, and achieving better segmentation results for fine targets.

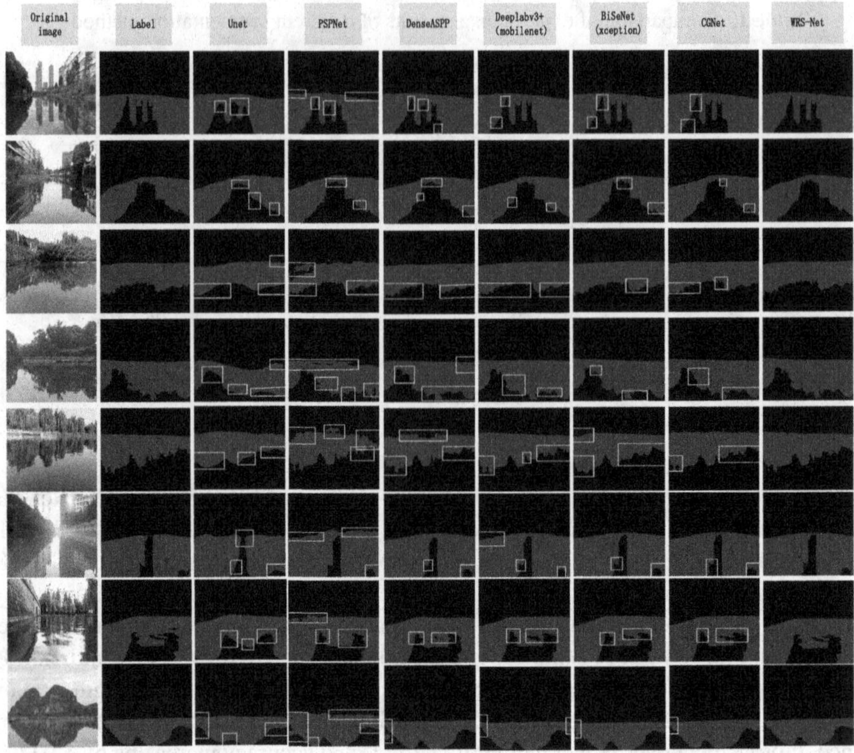

Fig. 6. Comparison with other reflection segmentation methods (Part I)

In conclusion, the WRS-Net proposed in this paper has enhanced the segmentation effect compared to other segmentation methods and demonstrated good performance in segmenting images of water reflection with more detailed contour margins.

5.2 Ablation Study

To determine the influence of the MSA and ICA modules on the overall performance of the WRS-Net, ablation experiments were conducted for each module and compared to the Deeplabv3+ network for analysis using the water surface reflection segmentation dataset generated by this study. Table 2 provides a comparison of the experimental results.

As shown in Table 2, the addition of the MSA module to the network led to the greatest improvement in the network's mIoU value, which increased by 4.11 percentage points; the ICA module led to the network's optimization in terms of its mIoU value, mPA value, and accuracy, which increased by 4.48 percentage points, 3.29 percentage points, and 2.24 percentage points, respectively.

The proposed WRS-Net has the greatest overall segmentation performance compared to the aforementioned networks, with a maximum improvement of 5.45 percentage points

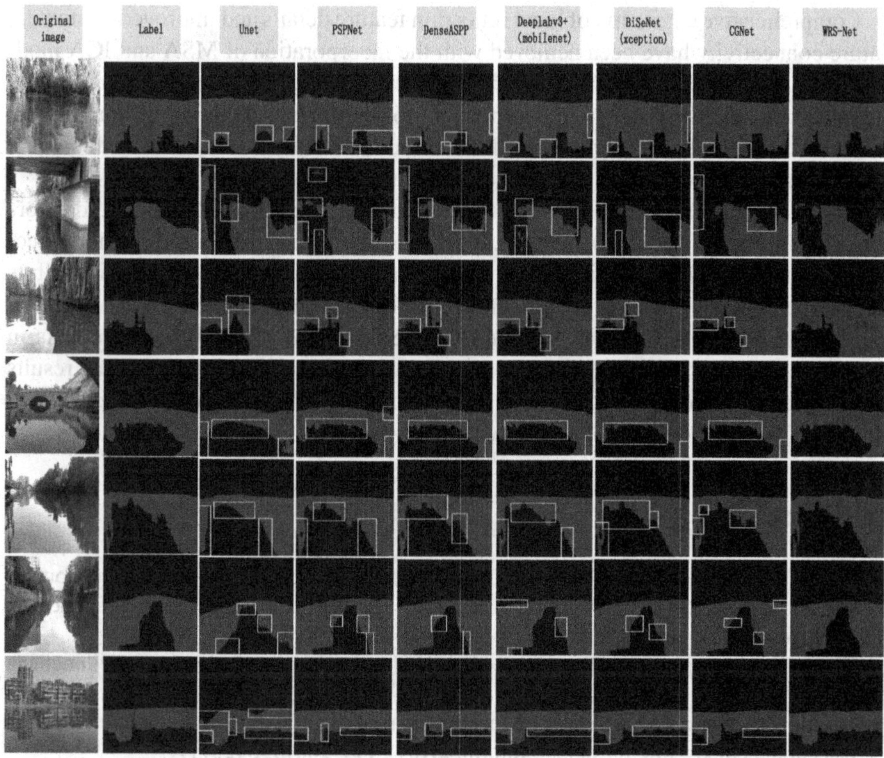

Fig. 7. Comparison with other reflection segmentation methods (Part II)

Table 2. Results of ablation experiments

Methods	mIoU	mPA	Precision
Deeplabv3+ (xception)	89.15%	93.18%	95.3%
MSA	93.26%	95.99%	97.12%
ICA	93.63%	96.47%	97.54%
WRS-Net	**94.60%**	**97.70%**	**97.96%**

and a minimum improvement of 2.66 percentage points in the mIoU value, as well as considerable improvements in the accuracy and mPA metrics.

6 Conclusion

Water reflection detection is very important for autonomous navigation systems for unmanned surface vessels. In this paper, an encoder-decoder structured semantic segmentation model named WRS-Net is proposed for end-to-end identification of reflections in surface images.

Comprehensive extraction of local reflection feature details and improved contextual feature connectivity have been achieved with the incorporation of MSA and ICA modules. In particular, cascaded MSA in the encoder part effectively reduces feature loss and enhances shallow features in complement to high-level semantic information; the ICA module in the decoder part consolidated the local reflection feature details while considering the global reflection distribution features.

The performance is confirmed by the distribution of various statistical indicators shown in Table 1–2. Moreover, Figs. 6, 7 demonstrate that the proposed method effectively segments surface reflections in various water environments. Specifically, it exhibits superior performance in ensuring reflection edge extraction and enhancing the accuracy of reflection identification. Our technique outperforms the state-of-the-art models in terms of overall segmentation performance of target pixels and shows promising results in edge feature identification, considering that mIoU, mPA, and average accuracy of 94.60%, 97.70%, and 97.96%, respectively.

References

1. Wang, W., Gheneti, B., Mateos, L.A., Duarte, F., Ratti, C., Rus, D.: Roboat: an autonomous surface vehicle for urban waterways. In: 2019 IE EE/RSJ International Conference on Intelligent Robots and Systems (IROS), pp. 6340–6347 (2019)
2. Wang, D.Z.: Detection of Water Reflection. Harbin Institute of Technology, Heilongjiang (2009)
3. Huang, P.P., Wang, J.H., Chen, C.F.: Experimental study of several water surface reflection detection methods. Micro Comput. Inform. **27**(09), 199–200+198 (2011)
4. Zhou, X.N., Hao, J.M., Chen, Y.: A study of a gray-scale histogram-based threshold segmentation algorithm. Digital Technol. Appl. 131 (2016)
5. Canny, J.: A computational approach to edge detection. J. IEEE Trans. Pattern Anal. Mach. Intell. **PAMI-8**(6), 679–698 (1986)
6. Loncaric, S.: A survey of shape analysis techniques. Pattern Recogn. **31**(8), 983–1001 (1998)
7. Basak, H., Kundu, R., Sarkar, R.: MFSNet: a multi focus segmentation network for skin lesion segmentation. Pattern Recogn. **128**, 108673 (2022)
8. Petzold, J., Wahby, M., Stark, F., et al.: If you could see me through my eyes: predicting pedestrian perception. In: 2022 8th International Conference on Control, Automation and Robotics (ICCAR), pp. 84–190. IEEE (2022)
9. Zhu, W., Wang, C.Y., Tseng, K.L.: Local-adaptive face recognition via graph-based meta-clustering and regularized adaptation. In: Proceedings of the IEEE/CVF Conference on Com puter Vision and Pattern Recognition, pp. 20301–20310 (2022)
10. Shelhamer, E., Long, J., Darrell, T.: Fully convolutional networks for semantic segmentation. IEEE Trans. Pattern Anal. Mach. Intell. **39**(4), 640–651 (2015)
11. Ronneberger, O., Fischer, P., Brox, T.: U-Net: convolutional networks for biomedical image segmentation. In: Navab, N., Hornegger, J., Wells, W.M., Frangi, A.F. (eds.) MICCAI 2015. LNCS, vol. 9351, pp. 234–241. Springer, Cham (2015). https://doi.org/10.1007/978-3-319-24574-4_28
12. Badrinarayanan, V., Kendall, A., Cipolla, R.: SegNet: a deep convolutional encoder – decoder architecture for image segmentation. IEEE Trans. Pattern Anal. Mach. Intell. **39**(12), 2481–2495 (2017)
13. Zhao, H., Shi, J., Qi, X., et al.: Pyramid scene parsing network. In: Proceedings of the IEEE Conference on Computer Vision and Pattern Recognition, pp. 2881–2890 (2017)

14. Chen, L.C., Papandreou, G., Kokkinos, I., et al.: Semantic image segmentation with deep convolutional nets and fully connected CRFs. arXiv preprint arXiv:1412.7062 (2014)
15. Chen, L.C., Papandreou, G., Kokkinos, I., et al.: Deeplab: Semantic image segmentation with deep convolutional nets, atrous convolution, and fully connected CRFs. IEEE Trans. Pattern Anal. Mach. Intell.. Pattern Anal. Mach. Intell. **40**(4), 834–848 (2017)
16. Chen, L.C., Papandreou, G., Schroff, F., et al.: Rethinking atrous convolution for semantic image segmentation. arXiv preprint arXiv:1706.05587 (2017)
17. Chen, L.-C., Zhu, Y., Papandreou, G., Schroff, F., Adam, H.: Encoder-decoder with atrous separable convolution for semantic image segmentation. In: Ferrari, V., Hebert, M., Sminchisescu, C., Weiss, Y. (eds.) ECCV 2018. LNCS, vol. 11211, pp. 833–851. Springer, Cham (2018). https://doi.org/10.1007/978-3-030-01234-2_49
18. Yang, M., Yu, K., Zhang, C., et al.: Denseaspp for semantic segmentation in street scenes. In: Proceedings of the IEEE Conference on Computer Vision and Pattern Recognition, pp. 3684–3692 (2018)
19. Huang, G., Liu, Z., Van Der Maaten, L., et al.: Densely connected convolutional networks. In: Proceedings of the IEEE Conference on Computer Vision and Pattern Recognition, pp. 4700–4708 (2017)
20. Zongwei Zhou, Md., Siddiquee, M.R., Tajbakhsh, N., Liang, J.: UNet++: a nested U-Net architecture for medical image segmentation. In: Stoyanov, D., Taylor, Z., Carneiro, G., Syeda-Mahmood, T., Martel, A., Maier-Hein, L., João Manuel, R.S., Tavares, A.B., Papa, J.P., Belagiannis, V., Nascimento, J.C., Zhi, Lu., Conjeti, S., Moradi, M., Greenspan, H., Madabhushi, A. (eds.) DLMIA/ML-CDS -2018. LNCS, vol. 11045, pp. 3–11. Springer, Cham (2018). https://doi.org/10.1007/978-3-030-00889-5_1
21. Ding, X., Guo, Y., Ding, G., et al.: ACNet: strengthening the kernel skeletons for powerful CNN via asymmetric convolution blocks. In: Proceedings of the IEEE/CVF International Conference on Computer Vision, pp. 1911–1920 (2019)
22. Wu, T., Tang, S., Zhang, R., et al.: CGNet: a light-weight context guided network for semantic segmentation. IEEE Trans. Image Process. **2020**(30), 1169–1179 (2020)
23. Zhu, L., Ji, D., Zhu, S., et al.: Learning statistical texture for semantic segmentation. In: Proceedings of the IEEE/CVF Conference on Computer Vision and Pattern Recognition, pp. 12537–12546 (2021)
24. Chen, L.C., Yang, Y., Wang, J., et al.: Attention to scale: scale-aware semantic image segmentation. In: Proceedings of the IEEE (2016)
25. Fu, J., Liu, J., Tian, H., et al.: Dual attention network for scene segmentation. In: Proceedings of the IEEE/CVF Conference on Computer Vision and Pattern Recognition, pp. 3146–3154 (2019)
26. Zhao, H., Zhang, Yi., Liu, S., Shi, J., Loy, C.C., Lin, D., Jia, J.: PSANet: point-wise spatial attention network for scene parsing. In: Ferrari, V., Hebert, M., Sminchisescu, C., Weiss, Y. (eds.) Computer Vision – ECCV 2018: 15th European Conference, Munich, Germany, September 8–14, 2018, Proceedings, Part IX, pp. 270–286. Springer International Publishing, Cham (2018). https://doi.org/10.1007/978-3-030-01240-3_17
27. Niu, R., Sun, X., Tian, Y., et al.: Hybrid multiple attention network for semantic segmentation in aerial images. IEEE Trans. Geosci. Remote Sens. **60**, 1–18 (2021)
28. Huang, Z., Wang, X., Huang, L., et al.: CCNet: criss-cross attention for semantic segmentation. In: Proceedings of the IEEE/CVF International Conference on Computer Vision, pp. 603–612 (2019)
29. Liu, Z., Lin, Y., Cao, Y., et al.: Swin transformer: hierarchical vision transformer using shifted windows. In: Proceedings of the IEEE /CVF International Conference on Computer Vision, pp. 10012–10022 (2021)
30. Paszke, A., Gross, S., Massa, F., et al.: Pytorch: an imperative style, high-performance deep learning library. In: Advances in Neural Information Processing Systems, 32 (2019)

Author Index

Q. Liu et al. (Eds.): PRCV 2023, LNCS 14426, pp. 507–509, 2024.
https://doi.org/10.1007/978-981-99-8432-9